Ad Hoc and
Sensor
Networks

Theory and Applications (2nd Edition)

T0178484

Ad Hoc and
Sensor
Networks

Theory and Applications (2nd Edition)

Carlos de Morais Cordeiro

Intel Corporation, USA

Dharma Prakash Agrawal

University of Cincinnati, USA

World Scientific

NEW JERSEY • LONDON • SINGAPORE • BEIJING • SHANGHAI • HONG KONG • TAIPEI • CHENNAI

Published by

World Scientific Publishing Co. Pte. Ltd.

5 Toh Tuck Link, Singapore 596224

USA office: 27 Warren Street, Suite 401-402, Hackensack, NJ 07601

UK office: 57 Shelton Street, Covent Garden, London WC2H 9HE

British Library Cataloguing-in-Publication Data
A catalogue record for this book is available from the British Library.

ISBN-13 978-981-4338-88-2
ISBN-10 981-4338-88-5
ISBN-13 978-981-4338-89-9 (pbk)
ISBN-10 981-4338-89-3 (pbk)

Printed in Singapore.

To my wife Wilma and our sons Matheus and Alex.

Carlos de Morais Cordeiro

To my wife Purnima.

Dharma Prakash Agrawal

Preface to the Second Edition

The area of ad hoc and sensor networks is growing at a much faster rate than any one's anticipation. To reflect such rapid advances, we have added three new chapters: Chapter 6 on Wireless Mesh Networks, Chapter 8 on Cognitive Radio and Networks, and Chapter 12 on Sensor Networks for Controlled Environment and Actuators. Besides this, the remaining chapters have been updated to reflect recent advances and changes in the area. Specifically, the Chapters on Wireless LAN, Wireless PAN, Applications of Sensor Networks, Sensor Networks Design Considerations, and Security in Ad Hoc and Sensor Networks have been thoroughly revised. Additional questions have been added in most of the chapters.

The authors are planning to prepare power point slides for all the chapters and will supply them to instructors if the book is adopted for a class. We hope, this new edition will continue to serve the readership in the best possible way.

Carlos de Morais Cordeiro
Dharma Prakash Agrawal

Preface to the First Edition

The intention of this textbook is to serve as the primary reference in the field of ad hoc and sensor networks for individuals with academic, industry, or military background. It targets not only researchers and engineers, but also those who would like to have a deep yet easy coverage of this growing field, and the current state of research in this area. It comes to fill in the gap of existing literature on ad hoc and sensor networks by providing a comprehensive coverage of the subject matter. This textbook has been written with great care to address the need of those who seek not only detailed knowledge of this important field, but also the breadth. After all, this area is poised to be a key component of future communication networks and likely to have an undaunted impact on our daily lives.

If there is one thing that we have learnt in all these years of research and development on ad hoc and sensor networks is that there is a major interdependence among various layers of the network protocol stack. Contrary to wired or even one-hop wireless (e.g., cellular or mobile) networks, the lack of a fixed infrastructure, the inherent mobility, and the underlying routing mechanism by ad hoc and sensor networks introduce a number of technological challenges that are very hard to be addressed within the boundaries of a single protocol layer. Despite of this clear fact, all existing edited textbooks on ad hoc and sensor networks often focus on a specific aspect of the technology in isolation, fail to provide critical insights on cross-layer interdependencies, and hence leave major questions in the minds of the readers.

Our experience in dealing with students, professionals, and researchers working on ad hoc and sensor networks have revealed the need for a textbook that covers the many interrelated aspects of these networks and which can also clearly pinpoint iterative interactions between different layers. The study of ad hoc and sensor networks is very peculiar and intriguing, and to be able to fully understand this area it is not only enough to understand specific solutions individually, but also their many interdependencies and cross-layer interactions. We are confident that this knowledge will allow readers to firmly grasp this topic, understand its intricacies, and stimulate creativity.

This is in essence the approach we take in this textbook. From the physical up to the application layer, we provide a detailed investigation of ad hoc and sensor networks to date. In addition, wherever applicable, the discussion of these topics is closely followed by their impact on other layers of the network protocol stack. With this explanatory model, we aim to provide the readers with not only the depth in understanding but also the breadth. The ultimate goal is to provide a superior experience that opens up new horizons as one move on from one chapter to another.

The organization of this textbook is based on the authors' long experience in academia and industry, dealing with students and professionals, where we feel that the easiest way to start this journey is through the routing layer. Technologies in this layer are often more easily absorbable so as to create a solid foundation for the follow-up subject areas. Therefore, after an introduction and overview of existing and future wireless communication systems in Chapter 1, we start with detailed technical discussions in Chapter 2 by examining unicast routing protocols and algorithms. To accommodate important new applications and improve the system performance of ad hoc and sensor networks, this is followed by the investigation of mechanisms for broadcasting, multicasting and geocasting in Chapter 3. Once all networking concepts are in place, it is time to move down in the protocol stack. In Chapters 4 and 5 we discuss the enabling technologies that are used at the physical and medium access control (MAC) layers of ad hoc and sensor networks. From IEEE 802.11 to IEEE 802.15, these chapters provide a detailed coverage of existing and forthcoming wireless technologies. Chapter 6 deals with directional antennas, which is a powerful way of increasing the capacity, connectivity, and covertness of ad hoc networks. This is the first textbook that deals with directional antennas from a networking perspective, concentrating on the MAC and routing issues when these types of antennas are in use. Next, we move up the stack to the transport layer and look at the many performance issues of the Transmission Control Protocol (TCP) over ad hoc networks, and discuss ways for improvements. Chapters 8 and 9 are fully dedicated to sensor networks and the unique characteristics and issues they face. As it shall be clear, sensor networks demand special treatment of certain issues which are inherently specific to them as compared to a generic ad hoc network. As both ad hoc and sensor networks are wireless, security becomes a critical component and is extensively discussed in Chapter 10. Finally, Chapter 11 investigates the increasingly important area of *all wireless networks* towards future fourth generation wireless systems and beyond. Among other things, we discuss the integration of heterogeneous wireless networks, such as cellular and wireless local area networks (LANs), with ad hoc and sensor networks, which will form the basis of the universal ubiquitous networking paradigm of the future. To ensure deep understanding of the subject, each chapter is accompanied by numerical questions and topics for simulation projects. Many of the exercises are open-ended and have been taken from open-book examination questions given to graduate students.

The authors are confident that the approach taken in this textbook together with its vast and extensive coverage of topics, will enable the readers to not only understand and position themselves in this hot area of ad hoc and sensor networks, but will also allow them to develop new capabilities, enhance skills, share expertise, consolidate knowledge and encourage further development of the area by identifying key problems, analyzing them and designing new and innovative solutions and applications.

Carlos de Morais Cordeiro
Dharma Prakash Agrawal

Acknowledgements for the Second Edition

The second edition of this book would not have been possible without the help from numerous individuals. Thanks are due to Weihuang Fu in collecting material and extensive inputs for the Chapter on Wireless Mesh Networks. Both Talmai Oliveira and Aparna Venkataraman helped in completing the page numbers for the index section.

A number of graduate students have read different versions of the drafts and helped improve the overall quality of this textbook. Our sincere thanks are due to Amitabh Mishra, Dushan Aththidiyavidanalage, Vaibhav Pandit, Nishan Weragama, Yang Chi Yang, Jung Hyun Jun, Sansit Sharma, Weihuang Fu, Sriram Narayanan, Narendra Katneni, and Avani Dalal.

Finally, we express our gratitude to the authorities of Intel Corporation and the University of Cincinnati for their encouragement in writing this second edition. The help by the staff of our publisher at the World Scientific Inc. is also very much appreciated.

Carlos de Morais Cordeiro
Dharma Prakash Agrawal

Acknowledgements for the First Edition

This book would not have been possible without the help from numerous individuals. A number of colleagues have read different versions of the drafts and helped improve the overall quality of this textbook. Our sincere thanks are due to Dave Cavalcanti, Hrishikesh Gossain, Anup Kumar, and Ashok Roy, for many helpful suggestions on different topics. Vivek Jain, Anurag Gupta, Anindo Mukherjee, Demin Wang, Yun Wang, and Qi Zhang provided useful material and references for some important concepts. Many thanks are also due to Torsha Banerjee, Ratnabali Biswas, Yi Cheng, Kaushik Chowdhury, Bing He, Nagesh Nandiraju, Lakshmi Santhanam and Haitang Wang for final proofreading of different chapters and identifying the index terms. Last, but not the least, our sincere thanks go to Abhishek Jain for going through the tiring process of proof-reading, updating diagrams, and performing endless painstaking corrections.

The authors would like to thank their families for enduring lonely evenings, weekends, and working vacations for the past two years while we have been busy writing this book. Finally, we express our gratitude to the authorities of Philips Research North America and the University of Cincinnati for their encouragement in writing this textbook. The help by the staff of our publisher at the World Scientific Inc. is also very much appreciated.

Carlos de Morais Cordeiro
Dharma Prakash Agrawal

Contents

Chapter 1

Introduction

1.1 Introduction

Over recent years, the market for wireless communications has enjoyed an unprecedented growth. Wireless technology is capable of reaching virtually every location on the surface of the earth. Hundreds of millions of people exchange information every day using pagers, cellular telephones, laptops, various types of personal digital assistants (PDAs) and other wireless communication products. With the tremendous success of wireless voice and messaging services, it is hardly surprising that wireless communication is beginning to be applied to the realm of personal and business computing. No longer bounded by the harnesses of wired networks, people will be able to access and share information on a global scale nearly anywhere one can think about.

Simply stating, a Mobile Ad hoc NETwork (MANET) [Agrawal2002, Cordeiro2002, Perkins2001] is one that comes together as needed, not necessarily with any support from the existing infrastructure or any other kind of fixed stations. We can formalize this statement by defining an ad hoc (ad-hoc or adhoc) network as an autonomous system of mobile hosts (MHs) (also serving as routers), connected by wireless links, the union of which forms a communication network modeled in the form of an arbitrary communication graph. This is in contrast to the well-known single hop cellular network model that supports the needs of wireless communication by installing base stations (BSs) as access points. In these cellular networks, communications between two MHs completely rely on the wired backbone and the fixed BSs. In a MANET, no such infrastructure exists and the network topology may dynamically change in an unpredictable manner since nodes are free to move.

As for the mode of operation, ad hoc networks are basically peer-to-peer multi-hop mobile wireless networks, where information

1

packets are transmitted in a store-and-forward manner from a source to an arbitrary destination, via intermediate nodes as shown in Figure 1.1. As the MHs move, the resulting change in network topology must be made known to the other nodes so that outdated topology information can be updated or removed. For example, as the MH2 in Figure 1.1 changes its point of attachment from MH3 to MH4 other nodes part of the network should use this new route to forward packets to MH2.

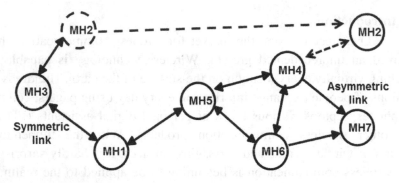

Figure 1.1 – A mobile ad hoc network (MANET)

Note that in Figure 1.1, and throughout this text, we assume that it is not possible to have all MHs within range of each other. In case all MHs are close-by, i.e., within radio range, there are no routing issues to be addressed. In real situations, the power needed to obtain complete connectivity may be, at least, infeasible, not to mention issues such as battery life and spatial reusability. Therefore, we are interested in scenarios where only few MHs are within radio range of each other. Figure 1.1 raises another issue of symmetric (bi-directional) and asymmetric (unidirectional) links. As we shall see later on, some of the protocols we discuss consider symmetric links with associative radio range, i.e., if (in Figure 1.1) MH1 is within radio range of MH3, then MH3 is also within radio range of MH1. This is to say that the communication links are symmetric. Although this assumption is not always valid, it is usually made because routing in asymmetric networks is a relatively hard task [Prakash1999]. In certain cases, it is possible to find routes that could avoid asymmetric links, since it is quite likely that these links imminently fail. Unless stated otherwise, throughout this text we consider symmetric links, with all MHs having identical capabilities and responsibilities.

The issue of symmetric and asymmetric links is one among the several challenges encountered in a MANET. Another important issue is that different nodes often have different mobility patterns. Some MHs are highly mobile, while others are primarily stationary. It is difficult to predict a MH's movement and pattern of movement. Table 1.1 summarizes some of the main characteristics [Cordeiro2002] and challenges in a MANET. A comprehensive look at the current challenges in ad hoc and sensor networking is provided later in this chapter.

Table 1.1 – Important characteristics of a MANET

Characteristic	Description
Dynamic Topologies	Nodes are free to move arbitrarily with different speeds; thus, the network topology may change randomly and at unpredictable times.
Energy-constrained Operation	Some or all of the nodes in an ad hoc network may rely on batteries or other exhaustible means for their energy. For these nodes, the most important system design optimization criterion may be energy conservation.
Limited Bandwidth	Wireless links continue to have significantly lower capacity than infrastructured networks. In addition, the realized throughput of wireless communications – after accounting for the effects of multiple access, fading, noise, and interference conditions, etc., is often much less than a radio's maximum transmission rate.
Security Threats	Mobile wireless networks are generally more prone to physical security threats than fixed-cable networks. The increased possibility of eavesdropping, spoofing, and minimization of denial-of-service type attacks should be carefully considered.

Wireless Sensor Network [Estrin1999, Kahn1999, Agrawal2002, Jain2005] is an emerging application area for ad hoc networks which has been receiving unprecedented attention. The idea is to use a collection of cheap to manufacture, stationary, tiny sensors to sense, and coordinate activities and transmit some physical characteristics about the surrounding environment to an associated BS or Sink Node. Once placed in a given environment, these sensors remain stationary, with power remaining a major driving issue behind protocols tailored since lifetime of the battery usually defines the sensor's lifetime. One of the most cited examples is the battlefield surveillance of enemy's territory, wherein a large number of sensors are dropped from a low-flying airplane or UAV so that activities on the ground can be detected and communicated. Other potential commercial fields include machinery prognosis, bio sensing, environmental monitoring and health of large bridges and structures.

1.2 The Communication Puzzle

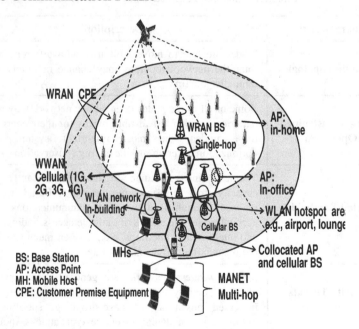

Figure 1.2 – The envisioned communication puzzle of 4G and beyond

Fourth-generation (4G) wireless networks, such as WiMax and LTE, are starting to be deployed around the world. These technologies are able to support Internet-like services, which are enabled through a seamless integration of many wireless networks with different transmission speeds and ranges interconnected through a high-speed backbone, as depicted in Figure 1.2. Fourth generation wireless networks include Wireless Personal Area Networks (Wireless PANs or WPANs for short), Wireless Local Area Networks (Wireless LANs or WLANs for short), Wireless Metropolitan Area Networks (Wireless MANs or WMANs for short), Wireless Regional Area Networks (Wireless RANs or WRAN for short) Wireless Local Loops (WLLs), Customer Premise Equipment (CPE), Cellular Wide Area Networks and Satellite Networks (see Figure 1.2).

These networks may be organized either with the support of a fixed infrastructure or in the form of a MANET [Cordeiro2003]. Usually, ad hoc networks are built upon the infrastructures provided by wireless LANs and PANs which are, in turn, supported through technologies such as the Institute of Electrical and Electronics Engineers (IEEE) 802.11 [IEEE-802.112007], High Performance Local Area Network (HIPERLAN) [HIPERLAN1999], Bluetooth [Bluetooth], IEEE 802.15 [IEEE-802.15] standards and so on. The widespread and integrated use of wireless networks will increase the usefulness of new wireless applications, especially multimedia applications deployment such as video-on-demand, audio-on-demand, voice over IP, streaming media, interactive gaming and so on.

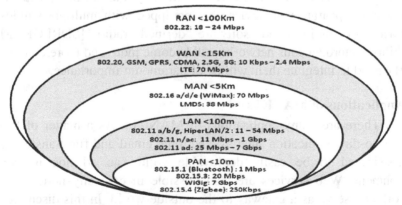

Figure 1.3 – The scope of various Wireless Technologies

LANs and Wide Area Network (WANs) are the original flavors of network design. The concept of "area" made good sense in early days, because a key distinction between a LAN and a WAN involves the physical distance spanned by the network. A LAN typically connects computers in a single building or campus, whereas a WAN generally covers large distances (states, countries, continents). A third category, the Metropolitan Area Networks (MANs), also fits into this distance-based scheme as it covers towns and cities. The fourth category, the Personal Area Network (PAN) has been designed to interact with highly mobile personal objects to share hardware and software resources. Recently, Regional Area Network (RAN) have been introduced [Cordeiro2005], which promises to provide coverage ranges in the order of tens of kilometers with applications in rural and remote areas. Figure 1.3 compares various wireless networks in terms of the popular standards, speeds, communication ranges and applications.

Since the infrastructure for building ad hoc networks are mostly within the framework of Wireless LANs and Wireless PANs, their scope given in Figure 1.3 is particularly useful. This is not to say, however, that the infrastructures provided by WMANs, Wireless WANs (WWANs), and WRANs, depicted in Figure 1.3, cannot interoperate with the ad hoc network. As a matter of fact, this problem is currently undergoing as to integrate MANETs with MANs and WWANs, where the infrastructure provided by these networks would serve as a backhaul to connect the MANET with the outside world (e.g., Internet). Furthermore, with the large scale appearance of devices are equipped with multiple wireless interfaces or multi band software defined radio [SDRFORUM] capability, heterogeneous networks will become more and more common and the need to integrate them will be of paramount importance.

1.3 Applications of MANETs

There are many applications of MANETs. As a matter of fact, any day-to-day application such as electronic email and file transfer can be considered to be easily deployable within an ad hoc network environment. Web services are also possible in case any node in the network can serve as a gateway to the outside world. In this discussion, we need not emphasize a wide range of military applications possible with ad hoc networks. Not to mention, the technology was initially developed keeping in mind the military applications, such as battlefield in an unknown territory where an infrastructure network is almost impossible to establish or maintain. In such situations, the ad hoc networks having self-organizing capability can be effectively used where other technologies either fail or cannot be deployed effectively. Advanced features of wireless mobile systems, including data rates compatible with multimedia applications, global roaming capability, and coordination with other network structures, are enabling new applications. Some well-known ad hoc network applications are:

- **Collaborative Work** – For some business scenarios, the need for collaborative computing might be more important outside office environments than inside a building. After all, it is often the case where people do need to have outside meetings to cooperate and exchange information on a given project;

- **Crisis-management Applications** – These arise, for example, as a result of natural disasters where the entire communications infrastructure is in disarray (for example, Tsunamis, hurricanes, etc.). Restoring communications quickly is essential. By using ad hoc networks, an infrastructure could be set up in hours instead of days/weeks required for wire-line communications;

- **Vehicular Area Network** – An ad hoc network is specifically useful in forming networks among different vehicles on the road and can propagate information like accidents, congestion, etc. It can also help determining close by facilities in the neighborhood such as gas station, restaurants, hospitals, and other facilities.

- **Personal Area Networking** – A personal area network (PAN) is a short-range, localized network where nodes are usually associated with a given person. These nodes could be attached to someone's cell phone, laptop, television set, and so on. In these scenarios, mobility is only a major consideration when interaction among several PANs is necessary, illustrating the case where, for instance, people meet in real life. Bluetooth [Haarsten1998] and WiGig [WiGigAlliance] are examples of technologies aimed at, among other things, supporting PANs by eliminating the need for wires between devices such as printers, cell phones, PDAs, laptop computers, headsets, television sets, and so on. Wireless PANs are discussed in Chapter 5.

1.4 Challenges

Ad hoc networking has been a popular field of study during the last few years. Almost every aspect of the network has been explored in some level of detail. Yet, no ultimate resolution to any of the problems is found or, at least, agreed upon. On the contrary, more questions have arisen. Similar to ad hoc networks, many aspects of sensor networks have also been explored but, contrary to ad hoc networks, there are many more issues which remain to be addressed.

This section outlines the major problems that ought to be addressed. The protocol dependent development possibilities are mostly omitted and the focus is on the "big picture", on the problems that stand in the way of having peer-to-peer connectivity everywhere in the future. The topics that need to be resolved are:

- Scalability;
- Quality of service;
- Client server model shift;
- Security;
- Interoperation with the Internet;
- Energy conservation; and
- Node cooperation.

Here, we plan to add on the approach presented in [Perkins2001, Penttinen2002], with several updates. The discussion here attempts to provide a thorough discussion of the future challenges in ad hoc and sensor networking.

1.4.1 Scalability

Most of the visionaries depicting applications which are anticipated to benefit from the ad hoc and sensor networking technology take scalability as granted. Imagine, for example, the vision of ubiquitous computing where networks can be of "any size". However, it is unclear how such large networks can actually grow. Ad hoc networks suffer from the scalability problems in capacity. To exemplify this, we may look into simple interference studies. In a non-cooperative network, where omni-directional antennas are being used, the throughput per node decreases at a rate $1/\sqrt{N}$, where N is the number of nodes [Gupta2000]. That is, in a network with 100 nodes, a single device gets, at most, approximately one tenth of the theoretical network data rate. This problem, however, cannot be fixed except by physical layer improvements, such as directional antennas which are discussed in Chapter 7.

If the available capacity sets some limits for communications, so do the protocols. Route acquisition, service location and encryption key exchanges are just few examples of tasks that will require considerable overhead as the network size grows. If the scarce resources are wasted with profuse control traffic, these networks may see never the day dawn. Therefore, scalability is a crucial research topic and has to be taken into account in the design of solutions for ad hoc and sensor networks.

1.4.2 Quality of Service

The heterogeneity of existing Internet applications has challenged network designers who have built the network to provide best-effort service only. Voice, live video and file transfer are just a few applications having very different requirements. Quality of Service (QoS)-aware solution is being developed to meet the emerging requirements of these applications. QoS has to be guaranteed by the network to provide certain performance for a given flow, or a collection of flows, in terms of QoS parameters such as delay, jitter, bandwidth, packet loss probability, and so on. QoS routing, discussed in Chapter 2, attempts to locate routes that satisfy given performance constraints and then reserve enough capacity for the flow. Despite the current research efforts in the QoS area, QoS in ad hoc and sensor networks is still an unexplored area. Issues of QoS robustness, QoS routing policies, algorithms and protocols with multiple, including preemptive, priorities remain to be addressed.

1.4.3 Client-Server Model Shift

In the Internet, a network client is typically configured to use a server as its partner for network transactions. These servers can be found automatically or by static configuration. In ad hoc networks, however, the network structure cannot be defined by collecting IP addresses into subnets. There may not be servers, but the demand for basic services still exists. Address allocation, name resolution, authentication and the service location itself are just examples of the very basic services which are needed but their location in the network is unknown and possibly even changing over time. Due to the infrastructureless nature of these networks and node mobility, a different addressing approach may be required. In addition, it is still not clear who will be responsible for managing various network services. Therefore, while there has been vast research initiatives in this area, the issue of shift from the traditional client-sever model remains to be appropriately addressed, although a lot of activity is going on within the Zero Configuration working group [ZeroConf] of the Internet Engineering Task Force (IETF) and also within the UPnP™ forum [UPnP] being considered in the context of the Digital Living Network Alliance [DLNA].

1.4.4 Security

Ad hoc and sensor networks are particularly prone to malicious behavior. Lack of any centralized network management or certification authority makes these dynamically changing wireless structures extremely vulnerable to infiltration, eavesdropping, interference, and so on. Security is often considered to be the major "roadblock" in the commercial application of this technology. Security is indeed one of the most difficult problems to be solved, but it has received only modest attention so far although considerable progress has been made as shown later in Chapter 9. The "golden age" of this research field can be expected to dawn only after the functional problems at the underlying layers have been agreed up on.

1.4.5 Interoperation with the Internet

It seems very likely that the most common applications of ad hoc networks require some Internet connectivity. However, the issue of defining the interface between the two very different networks is not straightforward. If a node has an Internet connection, it could offer Internet connectivity to the other nodes. This node could define itself as a default router and the whole network could be considered to be "single-hop" from the Internet perspective although the connections are physically over several hops. Recently, a practical solution for this problem was suggested in [Sun2002] that combines Mobile IP technology [Agrawal2002] with ad hoc routing so that the gateway node can be considered to be *foreign agent* as defined in Mobile IP.

1.4.6 Energy Conservation

Energy conservative networks are becoming extremely popular within the ad hoc and specially sensor networking research community. Energy conservation is currently being addressed in every layer of the protocol stack. There are two primary research topics which are almost identical: maximization of lifetime of a single battery and maximization of the lifetime of the whole network. The former is related to commercial applications and node cooperation issues whereas the latter is more crucial, for instance, in military environments where node cooperation is assumed. The goals can be achieved either by developing better batteries, or by making the network terminals to be more energy efficient. The first approach is likely to give a 40% increase in battery life in the near future (with Li-Polymer batteries) [Petrioli2001]. As to the device power consumption, the primary aspect of achieving energy savings is through low power hardware development such as variable clock speed CPUs, flash memory, and disk spindown [Jones2001]. However, from the networking point of view, our interest naturally focuses on the device's network interface, which is often the single largest consumer of power.

Energy efficiency at the network interface can be improved by developing transmission/reception technologies at the physical layer and by sensing inactivity at the application layer, but especially with specific networking algorithms. Much research has been carried out at the physical, medium access control (MAC) and routing layers, while little has been done at the transport and application layers. Nevertheless, there is still much more work need to be done.

1.4.7 Node (MH) Cooperation

Closely related to the security issues, the node cooperation stands in the way of commercial application of the technology. The fundamental question is why anyone should relay other people's data so as to receive the corresponding service from the others. However, when differences in amount and data priority come into picture, the situation becomes far more complex. Surely, a critical fire alarm box should not waste its batteries for relaying gaming data, nor should it be denied access to other nodes because of such restrictive behavior. Encouraging nodes to cooperate may lead to the introduction of billing, similar to Internet congestion control [MacKie-Mason1994]. Well-behaving network members could be rewarded, while selfish or malicious users could be charged higher rates. Implementation of any kind of billing mechanism is very challenging and is still wide open [Yoo2006].

1.4.8 Interoperation

The self-organization of ad hoc networks is a challenge when two independently formed networks come physically close to each other. This is an unexplored research topic that has implications on all levels of the system design. The issue is: what happens when two autonomous ad hoc networks move into same area. Surely they would be unable to avoid interfering with each other. Ideally, the networks would recognize the situation and be merged. However, the issue of joining two networks is not trivial; the networks may be using different synchronization, or even different MAC or routing protocols. Security also becomes a major concern. Can the networks adapt to the situation? For example; a military unit moving into an area covered by a sensor network could be such a situation; moving unit would probably be using different routing protocol with location information support, while the sensor network would have a simple static routing protocol.

Another important issue comes into picture when we talk about all wireless networks. One of the most important aims of recent research on all wireless networks is to provide seamless integration of all types of networks. This issue raises questions on how the ad hoc network could be designed so that they are compatible with, for instance, wireless LANs, third generation (3G) and 4G cellular networks. In Chapter 14 we discuss this complex issue and provide insights on the current status in this area.

1.5 Book Organization

The organization of this book follows a new approach which we find to be best suitable when discussing ad hoc networks. Unlike traditional networking books which adopt either a strict bottom-up or a top-down approach, our experience as educators, researchers and learners in the ad hoc and sensor networking arena has shown that such approaches are not suitable to understand how these networks really work. Ad hoc and sensor networks are very particular as there are many cross-layer interactions and one layer cannot be fully understood without at least knowing the basics of the others. Thus, employing a strict top-down or bottom-up approach is not appropriate. Here, we introduce a new explanatory model specifically designed to best understand all the aspects of these networks, from design to performance issues. We initiate the discussion by the network layer, which we believe is the best layer to kick-start the study in this area as it may be the one which requires minimum knowledge of the others. In addition, the network layer is often seen as perhaps "the easiest" to understand for both beginners and advanced people. Next, we move down and present solutions at the lower layers (i.e., physical and MAC) and, finally, upper layers (i.e., transport). We conclude this book by discussing the growing field of integration of heterogeneous wireless networks, where ad hoc and sensor networks are required to interoperate with other wireless networks such as cellular and wireless LANs.

The way this book is organized is depicted in Figure 1.4. In this figure, we clearly indicate which layers of the protocol stack are covered in which chapters. It is worth noticing that we do not have a separate chapter for the application layer. Again, this is due to the unique nature of ad hoc and sensor networks where the design choice of a solution is usually taken on the basis of supporting a particular category of applications. As in other areas of computer science, there is no one fits-all solution. Therefore, we take a different and novel approach where we believe that applications have to be discussed throughout the book, in a scattered manner, together with the associated solutions. This way, a reader can get the right perspective about the best suitability of a given solution to a specific application. In view of this, we organized this book as follows:

- **Chapter 2**: Here we introduce unicast routing over ad hoc networks. We provide a thorough discussion of the major unicast protocols, including proactive, reactive, position-based, and QoS routing. In addition, we present the broadcast storm problem in ad hoc networks and possible solutions.

- **Chapter 3**: In this chapter we present important issues of multicasting and geocasting in ad hoc networks, by discussing the applications, giving the motivation, and finally providing a comprehensive coverage of various proposed protocols.
- **Chapters 4 and 5**: These chapters deal with the most prominent and widely used MAC and physical layers for ad hoc networks, namely, the IEEE 802.11 for Wireless LANs and the IEEE 802.15 (including the Bluetooth technology) for Wireless PANs, respectively. These chapters provide a thorough discussion of these two standards and how they are used to support ad hoc networking. Here, we note that the Link Layer Control (LLC) sub-layer is not addressed in this book as it is standardized in the IEEE 802.2 to provide a uniform interface between the various network and MAC layer protocols.
- **Chapter 6**: Wireless Mesh Networks are seen to be the technology for the next generation multimedia wireless service that intelligently uses ad hoc network in connecting different routers so that traffic can be supported with just few routers having Internet access. Some of the important issues are discussed in this chapter.
- **Chapter 7**: In this chapter, we move on to a new and powerful way for increasing the capacity, connectivity, and covertness of ad hoc networks, namely, the use of directional antenna systems. We discuss directional antenna systems from the basic concept of the antenna model, going through the physical, MAC and network layers.
- **Chapter 8**: Cognitive radio networks is an up and coming area in wireless communication, which deals with the ability of radios to harvest unused frequency spectrum and reuse it on an opportunistic basis without causing harmful interference. We introduce cognitive radios and how it is poised to revolutionize wireless communications as we know it today.
- **Chapter 9**: The issue of TCP (Transmission Control Protocol) over ad hoc networks is covered in this chapter. Several aspects of TCP performance are analyzed, including the impact of node mobility, congestion window size, unfairness and the capture problem;
- **Chapter 10**: Here, we present a summary of various emerging applications of wireless sensor networks.
- **Chapter 11**: In this chapter we continue the discussion on sensor networks, with a comprehensive analysis of these networks including energy consumption, coverage and connectivity, MAC and network

layer, multipath routing, networks under constraints such as energy consumption and delay.

- **Chapter 12**: This chapter covers regularly deployed wireless sensor networks in the form of rectangular, triangular, and hexagonal schemes and provides comparison with random topology.
- **Chapter 13**: Security over ad hoc networks is discussed in this chapter. Initially, we show that security in ad hoc networks is a much harder task than in wired networks and motivate the need for security over these networks. Next, we delve into specifics of security over ad hoc networks including key management schemes, secure routing algorithms, cooperation, and intrusion detection systems.
- **Chapter 14**: In this chapter we cover the area of *all wireless networks*. More specifically, we discuss the integration of heterogeneous wireless technologies in the context of ad hoc and sensor networks, and the many issues involved at every layer of the protocol stack. We describe proposed integrated architectures and thoroughly compare them, as well as point out future directions for research.

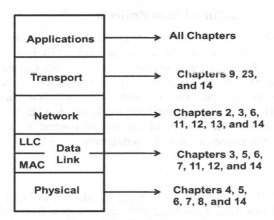

Figure 1.4 – Book organization

As we can see, the organization of this book follows a different approach, but we hope and believe this new approach facilitates better understanding of not only various protocols as separate entities, but also how they interact with each other. At the end of each chapter, we include a special section where we discuss the future directions and challenges in each particular area. These open problems are highlighted to allow the reader to think further into the matter and potentially pursue it as a future research topic.

Finally, we note that this book is intended for researchers and engineers in both industry and academia, as well as for anyone who would like to get a deeper understanding of this growing field of ad hoc and sensor networks, and the current state of research in this area. It is designed to provide a thorough discussion on the issues related to every layer of the protocol stack and is presented in Figure 1.4.

1.6 Conclusions and Future Directions

The topics covered in this book represent a significant portion of what is going on in academia, industry, military, and commercial networks. The vast and extensive material contained in this book will enable the readers to not only understand and position themselves in this hot area of ad hoc and sensor networks, but will allow them to develop new capabilities, enhance skills, share expertise, consolidate knowledge and further develop the area by analyzing and designing future solutions.

Homework Questions/Simulation Projects

Q. 1. There are many performance parameters considered in this chapter. Can you think of a rationale for allocating weights to these parameters? Explain clearly.

Q. 2. What are the other performance parameters you would like to consider besides the ones covered in this chapter? Explain clearly your rationale for selecting the same.

Q. 3. What are the requirements and service expectations in multimedia communication as compared to voice or text transfer using wireless devices? Explain clearly.

Q. 4. Design a problem based on any of the material covered in this chapter (or in references contained therein) and solve it diligently.

References

[Agrawal2002] D. P. Agrawal and Q.-A. Zeng, Introduction to Wireless and Mobile Systems, Brooks/Cole Publishing, 438 pages, August 2002, ISBN No. 0534-40851-6.

[Bluetooth] Bluetooth SIG, http://www.bluetooth.com/.

[Cordeiro2002] C. M. Cordeiro and D. P. Agrawal, "Mobile Ad Hoc Networking," Tutorial Presented in the 20th Brazilian Symposium on Computer Networks, Editors: Jose Rezende, Lucy Pirmez, and Luiz da Costa Carmo - ACS/NCE/UFRJ, CDD 004.6506 - pp. 125-186, May 2002.

[Cordeiro2003] C. M. Cordeiro, H. Gossain, R. L. Ashok, and D. P. Agrawal. "The Last Mile: Wireless Technologies for Broadband and Home Networks," Tutorial Presented in the 21th Brazilian Symposium on Computer Networks. Editors: Thais Batista, Carlos Ferraz, and Gledson Elias - UFRN/DIMAp, ISBN: 85-88442-51-5, pp. 119-178, May 2003.

[Cordeiro2005] C. Cordeiro, K. Challapali, D. Birru, and S. Shankar, "IEEE 802.22: The First Worldwide Wireless Standard based on Cognitive Radios," in IEEE International Conference on Dynamic Spectrum Access Networks (DySPAN), 2005.

[DLNA] Digital Living Network Alliance (DLNA), "http://www.dlna.org".

[Estrin1999] D. Estrin et al, "New Century Challenges: Scalable Coordination in Sensor Networks," ACM Mobicom, 1999.

[Gupta2000] P. Gupta and P. Kumar, "The capacity of wireless networks," IEEE Transactions on Information Theory, 46(2):388–404, March 2000.

[Haarsten1998] J. Haarsten, "Bluetooth – The Universal Radio Interface for Ad Hoc Wireless Connectivity," Ericsson Review (3), 1998.

[HIPERLAN1999] ETSI, "Broadband Radio Access Networks (BRAN); HIPERLAN type 2 technical specification; Physical (PHY) layer," August 1999.

[IEEE-802.112007] IEEE Std. 802.11. "IEEE Standard for Wireless LAN Medium Access Control (MAC) and Physical Layer (PHY) Specification," 2007.

[IEEE-802.15] IEEE 802.15 Working Group for WPAN, http://www.ieee802.org/15/.

[Jain2005] N. Jain and D.P. Agrawal, "Current Trends in Wireless Sensor Network Design," International Journal of Distributed Sensor Networks, Vol. 1, No. 1, 2005, pp. 101-122.

[Jones2001] C. Jones, K. Sivalingam, P. Agrawal, and J. Chen, "A survey of energy efficient network protocols for wireless networks," Wireless Networks, 7(4):343–358, September 2001.

[Kahn1999] J. M. Kahn, "New Century Challenges: Mobile Networking for Smart Dust," ACM Mobicom, 1999.

[MacKie-Mason1994] J. MacKie-Mason and H. Varian, "Pricing the Internet," In Public Access to the Internet, Prentice-Hall, New Jersey, 1994.

[Penttinen2002] A. Penttinen, "Research on Ad Hoc Networking: Current Activity and Future Directions," Networking Laboratory, Helsinki University of Technology, 2002.

[Perkins2001] C. E. Perkins, "Ad Hoc Networking," Addison-Wesley, ISBN: 0201-30976-9, 2001.

[Petrioli2001] C. Petrioli, R. Rao, and J. Redi, "Guest editorial: Energy conserving protocols," in ACM Mobile Networks and Applications (MONET), 6(3):207–209, June 2001.

[Prakash1999] R. Prakash, "Unidirectional Links Prove Costly in Wireless Ad Hoc Networks," In Proceedings of the Third International Workshop on Discrete Algorithms and Methods for Mobile Computing and Communications, August 1999, 15-22.

[SDRFORUM] Software Defined Radio (SDR) Forum, http://www.sdrforum.org/.

[Sun2002] Y. Sun, E. Belding-Royer, and C. E. Perkins, "Internet connectivity for ad hoc mobile networks," International Journal of Wireless Information Networks, special issue on Mobile Ad hoc Networks, 2002.

[UPnP] UPnPTM Forum, "http://www.upnp.org".

[Yoo2006] Y. Yoo and D. P. Agrawal, "Why it pays to be selfish in MANETs," IEEE Wireless Communications Magazine, 2006.

[ZeroConf] IETF zeroconf, Zero Configuration Networking (zeroconf), http://http://www1.ietf.org/html.charters/zeroconf-charter.html.

[WiGigAlliance] Wireless Gigabit Alliance (WiGig), http://www.wigig.org.

Chapter 2

Routing in Ad Hoc Networks

2.1 Introduction

A MANET environment, illustrated in Figure 2.1(a), is characterized by energy-limited nodes (Mobile Hosts (MHs)), bandwidth-constrained, variable-capacity wireless links and dynamic topology, leading to frequent and unpredictable connectivity changes. For example, assume in Figure 2.1(a) that MH S uses MH B to communicate with MH D. However, as MHs in a MANET are mobile, it may so happen that the route from node S to node D changes while in use, and now traverses nodes A and B as depicted in Figure 2.1(b). Therefore, traditional link-state and distance vector routing algorithms [Tanenbaum1996] are not effective in this environment. Numerous MANET routing protocols have been proposed, both under and outside the umbrella of the IETF MANET working group [MANET 1998]. We use the term MH and node interchangeably throughout the text.

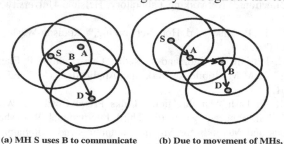

(a) MH S uses B to communicate   (b) Due to movement of MHs,
with MH D   S now uses A and B to reach D

Figure 2.1 – An example of a multi-hop MANET

Routing in a MANET depends on many factors including topology, selection of routers, request initiator's location, and specific underlying characteristics that could serve as a heuristic in finding the path quickly and efficiently. One of the major challenges in designing a routing protocol [Jubin1987] for MANETs is that a node at least needs to know the reachability information to its neighbors for determining a packet route, while the network topology can change quite often in a MANET.

Furthermore, as the number of network nodes can be large, finding route to a destination also requires frequent exchange of routing control information among the MHs. Thus, the amount of update traffic can be substantial, and it is even higher when nodes with increased mobility are present. The MHs can impact route maintenance overhead of routing algorithms in such a way that no bandwidth might be left for the transmission of data packets [Corson1996].

2.2 Topology-Based versus Position-Based Approaches

Routing over ad hoc networks can be broadly classified as *topology-based* or *position-based* approaches. Topology-based routing protocols depend on the information about existing links in the network and utilize them to carry out the task of packet forwarding. They can be further subdivided as being Proactive (or table-driven), Reactive (or on-demand), or Hybrid protocols. Proactive algorithms employ classical routing strategies such as distance-vector or link-state routing and any changes in the link connections are updated periodically throughout the network. They mandate that MHs in a MANET should keep track of routes to all possible destinations all the time. However, proactive protocols may not be appropriate in highly mobile MANETs. This may cause continuous use of a substantial fraction of the network capacity so that the routing information could be kept current. In addition, the quality of channels may change with time due to the shadowing and fast fading and may not be good to use even if there is no mobility [Lin2005].

On the other hand, reactive protocols employ a lazy approach whereby nodes only discover routes to destinations on-demand. In other words, reactive protocols adopt the opposite approach as compared to proactive schemes by finding a route to a destination only when needed. Reactive protocols often take substantially large delay in determining a route while consume much less bandwidth than proactive protocols. Another disadvantage is that in reactive protocols, even though route maintenance is limited to routes currently in use, it may still generate a significant amount of network control traffic when the topology of the network changes frequently. Lastly, packets en route to the destination are likely to be lost if the route in use changes.

Hybrid protocols combine local proactive and global reactive routing in order to achieve a higher level of efficiency and scalability. For example, a proactive scheme may be used for close by MHs only, while routes to distant nodes are found using reactive mode. Usually, but not always, hybrid protocols may be associated with some sort of hierarchy which can either be based on the neighbors of a node or on logical partitions of the network. The major limitation of hybrid schemes combining both strategies is that it still needs to maintain at least those paths that are currently in use. This limits the amount of topological changes that can be tolerated within a given time span.

Finally, position-based routing algorithms overcome some of the limitations of topology-based routing by relying on the availability of additional knowledge of the physical location. Typically, each or some of the MHs determine their own position through the use of the Global Positioning System (GPS) or some other type of positioning technique [Hightower2001]. The sender normally uses a *location service* to determine the position of the destination node, and to incorporate it in the packet destination address field. As we can see, position-based routing does not require establishment or maintenance of routes, but this usually comes at the expense of an extra hardware. As a further enhancement, position-based routing supports the delivery of packets to all nodes in a given geographical region in a natural way, and this is called *geocasting* which is discussed in the next chapter. In the following sections we elaborate on the most prominent protocols under each of these categories.

2.3 Topology-Based Routing Protocols

In this section we describe the protocols hereby termed as topology-based. We start with those employing proactive approach, followed by reactive ones, and hybrid schemes, and finally conclude with a comparison amongst them.

2.3.1 Proactive Routing Approach

In this section, we consider some of the important proactive routing protocols. The most important one is the destination-sequenced distance-vector (DSDV) [Perkins1994] which requires each node to broadcast routing updates periodically.

2.3.1.1 Destination-Sequenced Distance-Vector Protocol

In this routing protocol, every MH in the network maintains a routing table for all possible destinations within the network and the number of hops to each destination. Each entry is marked with a sequence number assigned by the destination MH. The sequence numbers enable the MHs to distinguish stale routes from new ones, thereby avoiding the formation of routing loops. Routing table updates are periodically transmitted throughout the network in order to maintain consistency in the tables.

To alleviate potentially large network update traffic, two possible types of packets can be employed: full dumps or small increment packets. A full dump type of packet carries all available routing information and can require multiple network protocol data units (NPDUs). These packets are transmitted less frequently during periods of occasional movements. Smaller incremental packets are used to relay only the information that has changed since the last full dump. Each of these broadcasts should fit into a standard-size NPDU, thereby decreasing the amount of traffic generated. The MHs maintain an additional table where they store the data sent in the incremental routing information packets. New route broadcasts contain the address of the destination, the number of hops to reach the destination, the sequence number of the information received regarding the destination, as well as a new sequence number unique to the broadcast. The route labeled with the most recent sequence number is always used. In the event that two updates have the same sequence number, the route with the smaller metric is used in order to optimize (shorten) the path. MHs also keep track of settling time of the routes, or the weighted average time that routes to a destination could fluctuate before the route with the best metric is received. By delaying the broadcast of a routing update by the length of the settling time, MHs can reduce network traffic.

Note that if each MH in the network advertises a monotonically increasing sequence number for itself, it may imply that the route just got broken. For example, MH B in Figure 2.1 decides that its route to a destination D is broken; it advertises the route to D with an infinite metric. This results in any node A, which is currently routing packets through B, to incorporate the infinite-metric route into its routing table until node A hears a route to D with a higher sequence number.

2.3.1.2 The Wireless Routing Protocol

The Wireless Routing Protocol (WRP) [Murthy1996] is a table-driven protocol with the goal of maintaining routing information among all nodes in the network. Each node in the network is responsible for maintaining four tables: Distance table, Routing table, Link-cost table, and the Message Retransmission List (MRL) table. Each entry of the MRL contains the sequence number of the update message, a re-transmission counter, an acknowledgment-required flag vector with one entry per neighbor, and a list of updates sent in the update message. The MRL records which updates in an update message ought to be retransmitted and neighbors need to acknowledge the retransmission.

MHs keep each other informed of all link changes through the use of update messages between the neighboring MHs and contains a list of updates (destination, distance to destination, and predecessor of destination), as well as a list indicating which MHs should acknowledge (ACK) the update. After processing updates from neighbors or detecting a change in a link, MHs send update messages to a neighbor. Similarly, any new paths are relayed back to the original MHs so that they can update their tables accordingly.

MHs learn about the existence of their neighbors from the receipt of acknowledgments and other messages. If a MH does not send any message for a specified time period, it must send a hello message to ensure connectivity. Otherwise, the lack of messages from the MH indicates the failure of that link and this may cause a false alarm. Whenever a MH receives a hello message from a new MH, it adds this new MH to its routing table and sends a copy of its routing table information to this new MH.

Part of the novelty in WRP stems from the way it achieves freedom from loops as nodes communicate the distance and second-to-last hop information for each destination in the network. WRP belongs to the class of path-finding algorithms with an important exception that it avoids the "count-to-infinity" problem by forcing each node to perform consistency checks on predecessor information reported by all its neighbors. This ultimately eliminates looping situations and provides faster route convergence if and when a link failure occurs.

2.3.1.3 The Topology Broadcast based on Reverse Path Forwarding Protocol

The Topology Broadcast based on Reverse Path Forwarding (TBRPF) protocol [Bellur1999] considers the problem of broadcasting topology information (including link costs and up/down status) to all nodes of a communication network. This information, together with a path selection algorithm, can be used by each node to compute preferred paths to all destinations, i.e., to perform routing based on link states. Most link-state routing protocols, including the Open Shortest Path First (OSPF) [Tanenbaum1996], are based on flooding. In these protocols, each link-state update is sent on every link of the network. Although flooding is useful in networks with high bandwidth links, it can consume a significant percentage of link bandwidth in MANETs where the network contains links with relatively low bandwidth.

The communication cost of broadcasting topology information can be reduced if the updates are sent along spanning trees. However, there is additional communication cost for maintaining these trees. The main concern here is whether the total communication cost is significantly less as compared to this additional cost. The TBRPF protocol is based on the *extended reverse-path forwarding* (ERPF) algorithm [Dalal1978], in which messages generated by a given source are broadcast in the reverse direction along the directed spanning tree formed by the shortest paths from all nodes to the source. ERPF assumes the use of an underlying routing algorithm by each node i in selecting the next node $p_i(v)$ along the shortest path to each destination (or broadcast source) v. The node $p_i(v)$ then becomes the *parent* of i on the broadcast tree rooted at source v. Each node informs its parent so that each parent becomes aware of its *children* for each source. A node i receiving a broadcast message originating from source v from its parent $p_i(v)$ forwards the message to its children for source v. ERPF is not reliable when the shortest paths can change due to the dynamic topology

[Dalal1978]. In fact, since ERPF is not reliable, the underlying routing algorithm should not depend on ERPF for topology broadcast.

TBRPF combines the concept of ERPF with the use of sequence numbers to achieve reliability, and the computation of minimum-hop paths based on the topology information received along the broadcast tree rooted at the source of the information. Since minimum-hop paths are computed, each source node broadcasts link-state updates for its outgoing links along a minimum-hop tree rooted at the source. Therefore, a separate broadcast tree is created for each source. The use of minimum-hop trees instead of shortest-path trees (based on link costs) results in less frequent changes in the broadcast trees and therefore less communication cost to maintain the trees.

TBRPF has the following chicken-egg paradox: it computes the paths for the broadcast trees based on the information received along the trees themselves. Thus, the correctness of TBRPF is not obvious. However, it is shown in [Bellur1999] that every MH knows the correct topology in finite time using TBRPF, if no topology changes occur for some time. TBRPF is a simple, practical protocol that generates less update/control traffic than flooding and is therefore especially useful in networks that have frequent topology changes and have limited bandwidth.

2.3.1.4 The Optimized Link State Routing Protocol

The Optimized Link State Routing (OLSR) protocol [Jacquet2001] is a proactive protocol based on the link state algorithm. In a pure link state protocol, all the links with neighboring nodes are declared and are flooded in the entire network. OLSR protocol is an optimization of a pure link state protocol for MANETs. First, it reduces the size of control packets: instead of all links, it declares only a subset of links amongst its neighbors which serves as its multipoint relay selectors (described next). Secondly, it minimizes flooding of this control traffic by using only the selected nodes, called multipoint relays, in diffusing its messages throughout the network. Apart from normal periodic control messages, the protocol does not generate extra control traffic in response to link failures or additions. The protocol keeps the routes for all the destinations in the network, hence it is beneficial for the traffic patterns with a large subset of MHs are communicating with each other, and the <source, destination> pairs are also changing with time. The protocol is particularly suitable for large and dense networks, as the optimization done using the multipoint relays works well in this context.

OLSR is designed to work in a completely distributed manner and thus does not depend upon any central entity. It does not require a reliable transmission for its control messages: each node sends its control messages periodically, and can therefore sustain a loss of some packets from time to time which happens due to collisions or other transmission problems in radio networks. In addition, OLSR does not need an in-order delivery of its messages: each control message contains a sequence number of most recent information, reordering can be done at the receiving end. OLSR protocol performs hop-by-hop routing, i.e., each node uses its most recent information to route a packet. Therefore, when a node is moving, its packets can be successfully delivered to it, if its speed is such that its movement could be followed in its neighborhood.

2.3.1.4.1 Multipoint Relays

The idea of multipoint relays [HIPERLAN1996] is to minimize the flooding of broadcast packets in the network by reducing duplicate retransmissions in the same region. Each MH in the network selects a set of neighboring MHs, to retransmit its packets and is called the multipoint relays (MPRs) of that node. The neighbors of any node N which are not in its MPR set receive the packet but do not retransmit it. Every broadcast message coming from these MPR Selectors of a node is assumed to be retransmitted by that node. This set can change over time and is indicated by the selector nodes in their hello messages.

Each node selects its multipoint relay set MPR among its one hop neighbors in such a manner that the set covers (in terms of radio range) all the nodes that are two hops away. The smaller is the multipoint relay set, the more optimal is the routing protocol. Figure 2.2 shows the multipoint relay selection around MH N. Multipoint relays are selected among the one-hop neighbors with a bi-directional link. Therefore, selecting the route through multipoint relays automatically avoids the problems associated with data packet transfer on unidirectional links.

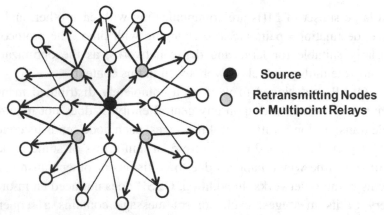

Figure 2.2 – Multipoint relays [Taken from Jacquet2001]

2.3.1.5 The Source Tree Adaptive Routing Protocol

Unlike most of the other proactive ad hoc routing approaches, the Source Tree Adaptive Routing (STAR) protocol [Garcia-Luna-Aceves1999] does not use periodic messages to update its neighbors. STAR is an attempt to create the same routing performance as the other proactive protocols and still be equal or better on bandwidth efficiency. To be able to do this, on demand route optimization has been put aside and the routes are allowed to be non-optimal to save bandwidth. However, STAR depends on an underlying protocol which must reliably keep track of the neighboring MHs. This could be implemented with periodic messages, but is not required. In addition to this, the link layer must provide reliable broadcasting, or else this feature will have to be implemented into STAR with an extra routing rule.

2.3.2 Reactive Routing Approach

In this section, we describe some of the most cited reactive routing protocols.

2.3.2.1 Dynamic Source Routing

The Dynamic Source Routing (DSR) [Broch1998, Johnson1996] algorithm is an innovative approach to routing in a MANET in which nodes communicate along paths stored in source routes carried by the data packets. It is referred to as one of the purest examples of an on-demand protocol [Perkins2001]. In DSR, MHs maintain route caches that contain the source routes which the MH is aware of. Entries in the route cache are continually updated as new routes are learned. The protocol consists of two major phases: route discovery and route maintenance. When a MH has a packet to send to some destination, it first consults its route cache to determine whether it already has a route to the destination. If it has a route to the destination, it will use this route to send the packet. If the MH does not have such an unexpired route, it initiates route discovery by broadcasting a route request packet containing the address of the destination, along with the source MH's address and a unique identification number. Each node receiving the packet checks whether it knows of a route to the destination. If it does not, it adds its own address to the route record of the packet and then forwards the packet along its outgoing links. To limit the number of route requests propagated on the outgoing links of a MH, a MH only forwards the route request if it has not yet seen the request.

A route reply is generated when the route request reaches either the destination itself, or an intermediate node that in its route cache contains an unexpired route to the destination. By the time the packet reaches either the destination or such an intermediate node, it contains a route record with the sequence of hops taken. Figure 2.3(a) illustrates the formation of the route as the route request propagates through the network. If the node generating the route reply is the destination, it places the route record contained in the route request into the route reply. If the responding node is an intermediate node, it appends its cached route to the route record and then generates the route reply. To return the route reply, the responding node must have a route to the initiator. If it has a route to the initiator in its route cache, it may use that route. Otherwise, if symmetric links (defined in Chapter 1) are supported, the node may reverse the route in the route record. If symmetric links are not supported, the node may initiate its own route discovery and piggyback the route reply on the new route request. Figure 2.3(b) shows the transmission of route record back to the source node.

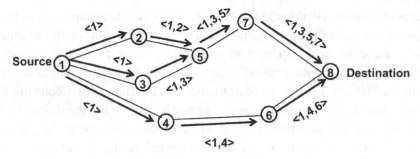

Figure 2.3(a) – Route discovery in DSR

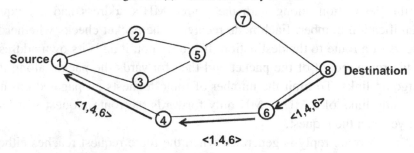

Figure 2.3(b) – Propagation of route reply in DSR

Route maintenance is accomplished through the use of route error packets and acknowledgments. Route error packets are generated at a node when the data link layer encounters a fatal transmission problem. When a route error packet is received, the hop in error is removed from the node's route cache and all routes containing the hop are truncated at that point. In addition to route error messages, acknowledgments are used to verify the correct operation of the route links. These include passive acknowledgments, where a MH is able to hear the next hop forwarding the packet along the route.

DSR also supports multi-path in its design as a built-in feature with no need for extra add-ons. This comes in very handy when a route fails, another valid route can be obtained from the route cache if one exists. In other words, route cache itself possesses multi-path capability by allowing the storage of more than one route to a destination.

The next routing protocol we consider is the Ad Hoc On-Demand Distance Vector (AODV [Perkins1999] which is basically a combination of DSDV and DSR.

2.3.2.2 The Ad Hoc On-Demand Distance Vector Protocol

The Ad Hoc On-Demand Distance Vector (AODV) routing protocol [Perkins1999] is basically a combination of DSDV and DSR. It borrows the basic on-demand mechanism of Route Discovery and Route Maintenance from DSR, plus the use of hop-by-hop routing, sequence numbers, and periodic beacons from DSDV. AODV minimizes the number of required broadcasts by creating routes only on-demand basis, as opposed to maintaining a complete list of routes as in the DSDV algorithm. Authors of AODV classify it as a pure on-demand route acquisition system since MHs that are not on a selected path, do not maintain routing information or participate in routing table exchanges. It supports only symmetric links with two different phases:

- Route Discovery, Route Maintenance; and
- Data forwarding.

When a source MH desires to send a message and does not already have a valid route to the destination, it initiates a path discovery process to locate the corresponding MH. It broadcasts a route request (RREQ) packet to its neighbors, which then forwards the request to their neighbors, and so on, until either the destination or an intermediate MH with a "fresh enough" route to the destination is reached. Figure 2.4(a) illustrates the propagation of the broadcast RREQs across the network. AODV utilizes destination sequence numbers to ensure all routes are loop-free and contain the most recent route information. Each node maintains its own sequence number, as well as a broadcast ID which is incremented for every RREQ the node initiates Together with the node's IP address, this uniquely identifies an RREQ. Along with the node's sequence number and the broadcast ID, the RREQ includes the most recent sequence number it has for the destination. Intermediate nodes can reply to the RREQ only if they have a route to the destination whose corresponding destination sequence number is greater than or equal to that contained in the RREQ.

During the process of forwarding the RREQ, intermediate nodes record in their route tables the address of the neighbor from which the first copy of the broadcast packet was received, thereby establishing a reverse path. If additional copies of the same RREQ are later received, they are discarded. Once the RREQ reaches the destination or an intermediate node with a fresh enough route, the destination/intermediate node responds by unicasting a route reply (RREP) packet back to the neighbor from which it first received the RREQ (Figure 2.4(b)). As the RREP is routed back along the reverse path, nodes along this path set up forward route entries in their route tables that point to the node from which the RREP came. Associated with each route entry is a route timer which causes the deletion of the entry if it is not used within the specified lifetime. Because the RREP is forwarded along the path established by the RREQ, AODV only supports the use of symmetric links.

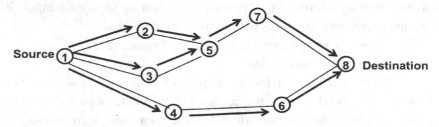

Figure 2.4(a) – Propagation of RREQ in AODV

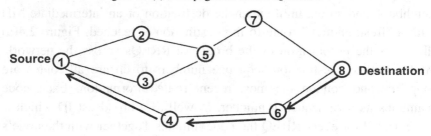

Figure 2.4(b) – Path taken by the RREP in AODV

Routes are maintained as follows. If a source MH moves, it is able to reinitiate the route discovery protocol to find a new route to the destination. If a MH along the route moves, its upstream neighbor notices the move and propagates a link failure notification message (an RREP with infinite metric) to each of its active upstream neighbors to inform them of the breakage of that part of the route. These MHs in turn propagate the link failure notification to their upstream neighbors, and so on until the source node is reached. The source MH may then choose to re-initiate route discovery for that destination if a route is still desired.

An important aspect of the protocol is the use of hello messages as periodic local broadcasts to inform each MH in its neighborhood to maintain local connectivity. However, the use of hello messages may not be required at all times. Nodes listen for re-transmission of data packets to ensure that the next hop is still within reach. If such a re-transmission is not heard, the node may use techniques to determine whether the next hop is within its communication range. The hello messages may also list other nodes from which a mobile node has recently heard, thereby yielding greater knowledge of network connectivity.

AODV is designed for unicast routing only, and multi-path is not supported. In other words, only one route to a given destination can exist at a time. However, enhancements have been proposed which extend the base AODV to provide multi-path capability, and it is known as Multi-path AODV (MAODV) [Marina2001].

2.3.2.3 Link Reversal Routing and TORA

The Temporally Ordered Routing Algorithm (TORA) [Park1997] is a highly adaptive loop-free distributed routing algorithm based on the concept of link reversal. It is designed to minimize reaction to topological changes. A key design concept in TORA is that it decouples the generation of potentially far-reaching control messages from the rate of topological changes. Such messaging is typically localized to a very small set of nodes near the change without having to resort to a complex dynamic, hierarchical routing solution. Route optimality (shortest-path) is considered of secondary importance, and longer routes are often used if discovery of newer routes could be avoided. TORA is also characterized by a multi-path routing capability.

Each node has a height with respect to the destination that is computed by the routing protocol. Figure 2.5 illustrates the use of the height metric. It is simply the distance from the destination node. TORA is proposed to operate in a highly dynamic mobile networking environment. It is source initiated and provides multiple routes for any desired source/destination pair. To accomplish this, nodes need to maintain routing information about adjacent (one-hop) nodes. The protocol performs three basic functions:

- Route creation,

- Route maintenance, and
- Route erasure.

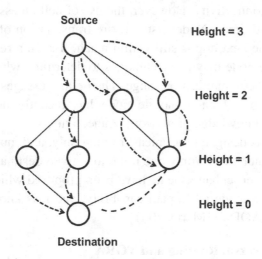

Figure 2.5 – TORA Height Metric

From each node to each destination in the network, a separate directed acyclic graph (DAG) is maintained. When a node needs a route to a particular destination, it broadcasts a QUERY packet containing the address of the destination for which it requires a route. This packet propagates through the network until it reaches either the destination, or an intermediate node having a route to the destination. The recipient of the QUERY then broadcasts an UPDATE packet, listing its height with respect to the destination. As this packet propagates through the network, each node that receives the UPDATE sets its height to a value greater than the height of the neighbor from which the UPDATE has been received. This has the effect of creating a series of directed links from the original sender of the QUERY to the node that initially generated the UPDATE. When a node discovers that a route to a destination is no longer valid, it adjusts its height so that it is a local maximum with respect to its neighbors and transmits an UPDATE packet. If the node has no neighbors of finite height with respect to this destination, then the MH attempts to discover a new route as described above. When a node detects a network partition, it generates a CLEAR packet that resets routing state and removes invalid routes from the network.

TORA is layered on top of IMEP, the Internet MANET Encapsulation Protocol [Corson997], which is required to provide reliable, in-order delivery of all routing control messages from a node to each of its neighbors, plus notification to the routing protocol whenever a link to one of its neighbors is created or broken. To minimize overhead, IMEP aggregates many TORA and IMEP control messages (which IMEP refers to as *objects*) together into a single packet (as an *object block*) before transmission. Each block carries a sequence number and a response list of other nodes from which an ACK has not yet been received, and only those nodes acknowledge the block when receiving it; IMEP retransmits each block with some period, and continues to retransmit it if needed for some maximum total period, after which TORA is notified of each broken link to unacknowledged nodes. For link status sensing and maintaining a list of a node's neighbors, each IMEP node periodically transmits a BEACON packet, which is answered by each node hearing it with a HELLO packet.

As we mentioned earlier, during the route creation and maintenance phases, nodes use the "height" metric to establish a DAG rooted at the destination. Thereafter, links are assigned a direction (upstream or downstream) based on the relative height metric of neighboring nodes as shown in Figure 2.6(a). When node mobility causes the DAG route to be broken, route maintenance becomes necessary to reestablish a DAG rooted at the same destination. As shown in Figure 2.6(b), upon failure of the last downstream link, a node generates a new reference level that effectively coordinates a structured reaction to the failure. Links are reversed to reflect the change in adapting to the new reference level. Timing is an important factor for TORA because the "height" metric is dependent on the logical time of a link failure; TORA assumes that all nodes have synchronized clocks (accomplished via an external time source such as the Global Positioning System). TORA's metric comprises of quintuple elements, namely:

- Logical time of a link failure,
- The unique ID of the node that defined new reference level,
- A reflection indicator bit,
- A propagation ordering parameter, and
- The unique ID of the node.

The first three elements collectively represent the reference level. A new reference level is defined each time a node loses its last downstream link due to a link failure. TORA's route erasure phase essentially involves flooding a broadcast clear packet (CLR) throughout the network to erase invalid routes. In TORA, oscillations might occur, especially when multiple sets of coordinating nodes concurrently detect partitions, erase routes, and build new routes based on each other (Figure 2.7). Because TORA uses inter-nodal coordination, its instability is similar to the "count-to-infinity" problem, except that such oscillations are temporary and the route ultimately convergences. Note that TORA is partially proactive and partially reactive. It is reactive in the sense that route creation is initiated on-demand. However, route maintenance is done on a proactive basis such that multiple routing options are available in case of link failures.

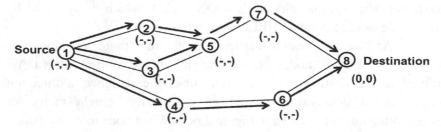

Figure 2.6(a) – Propagation of the Query Message

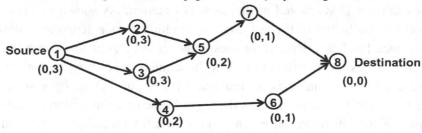

Figure 2.6(b) – Node's Height updated as a Result of the Update Message

2.3.3 Hybrid Routing Approach

Even though sometimes not explicit, most hybrid protocols do try to employ some sort of hierarchical arrangement (or pseudo hierarchy). Usually, this hierarchy can be based either on the neighbors of a node or in different partitions of the network. We now present some of the most referred hybrid routing protocols for MANETs.

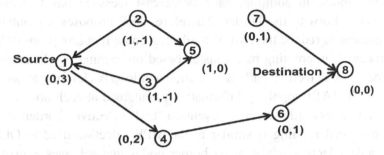

Figure 2.7 – Route Maintenance in TORA

2.3.3.1 Zone Routing Protocol

Zone Routing Protocol (ZRP) [Haas1998a, Haas1998b] is an example of hybrid reactive and proactive schemes. It limits the scope of the proactive procedure only to the node's local neighborhood, while the search being global throughout the network can be performed efficiently by querying selected nodes in the network, as opposed to querying all the nodes. A node employing ZRP proactively maintains routes to destinations within a local neighborhood, referred to as a routing zone and is defined as a collection of nodes with hop distance from the node in question is no greater than a parameter referred to as zone radius.

The construction of a routing zone requires a node to first know who its neighbors are. A neighbor is defined as a node that can communicate directly with the node in question and is discovered through a MAC level Neighbor discovery protocol (NDP). The ZRP maintains routing zones through a proactive component called the Intrazone routing protocol (IARP) which is implemented as a modified distance vector scheme. On the other hand, the Interzone routing protocol (IERP) is responsible for acquiring routes to destinations that are located beyond the routing zone. The IERP uses a query-response mechanism to discover routes on-demand. The IERP is distinguished from the standard flooding algorithm by exploiting the structure of the routing zone, through a process known as *bordercasting*. The ZRP provides this service through a component called Border resolution protocol (BRP).

Bordercast is more expensive than the broadcast flooding used in other reactive protocols. Nodes generally have many more border nodes than neighbors. In addition, each bordercast message has to traverse zone-radius hops to the border. Therefore, ZRP proposes a number of mechanisms to reduce the cost of bordercast route requests [Haas1998a]. Redundancy suppressing mechanisms based on caching overhead traffic include query detection, early termination and loop back termination.

The IARP topology information maintained at each node can be used for backward search prevention and selective bordercasting. Selective bordercasting is similar to the MPR selection used in OLSR; each node selects a subset of its border nodes that achieves equivalent coverage. The network layer triggers an IERP route query when a data packet is to be sent to a destination that does not lie within its routing zone. The source generates a route query packet, which is uniquely identified by a combination of the source node's ID and the request number. The query is then broadcast to all the peripheral nodes of the source by adding the node ID to the query. The sequence of recorded node IDs specifies an accumulated route from the source to the current routing zone. If the destination does not appear in the node's routing zone, the node bordercasts the query to its peripheral nodes. If the destination is a member of the routing zone, a route reply is sent back to the source, along the path specified by reversing the accumulated route. A node discards any route query packet for a query that it has previously encountered. An important feature of this route discovery process is that a single route query can return multiple route replies with the quality determined based on some metric. Then, the relative quality of the route can be used to select the best route. Route failure is detected proactively, in conjunction with the IARP and can be repaired locally. If necessary, a hop-limited local request can be used to repair the route, or a route error message can be set to re-initiate the route discovery from the source.

An adaptive and distributed configuration of each node's routing zone in ZRP provides a flexible solution [Samar2004]. This is possible by incorporating local characteristics such as local route information for global route discovery, etc. A substantial improvement is observed that enhances the network scalability and routing robustness.

2.3.3.2 Fisheye State Routing

The Fisheye State Routing (FSR) protocol [Iwata1999] introduces the notion of multi-level fisheye scope to reduce routing update overhead in large networks. Nodes exchange link state entries with their neighbors with a frequency that depends on distance to destination. From link state entries, nodes construct the topology map of the entire network and compute optimal routes. FSR tries to improve the scalability of a routing protocol by putting most efforts in gathering data on the topology information that is most likely to be needed soon. Assuming that nearby changes to the network topology are those most likely to matter, FSR tries to focus its view on nearby changes by observing them with the highest resolution in time and changes at distant nodes are observed with a lower resolution and less frequently. It is possible to interpret the FSR as the one blurring the sharp boundary defined by the ZRP model.

2.3.3.3 Landmark Routing (LANMAR) with Group Mobility

Landmark Ad Hoc Routing (LANMAR) [Pei2000] combines the features of FSR and Landmark routing. The key feature is the use of landmarks for each set of nodes which move as a group (e.g., a group of soldiers in a battlefield). Like FSR, nodes exchange link state only with their neighbors. Routes within Fisheye scope are accurate, while routes to remote groups of nodes are "summarized" by the corresponding landmarks. A packet directed to a remote destination, initially aims at the landmark; as it gets closer to destination it eventually switches to the accurate route provided by Fisheye. In the original wired landmark scheme [Tsuchiya1988], predefined hierarchical address of each node reflects its position within the hierarchy. Each node knows the routes to all the nodes within its hierarchical partition. Moreover, each node knows the routes to various "landmarks" at different hierarchical levels. Packet forwarding is consistent with the landmark hierarchy and the path is gradually refined from top-level hierarchy to lower levels as a packet approaches the destination.

LANMAR borrows the notion of landmarks [Tsuchiya1988] to keep track of logical subnets. A subnet consists of members which have a commonality of interests and are likely to move as a "group" (e.g., soldiers in the battlefield). A "landmark" node is elected in each subnet. The routing scheme itself is a modified version of FSR. The main difference is that the FSR routing table contains "all" nodes in the network, while the LANMAR routing table includes only the nodes within the scope and the landmark nodes. This feature greatly improves scalability by reducing routing table size and update traffic overhead. When a node needs to relay a packet, if the destination is within its neighboring scope, the address is found in the routing table and the packet is forwarded directly. Otherwise, the logical subnet field of the destination is searched and the packet is routed towards the landmark for that logical subnet. The packet, however, does not need to pass through the landmark. Rather, once the packet gets within the scope of the destination, it is routed directly.

The routing update exchange in LANMAR routing is similar to FSR. Each node periodically exchanges topology information with its immediate neighbors. In each update, the node sends entries within its fisheye scope. It also piggybacks a distance vector with size equal to the number of logical subnets and thus landmark nodes. Through this exchange process, the table entries with larger sequence numbers replace the ones with smaller sequence numbers.

2.3.3.4 Cluster-Based Routing Protocol

The Cluster-Based Routing Protocol (CBRP) [Jiang1998] is a partitioning protocol emphasizing support for unidirectional links. Clusters are defined by bi-directional links, but inter-cluster connectivity may be obtained via a pair of unidirectional links. Each node maintains two-hop topology information to define clusters. Each cluster includes an elected cluster head (CH), with which each member node has a bi-directional link. Clusters may be overlapping or disjoint; however, CH may not be adjacent. In addition to exchanging neighbor information for cluster formation, nodes must find and inform their CH(s) of the status of the "gateway" nodes, cluster members which can be reached from a node belonging to another cluster. Thus, each CH has knowledge of all the clusters with which it has bi-directional connectivity, possibly via a pair of unrelated unidirectional links. The latter are discovered by flooding adjacent CHs with a request for an appropriate link. When a source has no route to a destination, it forwards a route request to its CH. The cluster infrastructure is used to reduce the cost of disseminating the request. When a CH receives a request, it appends to the request packet its ID, as well as a list of (non-redundant) adjacent clusters, and rebroadcasts it. A neighboring node which is a gateway to one or more adjacent clusters, uncast the request to the appropriate CH.

When the request reaches the destination, it contains a loose source routing specifying a sequence of clusters. When the route reply is sent from the destination back to the source, each intermediate cluster head writes a complete source route into the reply, optimizing that portion of the route based on its knowledge of cluster topology. Therefore, routes need not pass through cluster heads. When the complete source route is received at the source, it is used for data traffic.

As with DSR, intermediate nodes may generate new routes to take advantage of improved routes or salvaged failed routes. Unlike DSR, only cluster-level (two-hop neighborhood) information may be used for this purpose: nodes do not attempt to cache network-scale topology information.

Table 2.1 – An overview of Protocol Characteristics

Routing Protocol	Route Acquisition	Flood for Route Discovery	Delay for Route Discovery	Multipath Capability	Upon Route Failure
DSDV	Computed a priori	No	No	No	Flood route updates throughout the network
WRP	Computed a priori	No	No	No	Ultimately, updates the routing tables of all nodes by exchanging MRL between neighbors
DSR	On-demand, only when needed	Yes, aggressive use of caching may reduce flood	Yes	Not explicitly, as the technique of salvaging may quickly restore a route	Route error propagated up to the source to erase invalid path
AODV	On-demand, only when needed	Yes, conservative use of cache to reduce route discovery delay	Yes	Not directly, however, multipath AODV (MAODV) protocol includes this support	Route error broadcasted to erase multipath
TORA	On-demand, only when needed	Usually only one flood for initial DAG construction	Yes, once the DAG is constructed, multiple paths are found	Yes	Error is recovered locally and only when alternative routes are not available
ZRP	Hybrid	Only outside a source's zone	Only if the destination is outside the source's zone	No	Hybrid of updating nodes' tables within a zone and propagating route error to the source

2.3.4 Comparison

Table 2.1 summarizes the main characteristics of some of the most prominent topology-based protocols discussed so far. The criteria used for comparison are self-explanatory and have been extensively covered in the previous sections.

2.4 Position-Based Routing

In this section we discuss some ad hoc routing protocols that take advantage of some sort of location information in the routing process [Mauve2001]. Before delving into the forwarding schemes, it is of paramount importance to discuss the principles and issues behind position-based routing, as well as to look into location services.

2.4.1 Principles and Issues

The philosophy of position-based routing is that it is necessary to determine the location of the destination before a packet can be sent. Generally, a location service takes this responsible. Existing location services can be classified according to how many MHs have the service. This can be either *some* specific nodes or *all* the network nodes. Moreover, each location server may maintain the position of *some* specific nodes or *all* the nodes in the network. In the following discussion on location services, four possible combinations are some-for-some, some-for-all, all-for-some, and all-for-all MHs. In position-based routing, the forwarding decision by a MH is essentially based on the position of a packet's destination and the position of the node's immediate one-hop neighbor. Clearly, the position of the destination is contained in the header of the packet. If a node happens to know an accurate position of the destination, it may choose to update the position of the packet before forwarding it. The position of the neighbors is typically learned through one-hop broadcasts. These beacons are sent periodically by all nodes and contain the position of the sending node.

Three main packet forwarding schemes can be defined:
- Greedy forwarding;
- Restricted directional flooding;
- Hierarchical approaches.

For the first two, a node forwards a given packet to one (greedy forwarding) or more (restricted directional flooding) one-hop neighbors that are located closer to the destination than the forwarding node itself. The selection of the neighbor in the greedy case depends on the optimization criteria of the algorithm. It is fairly obvious that both forwarding strategies may fail if there is no one-hop neighbor that is closer to the destination than the forwarding node itself. *Recovery strategies* that cope with this kind of failure are also discussed. The third forwarding strategy is to form a hierarchy in order to scale to a large number of MHs. In this chapter we investigate two representatives of hierarchical routing that use greedy forwarding for wide area routing and non-position based approaches for local area routing.

Figure 2.8 depicts the two main building blocks, namely, location service and forwarding strategy, that are required for position-based routing. In addition, we illustrate potential classification criteria for the various existing approaches.

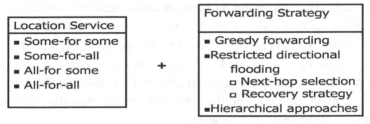

Figure 2.8 – Building blocks for position-based routing
[Taken from Mauve2001]

2.4.2 Location Services

In order to learn the current position of a specific node, help is needed from a location service. MHs register their current position with this service. When a node does not know the position of a desired communication partner, it contacts the location service and requests that information. In classical one-hop cellular networks, there are dedicated position servers (with well-known addresses) that maintain position information about the nodes in the network. With respect to classification, this is some-for-all approach as the servers are *some* specific nodes, each maintaining the position information about *all* MHs.

In MANETs, such centralized approach is viable only as an eternal service that can be reached via non-ad hoc means. There are two main reasons for this. First, it would be difficult to obtain the location of a position server if the server is a part of the MANET itself. This would represent a chicken-and-egg problem: without the position server it is not possible to get position information, but without the position information the server cannot be reached. Second, since a MANET is dynamic, it might be difficult to guarantee that at least one position server will be present in a given MANET. In the following, we concentrate on decentralized location services that are part of the MANET.

2.4.2.1 Distance Routing Effect Algorithm for Mobility

Within Distance Routing Effect Algorithm for Mobility (DREAM) framework [Basagni1998], each node maintains a position database that stores the location information about other nodes that are part of the network. As a consequence, it can be classified as an all-for-all approach. An entry in the position database includes a node identifier, the direction of and distance to the node, as well as a time value when this information has been generated. Obviously, the accuracy of such an entry depends upon its age. Each node running DREAM periodically floods packets to update the position information maintained by the other nodes. A node can control the accuracy of its position information available to other nodes in two ways:

- By changing the frequency at which it sends position updates. This is known as *temporal resolution*.
- By indicating how far a position update may travel before it is discarded. This is known as *spatial resolution*.

The temporal resolution of sending updates is coupled with the mobility rate of a node, i.e., the higher the speed is, more frequent the updates will be. The spatial resolution is used to provide accurate position information in the direct neighborhood of a node and less accurate information for far away nodes. The costs associated with accurate position information at remote nodes can be reduced since greater the distance separating two nodes is, slower they appear to be moving with respect to each other. Accordingly, the location information in routing tables can be updated as a function of the distance separating nodes without compromising the routing accuracy. This is called as the *distance effect* and is exemplified by Figure 2.9 where MH A is assumed stationary, while MHs B and C are moving in the same direction at the same speed. From node A's perspective, the change in direction will be greater for node B than for node C. The distance effect allows low spatial resolution areas far away from the target node, provided that intermediate hops are able to update the position information contained in the packet header. Based on the resulting routing tables, DREAM forwards packets in the *recorded direction* of the destination node, guaranteeing delivery by following the direction with a given probability.

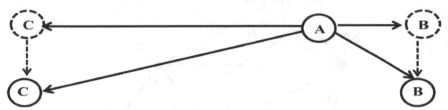

Figure 2.9 – The distance effect in DREAM [Taken from: Mauve2001]

2.4.2.2 Quorum-Based Location Service

The concept of quorum systems is quite popular in distributed systems and information replication in databases. Here, information updates (write operations) are sent to a subset (quorum) of available nodes, and information requests (read operations) are referred to a potentially different subset. When these subsets are designed such that their intersection is nonempty, it is ensured that an up-to-date version of the sought-after information can always be found.

In [Haas1999], this scheme is employed to develop a location service for MANETs. It is instructive to discuss this scheme through a sample network shown in Figure 2.10. A set of MHs is chosen to host position databases, and this is illustrated by nodes 1-6 in Figure 2.10. Next, a virtual backbone is constructed among the nodes of the subset by utilizing a non-position-based ad hoc routing algorithm. A MH sends position update messages to the nearest backbone node, which then chooses a quorum of backbone nodes to host the position information. In our example, node D sends its updates to node 6, which might then select quorum A with nodes 1, 2, and 6 to host the information. When a node S wants to obtain the position information, it sends a query to the nearest backbone node, which in turn contacts (through unicast or even multicast) the nodes of a (usually different) quorum. Node 4 might, for example, choose quorum B, consisting of nodes 4, 5, and 6 for the query. Since, by definition, the intersection of two quorum systems is nonempty, the querying node is guaranteed to obtain at least one response with the desired position information. It is important to timestamp position updates. If several responses are received, the one representing the most current position update is selected.

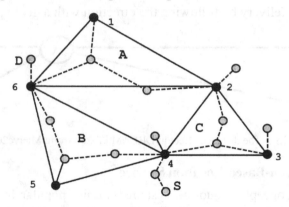

Figure 2.10 – Example of a quorum [Taken from Mauve2001]

An important trade-off in quorum-based position services is that larger the quorum set is, higher the cost for position updates and queries are, while larger the number of nodes in the intersection of two quorums will be. This improves resilience against unreachable backbone nodes. Several methods on how to generate quorum systems with desired properties can be found in [Haas1999]. The quorum-based position service can be configured to operate as all-for-all, all-for-some, or some-for-some approach, depending upon how the size of the backbone and the quorum is chosen. However, it will typically work as some-for-some scheme with the backbone being a small subset of all available nodes and a quorum being a small subset of the backbone nodes. Another work based on quorums in presented in [Stojmenovic1999a]. Here, position information for the nodes is propagated in a north-south direction. Whenever a node has to be contacted whose position is not known, position information is searched in east-west direction until the information is found.

2.4.2.3 Grid Location Service

The Grid Location Service (GLS) [Li2000, Morris2000] divides the area that contains the MANET into a hierarchy of squares. In this hierarchy, *n*-order squares contain exactly (*n*–1)-order squares, thus forming a *quad-tree*. Each node maintains a table of all other nodes within the local first-order square. The table is constructed with the help of periodic position broadcasts scoped to the area of the first order square. We present GLS with the assistance of Figure 2.11.

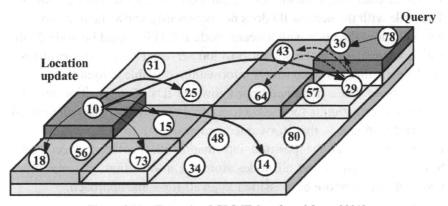

Figure 2.11 – Example of GLS [Taken from Mauve2001]

To determine where to store position information, GLS establishes a notion of *near* node IDs, defined as the least ID greater than a node's own ID. When node 10 in Figure 2.11 wants to distribute its position information, it sends position updates to the respective node with the nearest ID in each of the three surrounding first-order squares. Therefore, the position information is available at nodes 15, 18, 73, and at all nodes that are in the same first-order square as node 10 itself. In the surrounding three second-order squares, the nodes with the nearest ID are selected to host the node's position (nodes 14, 25, and 29 in the example of Figure 2.11). This process is repeated until the area of the MANET has been fully covered. The density of the position information for a given node decreases logarithmically with the distance from that node.

Now assume that node 78 wants to obtain the position of node 10. Firstly, it should locate a nearby node that knows about the position of node 10 which is node 29. While node 78 does not know that node 29 possesses the required position, it is able to discover this information. To understand how this process works, look at the position servers of node 29 which is stored in the three surrounding first-order squares at nodes 36, 43, and 64. Note that each of these nodes, including node 29, are also automatically the ones in their respective first-order square with the ID nearest to 10. Thus, there exists a "trail" of descending node IDs from each of the squares of *all* orders to the correct position server. Position queries for a node can now be directed to the node with the nearest ID of which the querying node knows. In our example, this would be node 36. The node with the nearest ID does not necessarily know the node sought, but will know the node with a nearer node ID. This would be node 29 in, which happens to be the sought position server. This process continues until a node that has the position information available is found.

Note that a node need not know the IDs of its position servers. Position information is forwarded to a certain position of each element in the quad-tree and is then forwarded progressively to nodes with closer IDs to ensure that the position information reaches the correct node. Since GLS requires that all nodes store the information on some other nodes, it can therefore be classified as an all-for-some approach.

2.4.2.4 Homezone

Two almost identical location services have been proposed independently in [Giordano1999, Stojmenovic1999b]. Both use the concept of a virtual *Homezone* where position information for a node is stored. By applying a well-known hash function to the node identifier, it is possible to derive the position C of the Homezone for a node. All nodes within a disk of radius R centered at C have to maintain position information for the node. Thus, as in the case of GLS, a position database can be found by means of a hash function on which sender and receiver agree without having to exchange information. If the Homezone is sparsely populated, R may have to be increased, resulting in increasing R for updates as well as for queries. Therefore the Homezone approaches are also all-for-some approaches.

2.4.3 Forwarding Strategies

In this section we describe the three major forwarding strategies employed in position-based routing.

2.4.3.1 Greedy Packet Forwarding

Using greedy packet forwarding, the sender of a packet includes an approximate position of the recipient in the packet. This information is gathered by an appropriate location service (e.g., described in the previous section). When an intermediate node receives a packet, it forwards the packet to a neighbor lying in the general direction of the recipient. Ideally, this process can be repeated until recipient has been reached.

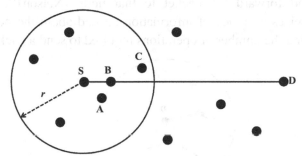

Figure 2.12 – Greedy packet forwarding strategies
[Taken from Mauve2001]

Typically, there are three different strategies a node can use to decide to which neighbor a given packet should be forwarded. These are illustrated in Figure 2.12, where node S and D denote the source and destination respectively. The circle with radius r indicates the maximum transmission range of node S. One intuitive strategy is to forward the packet to the node that makes the most progress towards (i.e., closest to) node D. In Figure 2.12, this would be node C. This strategy is known as *most forward within r* (MFR) [Takagi1984].

MFR may be a good strategy in scenarios where the sender of a packet cannot adjust the transmission signal strength to the distance between the sender and receiver. However, in [Hou1986] it is shown that a different strategy performs better than MFR in situations where the sender can adapt its transmitting power. In *nearest* with *forward progress*

(NFP), the packet is transmitted to the nearest neighbor of the sender which is in the direction of the destination. In Figure 2.12, this would be node A. If all nodes employ NFP, the probability of packet collisions is significantly reduced. Thus, the average progress of the packet is calculated as $p\ f(a,\ b)$ where p is the likelihood of a successful transmission and $f(a,\ b)$ is the progress of the packet being successfully forwarded from a to b. This is higher for NFP than for MFR. Another strategy for forwarding packets is compass routing, in which the neighbor closer to the straight line between sender and destination is selected [Kranakis1999]. In Figure 2.12, this would be node B. Compass routing tries to minimize the spatial distance a packet travels. Finally, it is possible to let the sender randomly select one of the nodes closer to the destination and forward the packet to that node [Nelson1984]. This strategy minimizes accuracy of information needed about the neighbors, thereby reducing the number of operations required to send a packet.

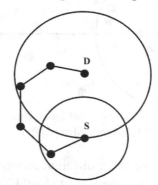

Figure 2.13 – Greedy routing failure

Unfortunately, greedy routing may fail to find a path between a sender and a destination, even though one does exist. This can be seen through Figure 2.13, where the circle around node D has the radius of the distance between nodes S and D, and circle around node S shows its transmission range. Note that there exists a valid path from node S to node D. The problem here is that node S is closer to the destination node D than any of the nodes in its transmission range. Greedy routing has therefore reached a local maximum from which it cannot recover. To counter this problem, it has been suggested that the packet should be forwarded to the node with the least backward progress [Takagi1984] if no node can be found in the forward direction. However, this raises the problem of looping, which cannot occur when packets are forwarded with positive progress toward the destination. Other studies [Hou1986] suggest not to forward packets that have reached a local maximum.

The face-2 algorithm [Bose1999] and the perimeter routing strategy of the Greedy Perimeter Stateless Routing Protocol (GPSR) [Karp2000] are two similar recovery approaches based on planar graph traversal. Both are performed on a per-packet basis and do not require nodes to store any additional information. A packet enters the recovery mode when it arrives at a local maximum. It returns to greedy mode when it reaches a node closer to the destination than the node where the packet entered the recovery mode. Planar graphs are graphs with non-intersecting edges. A set of nodes in a MANET can be considered a graph in which the nodes are vertices and an edge exists between two nodes if they are close enough to communicate directly with each other. The graph formed by a MANET is generally not planar, as shown in Figure 2.14 where the transmission range of each node contains all other nodes.

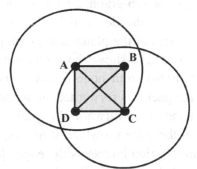

Figure 2.14 – An example of a non-planar graph
[Taken from Mauve2001]

In order to construct a connected planar sub-graph of the graph formed by the nodes in a MANET, a well-known mechanism is employed [Toussaint1980]: an edge between two nodes A and B is included in the graph only if the intersection of the two circles with radii equal to the distance between node A and B around those two nodes does not contain any other nodes. For example, in Figure 2.14 the edge between nodes A and C would not be included in the planar sub-graph since nodes B and D are contained in the intersection of the circles. It is important to realize that each node can locally make the decision as to whether an edge is within the planar sub-graph, since each node knows the position of all its neighbors.

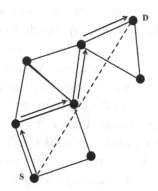

Figure 2.15 – Planar graph traversal
[Taken from Mauve2001]

Based on the planar sub-graph, a simple planar graph traversal is used to find a path toward the destination. The general concept is to forward the packet on faces of the planar sub-graph progressively closer to the destination. Figure 2.15 shows how this traversal is carried out when a packet is forwarded from node S toward node D on recovery mode. On each face, the packet is forwarded along the interior of the face by using the right hand rule: forward the packet on the next edge counterclockwise from the edge on which it arrived. Whenever the line between source and destination intersects the edge along which a packet is about to be forwarded, check if this intersection is closer to the destination than any other intersection previously encountered. If this is true, switch to the new face bordering the edge the packet is about to transverse. The packet is then forwarded counterclockwise to the edge it is about to be forwarded along before switching faces. This algorithm guarantees that a path will be found from the source to the destination if there exists one in the original non-planar graph.

The header of a packet contains additional information such as the position of the node where it entered recovery mode, the position of the last intersection that caused a face change, and the first edge traversed on the current face. Therefore, each node can make all routing decisions based only on the information about its local neighbors. This includes detection of an unreachable destination, when a packet traverses an earlier visited edge for the second time.

2.4.3.2 Restricted Directional Flooding

2.4.3.2.1 DREAM

In DREAM, the sender node S of a packet with destination node D forwards the packet to all one-hop neighbors that lie "in the direction of node D". In order to determine this direction, a node calculates the region that is likely to contain node D, called the *expected region* as depicted in Figure 2.16. Since this position information may be outdated, the radius r of the expected region is set to $(t_1 - t_0)v_{max}$, where t_1 is the current time, t_0 is the timestamp of the position information node S has about node D, and v_{max} is the maximum speed that a node may travel in the MANET. Given the expected region, the "direction towards node D" for the example in Figure 2.16 is defined by the line between nodes S and D and the angle ϕ. The neighboring nodes repeat this procedure using their information on node D's position. If a node does not have a one-hop neighbor in the required direction, a recovery procedure has to be started. This procedure is not part of DREAM specification.

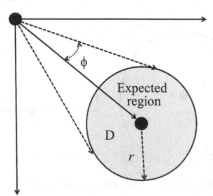

Figure 2.16 – Example of the expected region in DREAM
[Taken from Mauve2001]

2.4.3.2.2 Location-Aided Routing

The Location-Aided Routing (LAR) [Ko1998] protocol does not define a location-based routing protocol, but instead proposes the use of position information to enhance the route discovery phase of reactive ad hoc routing approaches, which often use flooding as a means of route discovery. Under the assumption that nodes have information about other node's positions, LAR uses this to restrict the flooding to a certain area and is similar to DREAM. LAR exploits location information to limit the scope of route request flood employed in protocols such as AODV and DSR and can be obtained through GPS. LAR limits the search for a route to so-called request zone, determined based on expected location of the destination node at the time of route discovery. Two concepts are important to understand how LAR works: Expected and Request Zones.

Let us first discuss what an Expected Zone is. Consider a node S that needs to find a route to node D. Assume that node S knows that node D was at location L at time t_0. Then, the *"expected zone"* of node D, from the viewpoint of node S at current time t_1, is the region expected to contain node D. For instance, if node S knows that node D travels with average speed v, then S may assume that the expected zone is the circular region of radius $v(t_1 - t_0)$, centered at location L (see Figure 2.17(a)). If actual speed happens to be larger than the average, then the destination may actually be outside the expected zone at time t_1. Thus, expected zone is only an estimate made by node S to determine a region that potentially contains D at time t_1.

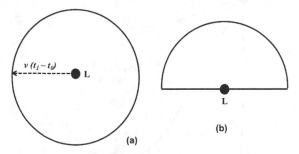

(a)

(b)

Figure 2.17 – Examples of *Expected Zone*

If node S does not know any previous location of node D, then node S cannot reasonably determine the expected zone (the entire region that may potentially be occupied by the ad hoc network is assumed to be the expected zone). In this case, LAR reduces to the basic flooding algorithm. In general, having more information regarding mobility of a destination node can result in a smaller expected zone as illustrated by Figure 2.17(b).

Based on expected zone, we can define the request zone. The proposed LAR algorithms use flooding with one modification. Node S defines (implicitly or explicitly) a *request zone* for the route request. A node forwards a route request *only if* it belongs to the request zone. To increase the probability that the route request will reach node D, the request zone should include the *expected zone*. Additionally, the request zone may also include other regions around the request zone.

Based on this information, the source node S can thus determine the four corners of the expected zone. For instance, in Figure 2.18 if the node I receives route request from another node, it forwards the request to neighbors, because it is within the rectangular request zone. However, when node J receives the route request, node J discards the request, as node J is not within the request zone (see Figure 2.18).

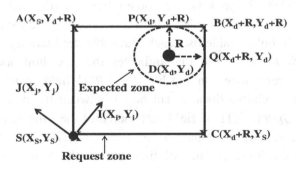

Figure 2.18 – LAR scheme [Taken from Ko1998]

This algorithm is called LAR scheme 1. The LAR scheme 2 is a slight modification to include two pieces of information within the route request packet: assume that node S knows the location $(X_d; Y_d)$ of node D at some time t_0 – the time at which route discovery is initiated by node S is t_1, where $t_1 \geq t_0$. Node S calculates its distance from location $(X_d; Y_d)$, denoted as $DIST_S$, and includes this distance with the route request message. The coordinates $(X_d; Y_d)$ are also included in the route request packet. With this information, a given node J forwards a route request forwarded by I (originated by node S), if J is within an expected distance from $(X_d; Y_d)$ than node I.

2.4.3.2.3 Relative Distance Micro-Discovery Ad Hoc Routing

The Relative Distance Micro-discovery Ad Hoc Routing (RDMAR) routing protocol [Aggelou1999], an adaptive and saleable routing protocol, is well suited in large mobile networks whose rate of topological changes is moderate. A key concept in its design is a typical localized route discovery in a very small region due to link failures near the change by using a flooding mechanism called Relative Distance Micro-discovery (RDM). To accomplish this, an iterative algorithm calculates an estimate of their RD given their previous RD, an average nodal mobility and information about the elapsed time since they last communicated. Based on the newly calculated RD, the query flood is then localized to a limited region of the network centered at the source node of the route discovery and with maximum propagation radius that equals to the estimated relative distance.

In RDMAR, packets are routed between the stations of the network by using routing tables which are stored at each station of the network. Each routing table lists each reachable destination j, it includes: the "D*efault Router*" field that indicates the next hop node through which the current node can reach j, the "*RD*" field which shows an estimate of the relative distance (in hops) between the node and j, the "*Time_Last_Update*" (TLU) field that indicates the time since the node last received routing information for j, a "*RT_Timeout*" field which records the remaining amount of time before the route is considered invalid, and a "R*oute Flag*" field which declares whether the route to j is active. RDMAR comprises of two main algorithms:

- Route Discovery – When an incoming call arrives at node i for destination node j and there is no route available, i initiates a route discovery phase. Here, there are two possible options for i; either to flood the network with a route query in which case the route query packets are broadcast into the whole network, or instead, limit the discovery in a smaller region of the network, if some kind of location prediction model for j can be established. In the latter case, the source of the route discovery, *i*, refers to its routing table in order to retrieve information on its previous relative distance with *j* and the time elapsed since *i* last received routing information for *j*. Let us designate this time as t_{motion}. Based on this information and assuming a moderate velocity, *Micro_Velocity*, and a moderate transmission range, *Micro_Range*, node *i* is then able to estimate its new relative distance to destination node *j* in terms of actual number of hops. To accomplish this, node *i* calculates the distance offset of *DST* (*DST_Offset*) during t_{motion}, and "adjusts" the result onto their previous relative distance (*RDM_Radius*).

- Route Maintenance – An intermediate node *i*, upon receipt of a data packet, first processes the routing header and then forwards the packet to the next hop. In addition, node *i* sends an explicit message to examine whether a bi-directional link can be established with the previous node. RDMAR does not assume bi-directional links but in contrast nodes *exercise* the possibility of having bi-directional links. If node *i* is unable to forward the packet because there is no route available or a forwarding error occurs along the data path as a result of a link or node failure, *i* may attempt a number of additional re-transmissions of the same data packet, up to a maximum number of retries. However, if the failure persists, node *i* initiates a Route Discovery procedure.

2.4.3.3 Hierarchical Routing

In traditional networks, the complexity of the routing algorithm handled by each node can be reduced tremendously by establishing some form of hierarchy. Therefore, it is a valid question to ask whether position-based routing for MANETs can also benefit from the use of hierarchy.

2.4.3.3.1 Terminodes Routing

One approach that combines hierarchical and position-based routing is a part of the Terminodes project [Blazevic2001] with two levels of hierarchy [Blazevic2000]. Packets are routed according to a proactive distance vector scheme if the destination is close (in terms of number of hops) to the sending node. For long distance routing, a greedy position-based approach is used. Once a long distance packet reaches the area close to the recipient, it continues to be forwarded by means of the local routing algorithm. It is shown [Blazevic2000] that the hierarchy can significantly improve ratio of successful packets and the routing overhead compared to conventional reactive ad hoc routing protocols. In order to prevent greedy forwarding for long distance routing from encountering a local maximum, sender includes a list of positions in the packet header which are then traversed to the sender. In Terminodes routing, the sender requests this information from nodes it is already in contact with and checks at regular intervals whether the path of positions is still valid or can be improved.

2.4.3.3.2 Grid Routing

A second method for position-based ad hoc routing containing hierarchical elements is proposed within the Grid project [GRIDPROJECT]. The location proxy technique described in [Couto2001] is similar to the Terminodes routing: a proactive distance vector routing protocol used at the local level, while position-based routing employed for long-distance packet forwarding. In Grid routing, however, the hierarchy is not only introduced to improve scalability but to have at least one position-aware node in each area to be used as proxies. Packets that are addressed to a position-unaware node therefore arrive at a position-aware proxy and are then forwarded according to the information of the proactive distance vector protocol. As a repair mechanism for greedy long-distance routing, a mechanism called Intermediate Node Forwarding (INF) is proposed [Couto2001]. If a forwarding node has no neighbor with forward progress, it discards the packet and sends a notification to the sender of the packet. The sender of the packet then chooses a single intermediate position randomly for a circle around the midpoint of the line between the sender and the receiver. Packets have to traverse that intermediate position. If the packet is discarded again, the radius of the circle is increased and another random position is chosen and is repeated until the packets are delivered to the destination, or until a predefined number has been attempted. Then the sender assumes that the destination is unreachable.

2.4.3.4 Other Position-based Routing

Effectiveness of all position-based routing depends on the accuracy of the location of the destination node. The GPS-based systems do not provide good accuracy inside the building and the surrounding area can be classified [Hatami2005] in the following five categories:

- Typical office environment with no line-of-sight (NLOS) with 50ns delay spread.
- Large open space with 100ns delay spread with NLOS.
- Large indoor or outdoor space with 150ns delay spread with NLOS.
- Large indoor or outdoor space with line-of-sight and 140ns delay spread.
- Large indoor or outdoor space with NLOS and 250ns delay spread.

The instantaneous received signal strength for a fixed location inside a building is observed to vary with time due to shadow, fading and multi-path reception. The closest neighbors' location is observed to provide good accuracy and reasonable performance under all categories. Existing location based routing schemes use the last known destination location to the source as the best zone estimate. Therefore, it is better to combine location-based routing with specific geographical points known as anchors [Blazevic2005] as selected by the source. These imaginary locations assist in routing and are selected based on those nodes that could possibly assist in path discovery or could be based on geographical node density maps at the source node.

2.4.4 Comparisons

In this section, we compare the location services and forwarding strategies previously described. One key aspect of this comparison is how the individual approaches behave with an increasing number of nodes in the MANET. For the remainder of this section, we assume that the density of nodes remain constant when the number of nodes increases. Therefore, the area covered by the MANET has to increase as the number of nodes increases.

2.4.4.1 Location Services

A comparison between different location-based routing is given in [Stojmenovic2002, Hatami2005]. Table 2.2 summarizes various location services using several different criteria, where n represents the number of nodes and c is a constant. The *type* criterion indicates how many nodes participate in providing location information and for how many other nodes each node is required to maintain location information. The *communication complexity* describes the average number of one-hop transmissions required to look up or update a node's position. The *time complexity* measures the average time it takes to perform a position update or position lookup. The amount of state required at each node to maintain the position of other nodes is indicated by the *state volume*. Some location services provide *localized information* by maintaining a higher density or better quality of position information near the position of the node. This may be important if the communication in a MANET is mainly local. The *robustness* of a location service is considered to be low, medium, or high, depending on whether it takes the failure of a single node, the failure of a small subset of all nodes, or the failure of all nodes to render the position of a given node inaccessible. The *implementation complexity* indicates how well the location service is understood and how complex it is to implement and test it. We note that this measure is highly subjective, while we have tried to be as fair as possible.

Table 2.2 – Comparison of Location Services (n=number of nodes; c=constant)
[Taken from: Mauve2001]

Criterion	DREAM	Qurom System	GLS	Homezone
Type	All-for all	Some-for-some	All-for-some	All-for-some
Communication complexity (update)	$O(n)$	$O(\sqrt{n})$	$O(\sqrt{n})$	$O(\sqrt{n})$
Communication complexity (lookup)	$O(c)$	$O(\sqrt{n})$	$O(\sqrt{n})$	$O(\sqrt{n})$
Time complexity (update)	$O(n)$	$O(\sqrt{n})$	$O(\sqrt{n})$	$O(\sqrt{n})$
Time complexity (lookup)	$O(c)$	$O(\sqrt{n})$	$O(\sqrt{n})$	$O(\sqrt{n})$
State volume	$O(n)$	$O(c)$	$O(\log(n))$	$O(c)$
Localized information	Yes	No	Yes	No
Robustness	High	Medium	Medium	Medium
Implementation complexity	Low	High	Medium	Low

DREAM is fundamentally different from other position services, as it requires all nodes to maintain position information about every other node. The communication complexity of a position update and the position information maintained by each node scales with O(n), while a position query requires only a local lookup, which is independent of the number of nodes. The time required to perform a position update in DREAM is a linear function of the diameter of the network, leading to a complexity of O($\sqrt{n}$). Due to the communication complexity of position updates, DREAM is the least scalable position service and, hence, is inappropriate for large-scale and general purpose MANETs. However, it is suitable for specialized applications since it is very robust and provides localized information in situations such as notifying an emergency.

The quorum system requires the same operations for position updates and position lookups. In both cases, a constant number of nodes (the quorum) must be contacted. Each of these messages has a communication and time complexity that depends linearly on the diameter of the network and thus scales with O($\sqrt{n}$). The state information maintained in the backbone nodes is constant, since an individual backbone is formed for a fixed number of nodes. The general robustness of the approach is medium, since the position of a node will become unavailable if a significant number of backbone nodes fail. However, the number of such nodes is a parameter that can be freely configured for the position service. Furthermore, the position information is kept spatially distributed and independent. Thus, the robustness seems to be higher than that of GLS or Homezone. A major drawback of the quorum system is its dependence on a non-position-based ad hoc routing protocol for the virtual backbone, which tremendously increases the implementation complexity and may compromise the scalability of this approach. However, both position services offered by GLS and Homezone can be thought of as special case of the quorum systems, thereby overcoming this drawback.

GLS and Homezone are similar to each other in that each node selects a subset of all available nodes as position servers. For Homezone, position updates and lookups need to be sent to the virtual home region (VHR). The average distance from that region depends linearly on the diameter of the network. Therefore, the communication and time complexity of Homezone is O($\sqrt{n}$). The state information is constant, as each node should have a constant number of position servers in its Homezone. Due to localized strategy of forwarding updates and lookups, communication and time complexity is just a constant factor larger than the Homezone and remains at O($\sqrt{n}$). The main tradeoff between GLS and Homezone is in providing localized information and in the implementation complexity. GLS benefits greatly if the communicating nodes are close to each other. But, the behavior of GLS in a dynamic environment and in the presence of node failures is more difficult to control than that of Homezone.

2.4.4.2 Forwarding Strategies

Table 2.3 – Comparison of Forwarding Schemes (n=number of nodes)
[taken from Mauve2001]

Criterion	GREEDY	DREAM	LAR	Terminodes	Grid
Type	Greedy	Restricted Directional flooding	Restricted Directional flooding	Hierarchical	Hierarchical
Communication complexity	O $\left(\sqrt{n}\right)$	O(n)	O(n)	O $\left(\sqrt{n}\right)$	O $\left(\sqrt{n}\right)$
Tolerable position inaccuracy	Transmission range	Expected region	Expected region	Short-distance routing range	Short-distance routing range
Requires all-for-all location service	No	Yes	No	No	No
Robustness	Medium	High	High	Medium	Medium
Implementation complexity	Medium	Low	Low	High	High

Table 2.3 presents a summary of various forwarding strategies and their evaluation criteria, where *n* represents the number of nodes. *Type* describes the fundamental strategy used for packet forwarding, while the *communication complexity* indicates the average number of one-hop transmissions required to send a packet from one node to another node with known position. The strategies need to tolerate different degrees of inaccuracy with regard to the position of the receiver

and is reflected by the *tolerable position* inaccuracy criterion. Furthermore, the forwarding *requires all-for-all location service* criterion indicates whether the forwarding strategy requires all-for-all location service in order to work properly. The robustness of an approach is high if the failure of a single intermediate node does not prevent the packet from reaching its destination. Its value is medium if the failure of a single intermediate node might lead to the loss of the packet but does not require the setup of a new route. Finally, the robustness is low if the failure of an individual node might result in packet loss and requires setting up a new route.

Greedy forwarding is efficient, with a communication complexity of $O(\sqrt{n})$, and is well suited for MANETs with a highly dynamic topology. The face-2 algorithm [Bose1999] and the perimeter routing of GPSR [Karp2000] are currently the most advanced recovery strategies. One drawback of the current greedy approaches is that the position of the destination needs to be known with an accuracy of a one-hop transmission range, or else the packets cannot be delivered. The robustness is medium, as the failure of an individual node may cause the loss of a packet in transit. However, it does not require setting up a new route as would be the case in topology-based routing protocols. Due to repair strategy like face-2 or perimeter routing, we consider the implementation efforts to be of medium complexity.

Restricted directional flooding, as in DREAM and LAR, has communication complexity of $O(n)$ and therefore does not scale well for large networks with a high volume of data transmissions. One difference between DREAM and LAR is that in DREAM, it is expected that intermediate nodes update the position of the destination when they have better information than the sender of the packet, while this is not the case in LAR. As a result, DREAM packet forwarding requires and makes optimal use of all-for-all location service, while LAR can work with any location service but does not benefit much form an all-for-all location service. Both approaches are very robust against the failure of individual nodes and position inaccuracy, and are very simple to implement. This qualifies them for applications that require high reliability and fast message delivery for very infrequent data transmissions.

Both Terminodes [Blazevic2001] and Grid [Couto2001] routing provide hierarchical approaches to position-based ad hoc routing. For long-distance routing, both use a greedy approach and therefore have characteristics similar to those of greedy forwarding. However, the use of non-position-based approach at the local level, make them tolerant to position inaccuracy, while being significantly more complex to implement. Grid routing allows position-unaware nodes to use position-aware nodes as proxies in order to participate as a MANET, while for Terminodes, a GPS-free positioning service has been developed. The probabilistic repair strategy proposed by Grid is simpler and requires less state information than that of Terminodes.

2.5 Other Routing Protocols

There are plenty of routing protocols for MANETs, and the most important ones have been covered in detail. However, below we describe some other routing protocols which employ optimization criteria different from the ones described earlier.

2.5.1 Signal Stability Routing

Unlike the algorithms described so far, the on-demand Signal Stability-Based Adaptive Routing protocol (SSR) [Dube1997] selects routes based on the signal strength (weak or strong) between nodes and a node's location stability. This route selection criterion of SSR has the effect of choosing routes that have "stronger" connectivity [Chlamtac1986]. Basically, SSR is comprised of two cooperative protocols, namely, the Dynamic Routing Protocol (DRP) and the Static Routing Protocol (SRP).

The DRP is responsible for the maintenance of Signal Stability Table (SST) and the Routing Table (RT). After processing the packet and updating the appropriate tables, DRP passes the packet to the SRP. The SRP of a node processes by passing the packet up the stack if it is the intended receiver, or looks up in the routing table for the destination and forwards the packet if it is not. If no entry is found in the routing table for the destination, a route search process is initiated. One difference between route-discovery-procedure used in SSR with respect to that employed in AODV is that route requests are only forwarded to the next hop in SSR if they are received over strong channels.

If there is no route reply received at the source within a specified timeout period, the source changes the PREF field in the packet header to indicate that weak channels have been accepted, as these may be the only links over which the packet can be propagated. When a failed link is detected in the network, route error packets are sent and another search process is initiated. The source also sends an erase message to notify all the nodes about the broken link.

2.5.2 Power Aware Routing

In this protocol, power-aware metrics [Singh1998, Jin2000] are used for determining routes in MANETs that reduces the cost/packet of routing packets by 5 - 30 percent over shortest-hop routing. Furthermore, using these new metrics ensures that mean time to node failure is increased significantly, while packet delays do not increase. A recent work [Lee2000] concentrates on selecting a route based the traffic and congestion characteristics in the network.

2.5.3 Associativity-Based Routing

This is a totally different approach in mobile routing. The Associativity-Based Routing (ABR) [Toh1997] protocol is free from loops, deadlock, and duplicate packets. In ABR, a route is selected based on a metric that is known as the degree of association stability. Each node periodically generates a beacon to signify its existence and this beacon causes their associativity tables to be updated A to derive long-lived routes for ad hoc networks. For each beacon received, the associativity tick of the current node with respect to the beaconing node is incremented. A high (low) degree of association stability may indicate a low (high) state of the node mobility. Associativity ticks are reset when the neighbors of a node or the node itself move out of the proximity. The three phases of ABR are:

- Route discovery;
- Route reconstruction (RRC);
- Route deletion.

The route discovery phase is accomplished by a broadcast query and await-reply (BQ-REPLY) cycle. A node desiring a route broadcasts a BQ message in search of MHs that have a route to the destination. All nodes receiving the query (that are not the destination) append their addresses and their associativity ticks with their neighbors along with QoS information to the query packet. A successor node erases its upstream node neighbors' associativity tick entries and retains only the entry concerned with itself and its upstream node. In this way, each resultant packet arriving at the destination contains the associativity ticks of the nodes along the route to the destination. If multiple paths have the same overall degree of association stability, the route with the minimum number of hops is selected. The destination then sends a REPLY packet back to the source along this path. Nodes propagating the REPLY mark their routes as valid. All other routes remain inactive, and the possibility of duplicate packets arriving at the destination is avoided.

RRC may consist of partial route discovery, invalid route erasure, valid route updates, and new route discovery, depending on which node(s) along the route move. Movement by the source results in a new BQ-REPLY process. When the destination moves, the immediate upstream node erases its route and determines if the node is still reachable by a localized query (LQ[H]) process, where H refers to the hop count from the upstream node to the destination. If the destination receives the LQ packet, it REPLYs with the best partial route; otherwise, the initiating node times out and the process backtracks to the next upstream node. Here, a RN message is sent to the next upstream node to erase the invalid route and inform this node that it should invoke the LQ[H] process. If this process results in backtracking more than halfway to the source, the LQ process is discontinued and a new BQ process is initiated at the source.

2.5.4 QoS Routing

All the routing protocols discussed so far have been proposed either for routing messages along the shortest available path or within some system-level requirement. Routing applications using these paths may not be adequate for applications which require QoS (e.g., real-time applications). In this section we overview some routing schemes that can support QoS in MANETs.

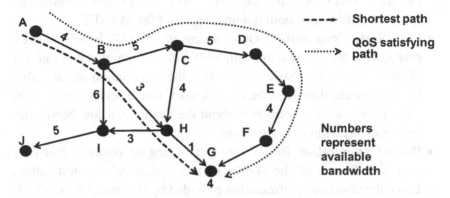

Figure 2.19 – A QoS routing example in a MANET

Figure 2.19 illustrates an example where nodes are labeled as A, B, C, ..., J. The numbers along each edge represent the available bandwidths of the wireless links. If we want to find a route from a source node A to a destination node G, conventional routing using shortest path (in terms of number of hops) as a metric, selects the route A-B-H-G. However, the QoS-based route selection process from node A to node G with a minimum bandwidth of 4 would use A-B-C-D-E-F-G as one possible path over the shortest path route A-B-H-G.

The QoS-aware path is determined within the constraints of bandwidth, minimal search, distance, and traffic conditions. To date, only a few QoS-aware routing protocols have been proposed for MANETs and we review the most prominent ones in the following sections.

2.5.4.1 Core Extraction Distributed Ad Hoc Routing

The Core Extraction Distributed Ad Hoc Routing (CEDAR) algorithm [Sinha1999] is a partitioning protocol proposed as a QoS routing scheme for small to medium size MANETs consisting of tens to hundreds of nodes. It dynamically establishes the core of the network, and then incrementally propagates the link states of stable high-bandwidth links to the core nodes. CEDAR has three key components:

- **Core Extraction**: A set of nodes is elected to form the core that maintains the local topology of the nodes in its domain, and also performs route computation. The core nodes are elected by approximating a minimum dominating set of the MANET.
- **Link State Propagation**: QoS routing in CEDAR is achieved by propagating the bandwidth availability information of stable links to all core nodes. The basic idea is that the information about stable high-bandwidth links can be made known to the nodes far away in the network, while information about the dynamic or low bandwidth links remains within the local area.
- **Route Computation**: Route computation first establishes a core path from the domain of the source to the domain of the destination. Using the directional information provided by the core path, CEDAR iteratively tries to find a partial route from the source to the domain of the furthest possible node in the core path, satisfying the requested bandwidth. This node then becomes the source of the next iteration.

In the CEDAR approach, the core provides an efficient and low-overhead infrastructure to perform routing, while the state propagation mechanism ensures availability of link-state information at the core nodes without incurring high overheads.

2.5.4.2 Incorporating QoS in Flooding-based Route Discovery

A ticket-based probing algorithm with imprecise state model has been proposed in [Chen1998] for discovering a QoS-aware routing path, by issuing a number of logical tickets to limit the amount of flooding (routing) messages. When a probing message arrives at a node, it may be split into multiple probes and forwarded to different next-hops with each child probe containing a subset of the tickets from their parents. When one or more probe(s) arrive(s) at the destination, the hop-by-hop path known and delay/bandwidth information can be used to perform resource reservation for the QoS-satisfying path.

In wired networks, a probability distribution can be calculated for a path based on delay and bandwidth information. In a MANET, however, building such a probability distribution is not suitable because wireless links are subject to breakage and state information is inaccurate. Therefore, a simple imprecise model has been proposed using the history and current (estimated) delay variations which is represented as a range of [*delay* - δ, *delay* + δ]. To adapt to the dynamic topology of MANETs, this algorithm allows different level of route redundancy. When a node detects a broken path, it notifies the source node which will reroute the connection through a new feasible path, and notifies the nodes along the old path to release the corresponding resources. Unlike the re-routing technique, the path-repairing technique does not find a completely new path. Instead, it tries to repair the path using local reconstructions.

Another approach for integrating QoS in the flooding-based route discovery process has been proposed in [Li2002]. This proposed positional attribute-based next-hop determination approach (PANDA) discriminates the next hop nodes based on their location or capabilities. When a route request is broadcast, instead of using a random broadcast delay, the receivers opt for a delay proportional to their abilities in meeting the QoS requirements of the path. The decisions at the receiver side are made on the basis of a predefined set of rules. Thus, the end-to-end path will be able to satisfy the QoS constraints as long as it is intact. A broken path will initiate the QoS-aware route discovery process.

2.5.4.3 QoS Support Using Bandwidth Calculations

An available bandwidth calculation algorithm for MANETs where time division multiple access (TDMA) is employed for communications is proposed in [Lin1999] that involves end-to-end bandwidth calculation and allocation. The source node can determine the resource availability for supporting the required QoS which is useful in call admission control. In wired networks, path bandwidth is the minimum available bandwidth of the links along the path. In time-slotted ad hoc networks, bandwidth calculation is much harder. We not only need to know the free slots along the path, but also need to assign the free slots at each hop. An example of Figure 2.20 illustrates where time slots 1, 2, and 3 are free between nodes A and B, and slots 2, 3, and 4 are free between nodes B and C. Assume node A wants to send some data to node C. Note that there will be collisions at node B if node A tries to use all three slots 1, 2, and 3 to send data to node B while node B is using one or both slots 2 and 3 to send data to node C. Thus, we have to somehow divide the common free slots 2 and 3 between the two links, namely, from node A to node B and from node B to node C.

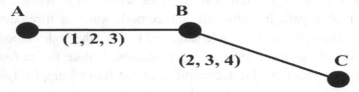

Figure 2.20 – A Bandwidth Calculation Example in a MANET

In TDMA systems, time is divided in slots which, in turn, are grouped into frames. Each frame contains two phases: control and data phases. During the control phase, each node takes turns to broadcast its information to all of its neighbors in a predefined slot. Hence, at the end of the control phase, each node has learned the free slots between itself and its neighbors. Based on this information, bandwidth calculation and assignment can be carried out in a distributed manner. Determining slot assignments while searching for the available bandwidth along the path is a NP-complete problem. Thus, a heuristic approach to tackle this issue has been proposed [Lin1999].

An on-demand QoS routing protocol using AODV has been designed for TDMA-based MANETs in [Zhu2002]. In this approach, a QoS-aware route reserves bandwidth from source to destination. In the route discovery procedure of AODV, a distributed algorithm is used to calculate the available bandwidth on a hop-by-hop basis. Route request messages with inadequate bandwidth are dropped by intermediate nodes. Only the destination node can reply to a route request message that has come along a path with sufficient bandwidth. The protocol can handle limited mobility by repairing broken paths. This approach is best applicable for small size networks or for short routes.

2.5.4.4 Multi-path QoS Routing

A multi-path QoS routing protocol has been introduced in [Liao2001] which is suitable for ad hoc networks with very limited bandwidth for each path, unlike other existing protocols for MANETs, which try to find a single path between the source and the destination, this algorithm searches for multiple paths for the QoS route. This protocol also adopts the idea of ticket-based probing scheme discussed earlier. Another rational for using multi-path routing is to enhance the routing resiliency by finding node/edge disjoint paths when link and/or node fail [Liang2005]. Another approach [Guo2005] is to use the extension of AODV to determine a backup source-destination routing path that could be used if the path gets disconnected frequently due to mobility or changing link signal quality. An analytical model has been developed [Guo2005] to justify having a backup path which can be easily piggybacked in data packets. Steps for immediate repairs of broken backup routes have also been suggested and extensive simulations have been done to validate the effectiveness of this scheme.

2.6 Conclusions and Future Directions

Routing is undoubtedly the most studied aspect of ad hoc networks. Yet, many issues remain open which deserve appropriate handling such as more robust security solutions, routing protocol scalability, QoS support, and so on. The integration of MANETs and infrastructure-based networks such as the Internet will be an important topic in wireless systems beyond 3G (discussed in Chapter 14). Also, efficient broadcasting schemes need to be examined carefully as it may be a serious roadblock to the scalability of ad hoc networks. One example is the assignment of IP addresses which is usually done by the use of Dynamic Host Configuration Protocol (DHCP) servers in fixed networks [Tanenbaum1996]. In MANETs, availability of such server many not be practical. As a result, nodes have to resort to some heuristic to obtain their IP addresses which may cause conflicts with other nodes' IP addresses. While the use of IP version 6 [Tanenbaum1996] may certainly help here due to its auto configuration capabilities, it is not a completely foolproof solution. It may be noted that the routing algorithms for MANETs are equally applicable to sensor networks [Agrawal2006] as basic characteristics of wireless sensor networks are similar to MANETs, except for low mobility, much larger number of sensor nodes and use of battery. Specific attributes of sensor networks, are considered in chapters 10, 11, and 12.

Homework Questions/Simulation Results

Q. 1. Ad hoc networks are special kinds of wireless network that does not have any underlying infrastructure. But, such networks are becoming increasingly important for both defense and civilian applications. Assuming a 60 x 60 grid connected ad hoc network is given to you, and address of each node is given by (i,j) with 0=<i/j<60. Node (k,l) need to communicate with mode (m,n).

    a. What path is followed if each node has information about their neighbors at distance d (number of hops)?

    b. What would be the size of the routing table?

    c. How many tables need to be updated if one node (p,q) is removed from the network?

    d. Route table/shortest route for cost(i,j) =f (i,j), e.g. cost (i,j)=j^2+i.

Q. 2. In Q. 1, node (k,l) need to communicate with node (m,n). Each node maintains a routing table of all those nodes which are at a maximum distance of *d* hops. What is the optimal value of *d* if maintaining 10 entries in a routing table is equivalent to one message transmission among adjacent nodes?

Q. 3. In the following ad hoc network, use one of the protocols to determine path from Node A to Node D:

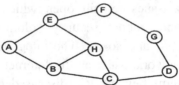

Example Ad hoc Network for Q.3

Q. 4. Repeat Q. 3 for another protocol and observe the difference.

Q. 5. Repeat Q. 3 for the following Ad hoc network:

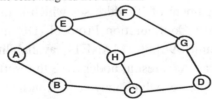

Example Ad hoc Network for Q. 5

Q. 6. In the following large ad hoc network, what will be impact on the path length if a new link can be established between MHs O and M? Explain carefully.

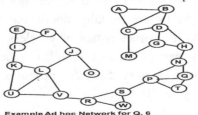

Example Ad hoc Network for Q. 6

Q. 7. Design a problem based on any of the material covered in this chapter (or in references contained therein) and solve it diligently.

References

[Aggelou1999] G. Aggelou and R. Tafazolli, "RDMAR: A bandwidth-efficient routing protocol for mobile ad hoc networks," in *Proceedings of ACM International Workshop on Wireless Mobile Multimedia (WoWMoM)*, August 1999.

[Agrawal2006] D. P. Agrawal and Q-A. Zeng, *Wireless and Mobile Systems*, Third Edition, Cengage, 2011.

[Basagni1998] S. Basagni, I. Chlamtac, V. R. Syrotiuk, and B. A. Woodward, "A distance routing effect algorithm for mobility (DREAM)," in *Proceedings of ACM/IEEE International Conference on Mobile Computing and Networking*, October 1998, 76-84.

[Bellur1999] B. Bellur and R. Ogier, "A Reliable, Efficient Topology Broadcast Protocol for Dynamic Networks," in *Proceedings of IEEE Infocom*, June 1999.

[Blazevic2000] L. Blazevic, S. Giordano, and J. Le Boudec, "Self Organized Terminode Routing," *Technical Report DCS/2000/040*, Swiss Federal Institute of Technology, 2000.

[Blazevic2001] L. Blazevic, L. Butty'an, S. Capkun, S. Giordano, J.-P. Hubaux, and J.-Y. Le Boudec, "Self-Organization in Mobile Ad Hoc Networks: The Approach of Terminodes," *IEEE Communications Magazine*, 2001.

[Blazevic2005] L. Blazevic, J.Y. La Boudec, and S. Giordano, "A location-based routing method for mobile ad-hoc Networks," IEEE Transactions on Mobile Computing, Vol. 4, No. 2, March/April 2005, pp. 97-110.

[Bose1999] P. Bose, P. Morin, I. Stojmenovic, and J. Urrutia, "Routing with Guaranteed Delivery in Ad Hoc Wireless Networks," in *Proceedings of 3rd ACM International Workshop in Discrete Algorithms and Methods for Mobile Computing and Communications*, 1999, pp. 48-55.

[Broch1998] J. Broth, D. B. Johnson, and D. A. Maltz, "The dynamic source routing protocol for mobile ad hoc networks," *Internet Draft*, 1998.

[Chen1998] S. Chen and K. Nahrstedt, "Distributed QoS routing with imprecise state information," in *Proceedings of IEEE ICCCN*, 1998.

[Chlamtac1986] L. Chlamtac and A. Lerner, "Link allocation in mobile radio networks with noisy channel," In *Proceedings of IEEE Infocom*, April 1986.

[Corson1996] M.S. Corson, J. Macker, and S. Batsell, "Architectural Considerations for Mobile Mesh Networking," In *Proceedings of the IEEE MILCOM*, October 1996.

[Corson1997] M.S. Corson and V.D. Park, "An Internet MANET Encapsulation Protocol (IMEP) Specification," *Internet-Draft*, November 1997.

[Couto2001] D. Couto and R. Morris, "Location Proxies and Intermediate Node Forwarding for Practical Geographic Forwarding," *Technical Report MIT-LCS-TR824*, MIT Lab. Computer Science, June 2001.

[Dalal1978] Y. Dalal and R. Metclafe, "Reverse Path Forwarding of Broadcast Packets," *Communications of the ACM*, Vol. 21, No. 12, December 1978, pp. 1040-1048.

[Dube1997] R. Dube, "Signal stability based adaptive routing for ad hoc mobile networks," In *Proceedings of IEEE Personal Communications*, February 1997, 36-45.

[Garcia-Luna-Aceves1999] J. J. Garcia-Luna-Aceves and M. Sphon, "Source Tree Adaptive Routing," *Internet Draft*, draft-ietf-manet-star-00.txt, work in progress, October 1999.

[Giordano1999] S. Giordano and M. Hamdi, "Mobility Management: The Virtual Home Region," *Technical Report*, October 1999.

[GRIDPROJECT] The grid project, http://www.pdos.lcs.mit.edu/grid.

[Guo2005] S. Guo, O. Yang, and Y. Shu, "Improving secure routing reliability in mobile ad hoc Networks," IEEE Transactions on Parallel and Distributed Systems, Vol. 16, No. 4, April 2005, pp. 362-373.

[Haas1998a] Z. Haas *et al.*, "The performance of query control schemes for the zone routing protocol," in *ACM SIGCOMM*, 1998.

[Haas1998b] Z. Haas and M. Pearlman, "The zone routing protocol (ZRP) for ad hoc networks," *Internet Draft*, 1998.

[Haas1999] Z. Haas and B. Liang, "Ad Hoc Mobility Management with Uniform Quorum Systems," *IEEE/ACM Transactions on Networking*, Vol. 7, No. 2, April 1999, pp. 228-240.

[Hatami2005] A. Hatami and K. Kahlavan, "A comparative performance evaluation of RSS-based positioning Algorithms used in WLAN Networks," Proceedings WCNC 2005.

[Hightower2001] J. Hightower and G. Borriello, "Location Systems for Ubiquitous Computing," *IEEE Computer*, Vol. 34, No. 8, August 2001, pp. 57-66.

[HIPERLAN1996] ETSI STC-RES10 Committee, "Radio Equipment and Systems: High Performance Radio Local Area Network (HIPERLAN) Type 1," Functional Specifications, June 1996, ETS 300-652.

[Hou1986] T.-C Hou and V. Li, "Transmission Range Control in Multihop Packet Radio Networks," *IEEE Transactions on Communications*, Vol. 34, No. 1, January 1986, pp. 38-44.

[IEEE-802.111997] IEEE Std. 802.11. "IEEE Standard for Wireless LAN Medium Access Control (MAC) and Physical Layer (PHY) Specification," June 1997.

[Iwata1999] A. Iwata, C.-C. Chiang, G. Pei, M. Gerla, and T.-W. Chen, "Scalable Routing Strategies for Ad Hoc Wireless Networks," *IEEE Journal on Selected Areas of Communications*, August 1999, 1369-1379.

[Jacquet2001] P. Jacquet, P. Mühlethaler, T. Clausen, A. Laouiti, A. Qayyum, L. Viennot, "Optimized Link State Routing Protocol for Ad Hoc Networks," in *Proceedings of IEEE INMIC*, Pakistan, 2001.

[Jiang1998] M. Jiang, J. Li, and Y. C. Tay, "Cluster based routing protocol (CBRP) functional specification," *Internet Draft*, 1998.

[Jin2000] K.T. Jin and D.H. Cho, "A MAC Algorithm for Energy-limited Ad Hoc Networks," in *Proceedings of Fall VTC 2000*, September 2000, 219-222.

[Johnson1996] D. Johnson and D. Maltz, "Dynamic Source Routing in Ad Hoc Wireless Networks," *Mobile Computing*, Kulwer Academic, 1996, 153-181.

[Jubin1987] J. Jubin and T. Truong, "Distributed Algorithm for Efficient and Interference-free Broadcasting in Radio Networks," in *Proceedings of IEEE Infocom*, January 1987, 21-32.

[Karp2000] B. Karp, "Geographic Routing for Wireless Networks," *Ph.D. Thesis*, Harvard University, 2000.

[Ko1998] Y.-B. Ko and N. H. Vaidya, "Location-aided routing (LAR) in mobile ad hoc networks," in *ACM MOBICOM*, November 1998.

[Kranakis1999] E. Kranakis, H. Singh, and J. Urrutia, "Compass Routing on Geometric Networks," in *Proceedings of Canadian Conference on Comp. Geo.*, Vancouver, August 1999.

[Lee2000] S.H. Lee and D.H. Cho, "A new adaptive routing scheme based on the traffic characteristics in mobile ad hoc networks," in *Proceedings of Fall IEEE VTC 2000*, September 2000, 2911-2914.

[Li2000] J. Li et al., "A Scalable Location Service for Geographic Ad Hoc Routing," in *Proceedings of ACM/IEEE International Conference on Mobile Computing and Networking (MOBICOM)*, 2000, pp. 120-130.

[Li2002] J. Li and P. Mohapatra, "PANDA: A Positional Attribute-Based Next-hop Determination Approach for Mobile Ad Hoc Networks," *Technical Report*, Department of Computer Science, University of California at Davis, 2002.

[Liang2005] W. Liang, Y. Liu, and X. Guo, "On-Line Disjoint path routing for network capacity Maximization in Ad Hoc Networks," WCNC 2005.

[Liao2001] W. H. Liao, Y. C. Tseng, S. L. Wang, and J. P. Sheu, "A Multi-Path QoS Routing Protocol in a Wireless Mobile Ad Hoc Network," in *Proceedings of IEEE ICN*, 2001.

[Lim2000] H. Lim and C. Kim, "Multicast tree construction and flooding in wireless ad hoc networks," In *Proceedings of the ACM International Workshop on Modeling, Analysis and Simulation of Wireless and Mobile Systems (MSWIM)*, 2000.

[Lin1999] C. R. Lin and J. S. Liu, "QoS Routing in Ad Hoc Wireless Networks," *IEEE Journal on Selected Areas in Communications*, Vol. 17, No. 8, August 1999

[Lin2005] J.H. Lin, Y.K. Kwok, and V.K.N. Lau, " A Quantitative comparison of Ad-Hoc routing protocols with and without channel adaptation," IEEE Transactions on Mobile Computing, Vol. 4, No. 2, March/April 2005, pp111=128.

[MANET1998] Mobile Ad Hoc Networks (MANET) Charter. Work in progress. http://www.ietf.org/html.charters/manet-charter.html, 1998.

[Marina2001] M. Marina and S. Das, "On-demand Multipath Distance Vector Routing in Ad Hoc Networks," *in the Proceedings of the International Conference for Network Protocols (ICNP)*, November 2001.

[Mauve2001] M. Mauve, J. Widmer, and H. Hartenstein, "A Survey on Position-Based Routing in Mobile Ad Hoc Networks," in *IEEE Network*, November/December 2001.
[Morris2000] R. Morris *et al.*, "Carnet: A Scalable Ad Hoc Wireless Network System," in *Proceedings of ACM SIGOPS European Workshop Beyond the PC: New Challenges for the Operating Systems*, September 2000.
[Murthy1996] S. Murthy and J.J. Garcia-Luna-Aceves, "An efficient routing protocol for wireless networks," in *ACM Mobile Networks and Applications Journal*, October 1996, 183-197.
[Nelson1984] R. Nelson and L. Kleinrock, "The Spatial Capacity of a Slotted Aloha Multihop Packet Radio Network with Capture," *IEEE Transactions on Communications*, Vol. 32, No. 6, June 1984, pp. 684-694.
[NS] Network Simulator (NS) version 2, http://www.isi.edu/nsnam/ns/index.html.
[Ogier2002] R. Ogier, "A Simulation Comparison of TBRPF, OLSR, and AODV," *SRI International*, July 2002.
[Park1997] V. D. Park and M. S. Corson, "A highly adaptive distributed routing algorithm for mobile and wireless networks," in *Proceeding of IEEE Infocom*, April 1997, 103-112.
[Pei2000] G. Pei, M. Gerla, and X. Hong, "Lanmar: Landmark routing for large scale wireless ad hoc networks with group mobility," in *Proceedings of ACM MobiHoc*, August 2000.
[Peng1999] W. Peng and X. Lu, "Efficient broadcast in mobile ad hoc networks using connected dominating sets," *Journal of* Software, Beijing, China, 1999.
[Perkins1994] C. E. Perkins and P. Bhagwat, "Highly dynamic destination-sequenced distance-vector routing (DSDV) for mobile computers," in *Computer Communication Review*, October 1994, 234-244.
[Perkins1999] C. E. Perkins and E. Royer, "Ad hoc on-demand distance vector routing," in *IEEE Workshop on Mobile Computing Systems and Applications*, February 1999, 90-100.
[Perkins2001] C. E. Perkins, *Ad Hoc Networking*, Addison-Wesley, ISBN: 0201309769, 2001.
[Samar2004] P. Samar, M.R. Pearlman, and Z. Haas, "Independent Zone Routing: An adaptive hybrid routing framework for ad-hoc wireless networks," IEEE Transactions on Networking , Vol. 12, No.4, Aug. 2004, pp 595-608.
[Singh1998] S. Singh, M. Woo, and C. S. Raghavendra, "Power-Aware Routing in Mobile Ad Hoc Networks," in *Proceedings of Mobihoc*, 1998, 181-190.
[Sinha1999] P. Sinha, R. Sivakumar, and V. Bharghavan, "CEDAR: a Core-Extraction Distributed Ad Hoc Routing algorithm," in *Proceedings of IEEE Infocom*, March 1999.
[Stojmenovic1999a] I. Stojmenovic, "A Routing Strategy and Quorum Based Location Update Scheme for Ad Hoc Wireless Networks," *Technical Report TR-99-09*, SITE, University of Ottawa, 1999.
[Stojmenovic1999b] I. Stojmenovic, "Home Agent based Location Update and Destination Search Schemes in Ad Hoc Wireless Networks," *Technical Report TR-99-10*, SITE, University of Ottawa, September 1999.
[Stojmenovic2002] I. Stojmenovic, "Position-based routing in Ad hoc Networks," IEEE Communications Magazine, July 2002, pp. 128-134.
[Takagi1984] H. Takagi and L. Kleinrock, "Optimal Transmission Ranges for Randomly Distributed Packet Radio Terminals," *IEEE Transactions on Communications*, Vol. 32, No. 3, March 1984, pp. 246-257.
[Tanenbaum1996] A. Tanenbaum, *Computer Networks*, Prentice Hall, ISBN 0-13-349945-6, 1996.
[Toh1997] C.-K. Toh, "Associativity based routing for ad hoc mobile networks," *Wireless Personal Communications*, March 1997.
[Toussaint1980] G. Toussaint, "The Relative Neighborhood Graph of a Finite Planar Set," *Pattern Recognition*, Vol. 12, No. 4, 1980, pp. 261-268.
[Tsuchiya1988] P.F. Tsuchiya, "The Landmark Hierarchy: a new hierarchy for routing in very large networks," In *Computer Communication Review*, Vol.18, No.4, Aug. 1988, 35-42.
[Vaidya2002] N. Vaidya, "Duplicate Address Detection in Mobile Ad Hoc Networks," in *ACM Mobihoc*, June 2002.
[Zhu2002] C. Zhu and M. S. Corson, "QoS Routing for Mobile Ad Hoc Networks," in *Proceedings of Infocom*, 2002.

Chapter 3

Broadcasting, Multicasting and Geocasting

3.1 Introduction

Recently, there has been an increasing interest in applications like online gaming, where multiple players residing at different locations participate in the same gaming session through their handheld portable devices. Consider a scenario with a user walking with a handheld device or waiting for a flight in airport terminal. He/She does not know about his/her neighbor, and switches on handheld device and tries to scan if someone would be interested in playing some game. This kind of "community centric" application is a major attraction in forthcoming data communication world. This is a typical MANET application, wherein interest is formed on demand by using portable devices.

As we have seen so far, there are many applications to MANETs such as electronic email and file transfer can be considered to be easily deployable within a MANET environment. Web services are also possible in case any node in the network can serve as a gateway to the outside world. We need not emphasize a wide range of possible military applications of MANETs as the technology was initially developed keeping them in mind. In such situations, the MANETs, having self-organizing capability, can be efficiently used where other technologies either fail or cannot be deployed effectively. Advanced features of wireless systems, including data rates compatible with multimedia applications, global roaming capability, and coordination with other network structures, are enabling new applications to be explored. Some of these diverse applications are characterized by a cooperative collaboration which is typical for the MANETs. Broadcasting, Multicasting and Geocasting are three enabling technologies which can add to liveliness of these applications.

Broadcasting is a common operation in many applications, e.g., graph-related and distributed computing problems. It is also widely used to resolve many network layer problems. In the particular case of a MANET where mobility is the rule and not the exception, broadcastings are expected to be performed more frequently (e.g., for paging a particular host, sending an alarm signal, and finding a route to a particular host such as in DSR, AODV, ZRP, and CBRP). Broadcasting may also be used in LAN emulation or serve as a last resort to provide multicast services in networks with rapid changing topologies. Therefore, broadcasting in a MANET is a basic service which needs deeper investigation and tuning.

Multicasting is the transmission of datagrams to a group of hosts identified by a single destination address and hence is intended for group-oriented computing [Agrawal2002]. In MANETs, multicasting can efficiently support a variety of applications that are characterized by close collaborative efforts. Multicasting could prove to be an efficient way of providing necessary services for these kinds of applications. If the group contains all the members of the network, then multicasting is changed to broadcasting. On the other hand, geocasting aims at delivering data packets to a group of nodes located in a specified geographical area (e.g., to broadcast emergency information within a mile radius of a fire, or to broadcast a coupon for coffee within a block of a Starbucks). Geocasting can be seen as a variant of the conventional multicasting problem, and distinguishes itself by specifying hosts as group members within a specified geographical region. In geocasting, the nodes eligible to receive packets are implicitly specified by a physical region; membership in a geocast group changes whenever a mobile node moves in or out of the geocast region [Boleng2001, Tseng2001].

Since broadcasting, multicasting and geocasting attack the issue of communication to a group of recipients, it is imperative to determine the best way to provide these services in an ad hoc environment by looking at broadcasting, multicasting and geocasting protocols simultaneously so that they could play important roles in their respective field. Therefore, if we can efficiently combine with multicasting and geocasting the features of MANET, it will be possible to realize a number of envisioned group oriented applications.

To, quantify which one is suitable and for what type of applications, it is necessary to investigate and discern on the applicability of existing ad hoc broadcast, multicast and geocast protocols. In this chapter, we provide a detailed description and comparison of broadcast, multicast and geocast protocols for MANETs. We also attempt to provide an insight into anticipated trends in the area and outline the approaches that are likely to play a major role in future, as well as point out open problems that need careful attention from the research community. It may be noted that there exists a large amount of literature for multicast in wired and infrastructured wireless networks and for a detailed investigation of them please refer to [Gossain2002]. Here, we are focusing only on multicasting over MANETs [Cordeiro2003].

3.2 The Broadcast Storm

Doing network-wide broadcasting in MANETs requires one device to broadcast the information to all its neighbors. For far-away devices, the message is rebroadcasted which could cause collision if multiple device broadcasts the same time and are in the neighborhood. This is also known as the broadcasting storm problem [Ni1999] and in this section we discuss ways to perform efficient rebroadcasting of messages. For the purpose of our discussion here, we assume that MHs in the MANET share a single common channel with carrier sense multiple access (CSMA) [Agrawal2002], but no collision detection (CD) or collision avoidance (CA) capability (e.g., the IEEE standard 802.11 [IEEE-802.11]). Synchronization in such a network with mobility is unlikely, and global network topology information is unavailable to facilitate the scheduling of a broadcast. Thus, one straightforward and obvious solution is to achieve broadcasting by flooding (for example, as it is done by mostly all MANET routing algorithms). Unfortunately, as we will see later, it is observed that redundancy, contention, and collision could exist if flooding is done blindly. Several problems arise in these situations including:

- As the radio propagation is omnidirectional and a physical location may be covered by the transmission ranges of several hosts, many rebroadcasts are considered to be redundant.
- Heavy contention could exist because rebroadcasting hosts are probably close to each other.

- As the RTS/CTS handshake is inapplicable for broadcast transmissions, collisions are more likely to occur as the timing of rebroadcasts is highly correlated.

3.2.1 Broadcasting in a MANET

A MANET consists of a set of MHs that may communicate with one another from time to time, and where no base stations are present. Each host is equipped with a CSMA/CA [Agrawal2002] transceiver. In such an environment, a MH may communicate with each other directly or indirectly. In the latter case, a multi-hop scenario occurs, where the packets originated from the source host are relayed by several intermediate MHs before reaching the destination. The broadcast problem refers to the transmission of a message to all other MHs in the network. The problem we consider has the following characteristics:

- **The broadcast is spontaneous**: Any MH can issue a broadcast operation at any time. For reasons such as the MH mobility and the lack of synchronization, preparing any kind of global topology knowledge is prohibitive.
- **The broadcast is frequently unreliable**: Acknowledgement mechanism is rarely used. However, attempt should be made to distribute a broadcast message to as many MHs as possible without putting too much effort. The motivations for such an assumption are:
  a. A MH may miss a broadcast message because it is off-line, it is temporarily isolated, or it experiences repetitive collisions.
  b. Acknowledgements may cause serious medium contention (and thus another "storm") surrounding the sender.
  c. In many applications, a 100% reliable broadcast is unnecessary.

In addition, we assume that a MH can detect duplicate broadcast messages. One way to do so is to associate a tuple (source ID, sequence number) with each broadcast message as in the case of DSR and AODV. Here, we focus on the flooding behavior in a MANET – the phenomenon where the transmission of a packet will trigger other surrounding MHs to transmit the same packet. We shall show that if flooding is used blindly, many redundant messages will be sent and serious contention/collision will be incurred.

A straightforward approach to perform broadcast is by flooding wherein a MH, on receiving a broadcast message for the first time, rebroadcast the message. But, associated limitations are discussed next.

3.2.2 Flooding-Generated Broadcast Storm

A simple broadcast will clearly have n transmissions in a network of n MHs. In a CSMA/CD network, drawbacks of flooding include:

- **Redundant rebroadcasts**: When a MH decides to rebroadcast a broadcast message to its neighbors, all its neighbors may already have the message.
- **Contention**: After a MH broadcasts a message, if many of its neighbors decide to rebroadcast the message, these transmissions may severely contend with each other.
- **Collision**: Because of the deficiency of backoff mechanism, lack of RTS/CTS handshake in broadcasts, and absence of collision detection (CD), collisions are more likely to occur.

Redundancy Analysis

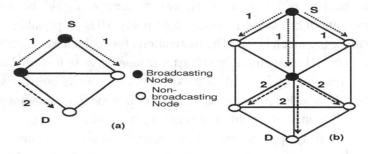

Figure 3.1 – Two optimal Broadcasts in a MANET. The links represent the connectivity among the nodes. Node S is the Source and node D is the "last" network node.

The main reason for redundancy is that radio signals from different transceivers may overlap with each other. Let us consider two examples where we denote node S is the source of the broadcast and node D as the "last" node to receive the broadcast. In Figure 3.1(a), it only takes two transmissions for node D to broadcast a message whereas normally, four transmissions will be carried out. Figure 3.1(b) presents a scenario where only two transmissions are sufficient for broadcast as opposed to seven transmissions generated.

As mentioned earlier, too many nodes start retransmitting the same message, which causes multiple copies of the same message being sent around and is referred to as the *broadcast storm problem*. Figure 3.2 exemplifies the broadcast storm problem, where node S initiates a route request to node D through a flooding. As we can see from the steps shown, flooding is highly redundant. Each node receives the route request *degree* times, and the route request propagates far beyond node D. Because nearby nodes will receive and rebroadcast the route request at nearly the same time, contention and collision will occur quite frequently. Some of them are explicitly shown in Figure 3.2.

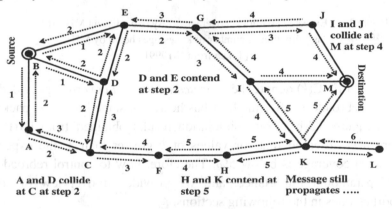

Figure 3.2 – The broadcast storm problem in a MANET
(13 nodes with an average degree of 2.6)

Let us assume that the total area covered by the radio signal transmitted by a transceiver is a circle of radius r. The areas covered by MHs S_1 and S_2 d distance apart are shown in Figure 3.3(a) by circles with radius r. The intersection of their coverage areas $INTC(d)$ is indicated by a shaded area. Upon hearing a packet for the first time, the *additional area coverage* provided by a MH which rebroadcasts the packet is equal to $\pi r^2 - INTC(d)$. When $d = r$, the additional coverage is maximum and is approximately equal to $0.61\pi r^2$. This is to say that a rebroadcast can provide only $0 \sim 61\%$ of additional coverage over what has been covered by the previous transmission. If we assume that a rebroadcasting host is randomly located within the transmitter's coverage, we can conclude through some calculation that the average additional coverage is $0.41\pi r^2$.

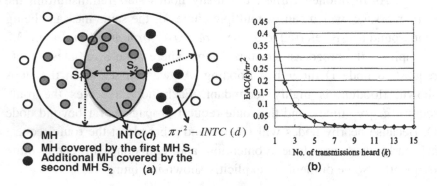

Figure 3.3 – Expected additional coverage (a) By two adjacent MHs, (b) $\dfrac{EAC\ (k)}{\pi\ r^2}$ as a function of number of k messages heard by MH
[Taken from Ni1999]

Let EAC(k) denote the *expected additional coverage* provided by a host's rebroadcast after the host has heard the same broadcast packet k times. Figure 3.3(b) shows simulation results obtained from [Ni1999] where it is observed that for values of k greater than 3, the expected additional coverage is below 5%. Therefore, how to control rebroadcast is of paramount importance, and we provide an overview of some important ones in the following sections.

3.2.3 Rebroadcasting Schemes

Recently, many proposals have addressed the broadcast storm problem with an aim of minimizing the number of retransmissions of a broadcast message while attempting to deliver a broadcast packet to each node in the network. A performance comparison of some of the schemes can be found in [Williams2002]. Before delving into the proposed solutions, it is important however to introduce common attributes of broadcast protocols considered here. We also note that the protocols proposed usually assume the IEEE standard 802.11 in the MAC layer.

3.2.3.1 Common Characteristics

In this section we describe common attributes of all broadcasting schemes. As we know, radio waves propagate at the speed of light. Therefore, when a source node transmits a broadcast packet, all of the source's neighbors will receive the transmission nearly at the same time, causing problems and are discussed next.

Design Considerations: Jitter and Random Delay Time

If all MHs are assumed to possess similar hardware and system loads, the neighbors will process the packet and rebroadcast it approximately at the same time, thereby causing packet collisions. To address this problem, broadcast protocols jitter the scheduling of broadcast packets from the network layer down to the MAC layer by some random amount of time. This jitter allows one neighbor to acquire the channel first, while other neighbors detect that the channel is busy. Many of the broadcasting protocols require a node to keep track of redundant packets received over a short time interval in order to determine whether or not to rebroadcast. This time interval, which is called as Random Delay Timer (RDT), is randomly chosen from a uniform distribution to accomplish two things. First, it allows nodes sufficient time to receive redundant packets. Second, the randomized scheduling mitigates packet collisions as discussed earlier.

An important design consideration in any broadcast protocol is the implementation of RDT. One approach is to send broadcast packets to the MAC layer after a short random time similar to the jitter. In this case, packets remain in the interface queue until the channel becomes clear for broadcast. While the packet is in the interface queue, redundant packets may be received, allowing the network layer to determine if rebroadcasting is still required. If the network layer protocol decides the packet should not be rebroadcasted, it informs the MAC layer to discard the corresponding packet. Based on the original work, the second approach is to make RAD (Random Delay Time) > jitter and that is how it differs from approach 1. During RDT, all redundant packets are assessed to decide on retransmission. After RDT expiration, the packet is either sent to the MAC layer or dropped. No attempts are made by the network layer to remove the packet after being sent to MAC layer.

Loop Prevention

None of the protocols discussed here require that a node rebroadcast a given packet more than one time. Thus, each broadcast protocol requires that nodes cache the original source node ID and the packet ID. This allows unique identification for each broadcast packet and assigns appropriate behavior upon reception of another broadcast packet.

3.2.3.2 Categories and Protocols

We now cover the broadcasting protocols for MANETs. In order to do that, we first categorize them into the following classes with increasing complexity:

- Simple flooding;
- Probability-based methods;
- Area-based methods; and
- Neighbor knowledge methods.

Simple flooding requires each node to rebroadcast all packets. *Probability-based methods* use the network topology to assign a probability to a node to rebroadcast. *Area-based methods* assume nodes have common transmission distances; and a node will only rebroadcast if this rebroadcast will likely provide sufficient additional coverage area. *Neighbor knowledge methods* maintain state on their neighborhood via hello packets which is used in deciding about rebroadcast.

3.2.3.2.1 Simple Flooding

Simple flooding [Ho1999, Jetcheva2001] starts off with a source node broadcasting a packet to all neighbors. The neighbors, upon receiving the broadcast packet, rebroadcast the packet exactly once and this continues until all reachable network nodes have received and rebroadcast the packet at least once. Flooding is proposed to achieve reliable broadcast in highly dynamic networks in [Ho1999].

3.2.3.2.2 Probability-Based Methods

Probabilistic Scheme

The *probabilistic scheme* [Ni1999] is similar to ordinary flooding, except that nodes only rebroadcast with a predetermined probability. In dense networks, it is much likely that multiple nodes share similar transmission coverage. Thus, having some random nodes not to rebroadcast saves network resources without harming its effectiveness. In sparse networks, there is less shared coverage and not all nodes will receive broadcast packets unless the probability parameter is high. When the probability is 100%, this scheme is similar to ordinary flooding.

Counter-Based Scheme

An inverse relationship is shown [Ni1999] between the number of times a packet is received at a node and the probability of this node's transmission being able to cover additional area on a rebroadcast and forms the basis of the *counter-based scheme*. Upon receipt of a previously unseen packet, the node initiates a counter with a value of one and sets a RDT. During the RDT, the counter is incremented by one for each redundant packet received. If the counter is less than a threshold value when the RDT expires, the packet is rebroadcast. Otherwise, it is simply dropped. Results reported in [Ni1999] show that threshold values above six relate to little additional coverage area being reached. The overriding features of the counter-based scheme are its simplicity and its inherent adaptability to local topologies. In other words, in a dense area of the network some nodes will not rebroadcast, whereas in sparse areas of the network all nodes will likely rebroadcast.

3.2.3.2.3 Area-Based Methods

Suppose a node receives a packet from a sender that is located only one meter away. If the receiving node rebroadcasts, the additional area covered by the retransmission is quite small. On the other hand, if a node is located at the boundary of the sender's radio coverage, then a rebroadcast would provide significant additional coverage area (to be more precise, up to 61% as discussed earlier). A node using an area-based method can try to estimate the additional coverage area based on all received redundant transmissions. The area-based methods only consider the coverage area of a transmission; while they do not consider the presence of nodes within this area.

Distance-Based Scheme

A node using the *distance-based scheme* compares the distance (e.g., through received signal strength indicator: RSSI) between itself and each of its neighbor nodes that has previously rebroadcast a given packet. Upon reception of a previously unseen packet, a RDT is initiated and redundant packets are cached. When the RDT expires, all source node locations are examined to determine if the distance between itself and any of its neighbor nodes is closer than a threshold distance value. If true, the node does not rebroadcast.

Location-Based Scheme

The *location-based scheme* [Ni1999] uses a more precise estimation of expected additional coverage area in deciding rebroadcast. In this method, each node must have the means to determine its own location (e.g., through GPS). Whenever a node originates or rebroadcasts a packet, it adds its own location to the header of the packet. Then, the node receiving a packet, calculates additional coverage area if it were to rebroadcast. If such an area is less than a threshold value, the node does not rebroadcast and all future receptions of the same packet are ignored. Otherwise, the node assigns a RDT before transmission. If the node receives a redundant packet during the RDT, it recalculates the additional coverage area and compares with the threshold. The area calculation and threshold comparison occur with all redundant broadcasts received, until the packet reaches its scheduled send time, or else it is dropped.

3.2.3.2.4 Neighbor Knowledge Methods

Flooding with Self Pruning

The simplest form of the neighbor knowledge methods is referred to as *flooding with self pruning* [Lim2000]. This protocol requires that each node have knowledge of its one-hop neighbors obtained via periodic hello packets. A node includes its list of known neighbors in the header of each broadcast packet and receiving node compares this list to the sender's neighbor list. If the receiving node would not reach any additional nodes, it refrains from rebroadcasting the packet. Otherwise, the node rebroadcasts the packet.

Scalable Broadcast Algorithm (SBA)

The Scalable Broadcast Algorithm (SBA) [Peng2000] requires that all nodes know about their neighbors within a two-hop radius. This neighbor knowledge coupled with the identity of the node from which a packet is received, allows a receiving node to determine if it would reach additional nodes by rebroadcasting. Two-hop neighbor knowledge is achievable via periodic hello packets; where each hello packet contains the source node's identifier (IP address) and the list of known neighbors.

Now suppose a node S_2 receives a broadcast data packet from node S_1. Since node S_1 is a neighbor, node S_2 can easily determine all the nodes which are simultaneously neighbor to both nodes S_1 and S_2. If node S_2 determines that its broadcast will cover additional neighbors not reached by node S_1's broadcast, node S_2 schedules the packet for transmission with a RDT. If, in the meantime, node S_2 receives a redundant broadcast packet from any another neighbor, node S_2 again determines if it can reach any new nodes by rebroadcasting. This process continues until either the RDT expires and the packet is sent, or the packet is dropped.

A method to dynamically adjust the RDT to network conditions is proposed in [Peng2000], where the RDT is calculated based on a node's relative neighbor degree. Specifically, each node searches its neighbor tables for the maximum neighbor degree of any neighboring nodes, say, d_{Nmax}, which can be obtained from the neighbors' broadcast packets. It then calculates a RDT based on the ratio of:

$$\left(\frac{d_{N\,max}}{d_{me}} \right)$$

where d_{me} is the number of current neighbors for the node. We note that this weighing scheme is greedy, as nodes with the largest number of neighbors usually broadcast before the others (smaller RDT).

Dominant Pruning

Dominant pruning also uses two-hop neighbor knowledge for routing decisions [Lim2000]. Unlike SBA, dominant pruning requires rebroadcasting nodes to proactively choose some or all of its one-hop neighbors as rebroadcasting nodes, and include their addresses as a part of the list in the broadcast packet header. Whenever a node receives a broadcast packet, it checks the header to see if its address is a part of the list. If so, it has to determine which of its neighbors should rebroadcast its packet so as to include them in the packet header. For that, it uses a Greedy Set Cover algorithm given the knowledge of which neighbors have already been covered by the previous sender's broadcast. One such algorithm [Lim2000] recursively chooses one-hop neighbors until it covers all of two-hop neighbors.

Multipoint Relaying

Multipoint relaying [Qayyum2000] is similar to Dominant Pruning in that rebroadcasting nodes are explicitly chosen by upstream senders. For example, say node A originates a broadcast packet. It has previously selected some, or in certain cases, all of its one-hop neighbors to rebroadcast all packets they receive from node A. The chosen nodes are called Multipoint Relays (MPRs) and each MPR is required to choose a subset of its one-hop neighbors as MPRs. The following algorithm for a node to choose its MPRs is suggested in [Qayyum2000]:

a. Find all two-hop neighbors that can only be reached by one one-hop neighbor. Assign those one-hop neighbors as MPRs.
b. Determine the resultant cover set (i.e., the set of two-hop neighbors that will receive the packet from the current MPR set).
c. From the remaining one-hop neighbors not yet in the MPR set, select that covers the most two-hop neighbors not in the cover set.
d. Repeat from step 2 until all two-hop neighbors are covered.

Multipoint relaying is described in detail as part of the OLSR protocol. In OLSR, hello packets include fields for a node to list the MPRs it has chosen. Clearly, the update interval for hello packets must be carefully chosen and optimized for given network conditions.

Ad Hoc Broadcast Protocol

The *Ad Hoc Broadcast Protocol* (AHBP) [Peng2002] utilizes an approach similar to Multipoint Relaying by designating nodes as a Broadcast Relay Gateway (BRG) within the broadcast packet header. BRGs are proactively chosen from each upstream sender which is a BRG itself. AHBP differs from Multipoint Relaying in three ways:

a. A node using AHBP informs one-hop neighbors of the BRG designation by a field in the header of each broadcast packet. This allows a node to calculate the most effective BRG set at the time a broadcast packet is transmitted. In contrast, Multipoint Relaying informs one-hop neighbors of MPR designation via hello packets.

b. In AHBP, when a node receives a broadcast packet and is listed as a BRG, the node uses two-hop neighbor knowledge to determine which neighbors also received the broadcast packet in the same transmission. These neighbors are considered already "covered" and are removed from the neighbor graph to choose next hop BRGs. In contrast, MPRs are not chosen considering the source route of the broadcast packet.

c. AHBP is extended to account for high mobility networks. Suppose node A receives a broadcast packet from node B, and node A does not list node B as a neighbor. In AHBP-EX (extended AHBP), node A will assume BRG status and rebroadcast the packet. Multipoint relaying could be similarly extended.

Dominating Set and Connected Dominating Set-Based Broadcasting

A dominating set (DS) is a subset of nodes in a graph such that all other remaining nodes are one hop away. So, if nodes of DS transmit a packet, all nodes in the network will receive a copy of the message, thereby achieving broadcasting. One such example is shown in Figure 3.4 (a) where S_4, S_7, S_{10}, S_{15}, and S_{17} form the DS of the graph.

A more calculation intensive algorithm for selecting DS - the *Connected Dominating Set (CDS)-Based Broadcast Algorithm* is described in [Peng1999] and is illustrated in Figure 3.4(b). Note that the nodes $S_4, S_5, S_7, S_9, S_{13}, S_{17}, S_{18}$, and S_{12} constituting CDS are in the form of a linear array. The idea behind CDS is once the packet that need to be broadcasted reaches one of the nodes of the DS, a successive retransmission by CDS nodes allows coverage of all CDS nodes and hence all nodes of the graph. Thus, CDS may have larger number of nodes as compared to DS; while it is still desirable as it makes broadcasting steps easier to define. A generic framework for distributed broadcast schemes in MANETs has been given in [Wu2004] which employs a dynamic self-pruning technique for changing a gateway node to a non-gateway one. Such a dominating-set-based broadcasting approach selects a subset of MHs to forward packets on behalf of other nodes, while all other nodes keep quiet. Finding DS or CDS of a graph is quite involved and many heuristics [He2009, Liu2010, Srinivasan2008] have been introduced to make the process simpler.

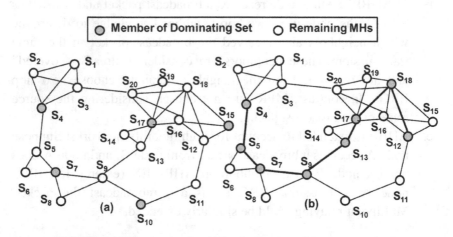

Figure 3.4 (a) Dominating Set (DS) of a Graph, and (b) Connected DS

A combination of gossiping and dominating set approach has been introduced in [Zhang2005a, Zhang2005b] by dividing MHs into four groups based on local information as follows:

- Group 1: Nodes with degrees larger than the degree of all neighboring nodes.
- Group 2: Nodes have a majority of neighbors with smaller degree than themselves.
- Group 3: Remaining nodes not belonging to groups 1, 2 and 4.
- Group 4: Nodes with degrees smaller than all the neighbors.

This process for the network of Figure 3.5 is illustrated in Table 3.1. The connectivity measurement and comparison with neighboring MHs provide a good idea about selecting nodes for retransmission. In any general MANET, once such grouping is done, the probability of using these groups as a message forwarder is assigned in a decreasing order as p_1, p_2, p_3 and p_4; p_1 for group 1 being the highest and p_4 for group 4 being the lowest. The idea behind selecting such groups is to ensure that the nodes with higher connectivity could possibly cover a larger number of newer nodes whenever broadcasting is needed. These values can be easily calculated as follows.

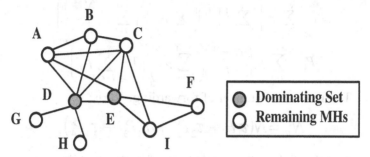

Figure 3.5 Grouping in an MANET

Table 3.1 Characteristics of Figure 3.5 in Terms of Connectivity

MH	A	B	C	D	E	F	G	H	I
Degree	4	3	5	6	5	2	1	1	3
Neighbors	B, C, D, E	A, C, D,	A, B, D, E, I	A, B, C, E, G, H	A, C, D, F, I	E, I	D	D	C, E, F
Connectivity compared to neighbors	Medium	Low	Medium	High	Medium	Low	Low	Low	Low
Group No.?	2	4	2	1	2	4	3	3	4

Let A be the area of the MANET, N be the number of MHs and r be the communication range of each mobile node with α being the fraction of the area covered, then $\alpha = \dfrac{\pi r^2}{A}$.

The average number of neighboring nodes $N_{neighbors}$, can be given by:

$$N_{neighbors} = (N-1)\,\alpha.$$

The probability p_i that a mobile node has I neighbors, can be given by:

$$p_i = \binom{N-1}{i}(1-\alpha)^{N-1-i}\,\alpha^i.$$

The probability p_1, p_2, p_3, and p_4 can be given by

$$P_1 = \sum_{i=1}^{N-1} p_i \binom{N-1}{i}\left(\sum_{j=1}^{i-1} p_j\right)^i\left(1-\sum_{j=1}^{i-1} p_j\right)^{N-1-i}.$$

$$P_2 = \sum_{i=1}^{N-1} P_i \left(\sum_{k=1}^{i/2} \left(\binom{i}{k} \left(\sum_{j=i}^{N-1} p_j \right)^k \left(\sum_{j=1}^{i-1} p_j \right)^{i-k} \right) \right).$$

$$P_3 = \sum_{i=1}^{N-1} P_i \left(\sum_{k=i/2}^{i} \left(\binom{i}{k} \left(\sum_{j=i}^{N-1} p_j \right)^k \left(\sum_{j=1}^{i-1} p_j \right)^{i-k} \right) \right).$$

$$P_4 = \sum_{i=1}^{N-1} P_i \binom{N-1}{i} \left(\sum_{j=i+1}^{N-1} p_j \right)^i \left(1 - \sum_{j=i+1}^{N-1} p_j \right)^{N-1-i}.$$

Then, the total number of forwarding, N_F, is:

$$N_F = N \left(\tau_1 P_1 + \tau_2 P_2 + \tau_3 P_3 + \tau_4 P_4 \right),$$

where τ_i is the forwarding probability for the mobile node in group i $(1 \le i \le 4)$. Such a scheme does not provide 100% coverage of all MANET nodes [Zhang2005c], but a good coverage and excellent saving are achieved under mobility and about 20% higher goodput is obtained than in the conventional AODV.

Lightweight and Efficient Network-Wide Broadcast

The Lightweight and Efficient Network-Wide Broadcast (LENWB) protocol [Sucec2000] also relies on two-hop neighbor knowledge obtained from hello packets. However, instead of a node explicitly choosing other nodes to rebroadcast, the decision is implicit. In LENWB, each node decides to rebroadcast based on the knowledge of which of its other one and two-hop neighbors are expected to rebroadcast. The information required for that decision is the knowledge of which neighbors have received a packet from the same source node, and which neighbors have a higher priority for rebroadcasting governed by the number of neighbors. The higher a node's degree is, the higher is its priority. It can proactively determine if all of its lower priority neighbors will receive those rebroadcasts.

3.3 Multicasting

In this section, we investigate the problem of multicasting in MANETs where the problem is to deliver a message to a subset of MANET MHs. We begin by understanding the hard task of multicasting to a group of MHs, together with the various issues behind the design and implementation of a multicast protocol for MANETs. Next, we study the existing multicast protocols for MANETs and show how different they are as compared to broadcasting.

3.3.1 Issues in Providing Multicast in a MANET

Well-established routing protocols do exist to offer an efficient multicasting service in conventional wired networks [Gossain2002]. Protocols for fixed networks fail to keep up with node movements and frequent topological changes. Besides, overhead of these protocols increases substantially with host mobility. The broadcast protocols cannot be used either as multicasting requires a selected set of nodes to receive the message do not consider whether a node belongs to a group or not. Rather, new protocols are being proposed and investigated which take issues such as locations of nodes belonging to a multicast group, and all associated topological changes. Moreover, MHs of a MANET run on batteries, the routing protocols must limit the amount of control information that is passed between nodes.

The majority of applications are in the areas where rapid deployment and dynamic reconfiguration are necessary and the wireline network is not available. These include military battlefields, emergency search and rescue sites, classrooms, and conventions where participants share information dynamically using their mobile devices. In addition, it is even more crucial to reduce the transmission overhead and power consumption in a wireless environment. Transient loops may form during reconfiguration of distribution structure (e.g., tree) as a result of mobility. Therefore, reconfiguration scheme should be kept simple to maintain low channel overhead. As we can see, providing an efficient multicasting over MANET faces many challenges including dynamic group membership and constant update of delivery path due to node movement. In the next sections, we cover the major protocols proposed so far and compare them under different criteria.

3.3.2 Multicast Routing Protocols

One straightforward way to provide multicast in a MANET is through flooding. With this approach, data packets are sent through out the MANET and every node that receives this packet broadcasts it to all its immediate neighboring nodes exactly once. It is suggested that in a highly mobile MANET, flooding of the whole network may be a viable alternative for reliable multicast. However, this approach has a considerable overhead as a number of duplicated packets are sent and packet collision does occur in a multiple-access based MANET.

We can classify the protocols into four categories based on how route to the members of the group is created:

- Tree-Based Approaches;
- Meshed-Based Approaches;
- Stateless Multicast; and
- Hybrid Approaches.

In the following we provide a description of the various multicast protocols in the above categories and compare them under several criteria.

3.3.2.1 Tree-Based Approaches

Most of the schemes for providing multicast in wired network are either source-based or shared tree-based. Different researchers have tried to extend the idea of tree-based approach to provide multicast in a MANET environment and many characteristics can be identified such as: a packet traverses each hop and node in a tree at most once; very simple routing decisions at each node; and the number of copies of a packet is minimized; such a loop-free tree structure represents shortest paths amongst nodes.

On the other hand, there are many issues that must be addressed in tree-based approaches. As mentioned earlier, trees provide a unique path between any two nodes. Therefore, having even one link failure could mean reconfiguration of the entire tree structure and could be a major drawback. In addition, multiple packets generated by different sources will require some consideration when utilizing multicast trees such that efficient routing can be established and maintained. Thus, it is common to consider the use of either a shared tree or establish a separate tree per each source. As highlighted in [Garcia-Luna-Aceves1999a], each approach has to deal with its own associated issues.

For separate source trees, each router (or node in case of MANETs) involved in multiple router groups must maintain a list of pertinent information for each group in which it is involved. Such management per router is inefficient and not scalable. On the other hand, for shared trees, there is a potential that packets may not only not traverse shorter paths, but in fact may be routed on paths with are much longer. While any scheme has positive and negative sides, simple structure coupled with ease of the approach has made multicast trees the primary method for realizing multicasting on the Internet.

3.3.2.1.1 Ad hoc Multicast Routing Protocol Utilizing Increasing Id-Numbers

Ad hoc Multicast Routing Protocol utilizing Increasing id-numberS (AMRIS) [Wu1998] is an on-demand protocol, which constructs a shared multicast delivery tree (see Figure 3.6) to support multiple senders and receivers. AMRIS dynamically assigns an id-number to each node in each multicast session. Based on the id-number, a multicast delivery tree – rooted at a special node with Sid (Smallest-ID) – is created and the id-number increases as the tree expands from the Sid. Generally, Sid is the source or the node that initiates a multicast session.

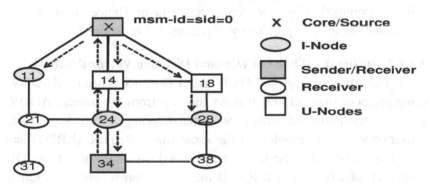

Figure 3.6 – AMRIS packet forwarding (X and 34 are sources, 11, 24, 28 are recipients) [Taken from Cordeiro2003]

The first step in AMRIS protocol operation is the selection of Sid. If there is only one sender for a group, the Sid is generally the source of the group. In case of multiple senders, a Sid is selected among the given set of senders. Once a Sid is identified, it sends a NEW-SESSION message to its neighbors. The content of this message includes

Sid's msm-id (multicast session member id) and the routing metrics. Nodes receiving the NEW-SESSION message generate their own msm-ids, which is larger than the msm-id of the sender. In case a node receives multiple NEW-SESSION messages from different nodes, it keeps the message with the best routing metrics and calculates its msm-ids. To join an ongoing session, a node checks the NEW-SESSION message, determines a parent with smallest msm-ids, and unicast a JOIN-REQ to its potential parent node. If parent node is already in the multicast delivery tree, it replies with a JOIN-ACK. Otherwise, the parent itself tries to join the multicast tree by sending a JOIN-REQ to its parent. If a node is unable to find any potential parent node, it executes a branch reconstruction (BR) process to rejoin the tree. BR consists of two sub-routines, namely, subroutines 1 (BR1) and 2 (BR2). The BR1 is executed when a node has potential parent node for a group. In case it does not find any potential parent node, BR2 is executed where the node broadcasts a JOIN-REQ which consists of a range field R to specif nodes till R hops. Upon link breakage, the node with larger msm-id tries to rejoin the tree by executing any of the BR mechanisms. AMRIS detects the link disconnection by a beaconing mechanism. Hence, until the tree is reconstructed, there is possibility of packets being dropped.

3.3.2.1.2 Multicast Ad hoc On-Demand Distance Vector Protocol

The Multicast Ad hoc On-Demand Distance Vector (MAODV) routing protocol [Royer1999] follows directly from the unicast AODV, and discovers multicast routes on-demand using a broadcast route discovery mechanism employing the same Route Request (RREQ) and Route Reply (RREP) messages that issued in the unicast AODV protocol. A MH originates a RREQ message when it wishes to join a multicast group, or when it has data to send to a multicast group without having a route to that group. Only a member of the desired multicast group may respond to a join RREQ. If the RREQ is not a join request, any node with a fresh enough route (based on group sequence number) to the multicast group may respond. If an intermediate node receives a join RREQ for a multicast group of which it is not a member, or if it receives a RREQ and it does not have a route to that group, it rebroadcasts the RREQ to its neighbors.

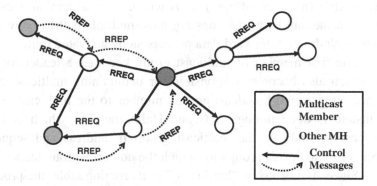

Figure 3.7 – Route Discovery in MAODV Protocol
[Taken from Cordeiro2003]

A node receiving a RREQ first, updates its route table to record the sequence number and the next hop information for the source node. This reverse route entry may later be used to relay a response back to the source. For join RREQs, an additional entry is added to the multicast route table and is not activated unless the route is selected to be part of the multicast tree. If a node receives a join RREQ for a multicast group, it may reply if it is a member for the multicast group's tree and its recorded sequence number for the multicast group is at least as great as that contained in the RREQ. The responding node then unicasts a RREP back to the source. As nodes along the path to the source receive the RREP, they add both a route table and a multicast route table entry for the node from which they received the RREP (see Figure 3.7).

When a source node broadcasts a RREQ for a multicast group, it often receives more than one reply. The source node keeps the received route with the greatest sequence number and shortest hop count to the nearest member of the multicast tree for a specified period of time, and disregards other routes. At the end of this period, it enables the selected next hop in its multicast route table, and unicasts an activation message (MACT) to this selected next hop which enables the entry for the source node. If this node is a member of the multicast tree, it does not propagate the message any further. Otherwise, it would have received one or more RREPs from its neighbors. It keeps the best next hop for its route to the multicast group. This process continues until the node that originated the chosen RREP (member of tree) is reached. The activation message ensures that the multicast tree does not have multiple paths to any node in the tree. Nodes only forward data packets along activated routes.

The first member of multicast group becomes a leader for that group which also becomes responsible for maintaining multicast group sequence number and broadcasting this number to the multicast group. This update is done through a Group Hello message which contains extensions that indicate the multicast group IP address and sequence numbers of all multicast groups for which the node is a group leader.

Since AODV keeps "hard-state" in its routing table, the protocol has to track actively and react to changes in this tree. If a member terminates its membership with the group, the multicast tree requires pruning. Links in the tree are monitored to detect link breakages and the node that is farther from the multicast group leader takes the responsibility to repair the broken link. If the tree cannot be reconnected, a new leader is chosen as follows. If the node initiating the route rebuilding is a multicast group member, it becomes the new group leader. On the other hand, if it was not a group member and has only one next hop for the tree, it prunes itself from the tree by sending its next hop a prune message. This continues until a group member is reached. Once separate partitions reconnect, a node eventually receives a Group Hello for the multicast group that contains group leader information different from the information it already has.

3.3.2.1.3 Lightweight Adaptive Multicast (LAM)

The Lightweight Adaptive Multicast (LAM) protocol [Ji1998] draws on the Core-Based Tree (CBT) protocol [Ballardie1993] and the TORA unicast routing algorithm in order to provide multicast services over MANETs. As illustrated in Figure 3.8, CBT originally designed for wired networks, has each a multicast server, or *core* that maintains a multicast group. Here, any node needs to communicate with a specific multicast group can query the core. LAM is based on the assumption that a tightly coupled unicast-multicast routing protocol is more suitable for MANETs than a multicast protocol. Although this coupling makes it less portable, it may be more efficient due to elimination of duplicated control functionality. LAM built on TORA unicast routing infrastructure, can provide multiple routes, and all MHs are globally ordered.

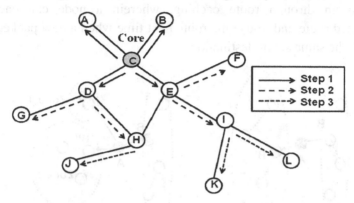

Figure 3.8 – Tree-based Multicasting with a Core

Similar to CBT, LAM builds a group-shared multicast routing tree for each multicast group centered at the CORE. Nodes in LAM maintain two variables, POTENTIAL-PARENT and PARENT, and two lists POTENTIAL-CHILD-LIST and CHILD-LIST. These potential data objects are used when the node is in a "join" or "rejoin" waiting state. To address the problem posed by having a single centralized core, Inter-core LAM (IC-LAM) is proposed [Ji1998]. IC-LAM is a tunnel-based protocol connecting multiple cores. By allowing multiple cores, IC-LAM avoids total group failure due to a single core failure.

3.3.2.1.4 Location Guided Tree Construction Algorithm for Small Group Multicast

The Location Guided Tree (LGT) [Chen2002] is a small group multicast scheme based on packet encapsulation. Multicast data is encapsulated in a unicast packet and transmitted only among the group nodes. It is based on the construction of two types of tree, location-guided k-array (LGK) tree and a location-guided Steiner (LGS) tree. The geometric location information of the destination nodes is utilized to construct the packet distribution tree without knowing the global topology of the network (Figure 3.9). It is assumed that longer the geometric distance is, more will be the network-level hops to reach the destination. Therefore, the algorithm attempts to construct a tree with short geometrical trees. The protocol also supports an optimization mechanism through route caching, wherein a node can cache the computed route and re-use the route next time when a new packet comes in with the same set of destinations.

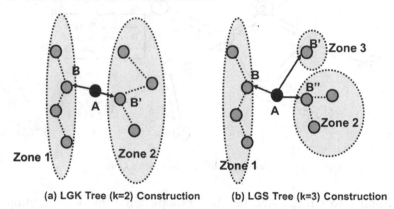

(a) LGK Tree (k=2) Construction   (b) LGS Tree (k=3) Construction

Figure 3.9 – Location Guided Tree Construction Algorithm
[Taken from Cordeiro2003]

In LGK (location-guided k-ary) tree approach, the sender first selects nearest k destinations as children nodes. The sender then groups the rest of the nodes to its k children as per the closeness of the geometric proximity. Once the group nodes are mapped, the sender forwards a copy of the encapsulated packet to each of the k children (sub destination list of group members) as destinations. Upon receiving this packet, each child runs the tree construction algorithm to further forward the packet to other destinations. The process stops when an in-coming packet has an empty destination list. In the LGS scheme, a Steiner tree is constructed, which uses the multicast group members as tree nodes. The protocol uses a hybrid mechanism for location/membership update, which includes in-band update and periodic update. If a node has no data packet to send for an extended period of time, it sends a periodic update with a null packet and its present geometric location.

3.3.2.1.5 Multicast Zone Routing

The Multicast Zone Routing (MZR) protocol [Devarapalli2002] is based on the hierarchical structure used by the ZRP unicast routing protocol. A ZRP network is partitioned into zones with a set of nodes. Each node computes its own zone that lies within a certain zone radius of the node. ZRP is described as a hybrid approach between the proactive and reactive routing protocols, where routing is proactive inside the zones (i.e., a unicast route is proactively maintained between every pair of nodes in a zone) and reactive between the zones (i.e., a route between two nodes in different zones is created when needed).

To create a zone, MZR node A broadcasts an ADVERTISEMENT message with a time-to-live (TTL) equal to a pre-configured ZONE-RADIUS. Node B within the zone radius decrements the TTL and forwards the message if appropriate. Node B makes an entry in its routing table for node A, with the last hop of the ADVERTISEMENT message as the next hop towards the destination. The distance is set to the hop count of the packet. Nodes that are ZONE-RADIUS hops away from node A become border nodes, and serve as a gateway between node A's zone and the rest of the network.

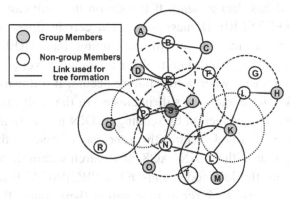

Figure 3.10 – A multicast tree extension through the tree network for MZR
[Taken from Devarapalli2002]

In the spirit of zone routing, MZR begins its search for a multicast tree within the zone before extending the search outward. When a source wants to start sending multicast traffic, it initiates the construction of a multicast tree. The source node sends a TREE-CREATE packet to each node in its zone. Nodes that want to receive multicast data respond with a TREE-CREATE-ACK packet back to the source of the multicast session. As the packet travels the tree, intermediate nodes mark in their routing tables the last hop of the TREE-CREATE-ACK as a downstream node. Tree creation continues in the new zone in the same way as described above and an example is shown in Figure 3.10 taken from [Devarapalli2002].

Routes in upon receipt of a TREE-CREATE-ACK, the border node unicasts a TREE-CREATE-ACK to the multicast source. This creates a link between the border node and the source. This sequence continues until every node in the network receives a TREE-CREATE message. When the source wants to extend the multicast tree to the entire network, it sends TREE-PROPAGATE to all boarder nodes in its zone. After receiving this, a boarder node sends a TREE-CREATE packet to all its zone boarders. If a node in the boarder's zone is interested in the session, it replies with a TREE-CREATE message. A boarder node receives a TREE-CTEATE-ACK.

MZR are updated through the use of TREE-REFRESH packets sent periodically by the source to its multicast receivers, indicating that the source still has data to send. If a node on the multicast tree fails to receive a TREEREFRESH message after a certain time, it deletes its multicast entry. It may be worth mentioning that TREE-REFRESH packets could be piggybacked on multicast data whenever possible. Zone routing performs well when a link failure occurs. If a downstream node detects a link failure and it is still interested in the multicast session, it initiates branch reconstruction by sending a JOIN packet to all the nodes within its zone. If a node within the zone has a route to the multicast source, it responds with a JOIN- ACK. If a search within the zone fails to produce a route, the lost node sends JOIN-PROPAGATE to its border nodes, which in turn, look for a route within their zones. If they find a route, they respond with a JOIN-ACK to the lost node. If not, they continue the search with JOIN-PROPAGATE to their border nodes. Essentially, if a route is not found within the lost node's zone, the search for a route is propagated throughout the entire network. However, if a route is found within the confines of the lost node's zone, the search is limited to those nodes and bandwidth is conserved.

Tree pruning is a relatively simple process. Any node N that wishes to leave a multicast group sends a PRUNE message to its upstream nodes. If node A is an upstream node of node N and node N is node A's only one-hop downstream node, node A will then stop forwarding multicast traffic. If node A does not want to receive multicast data itself and it does not have any other downstream nodes it sends a PRUNE message to its upstream node. This continues until the PRUNE reaches a node that wishes to receive multicast traffic or it reaches the source node. Nodes that wish to join an existing multicast session can perform a JOIN in the same way that a lost node does.

One advantage of the MZR protocol is that it creates a source specific, on-demand multicast tree with a minimal amount of routing overhead. Multicast zone routing attempts to reduce the amount of overhead incurred in route maintenance by preventing routing updates from spreading unnecessarily throughout the network. It would seem that the tradeoff in complexity and routing overhead incurred by the zone routing mechanism does not necessarily offset the advantages presented by MZR. The zone routing seems advantageous when attempting to route unicast packets, but not necessarily when creating multicast source trees.

Clearly, the hierarchical approach does not conserve bandwidth during the initial TREE-CREATE flood. In fact, MZR can introduce extra latency when a TREE-CREATE flood occurs and TREE-PROPAGATE messages are used. MZR requires that multicast tree being created beyond the source's immediate zone occur only after the intra-zone multicast tree has been created. Tree creation latency could be reduced if foreign zones did not have to wait for the source zone to complete its tree before creating their own multicast trees. The MZR algorithm does offer multicast re-joins within the zone. For example, if a node must look outside its zone for a new route the entire network is flooded. Only when a new route lays within ZONERADIUS hops from the lost node, the bandwidth is conserved. This situation may be common in the case of link failures, but not in the case that a node wishes to join an existing multicast tree for the first time. Instead, a node could simply send an initial JOIN message with a small TTL, i.e., on the order of a zone radius. If after a certain time this JOIN does not produce a valid route, it could resend the JOIN with a larger TTL.

3.3.2.1.6 Multicast Optimized Link State Routing

As in MZR, the Multicast Optimized Link State Routing (MOLSR) protocol [Jacquet2001] (an extension of the OLSR unicast routing protocol) creates a source specific multicast tree. Multicast-capable routers in an OLSR network periodically advertise their ability to route and build multicast routes with a MC_CLAIM message sent every MC_CLAIM_PERIOD seconds. As information in a MC_CLAIM message does not change over time, a relatively long MC_CLAIM_PERIOD should be used. Using this and the information provided by the OLSR's TC messages, MOLSR nodes can calculate shortest paths to every potential multicast source in the same manner, using only multicast-capable OLSR routers.

Similar to backward method used in MZR, a source advertises its intentions by broadcasting a SOURCE_CLAIM message to every node in the network. Before responding to the SOURCE_CLAIM, a multicast receiver first checks its multicast routing table. If an entry does not already exist, the node creates one and it sets the son address to the originator of CONFIRM_PARENT message. The node then sets the parent node to the next hop towards the multicast source using its multicast routing table, and sends a CONFIRM_PARENT to the parent node. If an entry does exist, the node simply updates the timer. In either case, if the node is a multipoint relay (MPR) node for the last hop, it forwards the SOURCE_CLAIM. When a node receives a CONFIRM_PARENT message, it checks its multicast routing table for an entry. If the entry does not exist, it creates one and sets the sons and parents to null. If the last hop of the CONFIRM_PARENT packet does not exist in the sons list, it adds and updates the son timer to SON_HOLD_TIME. If the son does exist, it simply updates the SON_HOLD_TIME.

A multicast source periodically sends a SOURCE_CLAIM message to every node in the network for two reasons. First, it informs multicast receivers that the source is still sending data and that all of the nodes in the multicast tree should update their multicast timers. Second, it allows unattached hosts to join the multicast group. If a node detects that the next hop entry towards the multicast source has changed, a node must inform the new entry that it is now a multicast parent. To do this, the node must send a CONFIRM_PARENT message to the node. If the old parent is reachable, the node may send a LEAVE message to the old parent. A node periodically sends out CONFIRM_PARENT messages to inform its parents that it still wishes to receive multicast traffic. Any MOLSR node with no sons, wanting to leave a multicast group sends a LEAVE message to its parent. The parent removes this node from the son list and, if the list becomes empty, it sends a LEAVE message to its parent. This continues until a node that wants to receive data or a node with at least two sons, is reached.

All the schemes discussed here assume that the nodes have multicast capability and does not consider a situation when this is not true. A case is presented in [Jacquet2001] when a node that does not have multicast capabilities sends multicast data. This is an interesting topic which deserves further investigation, and a proposal for a Wireless Internet Group Management Protocol (WIGMP) has been introduced in [Jacquet2001].

3.3.2.1.7 Other Protocols

The Associativity-Based Ad Hoc Multicast (ABAM) [Toh2000] is on-demand source-initiated association stability based multicast protocol by monitoring the link status of each node. ABAM deals with the network mobility on different levels according to varying mobility effects: branch repairs when the receiver moves, sub-tree repairs when a branching node moves, and full tree level repairs when the source node moves. Tree reconfiguration is required only when a link is broken, and a localized repair strategy comes into picture.

In contrast to local control in ABAM and MZR, On-demand Location-Aware Multicast (OLAM) protocol [Basagni2000] is a global method based on the expectation that each node is equipped with a positioning device such as GPS. With this assumption, each node can process and take a snapshot of the network topology and make up a multicast as a minimum spanning tree. This protocol does not use any distributed data structures or ad hoc routing protocol as foundation. Although it is shown in [Basagni2000] that OLAM's overhead is low and that it works well for varying mobility and group sizes, the GPS measurements may become a huge when the multicast tree is very large.

Because on-demand operation is driven by the presence of data packets instead of periodic or continuous control flooding, on-demand protocols have lower control overhead and react quickly to routing changes. In view of this, Adaptive Demand-Driven Multicast Routing (ADMR) [Jetcheva2001a] protocol attempts to reduce non-on-demand components. ADMR uses *tree flood* to enable packets to be forwarded following variant branches in the multicast tree. A multicast packet in ADMR floods within the multicast distribution tree only towards the group's receivers. The use of tree flood also increases the robustness of the tree structure. It also tends to scale well with group size and mobility. The Spiral-fat-tree-based On-demand Multicast (SOM) protocol [Chen2001] builds a spiral fat tree to increase the stability and redundancy of the tree structure as shown in Figure 3.11.

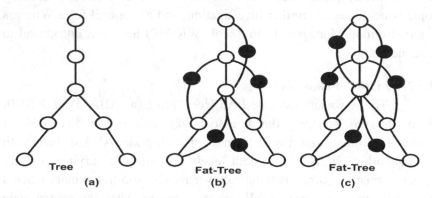

Figure 3.11 – The construction of spiral-fat-tree [Taken from Chen2001]

3.3.2.2 Mesh-Based Approaches

In contrast to tree-based approach, mesh-based multicast protocols may have multiple paths between any source and receiver pairs and are not necessarily the best suited for multicast in a MANET if the network topology changes frequently. In such an environment, mesh-based protocols seem to outperform tree-based proposals due to availability of alternative paths, which allow multicast datagrams to be delivered to all intended nodes even if links fail. For example, in Figure 3.8, if MHs C, D, E, and H are used as a group of cores, it changes in to a Mesh based approach.

The disadvantage of a mesh is an increase in data-forwarding overhead. The redundant forwarding consumes more bandwidth in the bandwidth constrained MANETs. Moreover, the probability of collisions is higher when a larger number of packets are generated. Therefore, one common problem mesh-based protocols have to consider is how to minimize the data-forwarding overhead caused by multiple flooding. As we shall see, different protocols attack this issue in different ways through the use of forwarding groups, cores, and so on. This section gives an overview of the mesh-based approaches.

3.3.2.2.1 On-Demand Multicast Routing Protocol

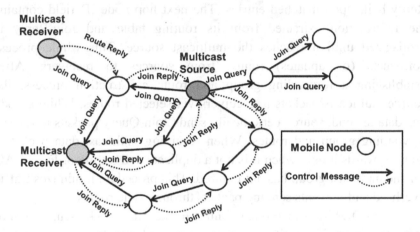

Figure 3.12 - Mesh creation in ODMRP Protocol
[Taken from Cordeiro2003]

On-demand Multicast Routing Protocol (ODMRP) [Gerla2000] is a mesh-based protocol, which employs a forwarding group concept (only a subset of nodes forwards the multicast packets). A soft state approach is taken in ODMRP to maintain multicast group members. No explicit control message is required to leave the group. The group membership and multicast routes are established and updated by the source on demand. When a multicast source has packets to send, but no route to the multicast group, it broadcasts a Join-Query control packet to the entire network. This Join-Query packet is periodically broadcasted to refresh the membership information and updates routes as depicted in Figure 3.12. When an intermediate node receives a Join-Query packet, it stores the source ID and the sequence number in its message cache to detect any potential duplicates. The routing table is updated with an appropriate node ID from which the message was received. If the message is not a duplicate and the TTL is valid, it is rebroadcasted.

When a Join-Query packet reaches a multicast receiver, it creates and broadcasts a Join-Reply to its neighbors. When a node receives a Join-Reply, it checks if the next hop node ID of one of the entries matches its own ID. If it does, the node realizes that it is on the path to the source and thus is a part of the forwarding group and sets the FG_FLAG (Forwarding Group Flag). It then broadcasts its own Join-Reply built upon matched entries. The next hop node ID field contains the information extracted from its routing table and Join-Reply is propagated until it reaches the multicast source. This whole process constructs (or updates) routes from sources to receivers. After establishing a forwarding group and route construction process, the source multicasts packets to receivers via selected routes. While a node has data to send, source periodically sends Join-Query packets to refresh forwarding group and routes. When receiving multicast data packet, a node forwards it only when it is not a duplicate and setting of FG_FLAG for the multicast group has not expired. This procedure minimizes traffic overhead and prevents sending packets through stale routes.

In ODMRP, no explicit control packets need to be sent to join or leave the group. If a multicast source wants to leave the group, it simply stops sending Join-Query packets. If a receiver no longer wants to receive from a particular multicast group, it does not send the Join-Reply. Nodes in the forwarding group are demoted to non-forwarding nodes if not refreshed (no Join-Replies received) before they timeout.

3.3.2.2.2 Core-Assisted Mesh Protocol

The Core-Assisted Mesh Protocol (CAMP) [Garcia-Luna-Aceves 1999b] supports multicasting by creating a shared mesh for each multicast group. Such meshes help in maintaining the connectivity to the multicast users, even in case of node mobility. It borrows concepts from CBT, but the core nodes are used for control traffic needed to join multicast groups. The basic operation of the CAMP includes building and maintaining the multicast mesh for each multicast group. Each router maintains a routing table (RT) built with the unicast routing protocol and is modified by CAMP when a multicast group needs to be inserted or removed. A router may update its multicast routing table (MRT) based on topological changes or messages received from its neighbors.

CAMP classifies the nodes in the network in three modes: simplex, duplex and non-member. A router joins a group in a simplex mode if it intends only to send traffic received from specific nodes or neighbors to the rest of the group, and does not intend to forward packets from the group. A duplex member forwards any multicast packets for the group, whereas a non-member node needs not to be in the multicast delivery mesh. CAMP uses a receiver-initiated method for routers to join a multicast group. If a router wishing to join a group has multiple duplex members of the multicast group, then it simply changes its MRT and directly announces to its neighbors that it's a new member for the multicast group using multicast routing update. If it has no neighbors that are members of the multicast group, it either propagates a join request to one of the multicast group "cores" or attempts to reach a member through expanding ring search [Perkins2003]. Any router that is a regular member of the multicast group and has received the join request is free to transmit a join acknowledgement (ACK) to the sending router. A router can leave a group if it has no hosts that are members of the group, and also it has no neighbors for whom it is an anchor, i.e., as long as they are not needed to provide efficient paths for the dissemination of packets in the multicast meshes for the groups. Cores are also allowed to leave multicast group if there are no routers using them as anchors.

CAMP ensures that the mesh contains all reverse shortest paths between a source and the recipients. A receiver node periodically reviews its packet cache in order to determine it has the reverse shortest path to the source. Otherwise, a HEARTBEAT message is sent to the successor in the reverse shortest path to the source. This HEARTBEAT message triggers a PUSH JOIN (PJ) message. If the successor is not a mesh member, the PJ forces the specific successor and all the routers in the path to join the mesh. CAMP does not use flooding and the requests only propagate to the mesh members. On the other hand, CAMP relies on an underlying unicast routing protocol to guarantee correct distances to all destinations within finite time.

3.3.2.2.3 Forwarding Group Multicast Protocol

Forwarding Group Multicast Protocol (FGMP) [Chiang1998] can be viewed as flooding with "limited scope", wherein the flooding is contained within a selected forwarding group (FG) nodes. FGMP makes innovative use of flags and an associated timer to forward multicast packets. When the forwarding flag is set, each node in FG forwards data packets belonging to a group G with flags on until the timer expires. This soft state approach of using timer works well in dynamically changing environments. FGMP uses two approaches to elect and maintain FG: FGMP-RA (Receiver Advertising) and FGMP-SA (Sender Advertising). In FGMP-RA, multicast receivers maintain a table with all receivers of the group and periodically announce their group membership by flooding. The nodes, which relay this message, store the next-hop to the sender. Multicast receivers join the group by sending replies to the sender. FGMP can be seen as a twin method to ODMRP, where their main difference relies on the way group meshes are established. Both FGMP and ODMRP do suffer from scalability problems due to flooding of control packets.

3.3.2.2.4 Other Protocols

In addition to the forwarding groups and cores used in the previously discussed mesh-based multicast protocols, a local routing scheme is proposed in the Neighbor Supporting ad hoc Multicast routing Protocol (NSMP) [Lee2000a] to lower the network load. In NSMP, there are two types of route discovery: flooding route discovery and local route discovery. In flooding route discovery, control packets flood the entire network in the initial route establishment or in repair of network partitions, while in local route discovery only a small number of nodes related to the multicast group are involved for routine path maintenance. In selecting a route, NSMP prefers a path with more existing forwarding nodes, which is supposed to increase the route efficiency. The neighboring nodes of the multicast group are important for mesh maintenance and are also used to limit the control messages to a small part of the nodes, thereby minimizing the frequency to flood the network. Result in [Lee2000a] reports that NSMP has decreased transmissions and reduced control overhead as compared to the ODMRP.

Intelligent On-Demand Multicast Routing Protocol (IOD-MRP) [Wang 2001] is a modified version of CAMP. It employs an on-demand receiver initiated procedure to dynamically build routes and maintain multicast group membership instead of using cores. Because the stale routing information in the network may make the routes to the cores unavailable, IOD-MRP discards the use of cores, thereby guaranteeing a node can join the mesh with a shorter path. IOD-MRP also proposed an intelligent mobility management procedure to handle the multicast mesh. In other words, the receiver compares the paths and determines which one is the best. This intelligent procedure can maintain and optimize the multicast mesh by monitoring the multicast traffic and learning about link states of the mesh. As a result, control messages due to flooding can be reduced significantly. IOD-MRP can guarantee that there is always a path between multicast senders and receivers. It is shown in [Wang2001] that IOD-MRP can often provide better results than CAMP.

Finally, the Source Routing-based Multicast Protocol (SRMP) [Laboid2001] applies the source routing mechanism defined by the DSR unicast protocol in a modified manner, decreasing the size of the packet header. SRMP obtains multicast routes on-demand through constructing a mesh (an arbitrary subnet) to connect group members providing robustness against mobility. This protocol minimizes the flooding scope and the criterion is to choose stable paths with enhanced battery life and operates in a loop-free manner and minimizing channel overhead. The mesh-based approach of SRMP avoids the drawbacks of multicast trees. SRMP outperforms other multicast protocols by providing available paths based on future prediction for links state. These paths also guarantee nodes stability with respect to their neighbors, strong connectivity between nodes, and better battery lifetime.

3.3.2.3 Stateless Approaches

Tree-based and mesh-based approaches have an overhead of creating and maintaining the delivery tree/mesh with time. In a MANET environment, frequent movement of MHs considerably increases the overhead in the delivery tree/mesh. To minimize the effect of such a problem, stateless multicast is proposed wherein a source explicitly mentions the list of destinations in the packet header. Stateless multicast approaches focus on small group multicast and assumes the underlying routing protocol to take care of forwarding the packet to the respective destinations based on the addresses contained in the header.

3.3.2.3.1 Differential Destination Multicast

Differential Destination Multicast (DDM) protocol [Ji2001] is meant for small-multicast groups operating in dynamic networks of any size. Unlike other routing protocols, DDM lets source to control multicast group membership. The source encodes multicast receiver addresses in data packets using a special DDM Data Header. This variable length destination list is placed in the packet headers, resulting in packets being self-routed towards the destinations using the underlying unicast routing protocol. It eliminates maintaining per-session multicast forwarding states at intermediate nodes and thus is easily scalable with respect to the number of sessions.

DDM supports two kinds of operating modes: "stateless" and "soft state". In stateless mode, the nodes along the data forwarding paths need not maintain multicast forwarding states. An intermediate node receiving a DDM packet only needs to look at the header to decide how to forward the packet. In the "soft-state" mode, based on in-band routing information, each node along the forwarding path remembers the destinations to which the packet has been forwarded last time and its next hop information. By caching this routing information at each node, the protocol need not list the entire destination in future data packets. In case changes occur, an upstream node only needs to inform its downstream nodes about the differences in the destination forwarding since the last packet; hence the name "Differential Destination Multicast".

At each node, there is one Forwarding Set (FS) for each multicast session, which records to which destinations this node forwards data. The nodes also maintain a Direction Set (DS) to record the particular next hop to which multicast destination data are forwarded. At the source node, FS contains the same set of nodes as the multicast Member List (ML). In the intermediate nodes, the FS is the union of several subsets based on the data stream received from upstream neighbors. Associated with each set FS_k, there is a sequence number SEQ(FS_k) which is used to record the last DDM Block Sequence

Number seen in a received DDM data packet from an upstream neighbor k. It helps to detect loss of data packet containing the forwarding set updates. At a given node, FS also needs to be partitioned into subsets according to the next hops for different destinations.

DDM supports two types of packet control and data packets, where the data packets may also contain control information. There are five types of control packets: JOIN, ACK, LEAVE, RSYNC, and CTRL_DATA. To join a multicast session, a receiver needs to unicast a JOIN message to the source for that session. The source updates its ML and replies with an ACK. After specified period of time, the source sets a POLL flag in the next outgoing data packet. To express their continued interest, multicast members need to unicast a JOIN message again to the source. A member can also leave the session by sending an explicit LEAVE message. CTRL_DATA is used to encapsulate multicast data to send it to a particular destination by using unicasting, while RSYNC message is used to synchronize the multicast destination address sets between a pair of neighboring nodes whenever the topology changes.

Both LGT and DDM are primarily meant to provide small group multicast. In DDM, the packet distribution tree is uncontrollable by upper layer transport and application layers, whereas in LGT the packet distribution tree is constructed explicitly with the flexibility of adding upper layer packet processing and routing. Additionally, DDM requires every node in the network to eventually participate in the packet forwarding, while in LGT only the nodes participating in the session need to cooperate.

3.3.2.3.2 DSR Simple Multicast and Broadcast Protocol

DSR Simple Multicast and Broadcast protocol (DSR-MB) [Jetcheva2001b] is designed to provide multicast and broadcast functionality in MANETs by utilizing flooding-based route discovery mechanism of DSR unicast protocol. Although derived from DSR, it can be implemented as a stand-alone protocol. If DSR has already been implemented, minor modifications are required to enable this protocol. This multicast and broadcast protocol is not intended as a general purpose multicast protocol while its applicability is mainly in environments characterized by very high mobility or by a relatively small number of nodes. In the former case, protocols relying on the establishment of multicast state perform inadequately because they are unable to track the rapid changes in topology. In the latter case, the overhead of keeping multicast state exceeds the overhead of flooding.

3.3.2.4 Hybrid Approaches

The protocols to provide multicast in MANETs discussed so far, either address efficiency or robustness but not both simultaneously. The tree-based approaches provide high data forwarding efficiency at the expense of low robustness, whereas mesh-based approaches lead to better robustness at the cost of higher forwarding overhead and increased network load. Thus, there is a possibility that a hybrid multicasting solution may achieve better performance by combining the advantages of both tree and meshed-based approaches. In this section, we explore the different hybrid approaches to enable ad hoc multicasting.

3.3.2.4.1 Ad hoc Multicast Routing Protocol

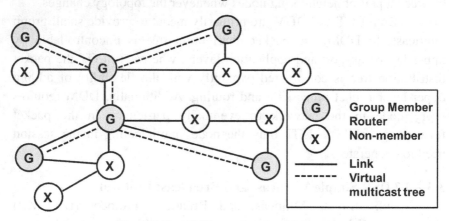

Figure 3.13 – AMRoute virtual multicast tree
[taken from Cordeiro2003]

The Ad hoc Multicast Routing Protocol (AMRoute) [Bommaiah1998] creates a bi-directional, shared tree by using only group senders and receivers as tree nodes for data distribution. The protocol has two main components: mesh creation and tree setup (see Figure 3.13). The mesh creation identifies and designates certain nodes as logical cores and these are responsible for initiating the signaling operation and maintaining the multicast tree to the rest of the group members. A non-core node only responds to messages from the core nodes and serves as a passive agent. The selection of logical core in AMRoute is dynamic and can migrate to any other member node, depending on the network dynamics and the group membership. To create a mesh, each member begins by identifying itself as a core and broadcasts JOIN_REQ packets with increasing TTL to discover other members. When a core receives JOIN_REQ from a core in a different mesh for the same group, it replies with a JOIN_ACK. A new bi-directional tunnel is created between the two cores and one of them is selected as core after the mesh merger. Once the mesh has been established, the core initiates the tree creation process. The core sends out periodic TREE_CREATE messages along all links incident on its mesh. Using unicast tunnels, the TREE_CREATE messages are sent only to the group members. Group members receiving non-duplicate TREE_CREATE message forwards it to all mesh links except the incoming one, and marks the incoming and outgoing links as a tree links. If a link is not going to be used as part of the tree, the TREE_CREATE is discarded and TREE_CREATE_NAK is sent back to incoming links. A member node, which wants to leave a group, can do so by sending a JOIN_NAK message to its neighboring nodes.

AMRoute employs the virtual mesh links to establish the multicast tree, which helps in keeping the multicast delivery tree the same even with the change of network topology as long as routes between core nodes and tree members exist via mesh links. The main disadvantage of this protocol is that it may have temporary loops and may create non-optimal trees in case of mobility.

3.3.2.4.2 Multicast Core-Extraction Distributed Ad Hoc Routing

The Multicast Core-Extraction Distributed Ad hoc Routing (MCEDAR) [Sinha1999] is a multicast extension to the CEDAR architecture. The main idea of MCEDAR is to provide the efficiency of the tree-based forwarding protocols and robustness of mesh-based protocols by combining these two approaches. It is worth pointing out that a source-based forwarding tree is created on a mesh. As such, this ensures that the infrastructure is robust and data forwarding occurs at minimum height trees. MCEDAR decouples the control infrastructure from the actual data forwarding in order to reduce the control overhead. The underlying unicast protocol, CEDAR, provides the core broadcasting for multicasting. The core is used for routing management and link state inspection. Also, the cores make up the mesh infrastructure which is referred to as an *mgraph*, and use joinIDs to perform the join operation.

As MCEDAR uses a mesh as the underlying infrastructure, it can tolerate a few link breakages without reconfiguration. The efficiency is achieved by using a forwarding mechanism on the mesh that creates an implicit route-based forwarding tree. As mentioned earlier, this ensures that the packets need to travel only the minimum distance in the tree.

3.3.2.4.3 Mobility-based Hybrid Multicast Routing

The Mobility-based Hybrid Multicast Routing (MHMR) protocol [An2001] is built on top of the mobility-based clustering infrastructure. In order to deal with the issues of scalability and stability, the structure is hierarchical in nature. The mobility and positioning information is provided via a GPS for each node. For a group of nodes, a cluster-head is chosen to manage and monitor the nodes in a cluster. A mesh structure is built based on all the current clusters. Thus, MHMR achieves high stability. This is followed by a tree structure built based on the mesh to ensure that the multicasting group achieves maximal efficiency. MHMR also provides a combination of proactive and reactive concepts which enable low route acquisition delay of proactive schemes while achieving low overhead of reactive methods.

It is interesting to note that cores are employed in both AMRoute and MCEDAR, as well as in many tree and mesh multicast algorithms. The use of cores has been shown to lower the control overhead. The use of cluster-heads has been proposed in MHMR. This has been shown to be a reasonable approach since dividing the nodes in an MANET into clusters seems to be a promising method in taking care of highly dynamic nodes. Hybrid methods can reveal themselves to be attractive as they can provide protocols that can address further robustness and efficiency. Though hybrid protocols have not been as deeply investigated as tree and mesh protocols, they are under development and recent results indicate its promising future.

3.3.3 Comparison

Table 3.2 – Comparison of ad hoc multicast routing protocols
[Taken from Cordeiro2003]

Protocol	Topology	Loop Free	Dependence on Unicast Protocol	Periodic Message	Control Packet Flooding Done/Required
Flooding	Mesh	Yes	No	No	No
AMRoute	Hybrid	No	Yes	Yes	Yes
AMRIS	Tree	Yes	No	Yes	Yes
MAODV	Tree	Yes	Yes	Yes	Yes
LAM	Tree	Yes	Yes	No	No
LGT-Based	Tree	Yes	No	Yes	No
ODMRP	Mesh	Yes	No	Yes	Yes
CAMP	Mesh	Yes	Yes	Yes	No
DDM	Stateless Tree	Yes	No	Yes	No
FGMP-RA	Mesh	Yes	Yes	Yes	Yes
FGMP-SA	Mesh	Yes	No	Yes	Yes
MCEDAR	Hybrid	Yes	Yes	Yes	Yes

The basic idea behind defining multicast routing protocol for MANET is to form path to the group members, with minimal redundancy and various algorithms described earlier, do attempt to achieve this goal using different mechanisms. The host mobility also influences the routes being selected and possibility of loop formation or the paths becoming non-optimal are important. It is also critical to know if the paths created are on demand, or optimal paths are determined once and updated periodically when needed. Another important consideration is if the control packets are flooded throughout the network, or it is limited to some nodes in the multicast delivery tree. Keeping this in mind, Table 3.2 compares different proposals to provide multicasting over MANETs using various metrics. A performance study of various multicast routing protocols can be found in [Lee2000b].

3.4 Geocasting

We now turn our attention to the problem of geocasting over MANETs. As we have mentioned earlier, geocasting is a variant of the conventional multicasting problem and distinguishes itself by specifying hosts as group members within a specified geographical region. In geocasting, the nodes eligible to receive packets are implicitly specified by a physical region and membership changes as mobile nodes move in or out of the region. The concept of geocast was first introduced in [Navas1997] as an Internet addition, not for MANET. GPS application in geographic messaging is described in [Navas1997], where it is discussed how to send packets to users who are located on a wired network within a particular polygon or circle defined by latitude and longitude.

In the future, it may be possible that GPS is deployed in almost every user terminal. With GPS, each node has its location readily available. Here, we assume that whenever a node in the geocast region receives a geocast packet, it floods the geocast packet to all its neighbors. In other words, flooding of geocast packets takes place within the geocast region. One effect of this assumption is that a geocast protocol works if at least one node in the geocast region receives the geocast packet. Lastly, the protocols presented here are assumed to use a *jitter* technique in order to avoid two packets colliding with each other by a broadcast. In other words, nodes offset transmissions by a random jitter to avoid their neighbors sending packets at the same time.

In this section, we classify existing geocast protocols into two categories: *data-transmission oriented protocols* and *routing creation oriented protocols*. Since all the nodes in the geocast region share information among each other by flooding, the difference between these two categories is how they transmit information from a source to nodes in the geocast region. Data-transmission oriented protocols use flooding or a variant of flooding to forward geocast packets from the source to the geocast region. Routing-creation oriented protocols create routes from the source to the geocast region via control packets. Both of these techniques eventually reach one or more nodes in the geocast region.

3.4.1 Geocast Routing Protocols

In this section, we discuss the main geocast routing protocols proposed for use in MANETs. We start with data-transmission oriented protocols, followed by the route creation oriented approaches.

3.4.1.1 Data-Transmission Oriented

Data-transmission oriented geocast protocols use flooding or a variant of flooding to forward data from the source to the geocast region - and are described here.

3.4.1.1.1 Location-Based Multicast

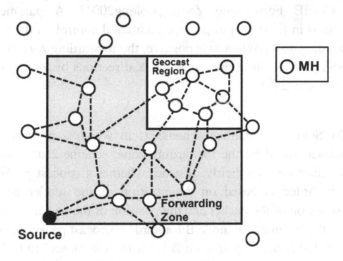

Figure 3.14 – A BOX forwarding zone [Taken from Boleng2001]

The Location-Based Multicast (LBM) protocol [Ko1999] extends the LAR unicast routing algorithm for geocasting. As we have seen, LAR is an approach to utilize location information to improve the performance (i.e., higher data packet delivery ratio and lower overhead) of a unicast routing protocol in a MANET. Similarly, the goal of LBM is to decrease delivery overhead of geocast packets by reducing the forwarding space for geocast packets, while maintaining accuracy of data delivery. The LBM algorithm is based on a flooding approach with a modification that a node determines whether to forward a geocast packet further via one of two schemes.

- **LBM Scheme 1**: When a node receives a geocast packet, it forwards to its neighbors if it is within a *forwarding zone*; otherwise, it discards the packet. Thus, how to define the forwarding zone becomes the key of this scheme [Boleng2001]. In Figure 3.14, the size of the forwarding zone is dependent on (i) the size of the geocast region, and (ii) the location of the sender. In a BOX Forwarding Zone, the smallest rectangle that covers both the source node and the geocast region. All the nodes in the forwarding zone forward data packets to their neighbors. Other kinds of forwarding zones are possible, such as the CONE Forwarding Zone [Boleng2001]. A parameter δ is discussed in [Ko1999] to provide additional control on the size of the forwarding zone. When δ is positive, the forwarding zone is extended in both positive and negative X and Y directions by δ (i.e., each side increases by 2δ).

- **LBM Scheme 2**: Unlike scheme 1, in which a geocast packet is forwarded based on the forwarding zone, scheme 2 does not have a forwarding zone explicitly. Instead, whether a geocast packet should be forwarded is based on the position of the sender node at the transmission of the packet and the position of the geocast region. That is, for the parameter δ, node B forwards a geocast packet from node A (originated at node S), if node B is "at least δ closer" to the center of the geocast region (Xc, Yc) than node A. In other words, $DISTA \geq DISTB + \delta$. We define (Xc, Yc) as the location of the geometrical center of the geocast region, and for any node Z, $DISTz$ denotes the distance of node Z from (Xc, Yc). In Figure 3.15 [Ko1999], node B will forward a geocast packet transmitted by node A since $DISTA \geq DISTB$ and $\delta = 0$. Node K will, however, discard a geocast packet transmitted by node B, since node K is not closer to (Xc, Yc) than node B. In brief, this protocol ensures that every packet transmission sends the packet closer to the destination.

As for the performance, the accuracy (i.e., ratio of the number of geocast group members that actually receive the geocast packets to the number of group members that were supposed to receive the packets) of both LBM schemes is comparable with that of flooding geocast packets throughout the network. However, the number of geocast packets transmitted is consistently lower for LBM than simple flooding.

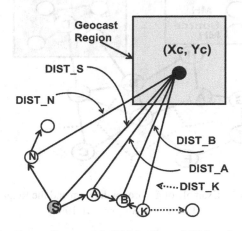

Figure 3.15 – Forwarding zone in LBM scheme 2 [Taken from Ko1999]

3.4.1.1.2 Voronoi Diagram Based Geocasting

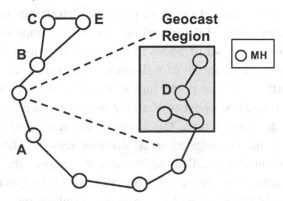

Figure 3.16 – An example of a problem in LBM
[Taken from Stojmenovic1999]

The goal of the Voronoi Diagram based Geocasting (VDG) protocol [Stojmenovic1999] is to enhance the success rate and decrease the hop count and flooding rate of LBM. It is observed that the forwarding zone defined in LMB may be a partitioned network between the source node and the geocast region, although a path between source and destination exists. An example of this is shown in Figure 3.16.

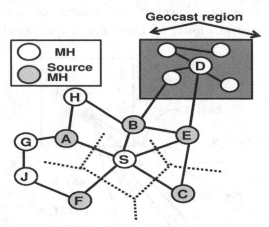

Figure 3.17 – Example of a Voronoi diagram and the Request Zone
[Taken from Stojmenovic1999]

In VDG, the definition of the forwarding zone of LBM has been modified. The neighbors of node A that are located within the forwarding zone in VDG are exactly those neighbors that are closest in the direction of the destination. This definition of precisely determines the expansion of the forwarding zone. This forwarding zone can be implemented with a *Voronoi diagram* for a set of nodes in a given node's neighborhood of a MANET. A Voronoi diagram of n distinct points (i.e., n neighbors) in a plane is a partition of the plane into n Voronoi regions, which, when associated with node A, consists of all the points in the plane that are the closest to A. In other words, the *Voronoi diagram model* is a model where every point is assigned to a Voronoi region. The subdivision induced by this model is called the *Voronoi diagram* of the set of nodes [Berg]. For example, in Figure 3.17 [Stojmenovic1999] five neighbors of source node S (A, B, C, E and F) carve up the plane into five Voronoi regions. The region associated with node A, consists of nodes G and H, since these two nodes are closer to node A than to any other node. The geocast region is the rectangle with the center D. In Figure 3.17, the Voronoi regions of nodes B and E intersect the geocast region; thus, only nodes B and E will forward geocast packets from node S.

Although there are not any simulations of the VDG algorithm, VDG reduces the flooding rates of LBM Scheme 1, as fewer packets should be transmitted. On the other hand, VDG may offer little improvement over LBM Scheme 2, as end results of the two protocols appears to be similar.

3.4.1.1.3 GeoGRID

Based on the unicast protocol GRID [Liao2001], the GeoGRID protocol [Liao2000] uses location information, which defines the forwarding zone and elects a special host (i.e, *gateway*) in each grid area responsible for forwarding the geocast packets. It is argued in [Liao2000] that the forwarding zone in LBM incurs unnecessary packet transmissions, and a tree-based solution is prohibitive in terms of control overhead. GeoGRID partitions the geographic area of the MANET into two-dimensional logical grids. Each grid is a square of size d x d (there arc tradc-offs in choosing a good value of d, as discussed in [Liao2000].) In GeoGRID, a gateway node is elected within each grid. The forwarding zone is defined by the location of the source and the geocast region. The main difference between GeoGRID, LBM and VDG is instead of every node in a forwarding zone transmitting data, only gateway nodes take this responsibility in GeoGRID. There are two schemes on how to send geocast packets in GeoGRID: *Flooding-Based GeoGRID* and *Ticket-Based GeoGRID*.

In Flooding-Based GeoGRID, only gateways in every grid within the forwarding zone rebroadcast the received geocast packets. Thus, gateway election becomes the key point of this protocol. In Ticket-Based GeoGRID, the geocast packets are still forwarded by gateway nodes, but not all the gateways in the forwarding zone forward every geocast packet. A total of $m+n$ tickets are created by the source if the geocast region is a rectangle of m x n grids. The source evenly distributes the $m+n$ tickets to the neighboring gateway nodes in the forwarding zone that are closer to the geocast region than the source. A gateway node that receives *X* tickets follows the same procedure as the one defined for the source. Consider the example in Figure 3.18 where node S begins with five (m=3, n=2) tickets. Node S may distribute two tickets to its neighboring nodes A and B, and one ticket to its neighbor node C, which are closer to the geocast region than node S. The philosophy is that each ticket is responsible for carrying one copy of the geocast packet to the geocast region. Hence, if a node is sent a geocast packet that it has seen before, it does not discard it. For example, if node C decides to give its ticket to node B in Figure 3.18, (i.e., node B receives a geocast packet from node C), node B will rebroadcast the packet. In other words, node B will transmit the geocast packet (at least) two times.

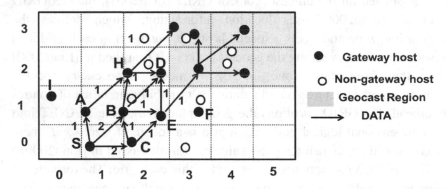

Figure 3.18 – A geocast example for the Ticket-Based GeoGRID protocol
[Taken from Liao2000]

Both the Flooding-Based GeoGRID and the Ticket-Based GeoGRID protocols need an efficient solution for the gateway election. Once this node is elected, it remains the gateway until it moves out of the grid. One problem of this selection process arises when another potential gateway roams closer to the physical center of the grid than the currently assigned gateway and cannot be elected as the gateway until the current gateway leaves the grid. To eliminate this possibility, multiple gateways could be temporally allowed to reside in a grid. In this situation, if a gateway hears a packet from another gateway at a location closer to the physical center of its grid, it silently turns itself into a non-gateway node and does not forward any further geocast packets. However, if the grid size is small, or the mobility of the node is low, this problem may not be severe. Another effective way of gateway election is via the concept of Node Weight [Basagni1999]. For example, we could assign the weight of a node as being inversely proportional to its speed. Flooding-Based GeoGRID and Ticket-Based GeoGRID have obvious advantages over LBM Scheme 1 and LBM Scheme 2, especially in dense networks. The two GeoGRID protocols should offer both higher accuracy and lower delivery cost than LBM and VDG due to the reduced number of transmitted packets.

3.4.1.2 Route Creation Oriented

As discussed in the previous chapter, flooding of packets may cause a *broadcast storming* effect, generating serious redundancy, contention and collision problems. In this section we introduce routing-creation oriented protocols, which create routes to transmit data from the source to the geocast region. One advantage of this kind of protocol is the reduced overhead in the transmission of data packets. One disadvantage is that it requires higher latency and control overhead to establish the routes.

3.4.1.2.1 GeoTORA

The goal of the GeoTORA protocol [Ko2000] is to reduce the overhead of transmitting geocast packets via flooding techniques, while maintaining high accuracy. The unicast routing protocol TORA is used by GeoTORA to transmit geocast packets to a geocast region. As TORA is a distributed routing protocol based on a "link reversal" algorithm, it provides multiple routes to a destination. Despite dynamic link failures, TORA attempts to maintain a destination-oriented directed acyclic graph such that each node can reach the destination. In GeoTORA, a source node essentially performs an *anycast* to any geocast group member (i.e, any node in the geocast region) via TORA. When a node in the geocast region receives the geocast packet, it floods the packet such that the flooding is limited to the geocast region. The accuracy of GeoTORA is high, but not as high as pure flooding or LBM.

3.4.1.2.2 Mesh-based Geocast Routing Protocol

The Mesh-based Geocast Routing (MGR) protocol [Boleng2001] uses a mesh for geocasting in an ad hoc environment in order to provide redundant paths between the source and the group members. Since the group members in a geocast region are in close proximity to each other, it is less costly to provide redundant paths from a source to a geocast region than to provide the redundant paths from a source to a multicast group of nodes that may not be in close proximity of each other. Instead of flooding geocast packets, the MGR Protocol tries to create redundant routes via control packets. First, the protocol floods JOIN-DEMAND packets in a forwarding zone. A JOIN-DEMAND packet is forwarded in the network until it reaches a node in the geocast region. By reversing the route of JOIN-DEMAND packet, this node unicasts a JOIN-TABLE packet back to the source. Thus, the nodes on the edge of the geocast region become a part of the mesh. Once the first JOIN-TABLE packet is received by the source, data packets can be sent to the nodes in the geocast region. Figure 3.19 shows an example of geocast communication via a mesh.

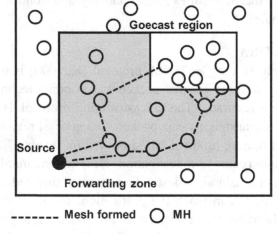

Figure 3.19 – A Mesh-based Geocast Routing protocol example
[Taken from Boleng2001]

Using the forwarding zone discussion from LBM, different forwarding zones have been evaluated to control the number of redundant paths in the mesh. A larger forwarding zone creates a larger mesh. Reducing the area of the forwarding zone reduces control overhead, network-wide data load, end-to-end delay, and network reliability. In addition, increasing the average node mobility leads to decreased network reliability.

In the data-transmission oriented category, GeoGRID appears to be an effective protocol. Both the Flooding-Based GeoGRID protocol and the Ticket-Based GeoGRID protocol may reduce the overhead of LBM and VDG, and especially in the dense network. So, let us look at their performance.

3.4.2 Comparison

As compared to data-transmission oriented protocols, the overhead of GeoTORA is small. However, the accuracy of GeoTORA is affected. Another potential problem with GeoTORA is that only one node receives a geocast packet from TORA which is then responsible for flooding through the geocast region. In the MGR protocol, multiple nodes in the geocast region will receive a geocast packet due to the redundant paths that are created between the source and the geocast region. While redundant paths will increase the overhead compared to a single path, the accuracy of redundant paths should also increase. In conclusion, there appears to be a trade-off between overhead and reliability. In other words, higher protocol overhead appears to provide better levels of reliability, while lower protocol overhead appears to provide lower levels of reliability.

3.5 Conclusions and Future Directions

As mentioned earlier, research in the area of broadcasting, multicasting and geocasting over MANETs is far from being exhaustive. Much of the effort so far has been on devising routing protocols to support effective and efficient communication between nodes. However, there are still many topics that deserve further investigation such as:

- **Scalability** – This issue is not only related to broadcasting, multicasting or geocasting in MANETs but also with the MANET itself. An obvious question is to what extent can an MANET grow? Can we design a multicast routing protocol for MANET, which is scalable with respect to number of members in the group, their mobility and other constraints posed by the MANET environment itself? Similarly, can we come up with a scalable location service and forwarding scheme to provide efficient geocasting services?

- **Applications for broadcast/multicast/geocast over MANETs** – Have we found a killer application? Does it exist or do we really need one? Although we talk about online gaming, military applications, environmental monitoring, and information dissemination in selected geographical areas, but what the potential commercial applications of MANETs are and if service providers can be convinced to support multicast and/or geocast, is still an open issue.

- **QoS** – This applies to broadcast, multicast and geocast. Is it feasible for bandwidth/delay-constrained multicast applications to run well in a MANET? Since MANET itself does not have a well-defined framework for QoS support yet, it may be difficult to address this task for some time.

- **Address configuration** – This has a lot to do with multicast services. Due to the infrastructureless nature of MANETs, a different addressing approach may be required. Special care needs to taken so that other groups should not reuse a multicast address used by a group at the same time. Node movement and network partitioning makes this task of synchronizing multicast addresses in a MANET really difficult.

- **Security** – How can the network secure itself from malicious or compromised hosts? Due to broadcast nature of the wireless medium, security provisioning has become more difficult. In the specific case of multicasting, further research is needed to investigate how to stop an intruder from joining an ongoing multicast session or stop a node from reception of session packets.

- **Power control** – How can battery life be maximized? Both source and core-based multicast approaches concentrate traffic on a single node. For example, in a stateless multicast the group membership is controlled by the source, which limits lifetime of its battery. It is still needed to investigate how to efficiently distribute traffic from a central node to other member nodes in a MANET. Similarly, efficient geocasting services can consume considerable amounts of energy. The research community is already looking into many of these questions; however, there is still a lot more work to be done.

Homework Questions/Simulation Projects

Q. 1. Why is multicasting not considered as restricted broadcasting in MANETs?

Q. 2. What are the limitations of "dominating sets" approach in achieving broadcasting in MANETs?

Q. 3. Consider a dominating set of size |D|. Prove that it is possible to connect D using at most 2D additional vertices.

Q. 4. When would you prefer multiple unicast over core-based routing?

Q. 5. Why is mesh-based multicasting better than core-based schemes? Explain clearly.

Q. 6. Multicasting is an important process in networking. Some routers in the network do have multicasting capability while others still do not support this feature. Assuming that there are 200 nodes network with no multicast capability and each multicast group consists of exactly 8 members and are randomly distributed in the network. What are different options do you have to perform multicasting and what are their relative advantages and disadvantages?

Q.7. Design a problem based on any of the material covered in this chapter (or in references contained therein) and solve it diligently.

References

[Agrawal2002] D. P. Agrawal and Q.-A. Zeng, "Introduction to Wireless and Mobile Systems," 3rd edition, Cengage Publishing, 2011.

[An2001] B. An and S. Papavassiliou, "A mobility-based hybrid multicast routing in mobile ad-hoc wireless networks," Proceedings of IEEE Milcom, 2001.

[Ballardie1993] T. Ballardie, P. Francis, and J. Crowcroft, "Core based trees (CBT): and architecture for scalable interdomain multicast routing," Proceedings of ACM SIGCOMM, 1993.

[Basagni1999] S. Basagni, "Distributed clustering for ad hoc networks," In Proceedings of the International Symposium on Parallel Architectures, Algorithms, and Networks (I-SPAN), pp. 310–315, 1999.

[Basagni2000] S. Basagni, I. Chlamtac, V. Syrotiuk, and R. Talebi, "On-demand location aware multicast (OLAM) for ad hoc networks," Proceedings of IEEE Wireless Communications and Networking Conference, 2000.

[Berg] M. D. Berg. Computational geometry. Algorithms and Applications.

[Boleng2001] J. Boleng, T. Camp, and V. Tolety, "Mesh-based geocast routing protocols in an ad hoc network," in Proceedings of the International Workshop on Parallel and Distributed Computing Issues in Wireless Networks and Mobile Computing (IPDPS), pp. 184–193, April 2001.

[Bommaiah1998] E. Bommaiah, M. Liu, A. McAuley, and R. Talpade, "AMRoute: Adhoc Multicast Routing Protocol," Internet-Draft, August 1998.

[Chen2001] Y.-S. Chen, T.-S. Chen, and C.-J. Huang, "SOM: Spiral-fat-tree-based on-demand multicast protocol in a wireless ad-hoc network," in Proceedings of International Conference on Information Networking, 2001.

[Chen2002] K. Chen and K. Nahrstedt, "Effective Location-Guided Tree Construction Algorithms for Small Group Multicast in MANET," in Proceedings of IEEE Infocom 2002.

[Chiang1998] C.-C. Chiang, M. Gerla, L. Zhang, "Forwarding group Multicast Protocol (FGMP) for Multihop, Mobile Wireless Networks," ACM/Kluwer Journal of Cluster Computing: Special Issue on Mobile Computing, Vol. 1, No. 2, 1998.

[Cordeiro2003] C. Cordeiro, H. Gossain and D. Agrawal, "Multicast over Wireless Mobile Ad Hoc Networks: Present and Future Directions," in IEEE Network, Special Issue on Multicasting: An Enabling Technology, Vol. 17, No. 1, January/February 2003.

[Devarapalli2002] V. Devarapalli, A. Selcuk, and D. Sidhu, "M2R: A Multicast protocol for mobile ad hoc networks, IETF, 2002.

[Garcia-Luna-Aceves1999a] J. J. Garcia-Luna-Aceves and E. Madruga, "A Multicast Routing Protocol for Ad-Hoc Networks," Proceedings of IEEE Infocom, March 1999.

128   *AD HOC & SENSOR NETWORKS*

[Garcia-Luna-Aceves1999b] J.J. Garcia-Luna-Aceves and E.L. Madruga, "The Core-Assisted Mesh Protocol," IEEE Journal on Selected Areas in Comm., pp. 1380-1394, August 1999.
[Gerla2000] M. Gerla, S.-J. Lee, and W. Su. "On-demand multicast routing protocol (ODMRP) for ad-hoc networks," Internet Draft, draft-ietf-manet-odmrp-02.txt, 2000.
[Gossain2002] H. Gossain, C. Cordeiro, and D. Agrawal, "Multicast: Wired to Wireless," IEEE Communications Magazine, Vol. 40, No. 6, pp. 116-123, June 2002.
[He2009] Hongmei He, Zhenhuan Zhu, and Erkki Mäkinen, "A neural network model to minimize the connected dominating set for self-configuration of wireless sensor networks Source," IEEE Transactions on Neural Networks , vol. 20, no. 6, June 2009, pp. 973-982.
[Ho1999] C. Ho, K. Obraczka, G. Tsudik, and K. Viswanath, "Flooding for reliable multicast in multi-hop ad hoc networks," In Proceedings of the International Workshop on Discrete Algorithms and Methods for Mobile Computing and Communication (DIALM), pp. 64-71, 1999.
[IEEE-802.11] IEEE Std. 802.11. "IEEE Standard for Wireless LAN Medium Access Control (MAC) and Physical Layer (PHY) Specification," June 1997.
[Jacquet2001] P. Jacquet, P. Minet, A. Laouiti, L. Viennot, T. Clausen, and C. Adjih, "Multicast Optimized Link State Routing," Internet Draft, draft-ietf-manetolsr-molsr-01.txt), November 2001.
[Jetcheva2001a] J. Jetcheva and D. Johnson, "Adaptive Demand-Driven Multicast Routing in Multi-hop Wireless Ad Hoc Networks," in Proceedings of the ACM International Symposium on Mobile Ad Hoc Networking and Computing, 2001.
[Jetcheva2001b] J. Jetcheva, Y.-C. Hu, D. Maltz, and D. Johnson, "A Simple Protocol for Multicast and Broadcast in Mobile Ad Hoc Networks," Internet Draft, draft-ietf-manet-simple-mbcast-00.txt, June 2001.
[Ji1998] L. Ji and M. S. Corson, "A Lightweight Adaptive Multicast Algorithm," IEEE Globecom, pp. 1036-1042, 1998.
[Ji2001] L. Ji and M S. Corson, "Differential Destination Multicast - A MANET Multicast Routing Protocol for Small Groups," Proceedings of IEEE Infocom, pp. 1192-1202, 2001.
[Ko1999] Y. Ko and N. H. Vaidya, "Geocasting in mobile ad hoc networks: Location-based multicast algorithms," Proceedings of the 2nd IEEE Workshop on Mobile Computing Systems and Applications (WMCSA), 1999.
[Ko2000] Y. Ko and N. H. Vaidya, "GeoTORA: A protocol for geocasting in mobile ad hoc networks," Proceedings of the International Conference on Network Protocols (ICNP), November 2000.
[Laboid2001] H. Laboid and H. Moustafa, "Source Routing-based Multicast Protocol (SRMP)," Internet Draft, draft-labiod-manet-srmp-00.txt, work in progress, June 2001.
[Lee2000a] S. Lee and C. Seoul, "Neighbor Supporting Ad Hoc Multicast Routing Protocol," Proceedings of the First Workshop on Mobile Ad Hoc Networking and Computing (Mobihoc), August 2000.
[Lee2000b] S.-J. Lee, W. Su, J. Hsu, M. Gerla, and R. Bagrodia, "A Performance Comparison Study of Ad Hoc Wireless Multicast Protocols," Proceedings of IEEE Infocom 2000, March 2000.
[Liao2000] W.-H. Liao, Y.-C. Tseng, K.-L. Lo, and J.-P. Sheu, "Geogrid: A geocasting protocol for mobile ad hoc networks based on grid," Journal of Internet Technology, Vol. 1, No. 2, pp. 23-32, 2000.
[Liao2001] W.-H. Liao, Y.-C. Tseng, and J.-P. Sheu, "Grid: A fully location-aware routing protocol for mobile ad hoc networks," Telecommunication Systems, Vol. 18, No. 1, pp. 37-60, 2001.
[Lim2000] H. Lim and C. Kim, "Multicast tree construction and flooding in wireless ad hoc networks," In Proceedings of the ACM International Workshop on Modeling, Analysis and Simulation of Wireless and Mobile Systems (MSWIM), 2000.
[Liu2010] Zhuo Liu, Bingwen Wang and Lejiang Guo, "A Survey on Connected Dominating Set Construction Algorithm for Wireless Sensor Networks," 2010, http://scialert.net/abstract/?doi=itj.2010.1081.1092.

[Navas1997] J. C. Navas and T. Imielinski, "Geocast – geographic addressing and routing," Proceedings of the ACM/IEEE International Conference on Mobile Computing and Networking (Mobicom), 1997.
[Ni1999] S-Y. Ni, Y-C. Tseng, Y-S. Chen, and J-P. Sheu, "The Broadcast Storm Problem in a Mobile Ad Hoc Network," in Proceedings of ACM/IEEE Mobicom, 1999.
[Peng1999] W. Peng and X. Lu, "Efficient broadcast in mobile ad hoc networks using connected dominating sets," Journal of Software, Beijing, China, 1999.
[Peng2000] W. Peng and X. Lu, "On the reduction of broadcast redundancy in mobile ad hoc networks," in Proceedings of ACM MobiHoc, 2000.
[Peng2002] W. Peng and X. Lu, "AHBP: An efficient broadcast protocol for mobile ad hoc networks," Journal of Science and Technology, Beijing, China, 2002.
[Perkins2003] C. E. Perkins, E. M. Royer and S. R. Das, "Ad Hoc On Demand Distance Vector Routing (AODV)," Internet RFC 3561, July 2003.
[Qayyum2000] A. Qayyum, L. Viennot, and A. Laouiti, "Multipoint relaying: An efficient technique for flooding in mobile wireless networks," Research Report No. 3898, INRIA, France, March 2000.
[Royer1999] E. M. Royer and C. E. Perkins, "Multicast operation of the ad-hoc on-demand distance vector routing protocol," Proceedings of ACM MOBICOM, pp. 207-218, August 1999.
[Sinha1999] P. Sinha, R. Sivakumar, and V. Bharghavan, "MCEDAR: Multicast Core-Extraction Distributed Ad hoc Routing," IEEE Wireless Communications and Networking Conference, September 1999.
[Srinivasan2008] A. Srinivasan and Jie Wu, "TRACK: A Novel Connected Dominating Set based Sink Mobility Model for WSNs," Proceedings of 17th International Conference on Computer Communications and Networks, 3-7 Aug. 2008, pp. 1-8.
[Stojmenovic1999] I. Stojmenovic, "Voronoi diagram and convex hull based geocasting and routing in wireless networks," Technical Report, University of Ottawa, TR-99-11, December 1999.
[Sucec2000] J. Sucec and I. Marsic, "An efficient distributed network-wide broadcast algorithm for mobile ad hoc networks," CAIP Technical Report 248, Rutgers University, September 2000.
[Toh2000] C.-K. Toh, G. Guichal, and S. Bunchua, "ABAM: on-demand associativity-based multicast routing for ad hoc mobile networks," Proceedings of IEEE VTC Fall Technology Conference, Vol. 3, 2000.
[Tseng2001] Y.-C. Tseng, S.-L. Wu, W.-H Liao, and C.-M. Chao, "Location awareness in ad hoc wireless mobile networks," Computer, Vol. 34, No. 6, pp. 46–52, 2001.
[Wang2001] K. Wang and C.-T. Chang, "An intelligent on-demand multicast routing protocol in ad hoc networks," Proceedings of International Conference on Information Networking, 2001.
[Williams2002] B. Williams and T. Camp, "Comparison of Broadcasting Techniques for Mobile Ad Hoc Networks," in Proceedings of ACM Mobihoc, 2002.
[Wu1998] C. W. Wu, Y.C. Tay, and C.-K. Toh, "Ad-hoc Multicast Routing protocol utilizing Increasing id-numberS (AMRIS) Functional Specification," Internet-Draft, November 1998.
[Wu2004] J. Wu and F. Dai, "A Generic Distributed Broadcast Scheme in Ad-hoc Wireless Networks," IEEE Transactions on Computers, Vol. 53, No. 10, Oct. 2004, pp 1343-1354.
[Zhang2005a] Q. Zhang and D.P. Agrawal, "Dynamic Probabilistic Broadcasting in MANETs," Journal of Parallel and Distributed Computing, Vol. 65, No. 2, Feb. 2005, pp 220-233.
[Zhang2005b] Q. Zhang and D.P. Agrawal, "Performance Evaluation of Leveled Probabilistic Broadcasting in MANETs and Wireless Sensor Networks," Special issue of Simulation, Transactions of the society of modeling and simulation international, Aug. 2005, vol. 81. No. 8, pp. 533-546.
[Zhang2005c] Q. Zhang, "Efficient Broadcasting in Mobile Ad-Hoc and Wireless sensor networks," Ph.D. Dissertation, University of Cincinnati, 2005.

Chapter 4

Wireless LANs

4.1 Introduction

During the last few years, the Internet has become a major driving force behind most of the new developments in the telecommunication networks field. The volume of packet data traffic has been growing at a much faster rate than the telephone traffic. Meanwhile, there has been a similar explosive growth in the wireless field. We have seen the rollout of four generations of wireless cellular systems, attracting end-users by providing efficient mobile communications. In addition, wireless technology has become an important component in providing networking infrastructure for data delivery. This revolution has been made possible by the introduction of new networking technologies and paradigms such as Wireless LANs (or WLANs like IEEE 802.11 [IEEE-802.111997]), Wireless PANs (or WPANs like Bluetooth [Bisdikian2001]), Wireless MANs (or WMAN like IEEE 802.16 [Eklund2002]), Wireless WANs (or WWANs like IEEE 802.20 [IEEE802.20www]), and Wireless RANs (or WRAN like IEEE 802.22 [Cordeiro2005]). In particular, WLANs are becoming very popular for indoor and outdoor mesh-based applications, mainly due to their flexible configuration, low installation and maintenance costs, and mobility support as compared to their traditional wired counterparts. The combination of both the growth of the Internet and the success of wireless networks suggest that the next trend will be an increasing demand for wireless access to Internet applications.

Although the MANET protocols discussed in the previous chapters can, in principle, be implemented over nearly any type of network, the dominant choice has been the ad hoc and mesh (or infrastructureless) modes offered by WLANs and WPANs technologies. Given the importance of these two new paradigms in wireless ad hoc communications, we need to investigate WLANs and WPANs in conjunction with ad hoc networking. For WLANs, the most well-known

130

representative is the IEEE 802.11 standard together with its various amendments [IEEE802.11www]. The best example representing short-range WPANs is an industry standard called Bluetooth [Bluetoothwww].

This chapter deals with the IEEE standard 802.11 for WLANs and all the circumventing design issues in detail, and how it can be used to enable ad hoc networking. We also present various research advances in the WLAN arena. Bluetooth and the IEEE 802.15 standards are discussed in the next chapter. We note that we focus on the ad hoc operating mode of these technologies, as this is the goal of this textbook.

4.2 Why Wireless LANs

Since the success of Xerox's Palo Alto Research Center's Ethernet project in early 1970's, the basic LAN technology has blossomed in both public and private sectors. Standard LAN protocols, such as inexpensive Ethernet, have brought networking to almost all computers from different organizations and explore the power of networking and collaborative distributed computing. However, until recently, LANs have been limited to hard-wired infrastructure within the building. Even with phone dial-ups, network nodes are limited to wired landline connections. Many network users, especially mobile users in businesses, medical profession, factories, and universities find many benefits from the added capabilities of wireless LANs [Goldberg1995].

A major motivation and flexibility provided by wireless LANs is the mobility and untethered connection when compared to the fixed conventional hardwired connections. The practical use of wireless networks is only limited by an individual's imagination. Medical professionals can obtain not only patient records, but also real-time vital signs and other reference data at the patient bedside. Factory workers can access parts without inconvenient or sometimes bothersome wired connections. Wireless connections for any real-time sensing system allow a engineer to diagnose and maintain the welfare of manufacturing equipment remotely, even in an environmentally-hostile environment. Warehouse inventories can be verified quickly and effectively by connecting wireless scanners to the inventory database. Even wireless "smart" price tags, completing with liquid crystal display (LCD) readouts, allow merchants to virtually eliminate discrepancies between stock-point pricing and scanned prices at the checkout lane.

In addition to increased mobility, wireless LANs offer increased flexibility. One can easily visualize a meeting in which employees use small computers and wireless links to share and discuss future design plans and products. Such an ad hoc network can be brought up and torn down in a very short time as needed, either around the conference table and/or around the world. Some car rental establishments already use wireless networks to help facilitate check-ins. Traders on Wall Street use wireless terminals to make market transactions. Increasing number of students in university campuses is accessing lecture notes and other course materials while wandering around the campus. Recently, wireless LAN has also been used to stream video between devices, such as from a laptop to a television set or from a handheld to a television set [WirelessDisplay2010]. Sometimes, it may even be economical to use a wireless LAN. For instance, in old buildings, the cost of asbestos cleanup or removal outweighs the cost of installing a wireless LAN solution.

4.3 Transmission Techniques

In this section we give an overview of the transmission technologies which have been developed for the many standards and products for wireless networks [Pahlavan2001]. We first introduce the wired technology, as they are important for understanding the wireless ones. Typically, we would like to transmit data with the highest possible data rate and with minimum level of signal power, smaller channel bandwidth, and reduced transmitter/receiver complexity. However, emphasis on these objectives varies according to the application requirements and medium for transmission. Finally, these objectives are often conflicting and the trade-offs decide the important factors.

4.3.1 Wired Networks

All data applications for wired networks, including LANs, employ very simple schemes for transmission over, for instance, twisted pair, co-axial cable, or optical fiber. The data received from upper layers are line coded (e.g., Manchester coded on Ethernet) and the voltages (or optical signals) are applied to the medium directly. These transmissions schemes are often referred to as *baseband transmission schemes*. In voice-band modems, Digital Subscriber Line (DSL), and coaxial cable model applications, the transmitted signal is modulated over a carrier. The amplitude, frequency, phase, or a combination of these is used to carry data. These digital modulation schemes are correspondingly called amplitude shift keying (ASK), frequency shift keying (FSK), phase shift keying (PHK), or quadrature amplitude modulation (QAM). In voice-band modems with the telephone channel passband of 300-3,300 Hz, carrier is chosen to be around 1,800 Hz which is the pass band center.

For DSL services, the spectrum is shifted away from the lower frequencies used for voice applications. Discrete multitone transmission, a form of Orthogonal Frequency Division Multiplexing (OFDM), is employed in DSL. In cable modems, modulation is employed to shift the spectrum of the signal to a particular frequency channel and thereby improve bandwidth efficiency of the channel by supporting higher data rates. In data networking industry, cable modems are referred to as broadband modems as they provide a much higher data rate than the voice-band modems. Specific impairments seen on the telephone channels are amplitude and delay distortion, phase jitter, frequency offset, and effects of nonlinearity. Many of practical design techniques of wired modems have been developed to deal with these limitations.

4.3.2 Wireless Networks

Popular digital wireless transmission techniques can be divided into four categories according to their applications. The first category is pulse transmission techniques employed mostly in Infrared (IR) applications and, in the impulse radio or ultra-wideband (UWB) transmission [UWBwww]. The second category is the basic modulation techniques widely used in Time Division Multiple Access (TDMA) cellular, as well as a number of mobile data networks such as GSM and EDGE. The third category is spread spectrum systems used in the Code Division Multiple Access (CDMA) and in wireless LANs operating in the unlicensed Industrial-Scientific-Medical (ISM) frequency bands. Finally, fourth category is multi-carrier modulation systems adopted in new generation of wireless LANs and in fourth generation cellular systems such as WiMax and Long Term Evolution (LTE).

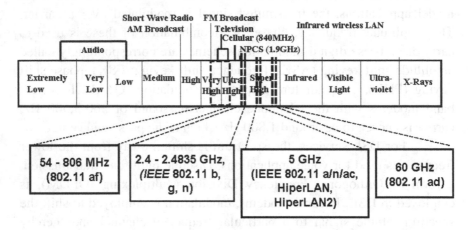

Figure 4.1 – Popular ISM bands in use by Wireless LAN Technologies

In 1985, the United States released the ISM frequency bands. These bands do not require licensing by the U.S. Federal Communications Commission (FCC). Figure 4.1 illustrates the ISM bands that are relevant to wireless LANs and which type of wireless LAN technology is allowed to operate in those bands. This unlicensed spectrum prompted most of the wireless LAN products to operate within ISM bands. For example, RF systems must confine the emitted spectrum to a band and are also limited to one watt of power. Microwave systems are considered very low power systems and must operate at 500 mW or less. In the context of wireless LANs, the three main propagation technologies used are Infrared, Microwave and Radio Frequency and are described below.

4.3.2.1 Infrared (IR)

Today, most of us are familiar with everyday devices that use IR technology such as remote controls for TVs, VCRs, DVD and CD players. IR transmission is categorized as a line-of-sight (LOS) wireless technology which means digital appliances must be in transmitter's direct line of sight. However, new diffused IR technologies that can work without LOS inside a room, and we expect to see these products in the very near future. IR networks can be implemented reasonably quickly; however, people walking or moisture in the air can weaken the signals. IR in-home technology is promoted by an international association of companies called Infrared Data Association (IrDA) [IrDAwww].

Infrared systems are simple to design and are inexpensive. They use the same signal frequencies as fiber optic links. IR systems detect only the amplitude of the signal and so interference is greatly reduced. These systems are not bandwidth limited and thus can achieve transmission speeds greater than other systems. Infrared transmission operates in the light spectrum and does not require a license from the FCC to operate. There are two conventional ways to set up an IR LAN. As infrared transmission can be aimed, the range could extend to a couple of kilometers and can be used outdoors. It also offers the highest bandwidth and throughput. The other way is to transmit omni-directionally and bounce the signals off of everything in every direction, which reduces the coverage to 30-60 feet at relatively cheaper price. The main drawback to an IR system is that the transmission spectrum is shared with the sun and other things such as fluorescent lights. Given the LOS requirement, IR signals cannot penetrate opaque objects. This means that walls, dividers, curtains, or even fog can obstruct the signal. Although not very popular, the earlier IEEE 802.11 standard also defines a physical layer for high-speed diffused IR employing pulse-position-modulation (PPM) utilizing a wavelength of 850nm–950nm and data rates of 1 and 2 Mbps.

4.3.2.2 Microwave

In compliance with FCC regulations, microwave (MW) systems operate at less than 500 mW of power. They use narrow-band transmission with single frequency modulation and are set up mostly in the 5.8 GHz band. The advantage to a MW system is higher throughput because they do not have the overhead involved with spread spectrum systems. RadioLAN [RadioLANwww] is an example of a system employing microwave technology.

4.3.2.3 Radio Frequency

Another main category of wireless technology is the RF which is more flexible, allowing consumers to link appliances that are distributed throughout the house. RF can be categorized as narrowband or spread spectrum. Narrowband technology includes microwave transmissions which are high-frequency radio waves that can be transmitted to distances up to 50 Km. Microwave technology is not suitable for local networks, but could be used to connect networks in separate buildings. Spread spectrum technology (SST) is one of the most widely used technologies and was developed during World War II to provide greater security for military applications. As it entails spreading the signal over a number of frequencies, spread spectrum technology makes the signal harder to intercept.

The main difference between the SST and traditional radio modem technology is that the transmitted signal in SST occupies a much larger bandwidth than the traditional radio modems with the signal bandwidth of the same order as the information signal at the baseband. Compared to UWB, occupied bandwidth by SST is still restricted enough so that the spread spectrum radio can share the medium with other spread spectrum and traditional radios using frequency division multiplexing [Pahlavan2001]. Two basic techniques for SST are: frequency hopping spread spectrum (FHSS) and direct sequence spread spectrum (DSSS).

4.3.2.3.1 Frequency Hopping Spread Spectrum

This technique splits the band into many small subchannels (e.g., 1-MHz). The signal then hops from subchannel to subchannel, transmitting short bursts of data for a set period of time, called dwell time. The hopping sequence must be synchronized at the sender and the receiver or else, the information is lost. The FCC requires that the band is split into at least 75 subchannels and that the dwell time is no longer than 400ms. Frequency hopping is less susceptible to interference because the frequency is constantly shifting which makes it extremely difficult to intercept. The whole band must be jammed in order to jam a frequency hopping system. These features are very attractive to agencies involved with law enforcement or military. Many FHSS LANs can be co-located if an orthogonal hopping sequence is used. Because the subchannels are smaller than in DSSS, a larger number of co-located LANs are possible with FHSS systems.

4.3.2.3.2 Direct Sequence Spread Spectrum (DSSS)

With DSSS, the transmission signal is spread over an allowed band and can be thought of as a two-stage modulation technique. In the first state, a random binary string, called the *spreading code*, is used to modulate the transmitted signal. The data bits are mapped (spread) to a pattern of "chips". In the second stage, the chips are transmitted over a traditional digital modulator. At the receiver, the chips are first demodulated and then passed through a correlator to map (dispreads) the chips back into data bits at the destination. The number of chips that represent a bit is called the *spreading ratio*. Higher the spreading ratio is, more is the resistance to interference for the signal. More bandwidth is available to the user with a lower spreading ratio. FCC dictates that the spreading ratio must be more than ten (typically, products have a spreading ratio of less than 20). For example, physical layer of the IEEE 802.11 standard employing DSSS requires a spreading ratio of eleven. The transmitter and the receiver must be synchronized with the same spreading code. If orthogonal spreading codes are used, then more than one LAN can share the same band. However, because DSSS systems use wide subchannels, the number of co-located LANs is limited by the size of those subchannels. Recovery is faster in DSSS systems because of the ability to spread the signal over a wider band.

As we know, the bandwidth of any digital system is inversely proportional to the duration of the transmitted pulse or symbol. Because the transmitted chips are much narrower than data bits, the bandwidth of the transmitted DSSS signal is much larger than systems without spreading. Therefore, the transmission bandwidth of DSSS is always wide, whereas FHSS is a narrowband system hopping over a number of frequencies in a wide spectrum. The DSSS systems provide a robust signal with better coverage area than FHSS. On the other hand, the FHSS can be implemented with much slower sampling rates, saving in the implementation costs and power consumption of the mobile units. These distinctions have guided the use of these systems in different technologies for WLANs and WPANs. For example, the IEEE 802.11b standard can be found in both DSSS and FHSS, while Bluetooth employs FHSS only, given its low power and low cost requirements.

4.4 Medium Access Control Protocol Issues

MAC protocols have been receiving considerable attention from both the industry and the academia. There are many issues that need to be addressed in order to design an efficient MAC protocol in a MANET environment [Royer2000]. Several MAC protocols can be employed for ad hoc networking such as IEEE 802.11 [IEEE802.112007], Bluetooth [Bluetoothwww] and HIPERLAN [HIPERLAN1996]. In this section, we discuss some fundamental issues that guide the design of MAC protocols for any wireless network. We note, however, that some of these issues pertain mostly to single channel MAC protocols such as the IEEE 802.11 which is discussed in detail later on in this chapter.

4.4.1 Hidden Terminal Problem

In Carrier Sense Multiple Access (CSMA) [Agrawal2002], every station (throughout this chapter, we use the terms node, mobile station, mobile host, and mobile terminal interchangeably) senses the carrier before transmitting, and the transmission is deferred if a signal is detected. Carrier sense attempts to avoid collisions by testing the signal strength in the vicinity of the transmitter. However, collisions occur at the receiver, not the transmitter; i.e., it is the presence of one or more interfering signals at the receiver that may lead to a collision. Since the receiver and the sender are typically not co-located, carrier sense does not provide the appropriate information for collision avoidance. An example illustrates this point in more detail. Consider the configuration depicted in Figure 4.2. Station A can hear B but not C, and station C can hear station B but not A (and by symmetry, we know that station B can hear both A and C). First, assume A is sending packets to B. When C is ready to transmit (perhaps to B or perhaps to some other station), it does not detect signal on carrier and thus commences transmission; this produces a collision at B. Station C's carrier sense did not provide the necessary information since station A was "hidden" from it. This is the classic "hidden terminal" scenario.

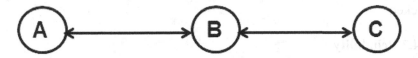

Figure 4.2 – Station B can hear both A and C, but A and C cannot hear each other

An "exposed" terminal scenario results if we assume that B is sending to A, rather than A sending to B. Then, when C is ready to transmit, it does detect carrier and therefore defers transmission. However, there is no reason to defer transmission to a station other than B since station A is out of C's range. Station C's carrier sense did not provide the necessary information since it was exposed to station B even though it would not collide or interfere with B's transmission. The point to note here is that carrier sense provides information about potential collisions at the sender, but not at the receiver. This information can be misleading when the configuration is distributed so that not all stations are within communication range of each other.

The solution to the hidden terminal problem has been proposed in the Medium Access with Collision Avoidance (MACA) protocol [Karn1990]. It consists of transmitting Request-to-Send (RTS) and Clear-to-Send (CTS) packets between nodes that wish to communicate. These RTS and CTS packets carry the duration of the data transfer of the communicating parties. Stations in the neighborhood not participating in the communication can overhear either RTS or CTS, can keep quiet for the duration of the transfer. Returning to our example of Figure 4.2, when node A wants to send a packet to node B, node A first sends a RTS packet to B. On receiving the RTS packet, node B responds by sending a CTS packet (provided node A is able to receive the packet). As a result of that, when node C overhears the CTS sent by B it keeps quiet for the duration of the transfer contained in the CTS packet. As for the exposed terminal problem, while in the IEEE 802.11 MAC layer there is almost no scheme to deal with it, the Medium Access with Collision Avoidance for Wireless (MACAW) protocol [Bharghavan1994] (based on MACA) solves this problem by having the source transmit a *data sending* control packet to alert exposed nodes of the impending arrival of an ACK packet.

4.4.2 Reliability

Wireless links are prone to errors. Packet error rates of wireless mediums are much higher than that of their wired counterparts. As a result, some protocols – which have been originally designed to work in wired world – suffer performance degradation when operating in a wireless environment. A classic example of this problem is TCP (which has been designed and fine-tuned for wired networks) that assumes transmission timer expiration as an indication of network congestion [Cordeiro2002a]. This event triggers the execution of TCP congestion control mechanisms which ultimately decreases the transmission rate, with the purpose of reducing the network congestion. As a matter of fact, this is often true in wired environment as the media is usually very reliable. However, in wireless environment this is often not the case as packet loss occurs every now and then due to effects such as multipath fading, interference, shadowing, distance between transmitter and receiver, etc. As a result, when a packet loss occurs in a TCP communication, the loss is erroneously assumed due to congestion and congestion control mechanisms are fired. There have been some proposals to cope up with such TCP behavior in wireless and mobile ad hoc networks [Cordeiro2002a, Liu2001] (Chapter 9 discusses the subject of TCP over ad hoc networks in detail).

As for the MAC protocol, a common approach to reduce packet loss rates experienced by upper layers is to introduce acknowledgment (ACK) packets. Returning to our earlier example of Figure 4.2, whenever node B received a packet from node A, node B sends an ACK packet to A. In case node A fails to receive the ACK from B, it will retransmit the packet. This approached is adopted in many protocols [Bharghavan1994]. As an example, the IEEE 802.11 Distributed Coordination Function (DCF) [Crow1997] uses RTS-CTS to avoid the hidden terminal problem and ACK to achieve reliability.

4.4.3 Collision Avoidance

The radios used in the wireless and mobile nodes for communication are half-duplex. This is to say that these radios are not able to transmit and receive at the same time and, thus, collision detection is not possible. To minimize collisions, wireless MAC protocols, such as CSMA with Collision Avoidance (CSMA/CA), often use collision avoidance techniques in conjunction with a carrier sense (be it physical or virtual) scheme. Collision avoidance is implemented by mandating that, when the channel is sensed idle, the node has to wait for a randomly chosen duration before attempting to transmit. This mechanism drastically decreases the probability that more than one node attempts to transmit at the same time, thereby avoiding any potential collision. Obviously, there will be cases where more than one node initiates transmission at the same time. In these cases, transmissions are corrupted, the ACK signal is not sent back by the receiver, and the corresponding nodes retry later on.

4.4.4 Congestion Avoidance

Congestion avoidance is a fundamental aspect in wireless MAC protocols. In IEEE 802.11 DCF, when a node detects the medium to be idle, it chooses a backoff interval between [0, CW], where CW is called contention window which usually has a minimum (CW_min) and maximum (CW_max) values. The node will count down the backoff interval and when it reaches zero, the node can transmit the RTS. In case the medium becomes busy while the node is still counting down the backoff interval, the countdown process is suspended.

To illustrate how DCF works, let us consider the example in Figure 4.3. In this figure, BO_1 and BO_2 are the backoff intervals of nodes 1 and 2, and we assume for this example that CW=31. As we can see from Figure 4.3, node 1 and node 2 have chosen a backoff interval of 25 and 20, respectively. Obviously, node 2 will reach zero five units of time earlier than node 1. When this happens, node 1 will notice that the medium became busy and freezes its backoff interval currently at 5. As soon as the medium becomes idle again, node 1 resumes its backoff countdown and transmits its data once the backoff interval reaches zero.

Similarly, upon node's 1 transmission, node 2 will freeze its backoff countdown process and resume it as soon as node 1 finishes its transmission. To a certain extent, collisions can be avoided by carrying out this procedure. Choosing a large CW leads to large backoff intervals and can result in larger overhead, since nodes have to carry out the countdown procedure. On the other hand, choosing a small CW leads to a larger number of collisions, and hence it is more likely for two nodes to count down to zero simultaneously.

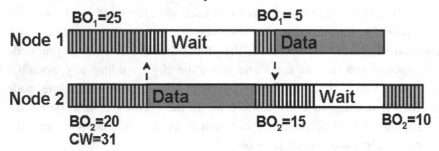

Figure 4.3 – Example of the Backoff Mechanism in DCF

4.4.5 Congestion Control

As mentioned earlier, the number of nodes attempting to transmit simultaneously may change with time. Therefore, some mechanism is desirable to manage congestion. In IEEE 802.11 DCF, congestion control is achieved by dynamically choosing the CW. When a node fails to receive CTS in response to its RTS, it assumes that congestion has built up, and hence doubles its contention window up to CW_max. When a node successfully completes its transmission, it resets its contention window to CW_min. This mechanism of dynamically controlling the CW is called Binary Exponential Backoff, since the CW increases exponentially with failed CTS indicated by its absence.

4.4.6 Energy Efficiency

Since many MH are operated by batteries, there is an increasing interest for MAC protocols that could conserve energy. The current proposals in this area usually suggest turning the radio off when it is not needed. IEEE 802.11 has a Power Saving (PS) mode whereby the Access Point (AP) periodically transmits a beacon, indicating which nodes have packets waiting for them. Each PS node wakes up periodically to receive the beacon transmitted by the AP. In case a node has a packet waiting for it, it sends a PS-POLL packet to the AP after waiting for a backoff interval in [0, CW_min]. Upon receipt of the PS-POLL packet, the AP transmits the data to the requesting node. Using this procedure, it is possible to extend the battery life of mobile nodes for a longer period of time. Later in this chapter, we discuss other ways by which energy can be conserved, including techniques such as transmission power control.

4.4.7 Other MAC Issues

The coverage of MAC protocol issues here is far from being exhaustive and many other issues need to be considered such as *fairness* which has many meanings with one of them being stations receiving equal bandwidth. This type of fairness is not easy to accomplish in the IEEE 802.11 MAC, since unfairness will eventually occur when one node in the same neighborhood backs off much more than some other node. MACAW's solution to this problem [Bharghavan1994] is to append the contention window value (CW) to packets a node transmits, so that all nodes hearing that CW, use it for their future transmissions. Since CW is an indication of the level of congestion in the vicinity of a specific receiver node, MACAW proposes maintaining a CW independently for each receiver. There are also other proposals such as Distributed Fair Scheduling [Vaidya2000] and Balanced MAC [Ozugur1998]. A final comment must be made on receiver-related issues in wireless MAC protocols. All protocols discussed so far are sender-initiated protocols. In other words, a sender always initiates a packet transfer to a receiver. The receiver might take a more active role in the process by assisting the transmitter in certain issues such as collision avoidance [Garcia-Luna-Aceves1999], and some sort of adaptive rate control [Holland 2001].

4.5 The IEEE 802.11 Standard for Wireless LANs

WLANs provide an excellent usage model for high-bandwidth consumers, and they are quite appealing for their low infrastructure cost and high data rates as compared to other wireless data technologies such as cellular or point-to-multipoint distribution systems. In June 1997, the IEEE approved the 802.11 standard (sometimes also referred to as Wi-Fi – for Wireless Fidelity) [IEEE802.112007, Nee1999, Cordeiro2002b] for WLANs, and in July 1997, the IEEE 802.11 has been adopted as a worldwide International Standards Organization (ISO) standard. The standard consists of three possible physical (PHY) layer implementations and a single common MAC layer supporting data rates of 1-Mb/s or 2-Mb/s. The alternatives for the PHY layer in the original standard include FHSS system using 2 or 4 Gaussian frequency-shift keying (GFSK) modulation, direct sequence spread spectrum (DSSS) system using differential binary phase-shift keying (DBPSK) or differential quadrature phase-shift keying (DQPSK) baseband modulation and IR physical layer.

Later in 1999, the IEEE 802.11b working group extended the IEEE 802.11 standard with the IEEE 802.11b addition and decided to drop the FHSS and use only DSSS. In addition, another working group, the IEEE 802.11a, significantly modified the PHY to add OFDM modulation, which effectively combines multicarrier, multisymbol, and multirate techniques. IEEE 802.11a can offer data rates up to 54 Mbps. Another amendment is the IEEE 802.11g, which is an extension to the IEEE 802.11 PHY standard, formally ratified in June 2003. The IEEE 802.11g provides the same maximum speed of 802.11a coupled with backwards compatibility with 802.11b devices by operating in the 2.4 GHz ISM band. Because of that, 802.11g is extremely popular nowadays. IEEE 802.11g compliant devices utilize OFDM modulation technology to achieve the higher data rates. These devices can automatically switch to CCK (Complementary Code Keying) modulation in order to communicate with the slower 802.11b and 802.11 compatible devices. Therefore, 802.11g PHY layer modulation can be seen as the union of the PHY layer modulations of both 802.11a and 802.11b.

The latest PHY amendment to the IEEE 802.11 standard is 802.11n, which was formally approved in October 2009. The major enhancement of 802.11n over the previous 802.11 PHYs is the use of Multiple Input Multiple Output (MIMO) and channel bonding up to 40 MHz. The use of MIMO and channel bonding coupled with an increase on the number of OFDM sub-carriers allows 802.11n to offer data rates nearing 600 Mbps.

Two other efforts are currently underway in IEEE 802.11 to further increase the PHY data rate as to meet the increasing demand of new usages and applications. The IEEE 802.11ac amendment aims at increasing the maximum data rates of the IEEE 802.11 standard to around 1 Gbps by employing wider channels of up to 160 MHz and taking advantage of multi-user MIMO. IEEE 802.11ac is expected to operate in the 5 GHz band only. The other amendment is IEEE 802.11ad, which is enhancing the IEEE 802.11 standard for operation in the 60 GHz ISM band. The current IEEE 802.11ad draft standard utilizes channel bandwidths of around 2 GHz, and is capable of reaching data rates up to 7 Gbps. Together, these amendments are future of Wi-Fi based technologies.

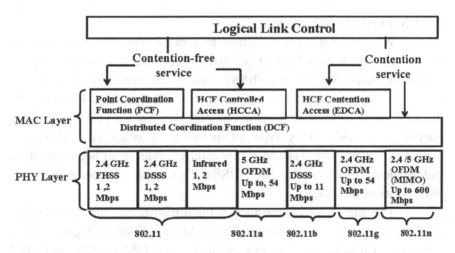

Figure 4.4 – A complete view of the Stack

Figure 4.4 gives a complete view of the IEEE 802.11 protocol stack with all the PHY layers until the latest approved IEEE 802.11n. As we can see from this figure, the MAC layer protocols are common across all standards while they are not always compatible at the PHY layer. With all these enhancements, ease of use, and customer satisfaction, the IEEE 802.11 standard is the most widely used WLAN technology today.

Therefore, in this section we investigate the IEEE 802.11 standard and describe the techniques underlying its PHY and MAC layers. Often viewed as the "brains" of the network, the 802.11 MAC layer uses an 802.11 PHY layer, such as 802.11a/b/g/n, to perform the tasks such as carrier sensing, transmission, and receiving of 802.11 frames. With regards to the MAC layer, the functional specifications are similar for all of the underlying 802.11 PHYs. We give special emphasis to the infrastructureless mode of operation, as this is typical to ad hoc networking. On the PHY side, we describe the physical layer specifications of the IEEE 802.11a/b standards, as the later is widely used nowadays and the former is finding increasing acceptance. The IEEE 802.11g PHY layer will be consistently analyzed as we describe the other PHYs. For more information on IEEE 802.11n, the reader can refer to [Paul2008].

4.5.1 Network Architecture

WLANs can be used either to replace wired LANs, or as an extension and the topology of an 802.11 network is shown in Figure 4.5(a). A Basic Service Set (BSS) consists of two or more wireless nodes, which establish communication after recognizing each other. In the most basic form, stations communicate directly with each other on a peer-to-peer mode. This type of network is often formed on an instantaneous basis, and is commonly referred to as an ad hoc network or Independent Basic Service Set (IBSS). This mode of operation is the main focus of this book.

The other form of operation is the infrastructure (or client/server) mode with the help of an Access Point (AP) as shown in Figure 4.5(b). AP forms a bridge between wireless and wired LANs while each BSS contains an AP. When an AP is present, all communications between stations or between a station and a wired network client go through the AP. AP's are not mobile, and form a part of the network infrastructure. MHs, on the other hand, are typically mobile and can roam between APs, thus requiring support to seamless coverage. A BSS in this configuration is said to be operating in the infrastructure mode. The Extended Service Set (ESS) shown in Figure 4.5(b) consists of a number of overlapping

BSSs connected together by means of a Distribution System (DS). Typically, the DS is an Ethernet LAN. Recently, however, wireless mesh networks (WMNs) have become increasingly popular as a DS [Akyildiz2005] where there is no need for wired connections amongst APs as covered in Chapter 6. APs are capable of wirelessly communicating with each other in a hierarchical fashion, which eliminates the need for any pre-existing wired infrastructure in WMNs and standardized are covered under the IEEE 802.11s task group.

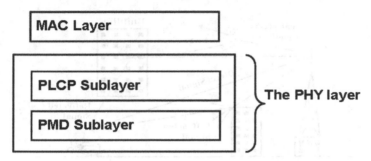

Figure 4.5 – Possible Network Topologies

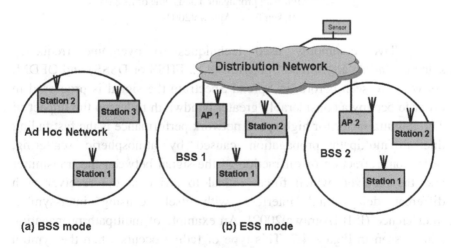

Figure 4.6 – The sublayers within the PHY

4.5.2 The Physical Layer

The PHY layer is the interface between the MAC and wireless media, which transmit and receive data frames over a shared wireless media (see Figure 4.6). The PHY provides three levels of functionality. Firstly, the PHY layer provides a frame exchange between the MAC and PHY under the control of the physical layer convergence procedure (PLCP) sublayer. Secondly, the PHY uses signal carrier and spread spectrum modulation to transmit data frames over the media under the control of the physical medium dependent (PMD) sublayer. Thirdly, the PHY provides a carrier sense indication back to the MAC to verify activity on the media.

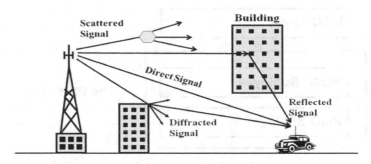

Figure 4.7 – Multipath propagation and some of its causes
[taken from: Agrawal2002]

Two commonly used techniques to overcome frequency selective fading are Spread Spectrum (e.g., FHSS or DSSS) and OFDM. As we have seen before, in Spread Spectrum the signal is processed in order to occupy a considerably greater bandwidth to lessen the impact of The limiting factor for high-speed network performance is the fast fading due to multipath propagation caused by atmospheric scattering, reflection, refraction or diffraction of the signal between the transmitter and the receiver, which forces signal to arrive at the receiver with different delays and interfere with itself causing inter-symbol interference (ISI) [Agrawal2002]. An example of multipath propagation can be seen in Figure 4.7. This type of fading occurs when the symbol time becomes much smaller than the channel delay spread, which makes it especially important as we increase the communication data rate. We will expose various techniques used in IEEE 802.11 to overcome the effect of fading frequency selective fading that will affect only a small part of the signal bandwidth. In OFDM, the data stream is split into a number of substreams, each having a bandwidth smaller than the coherence bandwidth of the channel that overcomes the frequency selective fading.

Understanding the meaning of some of the physical layer terminology is essential to grasp the intricacies of IEEE 802.11:

- GFSK (Gaussian Frequency Shift Keying) is a modulation scheme in which the data are first filtered by a Gaussian filter in the baseband, and then modulated with a simple frequency modulation. "2" and "4" represent the number of frequency offsets used to represent data symbols of one and two bits, respectively.

- DBPSK (Differential Binary PSK) is a phase modulation scheme using two distinct carrier phases for data signaling, providing one bit per symbol.

- DQPSK (Differential Quaternary PSK) is a type of phase modulation using two pairs of distinct carrier phases, in quadrature, to signal two bits per symbol. The differential characteristic of the modulation schemes indicates the use of the difference in phase from the last change or symbol to determine the current symbol's value, rather than any absolute measurements of the phase change.

Both the FHSS and DSSS modes are specified for operation in the 2.4 GHz ISM band, which has sometimes been jokingly referred to as the "interference suppression is mandatory" band as it is heavily used by various electronic products. The third physical layer alternative, which is not widely used, is an infrared system using near-visible light in the 850 nm to 950 nm ranges as the transmission medium.

At the forefront of the new WLAN options that would enable much higher data rates are three supplements to the IEEE 802.11 standard: 802.11a, 802.11b and 802.11g, as well as an ETSI standard HIPERLAN/2. Both 802.11a and HIPERLAN/2 have similar physical layer characteristics operating in the 5 GHz band and use the modulation scheme OFDM, but the MAC layers are considerably different. The focus of this section, however, is to discuss and compare the physical layer characteristics of the IEEE standards given that HIPERLAN/2 shares several of the same physical properties as 802.11a, and 802.11g PHY can be seen as a combination of 802.11a/b PHYs.

4.5.2.1 DSSS

The DSSS uses the 2.4 GHz frequency band as the RF transmission media. Data transfer over the media is controlled by the DSSS PMD sublayer as directed by the DSSS PLCP sublayer. The DSSS PMD takes the binary bits of information from the PLCP protocol data unit (PPDU) and transforms them into RF signals for the wireless media by using carrier modulation and DSSS techniques.

The IEEE 802.11 implements DSSS to fight frequency-selective fading. The IEEE 802.11b, approved by the IEEE in 1999, supports 5.5 and 11 Mb/s of higher payload data rates in addition to the original 1 and 2 Mb/s rates of IEEE 802.11. IEEE 802.11b also operates in the highly populated 2.4 GHz ISM band (2.40 to 2.4835 GHz), which provides only 83 MHz of spectrum to accommodate a variety of other products, including cordless phones, microwave ovens, other WLANs, and WPANs such as Bluetooth. This makes their susceptibility to interference a primary concern. To help mitigate interference effects, 802.11b designates an optional frequency agile or hopping mode using the three non-overlapping channels or six overlapping channels spaced at 10 MHz.

In DSSS, each bit in the original signal is mapped into a common pattern of chips in the transmitted signal, using a *pseudo-noise* (PN) sequence. This operation considerably enlarges the signal bandwidth and makes it more resistant to frequency selective fading. A PN sequence is a deterministic binary sequence that eventually repeats itself but that appears to be random (like noise). A single symbol of the PN sequence is called a *chip*.

4.5.2.1.1 Modulation

The DSSS PMD transmits the PLCP preamble and PLCP header at 1 Mbps using DBPSK. The MAC protocol data unit (MPDU) is sent at either 1 Mbps DBPSK or 2 Mbps DQPSK, depending upon the content in the signal field of the PLCP header.

4.5.2.1.2 Operating Channels and Power Requirements

Each DSSS PHY channel occupies 22 MHz of bandwidth. This allows for three non-interfering channels spaced 25 MHz apart in the 2.4

GHz ISM frequency band. This DSSS channel scheme is shown in Figure 4.8 with the corresponding channel nominations.

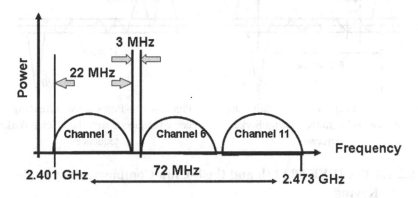

Figure 4.8 – DSSS non-overlapping channels

In addition to frequency and bandwidth allocations, transmit power is a key parameter that is regulated worldwide. The maximum allowable radiated emission for the DSSS PHY varies from region to region. Nowadays, the wireless manufacturers have selected 100 mW as the nominal RF transmits power level.

4.5.2.1.3 IEEE 802.11 and the 11-Chip Barker Sequence

In IEEE 802.11, the PN chosen for the DSSS PHY layer is the 11-chip Barker sequence [1, 1, 1, -1, -1, -1, 1, -1, -1, 1, -1]. This sequence has been selected as it has some very interesting autocorrelation properties which show some very sharp peaks when the transmitter and the receiver are synchronized. An example is shown in Figure 4.9 when we correlate the sequence '10' with this 11-chip Barker sequence. These peaks enable the receiver to lock on the strongest received signal, overcoming the 'echo' signals due to the multipath channel, as exemplified in Figure 4.10.

The first IEEE 802.11 standard used a symbol rate of 1 Mega-symbol per second (Msps) which yields a 11 MHz chipping rate with the Barker sequence, and is able to provide data rates of 1 Mbps (using DBPSK) and 2 Mbps (using DQPSK, where 2 bits are transmitted per symbol).

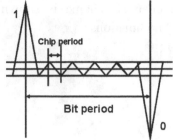

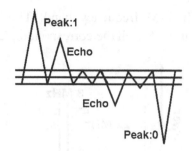

Figure 4.9 – Peaks when correlating the   Figure 4.10 – Peaks when correlating a
sequence '10' with the 11-chip Barker   received sequence with the 11-chip Barker
            sequence   sequence

4.5.2.1.4 The IEEE 802.11b and the 8-Chip Complementary Code Keying

IEEE 802.11b implements DSSS in an improved way which enables the enhanced data rates of 5.5 Mbps and 11 Mbps using symbol rates of 1.375 Msps (Million symbols per second) with an 8-chip Complementary Code Keying (CCK) modulation scheme. Instead of the Barker codes, CCK employs a nearly orthogonal complex code set called complementary sequences. The chip rate remains consistent with the original DSSS system at 1.375 Msps. The data rate varies to match channel conditions by changing the spreading factor and/or the modulation scheme.

To achieve data rates of 5.5 and 11 Mb/s, the spreading length is first reduced from 11 to eight chips. This increases the symbol rate from 1 Msps to 1.375 Msps. For the 5.5 Mbps bit rate with a 1.375 MHz symbol rate, it is necessary to transmit 4 bits/symbol (5.5 Mbps / 1.375 Mspss) and for 11 Mbps, an 8 bits/symbol. The CCK approach taken in the IEEE 802.11b keeps the QPSK spread spectrum signal and still provides the required number of bits/symbol, uses all but two of the bits to select from a set of spreading sequences and the remaining two bits to rotate the sequence. The selection of the sequence, coupled with the rotation, represents the symbol conveying the four or eight bits of data. For all the IEEE 802.11b payload data rates, the preamble and header are sent at the 1 Mbps to maintain compatibility with earlier versions.

4.5.2.2 FHSS

In FHSS, data transmission over the media is controlled by the FHSS PMD sublayer as directed by the FHSS PLCP sublayer. The FHSS PMD takes the binary bits of information from the whitened PLCP service data unit (PSDU) and transforms them into RF signals for the wireless media by using carrier modulation and FHSS techniques.

4.5.2.2.1 PSDU Data Whitening

Data whitening is applied to the PSDU before transmission to minimize bias on the data if long strings of 1's or 0's appear in the PSDU. The PHY stuffs a special symbol every 4 octets of the PSDU in a PPDU frame. A 127-bit sequence generator using the polynomial $S(x) = x^7 + x^4 + 1$ and 32/33 bias-suppression encoding algorithm are used to randomize and whiten the data.

4.5.2.2.2 Modulation

The IEEE 802.11 version released in 1997 uses two-level GFSK in the FHSS PMD to transmit the PSDU at the basic rate of 1 Mbps. The PLCP preamble and PLCP header are always transmitted at 1 Mbps. However, four-level GFSK is an optional modulation method defined in the standard that enables the whitened PSDU to be transmitted at a higher rate.

GFSK is a modulation technique used by the FHSS PMD, which deviates (shifts) the frequency either side of the carrier hop frequency depending on if the binary symbol from the PSDU is either a 1 or 0. A bandwidth bit period (Bt) = 0.5 is used. The changes in the frequency represent symbols containing PSDU information. For two-level GFSK, a binary 1 represents the upper deviation frequency from the hopped carrier, and a binary 0 represents the lower deviation frequency. The deviation frequency (fd) shall be greater than 110 KHz for IEEE 802.11 FHSS radios.

Four-level GFSK is similar to two-level GFSK and used to achieve a data rate of 2 Mbps in the same occupied frequency bandwidth. The modulator combines two binary bits from the whitened PSDU and encodes them into symbol pairs (10, 11, 01, 00). The symbol pairs generate four frequency deviations from the hopped carrier frequency, two upper and two lower. The symbol pairs are transmitted at 1 Mbps, and for each bit sent the resulting data rate is 2 Mbps.

4.5.2.2.3 Channel Hopping

A set of hop sequences is defined in IEEE 802.11 for use in the 2.4 GHz frequency band. The channels are evenly spaced across the band over a span 83.5 MHz. Hop channels differs from country to country. Channel hopping is controlled by the FHSS PMD. The FHSS PMD transmits the whitened PSDU by hopping from channel to channel in a pseudorandom fashion using one of the hopping sequences.

4.5.2.3 IR

The IR PHY is one of the three PHY layers supported in the IEEE 802.11 standard. The IR PHY differs from DSSS and FHSS because IR uses near-visible light as the transmission media. IR communication relies on the light energy, which is by line-of-sight or reflected off objects. The IR PHY operation is restricted to indoor environments and cannot pass through walls, such as DSSS and FHSS radio signals. Data transmission over the media is controlled by the IR PMD sublayer as directed by the IR PLCP sublayer.

4.5.2.3.1 Modulation

Table 4.1 – 4-PPM symbol map for 2 Mbps

Data bits	4-PPM symbol
00	0001
01	0010
11	0100
10	1000

The IR PHY transmits binary data at 1 and 2 Mbps using PPM modulation to reduce the optical power required of the Light Emitting Diode (LED) infrared source. The specific data rate is dependent upon the type of PPM. The modulation for 1 Mbps operation is 16-PPM, while it is 4-PPM for 2 Mbps. PPM is a modulation technique that keeps the amplitude, pulse width constant, and varies the position of the pulse in time. Each position represents a different symbol in time. For 2 Mbps operation, 4-PPM is used and two data bits are paired in the PSDU to form a 4-bit symbol map as shown in Table 4.1.

4.5.2.4 OFDM

While IEEE 802.11a has been approved in September 1999, new product development has proceeded much more slowly than IEEE 802.11b. This is due to the cost and complexity of implementation at the time of its approval. This standard employs over 300 MHz bandwidth in the 5 GHz unlicensed national information infrastructure (UNII) band. The spectrum is divided into three "domains," each having restrictions on the maximum allowed output power. The first 100 MHz in the lower frequency portion is restricted to a maximum power output of 50 mW. The second 100 MHz has a higher 250 mW maximum, while the third 100 MHz is mainly intended for outdoor applications and has a maximum of 1.0 W power output.

Employed in 802.11a/g/n, OFDM combines multicarrier, multisymbol, and multirate techniques, which require smart digital signal processing. The multicarrier operates by dividing the transmitted data into multiple parallel bit streams, each with relatively lower bit rates and modulating separate narrowband carriers, referred to as sub-carriers. The sub-carriers are orthogonal, so each can be received without interference from another. 802.11a specifies eight non-overlapping 20 MHz channels (regulations of specific countries may allow a larger or smaller number of channels to be used) in the lower two bands; each of these are divided into 52 sub-carriers (four of which carry pilot data) of 300-kHz bandwidth each. Four non-overlapping 20 MHz channels are specified in the upper band. The receiver processes the 52 individual bit streams, reconstructing the original high-rate data stream. This multicarrier technique has some important properties such as reducing multipath and allowing individual sub-carriers to be coded accordingly. This is also known as Coded OFDM (COFDM). As for 802.11g, the standard specifies the use of three channels so as to be backward compatible with 802.11b devices.

The multisymbol technique uses multiamplitude and multiphase modulation to increase the data rate. Four modulation methods are employed, depending on the data rate that can be supported by channel conditions between the transmitter and the receiver. These include BPSK, QPSK, 16-QAM, and 64-QAM signal constellations. QAM is a complex modulation method where data are carried in symbols represented by the phase and amplitude of the modulated carrier. 16-QAM has 16 symbols, each representing four data bits. 64-QAM has 64 symbols, each representing six data bits. Therefore, if the symbol rate for a constellation is 250 Kilo symbols per second (Ksps), the data rate for 16-QAM is (4 bits/symbol -> 250 Ksps) = 1 Mbps.

Table 4.2 – IEEE 802.11a data rate description

Data Rate (Mbit/s)	Modulation Type	Coding Rate (Convolution Encoding & Puncturing)	Coded bits per sub-carrier symbol	Coded bits per OFDM symbols	Data bits per OFDM symbol
6*	BPSK	1/2	1	48	24
9	BPSK	3/4	1	48	36
12*	QPSK	1/2	2	96	48
18	QPSK	3/4	2	96	72
24*	16-QAM	1/2	4	192	96
36	16-QAM	3/4	4	192	144
48	64-QAM	2/3	6	288	192
54	64-QAM	3/4	6	288	216
* Support for these data rates is required by the IEEE 802.11a standard					

Another approach to increase the data rate is to use a multirate modem, which provides one or more "fallback" modes of operation. The idea behind multirate technique is that if the modulation efficiency is increased (the number of bits per symbol is increased), the required signal-to-noise ratio (SNR) at the receiver also increases. For example, as the user moves away from the AP, the SNR reduces and the modem falls to a lower rate, providing reasonable error rates at lower values of the SNR. The data rates available in 802.11a are noted in Table 4.2, together with the type of modulation and the coding rate. Note that 802.11g also supports these data rates in addition to the date rates supported by the 802.11b standard. BPSK modulation is always used on four pilot sub-carriers. Although it adds a degree of complication to the baseband processing, 802.11a includes forward error correction (FEC) as a part of the specification. FEC, which does not present in 802.11b, enables the receiver to identify and correct errors occurring during transmission by sending additional data along with the primary transmission. OFDM has some very interesting properties: it can eliminate inter-symbol interference at no bandwidth cost (the total bandwidth of the transmitted signal stays roughly the same) while it does not require very complex signal processing (most is done with Fourier transforms). However, OFDM is very sensitive to frequency offsets and timing jitter and requires additional mechanisms to address these issues.

4.5.2.5 IEEE 802.11a/b/g PHY Comparison

Table 4.3 – Comparison among IEEE 802.11a/b/g

Characteristics	802.11a	802.11b	802.11g
Operating frequencies	5 GHz U-NII/ISM bands	2.4 GHz ISM band	2.4 GHz ISM band
Modulation techniques	OFDM	Barker Code/CCK	Barker Code/CCK/OFDM
Data rates (Mbps)	6, 9, 12, 18, 24, 36, 48, 54	1, 2, 5.5, 11	1, 2, 5.5, 11 6, 9, 12, 18, 24, 36, 48, 54
Slot time (µs)	9	20	20 9 (optional)
Preamble	OFDM	Long Short (optional)	Long/Short/OFDM

Table 4.3 summarizes the main differences between 802.11a/b/g WLAN systems. The 5 GHz band, employed by IEEE 802.11a, has received considerable attention, but a shorter wavelength is a drawback. Higher-frequency signals will have more trouble propagating through physical obstructions encountered in an office (walls, floors, and

furniture) than those at 2.4 GHz. An advantage of 802.11a is its intrinsic ability to handle delay spread or multipath reflection effects. The slower symbol rate and placement of significant guard time around each symbol, using a technique called cyclical extension, reduces the ISI (inter symbol interference) caused by multipath interference. (The last one-quarter of the symbol pulse is copied and attached to the beginning of the burst. Due to the periodic nature of the signal, the junction at the start of the original burst will always be continuous.). In contrast, 802.11b/g networks are generally range-limited by multipath interference rather than the loss of signal strength over distance.

Table 4.4 – Maximum transport level throughput in 802.11a/b/g

Technology	Maximum number of non-interfering channels	Maximum link rate (Mbps)	Theoretical maximum TCP rate (Mbps)	Theoretical maximum UDP rate (Mbps)
802.11b	3	11	5.9	7.1
802.11g (with 802.11b)	3	54	14.4	19.5
802.11g only	3	54	24.4	30.5
802.11a	19	54	24.4	30.5

When it comes to deployment of a wireless LAN, operational characteristics have been compared to those of cellular systems, where frequency planning of overlapping cells minimizes mutual interference, supports mobility as well as provides seamless channel handoff. The three non-overlapping frequency channels available for both IEEE 802.11b and 802.11g are at a disadvantage with respect to a greater number of channels available to 802.11a (over thirteen in US and up to nineteen in Europe depending on local regulations). The additional channels allow more overlapping access points within a given area while avoiding additional mutual interference and increasing the aggregate network capacity and the number of supported users.

The operating frequency is another important issue. Except for 802.11a, both 802.11b and 802.11g operate in the crowded 2.4 GHz band used by several others equipment such as Bluetooth devices, microwaves, cordless phones, garage door openers, and so on. This is seen as a major drawback, especially with regards to 802.11g which can have its throughput degraded by other co-channel technologies.

Another complicating factor for 802.11g is the backward interoperability requirement with 802.11b devices. To illustrate this point, consider Table 4.4 which depicts the maximum achievable throughputs for TCP and UDP for 802.11a/b/g for one particular setup [Atheros2003]. As we can see, the absence of 802.11b devices (802.11g-only environment) 802.11g throughput is equivalent to 802.11a. However, in a mixed mode of 802.11b/g environment, 802.11g devices have to adjust some properties (e.g., the slot time as shown in Table 4.3) as to be compatible to co-located 802.11b stations. This effectively reduces the data rate as 802.11g stations are now limited by slower 802.11b stations. In the worst case scenario, 802.11g performance may be as low as the slowest 802.11b device in the network. IEEE 802.11a, on the other hand, is not at all impacted as it operates in the 5 GHz band.

4.5.3 The MAC Layer

The responsibility of a MAC protocol is the arbitration of accesses to a shared medium among several end systems. In IEEE 802.11 this is carried out via an Ethernet-like stochastic and distributed mechanism: CSMA/CA [Agrawal2002]. The IEEE 802.11 protocol defines a multiple access network where all the devices using the same frequencies have to compete with each other to get access to the medium (the wireless channel). IEEE 802.11 specifies two medium access control protocols, Point Coordination Function (PCF) and Distributed Coordination Function (DCF). DCF is a fully distributed scheme which enables the ad hoc networking capabilities, whereas PCF is an optional centralized scheme built on top of the basic access method DCF of Figure 4.4.

4.5.3.1 The Hidden Terminal Problem

IEEE 802.11 MAC addresses this problem by adding two additional frames, the RTS and CTS discussed earlier. Here, the source sends a RTS and the destination replies with a CTS. Nodes overhearing the RTS and CTS suspend their transmissions for a specified time indicated in the RTS/CTS frames, as illustrated in Figure 4.11.

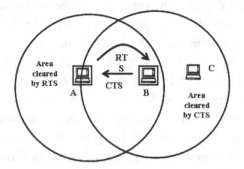

Figure 4.11 – RTS/CTS solve the hidden terminal problem

In the source station, a failure in the RTS/CTS handshake causes the RTS frame to be retransmitted and is treated as a collision. To prevent the MAC from being monopolized by consecutive retransmissions, there are retry counters and timers to limit the lifetime of a frame. RTS/CTS mechanism can be disabled by an attribute in the IEEE 802.11 management information base (MIB). The value of *dot11RTSThreshold* attribute defines the length of a frame that is required to be preceded by RTS and CTS frames. If the frame size is larger than this threshold, RTS/CTS are employed; otherwise the frame can be directly transmitted. The default value for this threshold is 128 octets. In addition, the RTS/CTS handshake can be disabled in the following situations:

- Low demand for bandwidth.
- Stations are concentrated in an area with all of them able to hear the transmissions of every other station.
- There is not much contention for the channel.

Note that for all practical purposes, an AP is heard by all stations in its BSS and will never have a hidden node.

4.5.3.2 The Retry Counters

There are two retry counters associated with every frame the MAC attempts to transmit: a short retry counter and a long retry counter.

The former is associated with short frames (i.e., frames with size less than *dot11RTSThreshold*), while the latter controls long frames. In addition to the counters, a lifetime timer is associated with every transmitted MAC frame. With this information, the MAC determines whether to cancel the frame's transmission and, hence discard it. Upon an unsuccessful transmission, the corresponding counter is incremented according to the frame size. When they reach the threshold defined in the MIB (i.e., *dot11ShortRetryLimit* and *dot11LongRetryLimit*), the frame is discarded.

4.5.3.3 Time Intervals

The IEEE 802.11 standard published in 2007 includes five time intervals through which both the DCF and PCF are implemented. Out of these five, two are defined by the PHY layer and the remaining by the MAC layer and are:

- The slot time defined in the PHY layer.
- The short interframe space (SIFS), defined by the PHY layer.
- The priority interframe space (PIFS).
- The distributed interframe space (DIFS).
- The extended interframe space (EIFS).

Basically, IFSs provide priority levels for channel access. In the IEEE 802.11b standard, the SIFS is the shortest interval (equal to 10µsec), followed by the slot time which is slightly longer (equal to 20µsec). The PIFS is equal to SIFS plus one slot time. The DIFS is equal to the SIFS plus two slot times. The EIFS is much larger than any of the other intervals. It is used by a station to set its NAV (Network Allocation Vector) when it receives a frame containing errors, allowing the possibility for ongoing MAC frame exchange to complete before the next transmission attempt. It is important to note that these values may change from standard to standard. For instance, in 802.11a the slot time value has been decreased (now equal to 9µsec), thereby supporting higher data rates. On the other hand, the IEEE 802.11g standard can use both IEEE 802.11a/b values for the slot time given its backward compatibility requirements.

4.5.3.4 Ranges and Zones

To understand the MAC operation and use of the various time intervals, it is of paramount importance to define terms such as transmission range, carrier-sensing range, carrier-sensing zone, and interfering range as depicted in Figure 4.12:

- **Transmission range**: This represents the range within which a MAC frame can be successfully received, provided there are no collisions at the receivers.

- **Carrier-sensing range**: The range within which a transmission can be detected is termed as carrier-sensing range. The *Carrier-sensing Zone (C-Zone)* is defined as the area where a signal can be detected, but it cannot be decoded hence indicating a busy medium. This is always larger than the transmission range, and may be more than two times its size [Kamerman1997, Sobrinho1999, Xu2001]. Given a particular transmit power level, the size of transmission and carrier-sensing range are often fixed (this is true at least for most existing Wireless LAN cards).

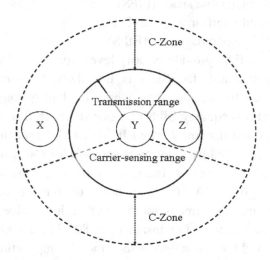

Figure 4.12 – Ranges and zones in IEEE 802.11

- **Interfering range**: This represents the range within which a node in receiving mode can be interfered by another transmission, leading to a collision at receiver. The interfering range depends upon many factors including the distance between transmitter and receiver, on the power level at which the packet is transmitted and also on the number of transmissions going on in a node's neighborhood [Cesana2003]. Hence, the size of interfering range may vary.

Figure 4.12 shows the transmission range, carrier-sensing range, and the C-Zone with respect to node Y. While in reality these ranges may not be circular, they are often assumed to be circular for the purpose of illustration. When node Y transmits a packet, node Z can receive it and decode it correctly since it is within node Y's transmission range. However, node X can only sense the signal and does not decode it correctly because it is located within node Y's C-zone. Note that we do not depict the interfering range in Figure 4.12 as it something that can vary significantly. Therefore, it cannot be easily visualized but can, however, be mathematically defined [Xu2002].

4.5.3.5 The Distributed Coordination Function (DCF)

DCF in IEEE 802.11 conducts two forms of carrier sensing: physical (by listening to the wireless shared medium) and virtual which uses the duration field included in the headers of RTS and CTS frames. These frames are utilized to set a station's NAV and can be used to determine the time when the source node would receive an ACK frame from the destination node. Using the duration information (see 4.5.3.7), nodes update their NAVs and the channel is considered to be busy if either physical or virtual carrier sensing (by the NAV) so indicates. Whenever NAV is zero, a station may transmit if the physical sensing allows. The area covered by the transmission range of the sender and the receiver is reserved for the data transfer, and hence other nodes cannot initiate transmission while communication is in progress. Given this fact, this region is hereby referred to as *silenced region*. By using the RTS and CTS handshake to silence the nodes in the silenced region, IEEE 802.11 is able to overcome, although not completely [Jung 2002] the hidden terminal problem [Fullmer1997, Moh1998].

As we have seen before, the IEEE 802.11 MAC protocol uses a backoff mechanism to resolve channel contention. When one station wants to send a frame, it senses the medium. If the medium is found idle for more than a DIFS period, then the frame can be transmitted. Otherwise, the transmission is deferred and the station uses an Exponential Random Backoff Mechanism by choosing a random backoff interval from [0, CW], where CW is called contention window. When the backoff counter reaches zero, the station transmits its frame. If collision occurs, the station doubles its CW, chooses a new backoff interval and tries retransmission. At the first transmission attempt, $CW=CW_{min}$ and is doubled at each retransmission up to CW_{max} as depicted in Figure 4.

This basic access mechanism of IEEE 802.11 can be extended by the RTS/CTS frame exchange, which reserves the channel before data transmission. When a station wants to send a frame with a size above a specified threshold (*dot11RTSThreshold*), it first sends a short control frame RTS to the destination station. After SIFS, the destination sends another short control frame CTS back to the source. The source then transmits its DATA frame after SIFS period, being sure that the channel is reserved for itself during all the frame duration. Both RTS and CTS frames carry the duration needed by the station and thus inform all stations how long the channel will be used. After the destination receives the DATA, it sends an ACK back to the source after SIFS period.

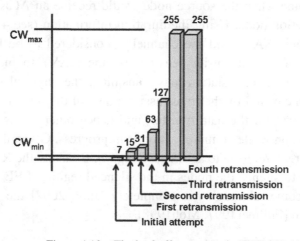

Figure 4.13 – The backoff procedure in IEEE 802.11

Figure 4.14 gives an example of channel access in the IEEE 802.11 MAC, showing how nodes within the transmission range and C-zone adjust their NAVs during RTS-CTS-DATA-ACK transmission. From this figure, we can see that nodes in transmission range correctly set their NAVs when receiving RTS or CTS. However, since nodes in the C-zone cannot decode the packet, they do not know the duration of the packet transmission. To prevent a collision with the ACK reception at the source node, nodes within the C-zone set their NAVs for the EIFS duration.

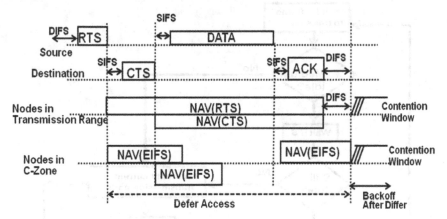

Figure 4.14 – Nodes in the transmission range and C-zone set their NAVs differently

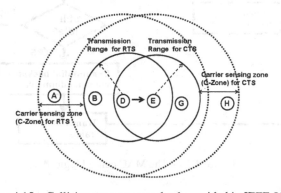

Figure 4.15 – Collisions are not completely avoided in IEEE 802.11

It is worth noticing that IEEE 802.11 does not completely prevent collisions due to hidden terminal, that is, nodes in the receiver's

C-zone but not in the sender's C-zone or transmission range, can cause a collision with the reception of a DATA packet at the receiver. For example, in Figure 4.15, suppose node D transmits a packet to node E. After the RTS-CTS handshake between nodes D and E, A and H will set their NAVs for EIFS duration. During D's DATA transmission, A defers its transmission because it senses D's DATA transmission. However, since node H does not sense any signal during D's DATA transmission, it assumes the channel to be idle (H is in E's C-zone, but not in D's).

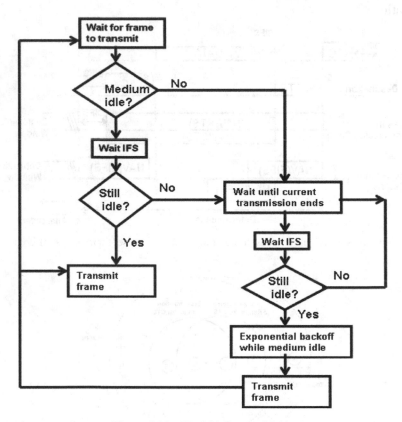

Figure 4.16 – The MAC control logic

Whenever H starts a new transmission, it can cause a collision with the reception of DATA at E. Since H is in E's C-zone, H can generate sufficient interference at node E to cause a collision with E's

DATA reception. Figure 4.16 depicts the overall MAC logic employed by the IEEE 802.11 and summarizes most of the discussion we have had so far.

4.5.3.6 The Point Coordination Function (PCF)

The PCF employs a poll and response protocol so as to eliminate the possibility of contention for the medium. Here, a point coordinator (PC) controls the medium access and is often co-located with the AP. In PCF, the PC maintains a polling list, and regularly polls the stations for traffic while also delivering traffic to the stations. The PCF is built over the DCF, and both of them operate simultaneously. However, the PCF uses PIFS instead of DIFS.

The PC begins a period of operation called the contention-free period (CFP), during which the PCF is operating. This period is called contention free because access to the medium is completely controlled by the PC. The CFP occurs periodically to provide a near-isochronous service to the stations, and alternates with a contention period where the normal DCF rules operate and all stations may compete for access to the medium. As per the standard, the contention period has to be long enough to contain at least one maximum length frame and its acknowledgement.

The CFP begins when the PC gains access to the medium, by using the normal DCF procedures and transmitting a Beacon frame which is required to be sent out periodically for the PC to compete for the medium. The traffic in the CFP consists of frames sent from the PC to one or more stations followed by the corresponding acknowledgements. In addition, the PC sends a contention-free-poll (CF-Poll) frame to those stations that have requested contention-free service. If the polled station has data to send, it responds to CF-Poll. For efficient medium utilization, it is possible to piggyback the acknowledgement and the CF-Poll onto data frames.

During the CFP, the PC ensures that the interval between transmissions to be no longer than PIFS so as to prevent a station operating under the DCF from gaining access to the medium. The NAV is what prevents stations from accessing the medium during the CFP.

The transmitted Beacon frames contain the information about the maximum expected length of the CFP. Finally, the PC announces the end of the CFP by transmitting a contention-free end (CF-End) frame. This frame resets the NAVs and stations begin DCF operation independently.

4.5.3.7 Framing

The MAC layer accepts MAC Service Data Units (MSDUs) from higher layers and adds appropriate headers and trailers to create MAC Protocol Data Units (MPDU). Optionally, the MAC may fragment MSDUs into several frames, hence attempting to increase the probability of each individual frame being delivered successfully. Table 4.5 depicts the general IEEE 802.11 MAC frame format [IEEE802.112007], where the sizes of the corresponding fields are in bytes. Altogether, the MAC frame comprises of ten fields:

Table 4.5 – The IEEE 802.11 MAC frame format

Octets: 2	2	6	6	6	2	6	2	0- 2324	4
Frame Control	Duration /ID	Address 1	Address 2	Address 3	Sequence Control	Address 4	QoS Control	Frame Body	FCS

- Frame control: this is comprised of 11 subfields, which encapsulate information such as protocol version, frame type and subtype, whether the frame is encrypted, power management and so on. Examples of frame type include control frames such as RTS and CTS and management frames.
- Duration/ID: the most common use of this field is to update the NAV at a station and is a key to the entire operation of the basic 802.11 access mechanism (DCF).
- Addresses: there are a total of four address fields, which depending on the situation can indicate: i) BSS Identifier (BSSID), which is the unique identifier for a particular BSS; ii) Transmitter Address (TA), which is the MAC address of the station that has transmitted the frame; iii) Receiver Address (RA), which is the MAC address of the destination node to which the frame has been sent; iv) Source Address (SA), which is the MAC address of the station that

originated the frame and can be different from the TA; and v) Destination Address (DA) which can be different than the RA because of the indirection.

- Sequence control: this field comprises of a 4-bit fragment number and a 12-bit sequence number. It allows a receiving station to eliminate duplicate received frames.
- QoS Control identifies the traffic category (TC) or traffic stream (TS) to which the frame belongs and various other QoS-related information about the frame that varies by frame type and subtype.
- Frame body is a variable length field which contains the information specific to the particular data or management frame.

Frame check sequence: generated using a CRC-32 polynomial.

4.5.4 Security

The first security scheme provided in the series of IEEE 802.11 standards is Wired Equivalent Privacy (WEP), specified as part of the 802.11b Wi-Fi standard and was originally designed to provide security level similar to wired LANs. The latter enjoys security and privacy due to the physical security mechanisms such as building access control. Unfortunately, physical security mechanisms do not prevent eavesdropping and unauthorized access in case of wireless communications. Therefore, WEP aimed at covering the lack of physical security akin to WLANs with security mechanisms based on cryptography. WEP suffers from various design flaws and some exposure in the underlying cryptographic techniques that seriously undermine its security claims. An amendment to 802.11 known as 802.11i was introduced, which has substantially improved on the security of the 802.11 standard.

4.5.4.1 WEP Security Mechanisms

WEP security mechanisms include simultaneous handling of data encryption and integrity for each frame as shown in Figure 4.17. First, an integrity check value (ICV) of the payload is computed using a cyclic redundancy check (CRC). The cleartext payload concatenated with the ICV is then encrypted using a bit-wise Exclusive-OR operation with a keystream. The keystream is a pseudorandom bit stream generated by the RC4 [Schneier1996] algorithm from a 40-bit secret key pre-appended with a 24-bit Initialization Value (IV). The resulting protected frame includes the cleartext frame header, the cleartext IV, the result of the encryption and a cleartext frame check sequence field.

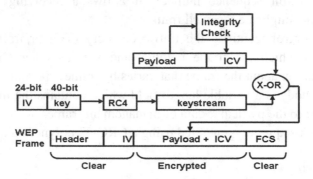

Figure 4.17 – WEP frame security mechanisms

The recipient of a WEP frame first generates keystream with RC4 using shared secret key and the IV value retrieved from the received frame. The resulting keystream is Exclusive-ORed with the encrypted field of the frame to decrypt the payload and the ICV. The integrity of the payload is then checked by comparing the integrity check computed on the cleartext payload with the ICV resulting from the decryption. The secret key can either be a default key shared by all the devices of a WLAN or a pair-wise secret shared only by two communicating devices. Since WEP does not provide any support for the exchange of pair-wise secret keys, the secret key must be manually installed on each device.

4.5.4.2 WEP Security Flaws

WEP suffers from many design flaws and some weaknesses in the way the RC4 cipher is used [Fluhrer2001, Stubblefield2001]. Data encryption in WEP is based on an approximation of the "one-time pad" [Schneier1996] algorithm that can guarantee perfect secrecy under some circumstances. Like WEP encryption, one-time pad encryption consists of the bit-wise Exclusive-OR between a binary plaintext message and a binary keystream as long as the message. The secrecy of the resulting cipher text is perfect, provided that each new message is encrypted with a different secret random keystream. The secrecy is not guaranteed when the keystream is re-used or its values can be predicted. Hence, a first class of attacks on WEP exploits possible weaknesses in WEP's keystream generation process that makes the secret keystream easily predictable or causes its re-use.

The first type of exposure is due to the likelihood of keystream re-use between a pair of communicating devices. Using the same secret key, the only variation in the input to the keystream generator is due to the variation in the IV. Since the IV is 24-bit value sent in a cleartext, the re-use of a keystream can be easily detected. The re-use of a keystream is also very likely because of the small set of possible IV values that can be exhausted in a few hours for busy traffic between two nodes. This type of exposure gets even worse if some care is not taken during the implementation of the standard: some products set the IV to a constant value (0 or 1) at the initialization of the encryption process for each frame sequence. The second type of exposure is due to the use of a 40-bit secret that is highly vulnerable to exhaustive search.

WEP data encryption is also exposed through an advanced attack that takes into account the characteristics of the RC4 algorithm [Stubblefield2001] and drastically reduces the set of possible keystream values based on the attacker's ability to recover the first byte of encrypted WEP payload. Another class of exposure on WEP is the data integrity mechanism using CRC in combination with one-time pad encryption using exclusive-or operation and is transparent with respect to modifications. As opposed to a cryptographically secure hash function, an integrity check computed with CRC yields predictable changes on the ICV with respect to single-bit modifications on the input message. Combining the transparency of Exclusive-OR with the predictable modification property of CRC, an attacker can flip bits on well-known positions of an encrypted WEP payload and on the corresponding positions of the encrypted ICV, so that the resulting cleartext payload is modified without the modification being detected by the recipient. It should be noted that the transparent modification of the WEP payload does not require the knowledge of the secret payload value since the attacker only needs to know the location of some selected fields in the payload to force the tampering of their value. Finally, another weakness of WEP is the lack of key management that could lead to a potential exposure for most attacks, exploiting manually distributed secrets shared by large populations.

4.5.4.3 The IEEE 802.11i Amendment: A New Security Scheme

To address the shortcomings of WEP, the IEEE 802.11 WG set up a special Task Group I (TGi) in charge of designing new security architecture as a part of the amendment called 802.11i. IEEE 802.11i, also known as Wi-Fi Protected Access (WPA) [WPAwww], proposes a long-term architecture based on the IEEE 802.1x standard, which itself is based on the IETF's Extensible Authentication Protocol (EAP). IEEE 802.1x has a flexible design supporting various authentication modes. The IEEE 802.11i consists of three major parts: Temporal Key Integrity Protocol (TKIP), counter mode cipher block chaining with message authentication codes (counter mode CBC-MAC) and IEEE 802.1x access control. TKIP primarily addresses the shortcomings of WEP and fixes its well-known problems, including small IV and short encryption keys. TKIP uses RC4, the same encryption algorithm as WEP to make it updateable from WEP, but it extends the IV from 24-bit to 48-bit in order to defend against the existing cryptographic attacks against WEP. Moreover, to cope up with brute force attacks, TKIP implements 128-bit encryption key to address the short-key problem of WEP. TKIP changes the way keys are derived and periodically rotates the broadcast keys so as to avoid the attack that is based on capturing large amount of data encrypted by the same key. It also adds a message-integrity-check function to prevent packet forgeries.

Counter mode CBC-MAC is designed to provide link layer data confidentiality and integrity. A new strong symmetric encryption standard, advanced encryption standard (AES) is deployed in which a 128-bit encryption key and 48-bit IV are used. AES is a block cipher where chunks (multiple bytes) of data are encrypted at once, as opposed to a stream cipher (like RC4) which handles encryption in order as the bits go through. With chunks of data encrypted at once, data is diffused within the block after encryption (rather than being allocated in a linear fashion, as in RC4) and hence it becomes much more difficult to predict the location of specific data within the encrypted stream. Contrary to TKIP, counter mode CBC-MAC has little resemblance to WEP and it is set to be a part of the second-generation WPA standard.

IEEE 802.1x is an authentication and key management protocol, which is designed for wired LANs, but has been extended to Wireless LANs. IEEE 802.1x authentication occurs when a client first joins a network. Then, authentication periodically recurs to verify the client has not been subverted or spoofed. IEEE 802.1x is a centralized, server-based authentication process where a MH sends an authentication request to an associated AP. The AP forwards the authentication information to a back-end authentication server via Remote Authentication Dial-In User Service (RADIUS) for verification. Once the verification process completes, the authentication server sends a response message to the AP that the client has been authenticated and network access should be granted. In 802.11i, the response message should contain the cryptographic keys sent to the client. After that, the AP transfers the mobile client to the authenticated state, hence allowing network access.

IEEE 802.1x is not a single authentication method. Rather, it utilizes EAP as its authentication framework. This means that 802.1x-enabled switches and APs can support a wide variety of authentication methods, including certificate-based authentication, smartcards, token cards, one-time passwords, etc. However, the 802.1x specification itself does not mandate any authentication methods. Since switches and APs act as a "pass through" for EAP, new authentication methods can be added without the need to upgrade the switch or AP, by adding software on the host and backend authentication server. Several common EAP methods have been defined in various IETF drafts or other industry documents, such as EAP-MD5, EAP-TLS, etc. While TKIP and counter mode CBC-MAC are still not implemented by most vendors, 802.11 x supports is already integrated into some operating systems.

In summary, TKIP/WPA provides enhanced security for existing infrastructure. Counter mode CBC-MAC protects the data integrity and confidentiality, and 802.11x presents a fully extensible authentication mechanism. Combining these techniques, 802.11i is significantly stronger than WEP. On the other hand, 802.11i requires changes to firmware and software drivers and may not be backward-compatible with some legacy devices and operating systems. A phased adoption process is anticipated because of the large amount of installed 802.11 devices.

4.5.5 The IEEE 802.11e MAC Protocol

The IEEE 802.11e [IEEE802.11e2001] is an amendment to the 802.11 standard for QoS provisioning. It prioritizes the radio channel access within a BBS of the IEEE 802.11 WLAN. A BSS that supports the new priority schemes of the 802.11e is referred to as QoS supporting BSS (QBSS). In order to effectively support QoS, the 802.11e MAC defines the Hybrid Coordination Function (HCF), which enhances the DCF mode in the IEEE 802.11 standard (see Figure 4.4).

The HCF is comprised of two parts: the Extended Distributed Channel Access (EDCA) and the HCF Controlled Channel Access (HCCA). Stations operating under the 802.11e are called QoS stations, and a QoS station which works as the centralized controller for all other stations within the same QBSS is called the *Hybrid Coordinator (HC)*. A QBSS is a BSS which includes an 802.11e-compliant HC and QoS stations. The HC typically resides within an 802.11e AP. In the following, we refer to an 802.11e-compliant QoS station simply as a station. Similar to DCF, the EDCA is the contention-based channel access mechanism of HCF. With 802.11e, there may still be the two phases of operation within a superframe, i.e., a CP and a CFP, which alternate over time continuously. The EDCA is used in the CP only while the HCF is used in both phases, thereby making this new coordination function hybrid.

4.5.5.1 The EDCA

The EDCA in 802.11e is the foundation for the HCF and the QoS is supported by the introduction of Traffic Categories (TCs). MSDUs are now delivered through multiple backoff instances within one station, wherein each backoff instance parameterized with TC-specific parameters. In the CP, each TC within the station contends for a transmission opportunity (TXOP) and independently starts a backoff after detecting the channel being idle for an Arbitration Interframe Space

(AIFS), which is at least equal to DIFS and can be increased individually for each TC. After waiting for AIFS, each backoff sets a counter to a random number drawn from the interval $[1; CW+1]$. The minimum size $(CW_{min}[TC])$ of the CW is another parameter dependent on the TC. Priority over legacy stations is provided by setting $CW_{min}[TC] < 15$ (in case of 802.11a PHY) and $AIFS = DIFS$. Figure 4.18 depicts the EDCA parameters.

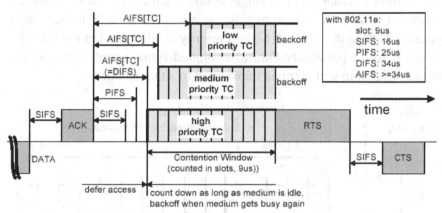

Figure 4.18 – Multiple parallel backoffs of MSDUs with different priorities

As in the legacy DCF, when the medium is determined busy before the counter reaches zero, the backoff has to wait for the medium being idle for AIFS again, before resuming the count down process. A major difference from the legacy DCF is that when the medium is determined as being idle for the period of AIFS, the backoff counter is reduced by one beginning the last slot interval of the AIFS period. Note that with the legacy DCF, the backoff counter is reduced by one beginning the first slot interval after the DIFS period.

After any unsuccessful transmission attempt, a new CW is calculated with the help of the persistence factor (PF), $PF[TC]$, and another uniformly distributed backoff counter out of this new, enlarged CW is drawn, so that the probability of a new collision is reduced. While in legacy 802.11, the CW is always doubled after any unsuccessful

transmission, 802.11e uses the PF to increase the CW differently for each
TC and is given by:

$$newCW \quad [TC \ ] \geq ((\ oldCW \quad [TC \ ] + 1) \cdot PF \ ) - 1$$

The CW never exceeds the parameter $CW_{max}[TC]$, which is the
maximum possible value for CW. A single station may implement up to
eight transmission queues realized as virtual stations inside a station,
with QoS parameters that determine their priorities. If the counters of
two or more parallel TCs in a single station reach zero at the same time, a
scheduler within the station avoids the *virtual collision*. The scheduler
grants TXOP to the TC with highest priority as illustrated in Figure 4.19.
There is still a possibility that a transmitted frame could collide at the
wireless medium with a frame transmitted by other stations.

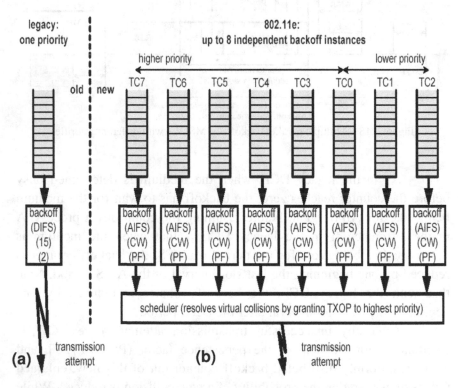

Figure 4.19 – Virtual backoff of eight Traffic Categories. (a) Legacy DCF, close to
EDCA with AIFS=34 μs , CWmin=15, PF=2; (b) EDCA with AIFS[TC]≥34 μs ,
CWmin[TC]=0-255, PF[TC]=1-16

Another important part of the 802.11e MAC is the *TXOP*. A TXOP is an interval of time when a station has the right to initiate transmissions, defined by a starting time and a maximum duration. TXOPs are acquired via contention (EDCA-TXOP) or granted by the HC via polling (polled TXOP). The duration of an EDCA-TXOP is limited by a QBSS-wide *TXOP* limit distributed in beacon frames, while polled TXOP is specified by the duration field inside the poll frame. However, although the poll frame is a new frame as part of the 802.11e, the legacy stations also set their NAVs upon receiving this frame. The prioritized channel access is realized with the QoS parameters per TC, which include AIFS[TC], CW_{min}[TC], and PF[TC], while CW_{max}[TC] is optional. In addition, there are discussions to introduce priority dependent EDCA-TXOP[TC]. The QoS parameters can be adapted over time by the HC, and is announced periodically via beacon frames. Protocol-related parameters are included in the beacon frame, which is transmitted at the beginning of each superframe.

4.5.5.2 The HCCA

HCCA extends the EDCA access rules. HC may allocate TXOPs to itself in order to initiate MSDU Deliveries whenever it desires, however, only after detecting the channel as being idle for PIFS (which is shorter than DIFS). To give the HC priority over the EDCA, AIFS must be longer than PIFS and can therefore not have a value smaller than DIFS.

During CP, each TXOP begins either when the medium is determined to be available under the EDCA rules, i.e., after AIFS plus backoff time, or when the station receives a special poll frame, the QoS CF-Poll from the HC. The QoS CF-Poll from the HC can be sent after a PIFS idle period without any backoff. Therefore, the HC can issue polled TXOPs in the CP using its prioritized medium access. During the CFP, the starting time and maximum duration of each TXOP is specified by the HC, which uses the QoS CF-Poll frames. Stations will not attempt to access the medium on its own during the CFP, so only the HC can grant TXOPs by sending QoS CF-Poll frames. The CFP ends after the time announced in the beacon frame or by a CF-End frame from the HC.

As a part of 802.11e, an additional random access protocol that allows fast collision resolution is defined. The HC polls stations for MSDU Delivery. For this, the HC requires information that has to be updated by the polled stations from time to time. Controlled contention is a way for the HC to learn which station needs to be polled, at what time, and for how much duration. The controlled contention mechanism allows stations to request the allocation of polled TXOPs by sending resource requests, without contending with other EDCA traffic. Each instance of controlled contention occurs during the controlled contention interval, which is started when the HC sends a specific control frame. This control frame forces legacy stations to set their NAV until the end of the controlled contention interval, thereby remaining silent during the controlled contention interval. The control frame defines a number of controlled contention opportunities (i.e., short intervals separated by SIFS) and a filtering mask containing the TCs in which resource requests may be placed. Each station with queued traffic for a TC matching the filtering mask chooses one opportunity interval and transmits a resource request frame containing the requested TC and TXOP duration, or the queue size of the requested TC. For fast collision resolution, HC acknowledges reception of request by generating a control frame with a feedback field so that requesting stations can detect collisions during controlled contention.

4.5.6 A Glimpse on Past and Present IEEE 802.11 Efforts

In addition to the various amendments to 802.11 discussed above, many others exist that extend the basic capabilities of the baseline 802.11 standard such as adding physical layer options, improving security, adding quality of service (QoS) features or providing better inter-operability. Table 4.6 describes the past (completed) and present (ongoing) efforts in the IEEE 802.11 WG in order to offer improved services for WLAN users. Up-to-date information on IEEE 802.11 efforts can always be found at [IEEE802.11www].

4.6 Enhancements to IEEE 802.11 MAC

Ever since the IEEE 802.11 standard has been released, research community has been working on enhancing both its PHY and MAC layers. However, research on the MAC layer seems to have been more intensive than in the PHY layer. In this section, we outline major recent and prominent advancements done in the MAC layer of IEEE 802.11.

Table 4.6 – Past and present IEEE 802.11 activities

Amendment	Status	Goal
802.11a	Completed	OFDM PHY in the 5GHz band with rates up to 54 Mbps.
802.11b	Completed	DSSS PHY in the 2.4GHz band with rates up to 11Mbps.
802.11d	Completed	Specifies MAC extensions for worldwide operation of 802.11 networks.
802.11e	Completed	Specifies MAC extensions to provide QoS support for WLAN applications such as video and voice.
802.11f	Completed	Recommended practice document that enables achieving interoperability between a multi-vendor WLAN networks.
802.11g	Completed	OFDM PHY in the 2.4GHz band with rates up to 54Mbps. Backward compatible with 802.11b.
802.11h	Completed	Specifies MAC extensions for so as to comply with European regulations for 5 GHz WLANs, which requires products to have transmission power control (TPC) and dynamic frequency selection (DFS).
802.11i	Completed	Specifies MAC extensions to improve the security of 802.11 networks. It is widely used in today's Wi-Fi products.
802.11j	Completed	Enables operation of 802.11 networks in Japan by complying with Japanese regulatory rules.
802.11k	Completed	Defines Radio Resource Measurement enhancements to provide mechanisms to higher layers for radio and network measurements.
802.11n	Completed	Defines extensions to the PHY and MAC for both 2.4GHz and 5GHz bands, supporting data rates of at least 100Mbps.
802.11p	Completed	Expands on conventional 802.11 wireless networking to allow for provisions that are specifically useful to automobiles, including communications in the automotive-allocated 5.9 GHz spectrum.
802.11r	Completed	Defines mechanisms speeding up the handoff between wireless access points, and is also known as "Fast BSS Transition".
802.11s	Ongoing	Developing an amendment to support extended set of service for wireless mesh networks.
802.11t	Completed	Defines a recommended practice for the evaluation of 802.11 wireless performance

Amendment	Status	Goal
802.11u	Completed	Specifies how 802.11 networks can work with other external networks, such as 3G/4G.
802.11v	Completed	Specifies techniques that allow wireless network management of client stations.
802.11w	Completed	Defines enhancements to the MAC that enable data integrity, data origin authenticity, replay protection, and data confidentiality for management frames.
802.11y	Completed	PHY and MAC extensions to support operation in the 3650–3700 MHz band in the US.
802.11z	Completed	MAC extensions defining a new mechanism known as Tunneled Direct Link Setup (TDLS), which enables client stations to communicate with each other directly.
802.11aa	Ongoing	MAC extensions for robust audio/video streaming in overlapping BSS environments.
802.11ac	Ongoing	Extensions to both PHY and MAC in the 5 GHz band to provide a maximum multi-station throughput of at least 1 Gbps and a maximum single link throughput of at least 500 Mbps.
802.11ad	Ongoing	Extensions to both PHY and MAC in the 60 GHz band to provide a maximum of at least 1 Gbps. The current draft standard supports a maximum data rate of 7 Gbps.
802.11ae	Ongoing	Defines mechanisms for prioritizing management frames using existing mechanisms for medium access.
802.11af	Ongoing	Defines modifications to both the PHY and MAC to meet the legal requirements for channel access and coexistence in the TV White Space (see Chapter 8).

4.6.1 Power Control

Power control is a determinant technique for energy conservation and thus is of fundamental importance for wireless ad hoc stations which rely on batteries. Besides energy saving, power control can also increase effective capacity of the network by enhancing spatial reuse of the wireless channel. Current research on power control MAC protocols focus on suitably varying transmit power in order to reduce energy consumption [Agarwal2001, Gomez2001, Jung2002, Wieselthier2000].

The power control strategies may be classified based upon the presence or absence of asymmetric links between nodes. In the context of IEEE 802.11 networks, link symmetry is assumed in its design while communication in asymmetric networks has been shown to be a relatively hard task [Narayanaswamy2002]. In the following subsections, we discuss two power control MAC protocols for use in MANETs.

4.6.1.1 The BASIC Protocol

As mentioned earlier, although power control can reduce energy consumption, it can also lead to asymmetry between nodes, that is, a given node A can reach a node B but node B cannot reach A. This is the result of the use of different power levels at different nodes. A clear drawback of this asymmetry is that it may result in increased number of collisions where certain nodes cannot sense ongoing low power transmissions, and hence transmit at a higher power level which ends up colliding with the current low power transmissions. Based on IEEE 802.11, the BASIC scheme aims at addressing this asymmetry issue by transmitting the RTS and CTS packets at maximum possible power level (p_{max}), while transmitting DATA and ACK at lowest power level needed to communicate ($p_{desired}$) [Agarwal2001, Gomez2001]. In this scheme, the RTS-CTS exchange is utilized to decide the transmission power for subsequent DATA and ACK packets.

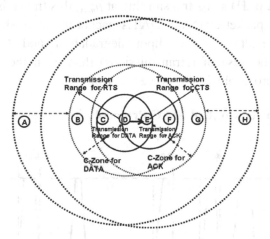

Figure 4.20 – The BASIC protocol

In this BASIC scheme, when nodes receive either a RTS or CTS packet (always transmitted at p_{max}), they set their NAVs for the duration of the DATA and ACK transmission. For instance, node D wants to transmit a packet to node E in Figure 4.20. When D and E transmit the RTS and CTS respectively, B and C receive the RTS, and F and G receive the CTS, so they will defer their transmissions for the duration of the D-E transmission. Since node A is in the C-zone of D, it will not be able decode the packets correctly but only sense the signal. Therefore, node A will set its NAV for EIFS duration whenever it senses the RTS transmission from D. Similarly, node H will set its NAV following CTS transmission from E.

In regular IEEE 802.11, when transmit power control is not used, the C-zone is the same for RTS-CTS and DATA-ACK since all packets are sent using the same power level (p_{max}). In the BASIC scheme, however, the transmission range for DATA-ACK is smaller than that of RTS-CTS whenever a source and destination pair decides to reduce the transmit power for DATA-ACK. Similarly, the C-zone for DATA-ACK is also smaller than that of RTS-CTS. Therefore, when nodes D and E in Figure 4.20 reduce their transmit power for DATA and ACK transmissions respectively, nodes A and H cannot sense the transmissions and thus consider the channel to be idle. When any of these nodes (A or H) starts transmitting at p_{max}, this transmission collides with the ACK packet at D and DATA packet at E. As shown in [Jung 2002], this results in throughput degradation and higher energy consumption (because of retransmissions) than even the regular IEEE 802.11 MAC protocol without power control.

4.6.1.2 The Power Control MAC Protocol

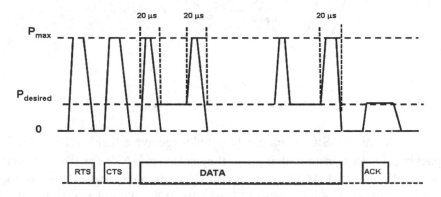

Figure 4.21 – Transmit power level transitions in PCM

To address the deficiency of the BASIC protocol, the PCM protocol has been proposed in [Jung2002]. Similar to the BASIC protocol, PCM transmits the RTS and the CTS packets at p_{max} and use the minimum power level (that is, $p_{desired}$) needed for communication for DATA and ACK. However, contrary to the BASIC scheme, eventual collisions with nodes in the C-zone are avoided by making the source node in a transmission periodically transmit the DATA packet at p_{max} so that nodes in the C-zone can sense the signal and set their NAVs accordingly.

Figure 4.21 illustrates the transmit power level transitions in PCM during a regular sequence of RTS-CTS-DATA-ACK transmission. As we can see, the source transmits the DATA at $p_{desired}$ and periodically employs p_{max}. In PCM, ACK packets are always transmitted at $p_{desired}$. With this modification to the BASIC protocol, nodes that can potentially interfere with the ongoing transmission will periodically sense the channel as busy, and defer their own transmission. Since nodes in the C-zone only defer for EIFS duration, the transmit power for DATA is increased once every EIFS duration. Also, the interval for which the DATA is transmitted at p_{max} should be larger than the time required for physical carrier sensing. In [Jung2002], it has been concluded that 20 μs is an appropriate value for such interval and is indicated in Figure 4.22. As we can see, PCM overcomes the deficiency of the BASIC scheme and can achieve throughput comparable to that of IEEE 802.11 [Jung2002], with less energy consumption. However, note that PCM, just like 802.11, does not prevent collisions completely. As discussed earlier, collisions with DATA being received by the destination can occur.

Despite of all this, PCM suffers from a drawback, namely, the inability to achieve spatial reuse (discussed in detail in the next section). Note that this shortcoming is also present in other existing power control MAC protocols, and is illustrated in Figure 4.23, where node A initiates a transmission to node B. A side effect of PCM strategy to periodically change the transmit power level during DATA transmission, is that the entire carrier-sensing range of node A is blocked. By blocked we mean that, in Figure 4.23, nodes C and D as well as nodes G and H that could eventually communicate with each other at such a power level that would not collide with the low power DATA and ACK transmission between A and B, are unable to do it so, given the periodic change in power level employed by node A. Considering that DATA transmission takes considerably longer (around two orders of magnitude [IEEE02.111997]) than RTS-CTS transmission, for the duration of the low power DATA-ACK transmission between nodes A and B no other communication can take place either in the transmission range or within the C-zone. Therefore, the increased channel capacity resulting from the low power DATA-ACK transmission cannot be reused. Note that, similar to PCM, this is also not possible in the BASIC scheme due to lack of coordination amongst nodes for channel spatial reuse, and because RTS and CTS are always transmitted at p_{max}.

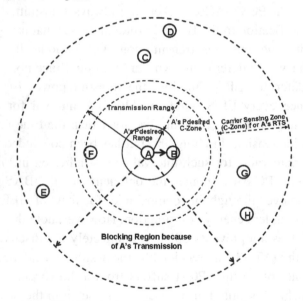

Figure 4.22 – PCM blocks the entire transmitter's carrier-sensing range

4.6.2 Spatial Reusability

In [Agrawal2003], the Spatial Reuse MAC (SRM) protocol is introduced which uses power control and employs a distributed form of transmission sneaking to accomplish appropriate spatial reuse of the channel. In other words, SRM tries to provide energy efficiency by employing power control while at the same time reusing the resulting additional channel capacity. SRM design is based on the fact that there are no specific mechanisms in existing power control MAC protocols that allows an efficient and coordinated reuse of the additional channel capacity resulted from using power control. These solutions either block the entire station's radio range as in PCM so as to prevent any collision, similar to the IEEE standard 802.11 [IEEE802.112007], or simply do not explore spatial reuse capability at all, and often consume higher energy than IEEE 802.11 without power control [Agarwal2001, Gomez2001, Jung2002, Karn1990, Pursley2000].

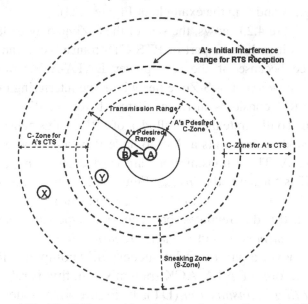

Figure 4.23 – The S-Zone and its relation to the other Ranges and Zones

The SRM protocol is similar to the BASIC scheme in that it transmits RTS and CTS at p_{max}, and DATA and ACK at $p_{desired}$. However, SRM implements a fully distributed *transmission sneaking technique* so as to enable channel spatial reuse. In short, transmission sneaking is a spatial reuse procedure by which a pair of nodes can communicate despite the ongoing transmission in its radio range, provided the low power sneaking DATA-ACK transmission does not collide with the ongoing data transmission. SRM is based on the concept of Sneaking Zone (S-Zone) depicted in Figure 4.23, where node B has a packet to send to node A. In this figure, if we assume that nodes B and A transmit RTS-CTS at full power (p_{max}) and DATA-ACK at $p_{desired}$, the S-Zone is defined as the area within the carrier sensing range of the RTS-CTS, where a transmission (called sneaking transmission) is possible without interfering with B-A's communication. It should be noted that this area is generally blocked in IEEE 802.11 as all the packets are transmitted at full power. PCM also blocks this zone due to its periodically increasing power level transmission. It can be shown that the size of the S-Zone is inversely proportional to the distance between the transmitter and receiver (e.g., A and B in the example of Figure 4.23).

As Figure 4.23 shows, the size of the S-Zone may be larger than the C-Zone as it includes a part of RTS-CTS transmission range, which becomes free because of the low power DATA-ACK transmission. Interestingly, now a part of S-Zone lies within the interfering range of A. Therefore, SRM considers different factors before starting a sneaking transmission. To illustrate the overall idea of SRM, let us consider Figure 4.25 where node A transmits a RTS to node B which, as a result, sends CTS back to A. These transmissions are carried out at p_{max}, while the DATA-ACK is transmitted at $p_{desired}$. Figure 4.24 depicts various ranges and zones of the RTS-CTS and DATA-ACK transmission between nodes A and B. SRM differentiates between two types of transmission: *Dominating Transmission* and *Sneaking Transmission*.

Whenever a pair of nodes successfully completes RTS-CTS handshake before a DATA-ACK transmission, this is referred to as the *Dominating Transmission* (DT). In Figure 4.25 nodes A and B have successfully completed RTS-CTS handshake and together with DATA-ACK transmission is called as the Dominating Transmission. Nodes A and B are called *Dominating Nodes* (DNs) of DT. Sneaking Transmission (ST): From Figure 4.25, we see that the pair of nodes C and D, E and F could eventually communicate with each other if they had the knowledge of the minimum power level required to communicate with each other. In this case, however, these pair of nodes cannot use RTS-CTS at p_{max} as they would collide with the current dominating transmission between A and B. The IEEE 802.11 standard allows   nodes   to   communicate   without   using   RTS-CTS

[IEEE802.112007] when the amount of data to be sent is less than the threshold *RTSThresh*. SRM utilizes this ability to directly transmit DATA without RTS-CTS in order to allow nodes C and D, E and F to communicate at low power despite the ongoing DT between nodes A and B. This transmission is defined as the *Sneaking Transmission* (ST), and the nodes involved as *Sneaking Transmitter* and *Sneaking Receiver*.

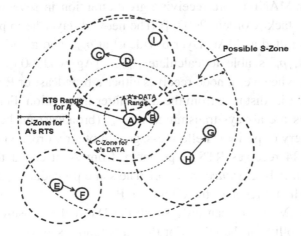

Figure 4.24 – Channel spatial reusability in the SRM protocol

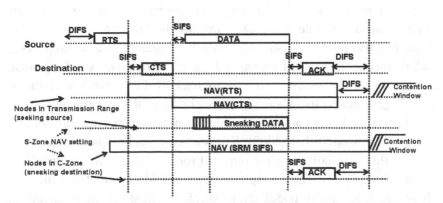

Figure 4.25 – Channel spatial Reusability in the SRM protocol (nodes within C-zone use SRM_EIFS for their NAVs)

In SRM, it is assumed that every node has access to a table with the minimum power level required to communicate with each of its neighbors. This may, however, be a reasonable assumption and has been considered [Agarwal2001]. One possible solution is to exchange Hello packets between neighboring nodes either at the MAC or at the network layer. Since many routing protocols already employ a form of hello packets to maintain network connectivity [Perkins2001], SRM follows a cross-layer design with the network layer assisting the MAC layer in the determination of various $p_{desired}$ amongst neighbor nodes (cross-layer solutions for MANETs are receiving great attention in several layers of the protocol stack [Cordeiro2002a]). The network layer hello packets are always transmitted as MAC layer broadcast at $p_{max,}$ and a node receiving this at level, p_r is able to calculate $p_{desired}$ [Agrawal2003]. With this information, whenever a node receives either a broadcast or RTS-CTS, it can determine its distance from the transmitter in question. For example, these packets are always transmitted at p_{max} (broadcast packets) during route discovery or periodic hello messages. In other words, when node B in Figure 4.24 receives RTS at power p_r from A, it can determine its distance from A by *distance* (p_{max}, p_r). The same procedure is carried out by node A when it receives the CTS from B.

In SRM, nodes can only sneak the DT if they ensure that their ST will not collide at the DNs. For that, a potential sneaking node needs to determine the amplitude of its ST. In other words, nodes in the C-Zone estimate both the transmission range and carrier sensing range of their potential sneaking node and make sure that the DNs are outside of this range. Finally, note in Figure 4.25 that the sneaking source employs a backoff mechanism called *sneaking backoff*. Before any sneaking node tries to sneak the medium, it has to backoff for a random duration between [20, 20×N] μs, where N is an estimate of the average number of neighbors a node has, and is dynamically obtained through the routing protocol [Johnson2001, Li2002]. The sneaking backoff is a multiple of 20 μs as this is usually the time required for a node to sense the medium activity [Jung 2002]. This is implemented in SRM to provide for the case where multiple nearby nodes try to sneak the medium simultaneously, hence causing collisions. This way, a node can interrupt its sneaking transmission if it detects the medium has become busy and returns to regular IEEE 802.11. Sneaking may be tried again in the next DT only.

4.6.3 QoS Provisioning

Several MAC schemes aimed at providing QoS guarantees have been proposed for wireless networks. However, these MAC protocols, in general, rely on a centralized control which is only viable for single hop wireless networks. In multi-hop wireless networks, a fully distributed scheme is needed. The discussion here applies to provisioning of QoS over the original IEEE 802.11 MAC, and not the IEEE 802.11e standard presented earlier which has been designed from scratch for QoS support. Many QoS sensitive applications will eventually need to run over the original IEEE 802.11 MAC, hence demanding some sort of QoS support. Therefore, in this section we look at proposals for QoS provisioning over multi-hop ad hoc networks.

4.6.3.1 An Extension to the IEEE 802.11 DCF

IEEE 802.11 DCF is a good example of a best-effort type MAC protocol. It has no notion of service differentiation and no support for real time traffic. A scheme to extend the IEEE 802.11 DCF is proposed in [Veres2001] with the ability to support at least two service classes: premium service (i.e., high priority) and best-effort. Traffic of premium service class is given lower values for congestion window than those of best-effort traffic. If packets of both types collide, the packet with smaller congestion window value is more likely to access the medium earlier.

4.6.3.2 The Black Burst Contention Scheme

The Black Burst (BB) contention scheme [Sobrinho1999] avoids packet collision in a very distinctive way, while at the same time solving the packet starvation problem. Packets from two or more flows of the same service class are scheduled in a distributed manner with fairness guarantees. Nodes contend for the medium after it has been idle for a period longer than the interframe space. Nodes with best-effort traffic and nodes with real-time traffic use different interframe space values to provide higher priority. The BB scheme can be added to any CSMA/CA type of protocol in the following way. Right before sending their packets when the medium remains idle long enough, real-time nodes first contend for transmission by jamming the media with pulses of energy, which a called BBs.

The novelty of this scheme is that each contending node is using a BB with different length, where the length of each BB is an integral number of black slots. The number of slots that forms a BB is an increasing function of the contention delay experienced by the node, measured from the instant when an attempt to access the channel has been scheduled until the node starts the transmission of its BB. Following each BB transmission, a node senses the channel for an observation interval. Since different nodes contend with BBs of different length, each node can determine without ambiguity whether its BB is of longest duration. Therefore, only the winner is produced after this contention period who will then successfully transmits its real-time data. BB contention ensures that real-time packets are transmitted without collisions and with priority over best-effort packets.

4.6.3.3 The MACA/PR Protocol

The Multi-hop Access Collision Avoidance with Piggyback Reservation (MACA/PR) protocol [Lin1997] provides guaranteed bandwidth support (via reservation) for real-time traffic. It establishes real-time connections over a single hop only, however it works with QoS routing algorithm and a fast reservation setup mechanism. The first data packet in the real-time stream makes reservations along the path. A RTS/CTS handshake is used on each link for this first packet in order to make sure that it is transmitted successfully. Both RTS and CTS specify how long the data packet will be. Any station near the sender which hears the RTS will defer long enough so that the sender can receive the returning CTS. Any node near the receiver which can hear the CTS will avoid colliding with the following data packet. As we have mentioned before, in MACA/PR the RTS/CTS handshake is employed only for the first packet to setup reservations. Subsequent packets do not use RTS/CTS.

When the sender transmits the data packet, it schedules the next transmission time after the current data transmission and piggybacks the reservation in the current data packet. Upon receiving the data packet correctly, the intended receiver enters the reservation into its reservation table and returns an ACK. The neighbors hearing the data packet can learn about the next packet transmission time. Similarly, neighbors at the receiver side which hear the ACK, avoid transmitting at the same time when the receiver is scheduled to receive the next packet. Note that ACK serves as renewing of reservation rather than for recovering from packet loss. If fact, if the ACK is not received, the packet is not retransmitted. Instead, if the sender consecutively fails to receive ACK for a certain number of transmissions, it assumes that the link is not satisfying the bandwidth requirement and notifies the upper layer, i.e., the QoS routing protocol. Hence, the "reservation ACK" serves as a "protector" for the given time window, and also as a mechanism to inform the sender if something is wrong with the link.

4.7 Conclusions and Future Directions

The low cost of wireless LANs has led to a tremendous growth of its worldwide use. Nowadays, we can find wireless LANs networks in nearly all enterprise environments, in many homes, hotspots, airport lounges, among others. In the near future, the use of wireless LANs will be as common as it is the use of cell phones nowadays. In homes, having a wireless LAN will be as natural as having a microwave, a coffee machine, a rice cooker, and so on. Wireless LANs are finding new applications in homes and enterprises. Today, there is a high demand for the efficient support of multimedia applications over wireless LANs. Needless to say that security is also a major concern. Obviously, these are just a few examples of some of the issues that have to be handled. With the advent of multi-user multiple-input multiple-output (MIMO) systems and multi-gigabit 802.11 networks operating in the 60 GHz band, a new revolution is expected in the wireless LAN arena. As we saw earlier in this chapter, the IEEE is indeed addressing some of these aspects while the problem space is much larger. The efficient utilization of the scarce radio resource is also an existing concern which needs more investigation, and cognitive and spectrum agile radios are attempts to address this issue. Finally, integration of wireless LANs into the next generation heterogeneous networks beyond 3G is also a very hot topic.

Now that we have covered wireless LANs, it is time to move to an area which is experiencing a tremendous growth in interest and applications, namely, wireless PANs. In the next chapter, we investigate wireless PANs and their enabling technologies in detail.

Homework Questions/Simulation Projects

Q. 1. A University building has a number of access points (APs) uniformly distributed in 2-D space and are to be accessed by students using laptops. The coverage range of each AP is 40 m. A group of 10-students need to work concurrently on different parts of a group project and they decided to exchange information using wireless access to APs. But, they found that, in spite of all the efforts, some areas in the building remain uncovered. The students decided to use dual-port radios and connect laptops in uncovered areas using an extension of Zigbee-based ad hoc network to a laptop connected to AP. The Zigbee devices have a range of 8 m and the students are equally distributed in an area of 120m x 120m.

   a. How many students access a single AP concurrently?
   b. Assuming each student is moving at a speed of 2 Km per hour, what is the probability that a student will have a handoff from one AP to another?
   c. To access any member of ad hoc network, a routing table needs to be formed at each AP. What will be the size of such a routing table?
   d. If the laptop of an ad hoc network serving as a gateway to an AP is also mobile, what will be the impact on the performance?
   e. If different devices have different mobility, how frequently do you need to update the routing table at each AP?

   Assume any relevant parameters. Validate your analytical results with appropriate simulation.

Q. 2. Consider two nodes connected by an IEEE 802.11b 11 Mbps wireless link. Assuming user payloads of 200, 600, and 1500 bytes, calculate the maximum user throughput over this wireless link. Assume that that the wireless link is collision and error-free, and that RTS/CTS control frames are not employed.

   IEEE 802.11b parameters are as follows: PLCP preamble and header is 24 bytes, MAC/LLC header and trailer (FCS) length is 42 bytes, ACK frame is 14 bytes, SIFS is 10μs, DIFS is 50μs and average backoff time between transmissions is 310μs (on average, Minimum Contention Window time/2). Assume that PLCP and control packets are always transmitted at the basic rate of 1Mbps.

   Requires solving discrete Markov chain model. (Hint: Look at the work done by G. Bianchi, " Performance Analysis of the IEEE 802.11 Distributed Coordination Function," IEEE JSAC Vol. 18, No. 3, March 2000)

Q. 3. As we have seen in this chapter, the IEEE 802.11 standard has been designed primarily for indoor environments. This is reflected in many of the design choices this standard has made in its physical and MAC layers. If you were to change the IEEE 802.11 protocol to operate in outdoor environments (up to 5 Km range), what changes would you make? (Tip: factors such as range and propagation need to be considered.)

Q. 4. Design a problem based on any of the material covered in this chapter (or in references contained therein) and solves it diligently.

References

[Agarwal2001] S. Agarwal, S. Krishnamurthy, R. Katz, and S. Dao, "Distributed Power Control in Ad-hoc Wireless Networks," *IEEE PIMRC,* Vol. 2, pp. F-59 – F-66, 2001.

[Agrawal2002] D. Agrawal and Q.-A. Zeng, *Introduction to Wireless and Mobile Systems,* Brooks/Cole Publishing, 438 pages, August 2002, ISBN No. 0534-40851-6.

[Agrawal2003] D. Agrawal, C. Cordeiro, and H. Gossain, "A Spatial Reuse Enabling Power Control MAC Protocol for Wireless Ad Hoc Networks," University of Cincinnati Intellectual Property Office No. 103-038, 2003.

[Akyildiz2005] I. F. Akyildiz, X. Wang, and W. Wang, "Wireless Mesh Networks: A Survey,'" *Elsevier Computer Networks Journal,* March 2005.

[Anastasi1998] G. Anastasi, L. Lenzini, and E. Mingozzi "MAC Protocols for Wideband Wireless Local Access: Evolution Toward Wireless ATM", *IEEE Personal Communications,* Oct. 1998, pp. 53–64.

[Atheros2003] Atheros Communications, "802.11 WLAN Performance," *White Paper,* 2003.

[Bharghavan1994] V. Bharghavan, A. Demers, S. Shenker, and L. Zhang, "MACAW: A media access protocol for wireless LANs," in *ACM SIGCOMM,* August 1994, pp. 212-225.

[Bisdikian2001] C. Bisdikian, "An Overview of the Bluetooth Wireless Technology," *IEEE Communications Magazine,* December 2001, pp.: 86-94.

[Bluetoothwww] Bluetooth SIG, "Bluetooth Specification," http://www.bluctooth.com.

[Cesana2003] M. Cesana, D. Maniezzo, P. Bergamo, M. Gerla, "Interference Aware (IA) MAC: an Enhancement to IEEE 802.11b DCF," *Proceedings of the IEEE VTC Fall,* 2003.

[Cordeiro2002a] C. Cordeiro, S. Das, and D. Agrawal, "COPAS: Dynamic Contention-Balancing to Enhance the Performance of TCP over Multi-hop Wireless Networks," in *Proceedings of IEEE IC³N,* October 2002.

[Cordeiro2002b] C. Cordeiro and D. Agrawal, "Mobile Ad Hoc Networking," *Tutorial/Short Course in 20th Brazilian Symposium on Computer Networks,* pp. 125-186, May 2002, http://www.ececs.uc.edu/~cordeicm/.

[Cordeiro2005] C. Cordeiro, K. Challapali, D. Birru, and S. Shankar, "IEEE 802.22: The First Worldwide Wireless Standard based on Cognitive Radios," in *IEEE International Conference on Dynamic Spectrum Access Networks (DySPAN),* 2005.

[Crow1997] B. P. Crow, I. Wadjaja, J. G. Kim, and P. T. Sakai, "IEEE 802.11 Wireless Local Area Networks," In *IEEE Communications Magazine,* September 1997, pp. 116-126.

[Ebert2000] J.-P. Ebert, B. Stremmel, E. Wiederhold, and A. Wolisz, "An Energy-efficient Power Control Approach for WLANs," *Journal of Communications and Networks (JCN)*, Vol. 2, No. 3, pp. 197-206, September 2000.

[Eklund2002] C. Eklund, R. Marks, K. Stanwood, and S. Wang, "IEEE Standard 802.16: A Technical Overview of the WirelessMAN™ Air Interface for Broadband Wireless Access," in *IEEE Communications Magazine*, June 2002.

[Fluhrer2001] S. Fluhrer, I. Mantin, and A. Shamir, "Weaknesses in the key scheduling algorithm of RC4," *Eighth Annual Workshop on Selected Areas in Cryptography*, August 2001.

[Fullmer1997] C. L. Fullmer, J. J. Garcia-Luna-Aceves, "Solutions to Hidden Terminal Problems in Wireless Networks," in *ACM SIGCOMM*, 1997.

[Garcia-Luna-Aceves1999] J. J. Garcia-Luna-Aceves, "Reversing the collision-avoidance handshake in wireless networks," in *ACM MOBICOM*, August 1999.

[Goldberg1995] L. Goldberg, "Wireless LANs: Mobile Computing's Second Wave," *Electronic Design*, June 1995.

[Gomez2001] J. Gomez, A. Campbell, M. Naghshineh, and C. Bisdikian, "Conserving Transmission Power in Wireless Ad Hoc Networks," *ICNP*, pp.: 24-34, November 2001.

[HIPERLAN/21999a] ETSI, "Broadband Radio Access Networks (BRAN); HIPERLAN type 2 technical specification; Physical (PHY) layer," August 1999.

[HIPERLAN/21999b] ETSI, "Broadband Radio Access Networks (BRAN); HIPERLAN Type 2; Data Link Control (DLC) Layer; Part 1: Basic Transport Functions," December1999.

[Holland2001] G. Holland, N. H. Vaidya, and P. Bahl, "A rate-adaptive MAC protocol for wireless multi-hop networks," in *ACM MOBICOM*, July 2001.

[IEEE802.11www] IEEE 802.11 Working Group for WLAN, http://www.ieee802.org/11/.

[IEEE802.112007] IEEE Std. 802.11. "IEEE Standard for Wireless LAN Medium Access Control (MAC) and Physical Layer (PHY) Specification," June 1997.

[IEEE802.11e2001] IEEE 802.11 WG, Draft Supplement to Standard for Telecommunications and Information Exchange Between Systems - LAN/MAN Specific Requirements - Part 11: Wireless Medium Access Control (MAC) and physical layer (PHY) specifications: Medium Access Control (MAC) Enhancements for Quality of Service (QoS), IEEE 802.11e/D2.0, Nov. 2001.

[IEEE802.15www] IEEE 802.15 Working Group for WPAN, http://www.ieee802.org/15/.

[IEEE802.20www] IEEE 802.20 Working Group for WWAN, http://www.ieee802.org/20/.

[IrDAwww] Infrared Data Association, "IrDA Standards," http://www.irda.org/standards/standards.asp, April 2000.

[Johnson2001] D. Johnson, D. Maltz, Y.-C. Hu, and J. Jetcheva, "The dynamic source routing protocol for mobile ad hoc networks (DSR)," IETF Internet-Draft, November 2001.

[Jung2002] E.-S. Jung and N. Vaidya, "A Power Control MAC Protocol for Ad Hoc Networks," in *ACM Mobicom*, 2002.

[Kamerman1997] A. Kamerman and L. Monteban, "WaveLAN-II: A High-Performance Wireless LAN for the Unilicensed Band," *Bell Labs Tech. Journal*, 2(3), 1997.

[Karn1990] P. Karn, "MACA - A new channel access method for packet radio," in *Proceedings of ARRL/CRRL Amateur Radio 9th Computer Networking Conference*, September 1990.

[Khun-Jush1999] J. Khun-Jush *et al.*, "Structure and Performance of HIPERLAN/2 Physical Layer," *IEEE VTC*, Fall 1999.

[Lettieri1998] P. Lettieri and M. Srivastava, "Adaptive Frame Length Control for Improving Wireless Link Throughput, Range, and Energy Efficiency," In *IEEE INFOCOM*, March 1998.

[Lin1997] C. R. Lin and M. Gerla, "Asynchronous Multimedia Multihop Wireless Networks," in *IEEE Infocom*, 1997.

[Liu2001] J. Liu and S. Singh, "ATCP: TCP for mobile ad hoc networks," In *IEEE J-SAC*, vol. 19, no. 7, pp. 1300–1315, July 2001.

[Moh1998] W. Moh, D. Yao, and K. Makki, "Wireless LAN: Study of hidden terminal effect and multimedia support," in *Proceedings of Computer Communications and Networks*, pp.422-431, 1998.

[Narayanaswamy2002] S. Narayanaswamy, V. Kawadia, R. Sreenivas, and P. Kumar, "Power Control in Ad-Hoc Networks: Theory, Architecture, Algorithm and Implementation of the COMPOW protocol," In *European Wireless 2002*, February 2002.

[Nee1999] R. Van Nee, G.A. Awater, M. Morikura, H. Takanashi, M.A. Webster, and K.W. Halford, "New High-Rate Wireless LAN Standards," *IEEE Communications Magazine*, vol. 37, no.12, December 1999, pp. 82–88.

[Ozugur1998] T. Ozugur, M. Naghshineh, P. Kermani, C. M. Olsen, B. Rezvani, and J. A. Copeland, "Balanced media access methods for wireless networks," in *ACM MOBICOM*, October 1998.

[Pahlavan2001] K. Pahlavan and P. Krishnamurthy, *Principles of Wireless Networks: A Unified Approach*, Prentice Hall, 608 pages, ISBN: 0130930032, 2001.

[Paul2008] T. Paul and T. Ogunfunmi, "Wireless LAN comes of age: Understanding the IEEE 802.11n amendment," *IEEE Circuits and Systems Magazine*, March 2008.

[Perkins2001] C. Perkins, E. Royer, and S. Das, "Ad Hoc On Demand Distance Vector Routing (AODV)," *Internet Draft*, March 2001 (work in Progress).

[RadioLANwww] RadioLAN Marketing Group, http://www.radiolan.com.

[Royer2000] E. M. Royer, S-J. Lee, and C. E. Perkins, "The Effects of MAC Protocols on Ad hoc Communication Protocols," *Proc. of IEEE WCNC 2000*, Sept. 2000.

[Schneier1996] B. Schneier, *Applied Cryptography*, John Wiley & Sons Inc., 2nd Ed., 758 pages, 1996.

[Sobrinho1999] J. Sobrinho and A. Krishnakumar, "Quality-of-Service in Ad Hoc Carrier Sense Multiple Access Wireless Networks," *J-SAC*, 17(8): 1353-1368, 1999.

[Stubblefield2001] A. Stubblefield, J. Ioannidis, and A. Rubin, "Using the Fluhrer, Mantin, and Shamir Attack to Break WEP," *AT&T Labs Technical Report TD-4ZCPZZ*, Revision 2, August 2001.

[UWBwww] Ultra-Wideband Working Group, "http://www.uwb.org".

[Vaidya2000] N. H. Vaidya, P. Bahl, and S. Gupta, "Distributed fair scheduling in a wireless LAN," in *ACM MOBICOM*, August 2000.

[Veres2001] A. Veres, A. T. Campbell, M. Barry, and L. H. Sun, "Supporting Service Differentiation in Wireless Packet Networks Using Distributed Control," *IEEE Journal on Selected Areas in Communications*, October 2001.

[Wieselthier2000] J. Wieselthier, G. Nguyen, and A. Ephremides, "On the Construction of Energy-Efficient Broadcast and Multicast Trees in Wireless Networks," *IEEE INFOCOM*, March 2000.

[WirelessDisplay2010] "Intel Wireless Display (WiDi): The Hottest Sleeper Technology", http://www.pcmag.com/article2/0,2817,2357919,00.asp, Jan. 2010.

[WPAwww] WPA (Wi-Fi Protected Access), *Wi-Fi Alliance*, http://www.wi-fi.com/.

[Xu2002] K. Xu, M. Gerla, and S. Bae, "How Effective is the IEEE 802.11 RTS/CTS Handshake in Ad Hoc Networks?," In IEEE Globecom, November 2002.

Chapter 5

Wireless PANs

5.1 Introduction

Introduction of Wireless PANs (WPANs) has caused the latest revolution in the wireless technology. WPANs are short to very short-range (from a couple centimeters to a couple of meters) wireless networks that can be used to exchange information between devices within the reach of an individual. WPANs can be used to replace cables between computers and their peripherals; to share multimedia content amongst devices; to build an infrastructure for sensor networking applications; or to establish various location aware services. The best example representing WPANs is the industry standard Bluetooth [Agrawal2002, Bluetoothwww, Cordeiro2002a], which these days can be found in many consumer electronics such as cell phones, PDAs, wireless headsets, wireless mouse and wireless keyboard. The IEEE 802 has also established the IEEE 802.15 Working Group (WG) for WPANs [IEEE-802.15www], which standardizes protocols and interfaces for WPANs. Altogether, the IEEE 802.15 family of standards is formed of seven Task Groups (TGs) and is described later in this chapter. Other less popular examples of WPAN technologies include Spike [Spikewww], IrDA [IrDAwww], and in the broad sense HomeRF [Negus2000].

A key feasibility issue of WPANs is the inter-working of wireless technologies to create heterogeneous wireless environment. For instance, WPANs and WLANs will enable an extension of the 3G/4G cellular networks (e.g., UMTS, cdma2000, WiMax, LTE) into devices without direct cellular access. Moreover, devices interconnected in a WPAN may be able to utilize a combination of 3G/4G access and WLAN access by selecting the access that is best at a given time. In such networks, 3G/4G, WLAN and WPAN (and many other) technologies do not compete, but complement each other by enabling a user to select the best service for their purpose. Figure 5.1 shows the operating space of the various IEEE 802 WLAN and WPAN standards and other activities still in progress (for a broader comparison perspective of the various IEEE 802 technologies, please refer to Figure 1.3).

Given the importance within the WPAN operating space, availability of devices and intensive research activities, we first present some of its applications. We then introduce the Bluetooth WPAN technology and provide an overview of its standard as defined by the Bluetooth Special Interest Group (SIG). The IEEE 802.15 family of protocols is discussed next, followed by a thorough comparison of the various WPAN technologies.

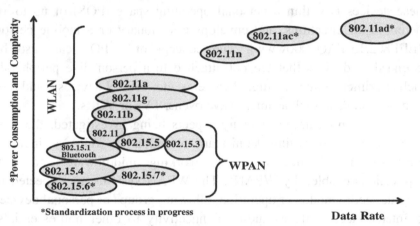

Figure 5.1 – The scope of various WLAN (802.11x) and WPAN (802.15.x, Bluetooth) Standards

5.2 Why Wireless PANs

The concept of Personal Area Networks (PANs) was first demonstrated by IBM researchers in 1996 that utilized human body to exchange digital information. Engineers used picoamp currents through the body at very low frequencies of around 1 MHz. The low power and frequencies prevented eavesdropping and interference to neighboring PANs, and the system enabled two researchers to electronically exchange a business card with a handshake [Zimmerman1996].

IBM engineers created a way to communicate between body-borne appliances by using the human body as a channel. The only limitation was that some form of human contact between devices is required which may not always be desirable or possible. To get around this problem of human contact, other alternatives such as infrared (IR) or far-field (radio) communications have been considered. Using wireless methods such as IR or radio frequency (RF) for PANs is known as WPAN which are typically smaller, operate on battery power, and are either worn on a human body or carried personally. The main design goal of WPANs is to allow either stationary or moving devices that are in close proximity to communicate and exchange information without wires. WPAN should allow devices to create or provide data/voice access points, personal ad hoc connectivity. The operating range for these devices is within a personal operating space (POS) of up to 10 meters in all directions, and envelops a stationary or a mobile person [IEEE802.15-FAQ2000www]. The concept of a POS can also be extended to devices that are not attached to a person, like peripherals such as printers, scanners, digital cameras, etc. WPAN devices could also be more mundane such as microwave ovens, TVs or VCRs.

Currently, amongst many aspects being investigated, the four research areas receiving significant attention in WPANs include: standards, interference issues, networking middleware and new applications enabled by WPANs. The WPAN systems are expected to provide secure modes of operation, allowing groups of personal devices to interconnect while excluding connectivity to other non-essentials. They should not affect the primary function, the form factor and power consumption of the devices in which they are embedded. As WPANs primarily use the license-free radio frequencies (e.g., ISM band), they have to coexist with other RF technologies that make use of these frequencies. A WPAN is functionally similar to a WLAN, while differs in terms of power consumption, coverage range, data rate and the cost.

5.3 The Bluetooth Technology

Bluetooth (or simply BT) has been a topic of considerable buzz in the telecommunications industry for the last decade. Bluetooth is named after a 10th-century Viking king known for his success in uniting

Denmark and Norway during his rule around 960 AD. Just as King Harald Bluetooth is known for uniting different Christian community, today's Bluetooth-enabled devices promise to unite different digital devices. Nowadays, the Bluetooth technology is the most prominent example of a WPAN.

Bluetooth is a low cost and short-range radio communication standard that was introduced as an idea in Ericsson Laboratories [Ericssonwww] back in 1994. Engineers envisioned a need for a wireless transmission technology that would be cheap, robust, flexible, and consume low power. The basic idea of cable replacement with possibility of many extensions was picked-up quickly and over 2500 companies joined the Bluetooth SIG [Bluetoothwww] and has been chosen to serve as the baseline of the IEEE 802.15.1 standard for WPANs, supporting both synchronous traffic such as voice, and asynchronous data communication.

5.3.1 History and Applications

The Bluetooth technology came to light in May 1998 and since then the Bluetooth SIG has steered the development of Bluetooth, including both protocols and applications profiles. The Bluetooth SIG is an industry group consisting of leaders in the telecommunications and computing industries such as Ericsson, IBM, Intel, Microsoft, Motorola, Nokia, Broadcom and Toshiba, and many more companies. Bluetooth wireless technology has become a de facto standard, as well as a specification for small-form factor, low-cost, short-range ad hoc radio links between mobile PCs, mobile phones and other portable devices. Among others, the Bluetooth SIG suggests five applications that provide a good illustration of the capabilities of the Bluetooth specification [BluetoothSpec], a three-in-one phone, an Internet bridge, an interactive conference, a headset, and an automatic synchronizer.

The three-in-one phone is a phone that can operate over a fixed-line phone when within the home, a mobile phone when outside the home, or as a walkie-talkie with another Bluetooth-enabled device when within its range. The Internet bridge allows a mobile computer to interact with another device within Bluetooth range. The interactive conference allows sharing of documents among several computers during a live conference. A Bluetooth-enabled headset can connect to any Bluetooth-enabled device that requires voice input or can provide sound, such as a wireline phone, mobile phone, or a music player. An automatic synchronizer is an application that allows multiple devices, such as desktop computers, laptops, PDAs, and/or mobile phones to synchronize with each other such that appointments and contact information available in the different devices matches.

Below we illustrate some other application areas where Bluetooth networks could be explored.

- Consumer– Wireless PC peripherals, smart house wireless PC peripherals, smart house integration, etc.
- Games– Controllers, virtual reality, etc.
- Professional– PDAs, cell phones, desktops, automobiles, etc.
- Services– Shipping, travel, hotels, etc.
- Industry– Delivery (e.g., scanners, printers), assembly lines, inspections, inventory control, etc.
- Sports training – Health sensors, monitors, motion tracking, etc.

Bluetooth has a tremendous potential in moving and synchronizing information in a localized setting – a natural phenomenon of human interaction, as we do business transactions and communicate more frequently with the people who are close by as compared to those who are far away.

5.3.2 Technical Overview

The Bluetooth classic specification (version 1.1) describes radio devices designed to operate over very short ranges – on the order of 10 meters – or optionally a medium range (100 meters) radio link capable of voice or data transmission to a maximum capacity of 720 kbps per channel (with a nominal throughput of 1 Mbps). The goals of the specification were to describe a device that is simple and robust, consumes little power, and particularly emphasizing it to be very inexpensive to produce.

Radio frequency operation is in the unlicensed ISM band at 2.4 to 2.48 GHz, using a frequency hopping spread spectrum (FHSS), full-duplex signal at up to 1600 hops/seconds. The signal hops among 79 frequencies at 1 MHz intervals to try to give a high degree of interference immunity from other external resources. This is crucial due to the number of electronic products sharing this frequency range. However, even these measures have not been sufficient to keep Bluetooth from suffering from both external interference and interference generated by its own devices [Cordeiro2003a]. In Bluetooth, RF output is specified as 0 dBm (1mW) in the 10m range version and -30 to +20 dBm (100mW) in the longer range version. The Bluetooth specifications are divided into two parts:

- *The Core* – This portion specifies components such as the radio, base band (medium access), link manager, service discovery protocol, transport layer, and interoperability with different communication protocols.
- *The Profile* – The Profile portion specifies the protocols and procedures required for different types of Bluetooth applications.

The Bluetooth specification covers details of the physical and data layers of the communication link. It should be noted that the strict partitioning of the different layers of the typical protocol stack defined by the Open System Interconnection (OSI) model [Tanenbaum1996] is losing its significance in wireless implementations [Cordeiro2002b]. It is sometimes in the best interest of an application to be aware of the current condition of the physical layer interface, thereby making the partitioning somewhat fuzzy.

5.3.2.1 Ad Hoc Networking

Whenever a pair or small group of Bluetooth devices come within radio range of each other, they can form an ad hoc network without requiring any infrastructure. Devices are added or removed from the network dynamically. Thus, they can connect to or disconnect from an existing network at will and without interruption to the other participants. In Bluetooth, the device making the communication initiative to another device assumes the role of a *master*, while the recipient becomes a *slave*.

5.3.2.1.1 The Piconet

The basic architectural unit of a Bluetooth is a *Piconet*, composed of one master device and up to seven active *slave devices*, which can communicate with each other only through their master. All the nodes in a Piconet share the same frequency hopping sequence using a slotted time division duplex (TDD) scheme with the maximum bandwidth of 1 Mbps. Figure 5.2 gives an example of a Piconet.

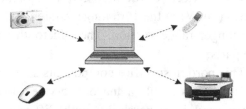

Figure 5.2 – A Bluetooth Piconet

Bluetooth Piconet is centered on the master device which controls a channel between all the slave devices. To become a master, a device requests a connection with another device: if the paged device accepts the link, the calling device becomes a master for that link and the responding device becomes a slave. Every Bluetooth device is exactly the same except for a 48-bit device identifier (BD_ADDR). Since all the traffic in Bluetooth has to go through the master device, it has full control over the communication within its own Piconet. According to stringent TDD schemes, a slave device is allowed to transmit in a slot only under the following three conditions:

- When the master has polled the slave, asking if it wants to send a message, as in the case of Asynchronous Connectionless (ACL) link (to be defined in the next sections);
- When a master sends a broadcast packet in the preceding slot; and
- The slave already has a reservation for that slot, as in the case of Synchronous Connection Oriented (SCO) links (to be defined in the next sections).

Besides the active slaves in a Piconet, additional devices can be connected in a parked state in which they listen but do not communicate. When they want to participate, they are swapped with one of the active devices. With this method, up to 255 devices can be virtually connected to the Piconet. If the acting master leaves the Piconet, one of the slaves assumes its role. During Piconet formation, the master allocates an *active member address* (am_address) to all the devices, and uses these addresses for communication over the Piconet. Also, each Piconet uses a different Frequency Hopping Sequence (FHS) in order to reduce interference with other nearby Piconets, i.e., reduce the inter-Piconet interference.

5.3.2.1.2 The Scatternet

To increase the number of devices in the network, a *scatternet* architecture consisting of several Piconets has been proposed and is shown in Figure 5.3, which depicts a scatternet comprised of three Piconets. Since scatternets span more than a single Piconet, a few nodes act as bridges (see Figure 5.3) responsible for relaying packets across Piconet boundaries. In this configuration, each Piconet is identified by its individual FHS. A bridge device usually participates in more than one Piconet, but can only be active in one at a time. A device can be a slave in several Piconets, but act as master in only one. Given that different Piconets may have different FHS, the bridge device selects the required master identity during communication between Piconets in order to synchronize with the FHS of the corresponding Piconet.

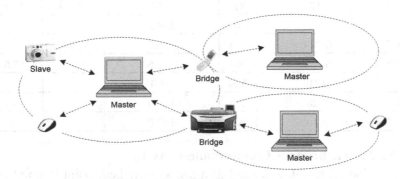

Figure 5.3 – A Bluetooth Scatternet

5.3.2.2 Voice and Data Transmission

The Bluetooth specification defines two different types of links for data and voice applications. They are:

- Synchronous Connection Oriented (SCO) link.
- Asynchronous Connectionless (ACL) link.

In the following subsections, we discuss each of these two links.

5.3.2.2.1 Synchronous Connection Oriented (SCO)

The SCO is a symmetric, point-to-point link between the master and one slave. Usually, the SCO link is used for audio applications with strict Quality of Service (QoS) requirements. Given its QoS demands, Piconet master reserves slots for SCO links so that it can be treated as a circuit switched link. There is no re-transmission of SCO packets (to be soon defined) while they are transmitted at predefined regular intervals (hereafter referred to as T_{SCO}). Each voice channel supports a 64 Kbps synchronous (voice) simplex (i.e., between the master and a slave) channel. Considering its nominal 1 Mbps Piconet bandwidth and the 64 Kbps requirement for a SCO connection, a Bluetooth Piconet can support up to three simplex SCO links (when using HV3 packets) so as to meet the required QoS needs. This can be concluded based on Table 5.1.

Table 5.1 – Bluetooth packet types

Type	User Payload (bytes)	FEC	Symmetric (Kbps)	Asymmetric (Kbps)	
DM1	0-17	Yes	108.0	108.8	108.8
DH1	0-27	No	172.8	172.8	172.8
DM3	0-121	Yes	256.0	384.0	54.4
DH3	0-183	No	384.0	576.0	86.4
DM5	0-224	Yes	286.7	477.8	36.3
DH5	0-339	No	432.6	721.0	57.6
HV1	0-10	Yes	64.0	-	-
HV2	0-20	Yes	128.0	-	-
HV3	0-30	No	192.0	-	-

5.3.2.2.2 Asynchronous Connectionless (ACL)

The ACL link is treated as a packet switched, point to point and point to multipoint data traffic link. Master maintains one ACL link with each active slave over which upper layer connection can be established and re-transmission is employed only when data integrity is necessary. In addition to ACL and SCO packets, the master and slaves exchange short POLL and NULL packets. POLL messages are sent by the master and require an acknowledgment (ACK), while NULL messages can be sent either by the master or the slave and do not require any ACK.

5.3.2.3 Bluetooth Packets

Bluetooth defines types of packets, and information can travel in these packet types only. Bluetooth uses 1, 3 and 5 slot packets as depicted in Figure 5.4. A TDD scheme divides the channel into 625 μsec slots at a 1 Mb/s symbol rate. As a result, at most 625 bits can be transmitted in a single slot. However, to change the Bluetooth device from transmit state to receive state and tune to the next frequency hop, a 259 μsec turn around time is kept at the end of the last slot. This results in reduction of effective bandwidth available for data transfer.

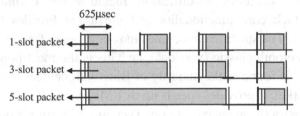

Figure 5.4 – Packet Transmission in Bluetooth

Table 5.1 summarizes the available packet types and their characteristics. Bluetooth employs HVx (High-quality Voice) packets for SCO transmissions and DMx (Data Medium-rate) or DHx (Data High-rate) packets for ACL data transmissions, where x = 1, 3 or 5. In case of DMx and DHx, x represents the number of slots a packet occupies as shown in Figure 5.4, while in the case of HVx, it represents the level of Forward Error Correction (FEC).

5.3.2.4 Connection Set Up

Connection setup in Bluetooth is as shown in Figure 5.5. First, each node discovers its neighborhood. This process is called *inquiry*. The execution of the inquiry procedure is not mandatory, and can be done when a node wants to refresh the information about its neighborhood. For two devices to discover each other, while one of them is in INQUIRY state the other has to be in INQUIRY SCAN. The node in INQUIRY SCAN responds to the INQUIRY. This way the node in INQUIRY state notices the presence of the node in INQUIRY SCAN. When the devices want to build up a connection, they begin the *page* procedure. Similar to the inquiry phase, there are two states: PAGE and

PAGE SCAN. When one of the nodes wants to build up a connection to the other node, it enters in the PAGE state. When the other node enters PAGE SCAN state, the connection setup is concluded.

Figure 5.5 – Connection establishment in Bluetooth

5.3.3 The Bluetooth Specifications

The Bluetooth specification includes the definition of the Protocol Stack core functionality and the usage Profiles for different applications. The specification comes in four primary flavors:

- Bluetooth classic (version 1.1): This is the most popular version and is found in the majority of Bluetooth devices available in the market. It provides speeds up to 1 Mbps.
- Bluetooth enhanced data rate (version 2.1): Builds on version 1.1 and increases the data rate up to 3 Mbps. Over time, this version of Bluetooth will become predominant in the marketplace.
- Bluetooth high-speed (version 3.0): This version of Bluetooth includes support of 802.11 as a high-speed channel. Depending on the usage profile and the availability of a 802.11 radio on the same device, the 802.11 can be turned on and off to support fast data transfers.
- Bluetooth low energy (version 4.0): this is the newest member to the Bluetooth family, and aims at significantly decreasing the power consumption of Bluetooth classic. Applications include sensors and actuators, mainly in the health and wellness arena.

In the following, we provide a brief overview of the Bluetooth protocol stack.

5.3.3.1 Bluetooth Protocol Stack

Figure 5.6 shows the layered structure of the Bluetooth protocol stack. The stack defines all layers unique to the Bluetooth technology. The Bluetooth specification only defines the Physical and the Data Link layers of the OSI Protocol Stack. The application layer shown in Figure 5.6 actually includes all the upper layers (IP, Transport, Application) sitting on the RFCOMM and the SDP. These layers are not themselves part of the stack and this host stack are typically handled in software. They communicate with lower layers via the Host Controller. The lower layers (RF, Baseband and LMP) are primarily built in hardware modules.

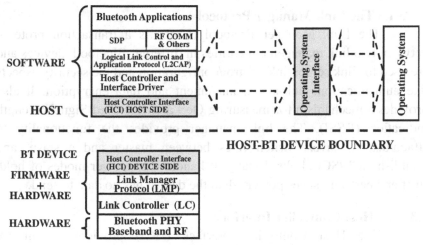

Figure 5.6 – The Bluetooth Protocol Stack

5.3.3.1.1 The Radio

The radio layer, which resides below the Baseband layer, defines the technical characteristics of the Bluetooth radios. It is the lowest layer in Bluetooth protocol stack and it defines the requirements of Bluetooth transceivers operating in unlicensed ISM band. Currently, many other wireless devices operate in this band and this creates interference. Bluetooth mitigates this effect using FHSS. It also uses FEC to reduce the impact of noise on longer links. It has a nominal range of 10 meters at a 0dBm (1 mW) power setting which can be extended up to 100 meters on a 20 dBm (100 mW) power setting. It uses a Binary Frequency Shift Keying (BFSK) modulation technique which represents a binary 1 as a negative frequency deviation.

5.3.3.1.2 The Baseband

The baseband defines the key procedures that enable devices to communicate with each other by incorporating the MAC procedures. It defines how Piconets are created, and also determines the packet formats, physical-logical channels and different methods for transferring voice and data. It provides link set-up and control routines for the layers above. Additionally, the baseband layer provides lower level encryption mechanisms for security.

5.3.3.1.3 The Link Manager Protocol

The Link Manager Protocol (LMP) is a transaction protocol between two link management entities in different Bluetooth devices and are used for link setup, link control/configuration and the security aspects like authentication, link-key management and data encryption. It also provides a mechanism for measuring QoS and Received Signal Strength Indication (RSSI). The link manager provides the functionality to attach/detach slaves, switch roles between master and a slave, and establish ACL/SCO links. Finally, it handles low power modes of hold, sniff and park and saves power when the device has no data to send.

5.3.3.1.4 Host Controller Interface

The Host Controller Interface (HCI) provides a uniform command interface to the baseband and the LMP layers, and also to the H/W status and the control registers (i.e., it gives higher-level protocols the possibility to access lower layers). The transparency allows the HCI to be independent of the physical link between the module and the host. The host application uses the HCI interface to send command packets to the Link Manager, such as setting up a connection or starting an inquiry. The HCI residing in firmware, implements commands for accessing baseband, LMP and the hardware registers, as well as for sending messages upward to the host.

5.3.3.1.5 The Logical Link Control and Adaptation Protocol

The Logical Link Control and Adaptation Protocol (L2CAP) layer shields the specifics of the lower layers and provides interface to higher layers. The L2CAP layer supports the higher level protocol multiplexing, packet segmentation and reassembly and QoS maintenance, thereby eliminating the concept of master and slave devices by providing a common base for data communication.

5.3.3.1.6 The RFCOMM

RFCOMM is a simple transport protocol that emulates serial port over the L2CAP protocol, and is intended for cable replacement. It is used in applications that would otherwise use a serial port.

5.3.3.1.7 The Service Discovery Protocol

The Service Discovery Protocol (SDP) is defined to find availability of services for each Bluetooth entity. The protocol should be able to determine properties of any future or present service with an arbitrary complexity. This is a very important part of Bluetooth technology since the range of services available is expected to grow rapidly as developers keep bringing out new products.

5.3.3.2 The Bluetooth Profiles

A profile is defined as a combination of protocols and procedures that are used by devices to implement specific services. For example, the "headset" profile uses AT Commands and the RFCOMM protocol and is one of the profiles used in the "Ultimate Headset" usage model. Profiles are used to maintain interoperability between devices (i.e., all devices conforming to a specific profile will be interoperable), which is one of the Bluetooth's primary goals. Bluetooth has defined dozens of profiles addressing different applications.

5.3.4 Piconet Synchronization and Bluetooth Clocks

Every Bluetooth unit has an internal clock called the native clock (CLKN) and a Bluetooth clock is derived from this free running native clock. For synchronization with other units, offsets are added to the native clock to obtain temporary Bluetooth clocks (CLK), which are mutually synchronized. When a Piconet is established, the master's native clock is communicated to all its slaves to generate the offset value. The Master keeps an exact interval of M*625μsec (where M is an even, positive integer greater than 0) between consecutive transmissions. The slave's Rx timing is adjusted with any packet sent in the master-to-slave slot, whereas the slave's Tx timing is adjusted based on the most recent slave Rx timing. As shown in Figure 5.7, every unit participating in a Piconet uses the derived clock (CLK), for all timing and scheduling activities in the Piconet. For a master, the offset is zero; hence CLK and CLKN are identical.

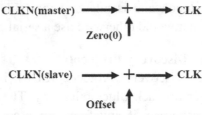

Figure 5.7 – CLK derivation in Bluetooth

5.3.5 Master-Slave Switch

The current Bluetooth specification provides means for a Master-Slave role switch (from now on referred to simply as M/S switch). This procedure is desirable on occasions such as:

- When a unit paging the master of an existing Piconet wants to join this Piconet.
- When a slave in the existing Piconet wants to set up a new Piconet involving itself as a master, and the current master as a slave.
- When a slave wants to fully take over an existing Piconet as a new master.

M/S switching is satisfied in three steps, namely TDD Switch, Piconet Switch for the previous master, and Piconet Switch for the remaining slaves. Consider a scenario when the Master device (B) wants to switch roles with a slave device (A). The details of messages exchanged and their significance are given in Table 5.2.

The time to perform the TDD Switch is the average amount of time between the instant the master decides to switch the Tx/Rx slot till the Tx/Rx slot is actually changed. It depends on the frequency with which the slave involved in the role switch is polled. The average time taken in an M/S switch ranges from 63 ms to 200 ms [BluetoothSpecwww]. Due to the substantial delay for an M/S switch, there is a perceivable period of silence for data or audio transmissions. A faster, delay bounded, predictive role switching is desirable and we introduce an improvement technique called Pseudo Role Switching later in this chapter.

Table 5.2 – Steps in Master (B)/Slave (A) role switching

Step	Message type	Direction	Purpose
1	LMP	B to A	Request to switch the Role (with Switch Instant)
2	LMP	A to B	Role Switch accepted (along with Slot Offset information)
3			TDD Switch at Switch Instant (max(32, 2*T-Poll))
4	FHS	A to B	Contains am_addr for B and FHS sequence
	ID	B to A	Acknowledgement
	Poll	A to B	Verify the Role Switch (with timer on)
	Null	B to A	Acknowledgement
5	LMP	A to others	Slot difference between new and old Piconet master
	FHS	A to others	am_addr (may be the same) and FHS sequence
	ID	Others to A	Acknowledgement
	Poll	A to others	Verify the Role Switch
	Null	Others to A	Acknowledgement
6	LMP	A to others	LMP connection establishment
	LMP	Others to A	Acknowledgement
7	L2CAP	A to others	L2CAP connection establishment
	L2CAP	Others to A	Acknowledgement

5.3.6 Bluetooth Security

Security is an important issue for any wireless technology. Bluetooth employs several layers of data encryption and user authentication measures. Bluetooth devices use a combination of the Personal Identification Number (PIN) and a Bluetooth address to identify other Bluetooth devices. Data encryption can be used to further enhance the degree of Bluetooth security [MobileInfowww]. Bluetooth uses transmission scheme that provides a self-level security. FHSS alleviates interference as the radio hops between the channels at a speed of 1600 hops per second. This provides some level of security on data transmission as it makes it harder to eavesdrop. In addition, the low power transmissions prevent the radio signals from propagating too far while additional protection is possible by encryption.

5.3.6.1 Security Architecture

Figure 5.8 presents a high level overview of the Bluetooth architecture together with the security components. The security manager stores information about the security of the services and the devices. On acceptance of the access or disconnection, it requires authentication and encryption if needed. The security manager also initiates setting up a trusted relationship and pairing, and handles the PIN code from the user.

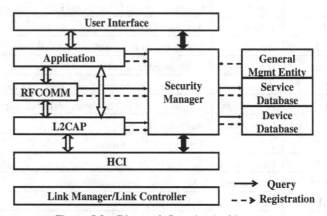

Figure 5.8 – Bluetooth Security Architecture

5.3.6.1.1 Security Levels

Bluetooth has several different security levels that can be defined for associated devices and services. All the devices obtain a status whenever they connect for the first time to some other device.

Device Trust Level

The devices can have two levels of trust: trusted and untrusted. The trusted level requires a fixed and trusted relationship and it has unrestricted access to all the services as, the device has to be previously authenticated. The untrusted device does not have fixed relationship and its access to services is limited. A new device is labeled as unknown device and is always considered untrusted.

Security Modes

In Bluetooth Generic Access Profile, there are three different security modes as follows:

- **Security Mode 1**: A device will not initiate any security procedure. This is a non-secure mode.
- **Security Mode 2**: A device does not initiate security procedures before channel establishment on L2CAP level. This mode allows different and flexible access policies for applications, especially when running with different security requirements in parallel. This is a service level enforced security mode.
- **Security Mode 3**: A device initiates security procedures before the link set-up on LPM level is completed. This is a link level enforced security mode.

Security Level of Services

When the connection is set up, there are different levels of security which the user can choose from. The security level of a service is defined by three attributes:

- **Authorization required**: Access is only granted automatically to trusted devices or untrusted devices after an authorization procedure;
- **Authentication required**: Before connecting to an application, remote device must be authenticated;
- **Encryption Required**: The link must be changed to encrypted mode, before access to the service is possible.

On the lowest level, the services can be set to be accessible to all devices. Usually, there is a need for some restrictions so the user can set the service and, hence, it need to be authenticated. When highest level of security is needed, the service can require both authorization and authentication and such a trusted device has access to the services. But, an untrusted device needs manual authorization.

5.3.6.1.2 Link Layer

In each device there are four entities used for security at the link level:

- The 48-bit public Bluetooth device address (BD_ADDR) is unique for each Bluetooth device and is defined by the IEEE.
- The private authentication key which is a 128-bit random number used for authentication purposes.
- The private encryption key varying from 8 through 128 bits in length is used for encryption.

- A 128-bit frequently changing random or pseudo-random number that is generated by the Bluetooth device itself.

BD_ADDR is used in the authentication process. When a challenge is received, the device has to respond with its own challenge that is based on the incoming challenge, its BD_ADDR and a link key shared between two devices. Other devices' BD_ADDRs are stored in the device database for further use.

5.3.6.2 Key Management

To ensure secure transmission, several types of keys are used in the Bluetooth system. The most important is the link key, which is used between two Bluetooth devices for authentication purposes. Using the link key, an encryption key is derived which secures the data of the packet and is regenerated for every new transmission.

5.3.6.2.1 Link Key

All security transactions between two or more parties are handled by a 128-bit random number called the link key and is used in the authentication process. Lifetime of a link key depends on whether it is semi-permanent or temporary. A semi-permanent key can be used after current session is over to authenticate Bluetooth units. A temporary key lasts only until current session is terminated and cannot be reused. Temporary keys are commonly used in point-to-multipoint connections, where the same information is transmitted to several recipients. Four keys are used to cover different types of applications.

The unit key, K_A, is derived at the installation of the Bluetooth device from a unit A. The combination key, K_{AB}, is generated for each new pair of Bluetooth devices. It is used when further security is needed. The master key, K_{MASTER}, is a temporary key used whenever the master device wants to transmit information to more than one device at once. The initialization key, K_{INIT}, is used in the initialization procedure. This key protects initialization parameters, and is used when there are no unit or combination keys. It is formed from a random number, an L-octet PIN code, and the BD_ADDR of the claimant unit.

5.3.6.2.2 Encryption Key

The encryption key is generated from the current link key, a 96-bit Ciphering Offset Number (COF) and a 128-bit random number. The COF is based on the Authenticated Ciphering Offset (ACO), which is generated during the authentication process and is discussed later on (see Figure 5.9). When the Bluetooth Link Manager activates the encryption, the encryption key is generated. It is automatically changed every time the Bluetooth device enters the encryption mode. The purpose of separating the authentication key and encryption key is to facilitate the use of a shorter encryption key without weakening the strength of the authentication procedure.

5.3.6.2.3 PIN Code

The PIN is a number to enhance the security of the system and can be either fixed or selected by the user. The length of the PIN code can vary between 1 and 16 octets. A regular 4-digit code is enough for some applications, but enhanced security requirements may need longer codes.

5.3.6.2.4 Key Generation and Initialization

Key exchange procedure is carried out separately for each unit during the initialization phase. All initialization procedures consist of the following five parts:
- Generation of an initialization key.
- Authentication.
- Generation of link key.
- Link key exchange.
- Generation of encryption key in each device.

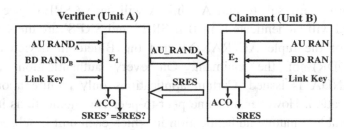

Figure 5.9 – Challenge-response scheme in Bluetooth

After this procedure, either connection is established or torn down.

5.3.6.3 Authentication

Authentication starts by issuing a challenge to another device which, in turn, sends a response back which is based on the received challenge, the recipient's BD_ADDR and link key shared between the devices. Once this procedure is successfully completed, authentication and encryption may be carried out. Without knowing the PIN, one unit cannot logon to another unit if authentication is activated. To make matters easier, the PIN can be stored somewhere inside the unit (e.g., Memory, Hard Drive, etc.), so if a connection is to be established, the user may not have to manually type in the PIN. Bluetooth uses a challenge-response scheme in which a claimant's knowledge of a secret key is checked through a 2-move protocol using symmetric secret keys, and is depicted in Figure 5.9. As a side product, the ACO is computed and stored in both devices and is later used to generate data encryption key that can be used between pair of devices.

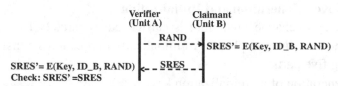

Figure 5.10 – Challenge-response for the symmetric key system

In Figure 5.9, unit A sends a random input, denoted by AU_RAND_A (a random number), with an authentication code, denoted by E1, to unit B. Unit B then calculates SRES as shown in Figure 5.10, and returns the result to unit A. Unit A will derive SRES' (see Figure 5.10) and will authenticate Unit B if SRES and SRES' are the same. E1 consist of the tuple AU_RANDA and the Bluetooth device address (BD_ADDR) of the claimant. On every authentication, a new AU_RANDA is issued. Certain applications only require a one-way authentication. However, in some peer-to-peer communications it might be preferable a mutual authentication in which each unit is subsequently the challenger (verifier) in two authentication procedure. The Link

Manager is responsible to determine in which direction(s) the authentication(s) has to take place.

5.4 Enhancements to Bluetooth

There is a vast literature on Bluetooth research. While initial focus was on interference analysis and mitigation, subsequent works have looked at various aspects of the technology, including Piconet and scatternet scheduling, scatternet formation algorithms, traffic engineering, QoS support, improving device discovery procedure, IP over Bluetooth, among others. In this section we give an overview of current research on Bluetooth which aims at either enhancing the current protocol performance or enabling it to support additional services.

5.4.1 Bluetooth Interference Issues

The 2.4 GHZ ISM radio frequency band is a broad, free and unlicensed spectrum space for used in microwave ovens, cordless phones, remote controllers, as well as Bluetooth and IEEE 802.11b/g (discussed in Chapter 4) devices. Therefore, all of these inventions have potential of interfering with each other [Cordeiro2002c, Derfler2000, Chiasserini2003]. In this section, we confine our discussion to interference in the 2.4 GHz band between Bluetooth and IEEE 802.11.

Bluetooth uses much lower transmission power than IEEE 802.11. Thus, powerful IEEE 802.11 devices may overwhelm its signal [Derfler2000]. To address this issue, the Task Group 2 within the IEEE 802.15 working group was established to improve the coexistence of the two standards. According to this working group, interference between IEEE 802.11 and Bluetooth causes a severe degradation in the overall system throughput when the distance between the interfering devices is within 2 meters. Between 2 and 4 meters, a slightly less significant degradation is observed [IEEEP802.15-145r12001].

Therefore, it is of paramount importance to analyze the impact of interference when Bluetooth and IEEE 802.11 devices operate in the same vicinity as well as when multiple Bluetooth Piconets are co-located. From now on, we refer to the interference generated by IEEE 802.11 devices over the Bluetooth channel as *persistent interference* [Cordeiro2002c], while the presence of multiple Piconets in the vicinity creates interference referred to as *intermittent interference* [Cordeiro2004b] (due to the frequency-hoped nature of the Bluetooth radio that generates interference in an intermittent fashion). Therefore, integrated solutions tackling both persistent and intermittent interference are of major interest. Obviously, it is sometimes possible to combine separate solutions into a single integrated scheme.

5.4.1.1 IEEE Efforts to Ensure Coexistence

The Bluetooth SIG and the task group 2 within the IEEE 802.15 working group defined mechanisms and recommended practices to ensure the coexistence of Bluetooth and Wi-Fi networks. In this context, *coexistence* is defined as the ability of one system to perform a task in a given shared environment where other systems may or may not be using the same set of rules [IEEE802.15.22000www]. These practices fall into two categories: collaborative and non-collaborative. In the following subsections we describe proposed mechanisms in each of this category.

Collaborative Mechanisms

A collaborative coexistence mechanism is defined where a WPAN and a WLAN communicate and collaborate to minimize mutual interference. Collaborative techniques require that a Wi-Fi device and a Bluetooth device be collocated (e.g., located in the same laptop or phone). TDMA (Time Division Multiple Access) techniques [IEEEP802.15-340r02001www, IEEEP802.15-300r12001www] allow Wi-Fi and Bluetooth to alternate transmissions. MEHTA [IEEEP802.15-300r12001www] (a Hebrew word for "Conductor") is a technique for managing packet transmission requests which grants permission to transmit a packet based on parameters like signal strength and difference between IEEE 802.11b and Bluetooth center frequencies. In addition, it can support SCO links. In conjunction with MEHTA, deterministic frequency nullifying [IEEEP802.15-364r2001www] mechanism is used that inserts a 1 MHz-wide null in the 22 MHz-wide IEEE 802.11b carrier that coincides with the current Bluetooth center frequency.

Non-Collaborative Mechanisms

A non-collaborative coexistence mechanism is one wherein there is no method for the WPAN and WLAN to communicate. There are several possible non-collaborative techniques. Adaptive packet selection and scheduling [IEEEP802.15-316r02001www] is a Bluetooth MAC-level enhancement that utilizes a frequency usage table to store statistics on channels that encounter interference. This table can subsequently be accessed by packet scheduling algorithms for transmissions to occur only when a good channel is made for the hop. Finally, adaptive frequency hopping (AFH) [IEEEP802.15-366r12001www] classifies channels and alters the regular hopping sequence to avoid interfering channels. This technique is widely used and will be described in more details.

5.4.1.2 Inter-Piconet Interference

With increasing scalability requirement, the number of co-located Piconets will eventually be so large that they will start interfering with each other (also called Intermittent Interference). The FHSS technique with 79 channels employed by Bluetooth will no longer suffice to keep interference at a desired minimum level, and presence of multiple Piconets create interference on the signal reception. Therefore, not only it is important to qualify and quantify such interference, but it also crucial to propose new ways to mitigate such negative effects. In [Cordeiro2003a], the impact of Intermittent Interferences on Piconet performance has been considered and this study serves as a basis for future work in this area.

Table 5.3 – DHx Throughput With/Without Interference (In Kbps)
[taken from Cordeiro2003a]

Packet Type	Ideal Conditions	Without Interference	With Interference
DH1	172.80	166.66	120.78
DH3	384.00	373.32	329.40
DH5	432.60	417.24	373.32

Table 5.3 gives a summary of the average throughput values obtained with the use of DHx packets with and without the presence of interference. A quick study of Table 5.3 indicates that the results are in line with the ideal ones when there is no interference. In the presence of interference, a drop of more than 30% in throughput is observed in DH1

links and lower throughput is experienced in all the cases, reinforcing a need for tailoring applications closer to these working conditions.

5.4.1.3 The Interference-Aware Packet Segmentation Algorithm

The Bluetooth standard defines various packet types to adjust according to different application requirements. Those range from single unprotected 1-slot packet to FEC (Forward Error Correction) encoded 5-slot packets. Ideally, the adaptation layer should choose the best suitable packet for transmission based both on the application requirements and the wireless channel condition. Furthermore, this choice cannot be static for the entire message due to dynamic nature of the error rate.

Motivated by these issues, [Cordeiro2002c] proposes an interference-aware algorithm called IBLUES (Interference-aware BLUEtooth Segmentation) to dynamically switch between Bluetooth packet types as packet error rates increases or decreases. This algorithm relies on the fact that interference is directly proportional to the packet error rate in Bluetooth (actually, it is not only interference that is taken into account in deciding the best suitable packet type). The rationale behind this algorithm is that a large packet outperforms a small packet in a low error rate channel (i.e., low interference level) since it possesses low overhead. On the other hand, small packets are best suitable when a channel has a high error rate (high interference level). In order to devise an accurate switching mechanism, it is mandatory to develop a combined model that takes into consideration interference from an environment consisting of multiple Piconets and IEEE 802.11b devices.

5.4.1.4 Overlap Avoidance Schemes

Two mechanisms, called OverLap Avoidance (OLA) schemes have been proposed in [Chiasserini2003] which are based on the traffic scheduling techniques at the MAC layer. The first mechanism, denoted as Voice OLA (V-OLA), is to be performed for the IEEE 802.11b in the presence of a Bluetooth voice (SCO) link. This scheme avoids overlap in time between the Bluetooth SCO traffic and IEEE 802.11b packets by performing a proper scheduling of the traffic transmissions at the IEEE 802.11b stations. In a Bluetooth network, each SCO link occupies FH/TDD channel slots according to a deterministic pattern. Therefore, an

IEEE 802.11b station shall start transmitting when the Bluetooth channel is idle and adjust the length of the WLAN packet so that it fits between two successive Bluetooth transmissions.

The second algorithm, denoted by data OLA (D-OLA), is to be performed at the Bluetooth system in case of a Bluetooth data link. As discussed before, the length of a Bluetooth data packet can vary from 1 to 5 time slots. In case of multi-slot transmissions, packets are sent by using a single frequency hop which is the hop corresponding to the slot at which the packet started. The key idea of the D-OLA scheme is to use a variety of packet length that characterizes the Bluetooth system so as to avoid frequency overlap between Bluetooth and IEEE 802.11b transmissions. Within each interfering Piconet, the D-OLA algorithm instructs the Bluetooth master device to schedule data packets with an appropriate slot duration so as to skip the frequency locations of the hopping sequence that are expected to use IEEE 802.11b band. An advantage of the OLA schemes is that they do not require a centralized packet scheduler. On the other hand, they do require changes in both the IEEE 802.11b standard and the Bluetooth specifications.

5.4.1.5 BlueStar: An Integrated Solution to Bluetooth and 802.11

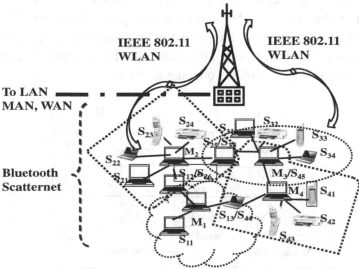

Figure 5.11 – The BlueStar architecture [Taken from Cordeiro2004b]

As we have so far discussed, it is most likely that Bluetooth devices and IEEE 802.11 WLAN stations operating in the same 2.4 GHz ISM frequency band should be able to coexist as well as cooperate with each other, and access each other's resources. These technologies are complementary to each other and such an integrated environment could be envisioned to allow Bluetooth devices to access the WLAN, and the Internet (integration of heterogeneous network is covered in Chapter 14). These cooperative requirements have lead to the BlueStar architecture [Cordeiro2004b], whereby few selected Bluetooth devices, called Bluetooth wireless gateways (BWG) are also members of a WLAN, thus empowering low-cost, short-range devices to access the global Internet infrastructure through the use of WLAN-based high-powered transmitters. This architecture is illustrated in Figure 5.11. Obviously, it is also possible that Bluetooth devices might access the WAN through a 3G/4G cellular infrastructure like UMTS, WiMax or LTE.

To combat both intermittent and persistent interference and provide effective coexistence, a unique hybrid approach of AFH, introduced earlier, and a new mechanism called Bluetooth carrier sense (BCS) are employed in BlueStar. AFH seeks to mitigate persistent interference by scanning the channels during a monitoring period and labeling them as "good" or "bad", based on whether the packet error rate (PER) of the channel is below or above a given threshold. BCS takes care of the intermittent interference by mandating that before any Bluetooth packet transmission, the transmitter has to sense the channel to determine the presence of any ongoing activity.

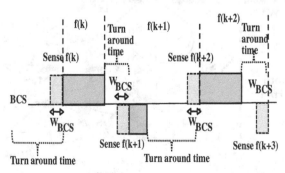

Figure 5.12 – Carrier sensing mechanism in Bluetooth

As shown in Figure 5.12, this channel sensing is performed during the turnaround time of the current slot, and it does not require any changes to the current Bluetooth slot structure. As we can see, BlueStar can be classified as a non-collaborative solution in the sense that the Bluetooth and the WLAN system operate independently, with no exchange of information. As shown in [Cordeiro2004b], the BlueStar architecture approximately doubles the performance of a regular Bluetooth. One disadvantage of Bluestar is that it requires considerable changes to Bluetooth specifications such as carrier sensing.

5.4.2 Intra and Inter Piconet Scheduling

The traffic in a Piconet is coordinated by the master device, and no two slaves can communicate without master's intervention. In other words, no slave is allowed to transmit without previously being polled by the master device as per some scheduling algorithm schemes and are referred to as intra-Piconet scheduling.

When the communication spans more than one Piconet over a scatternet, the scheduling does take a different perspective. It is now necessary for the master to synchronize (i.e., schedule) the presence of its bridge nodes in its Piconet. As the bridge nodes may "randomly" switch amongst Piconets, the master nodes of the corresponding Piconets may find themselves in a situation when the bridge is being polled while it is currently participating in some other Piconet. Therefore, appropriate protocols are necessary for master nodes to negotiate with a bridge node its presence in Piconets. This is referred to as inter-Piconet scheduling.

5.4.2.1 Intra-Piconet Scheduling

Bluetooth polling differs from classical polling in that the transmission from the master to a slave is always combined with the corresponding slave to master transmission. Therefore, the master has only partial status knowledge of slaves' queue states, while it only knows its own queues. Thus, classical polling models cannot be directly used, while they can still be used as benchmarks. In a Piconet, a polling system model can be adopted to describe the Bluetooth MAC operation. As seen earlier, Bluetooth supports two types of links between a master and a slave: SCO and ACL links. In case of an SCO link, the master has to poll the slave at regular intervals, given the stringent requirements of this type of traffic. Therefore, for SCO links, the master device does not have much freedom to use one or another scheduling algorithm. In case of an ACL link, however, polling can be performed in many different ways. Given this, we now focus on scheduling for ACL links only.

The Limited and Weighted Round Robin Scheme (LWRR)

Limited and Weighted Round Robin (LWRR) adopts a weighted round robin algorithm with weights being dynamically changed according to the observed status of the queues. In other words, LWRR considers the activeness of the slaves. Initially, each slave is assigned a weight, say W, which is reduced by one each time a slave is polled and no data is exchanged. Therefore, the slave misses as many chances in the polling cycle as is the difference between its current weight and W. The lowest value that a slave's weight can achieve is one, meaning that it has to wait a maximum of W-1 cycles to send data. Anytime there is a data exchange between a slave and master, the weight of the slave is reset to the W value.

On the down side, LWRR has some disadvantages, which is mainly due to the constantly changing number of slots during a polling cycle. First, an inactive slave may need to wait for a long time to get a chance to exchange data packets if the preceding polling cycle had a large number of slots. This can lead to a large delay for an idle slave. Second, an idle slave may be frequently polled if the previous polling cycles had a small number of slots. This may, therefore, reduce the efficiency of the system. LWRR scheme is just a very simple way of introducing the notion of "activeness" of slaves. In addition, the computation is not complex and required storage space is small.

The Pseudo-Random Cyclic Limited slot-Weighted Round Robin (PLsWRR)

The Pseudo-Random Cyclic Limited slot-Weighted Round Robin (PLsWRR) is based on two main properties:
* It attempts to distinguish between slaves on the basis of their "activeness", i.e., according to their traffic history. PLsWRR reduces

the rate of polling by not polling them for a certain number of slots. This lowers the bounds for maximum time between polls to a slave.

• The order in which slaves are polled in each cycle is determined in a pseudo-random manner so as to improve fairness. PLsWRR scheme has been shown to provide a certain degree of fairness and perform well on scenarios with different traffic sources like TCP and CBR.

PLsWRR is comprised of two main parts: a Pseudo-Random Cycle of Polling and a Limited slot-weighted Round Robin (PLsWRR) scheme. In the Pseudo-Random Cycle of Polling, the master keeps a separate buffer for each destination slave. For example, assume a scenario where slaves 2 and 4 are both transmitting to a slave 6. With plain Round Robin, slave 2 would have a higher chance to transmit a packet and get a buffer at the master as opposed to slave 4. Pseudo-random ordering attempts to break this unfairness between connections of slaves 2 and 4.

In the LsWRR scheme, each slave is assigned a slot-weight equal to Max-Slot-Priority (MSP) and is changed dynamically according to the outcome of previous poll, reduced if no data exchanged and increased otherwise. The lowest value a slot-weight can take is equal to 1. If a slave has skipped as many slots as the difference *MSP – current-slot-weight*, the master decides to poll the slave. Similarly, an active slave cannot be polled beyond a maximum number of times per cycle. LsWRR employs the number of slots as opposed to the number of cycles (as in LWRR) as a means to reduce the number of polls to less active slaves. LsWRR guarantees that a slave waits for a maximum of *MSP* slots until they get a chance to be polled again. Clearly, this makes the behavior more reliable as compared with LWRR, where the slave waits a bounded number of cycles but the length of these cycles may vary.

The Fair Exhaustive Polling (FEP)

In [Johansson1999], the Fair Exhaustive Polling (FEP) can be viewed as a combination of strict round robin polling and exhaustive polling that poll slaves not having to send anything. FEP achieves this by introducing two complementary states, namely, the active state and the inactive state, and also by associating a weight with each slave. In FEP, a polling cycle starts with the master moving all the slaves to active state, and then initiating one of several possible polling sub cycles in a round robin fashion. One distinguishing feature of FEP is that the master performs the task of packet scheduling for both the downlink (master to slave) and uplink (slave to master) flows. However, the master has only limited knowledge of the arrival processes at the slaves, which means that the scheduling of the uplink flows has to be based on the feedback it obtains when polling the slaves. A slave is moved from the active state to inactive state when two conditions are fulfilled. First, the slave has no packet to send. Second, the master has no packet to be sent to the corresponding slave. A slave is moved to the active state when the master has some packet destined for it.

This is an iterative process and continues until the active state is emptied (the exhaustive part of the algorithm), and when this happens, a new polling cycle starts. It is important to note that the exhaustive part of FEP is different from ERR. To implement fairness into FEP, a polling interval of a slave is added to some predetermined maximum time. This means that a slave, whose maximum polling interval timer has expired, is moved to the active state and is therefore polled in the next polling sub-cycle. This maximum polling interval is used by FEP to ensure that an inactive slave is regularly polled, and thus ensure if it has become active.

The Predictive Fair Poller (PFP)

The Predictive Fair Poller (PFP) [Yaiz2002] is a polling scheme that takes both efficiency and fairness into account similar to LWRR and FEP. For each slave, it predicts whether data is available or not while keeping track of the fairness. Based on these two aspects, it decides which slave to poll next. In the best effort case, PFP estimates the fair share of resources for each slave and keeps track of the fractions of these fair shares that each slave has been given. PFP distinguishes two types of traffics: the best effort and the QoS-based. For best effort traffic, PFP keeps track of both the fairness based on the fractions of fair share and the predictions, and thus can guarantee to poll the best effort traffic in a fair and efficient manner. In the QoS-based case, QoS requirements are negotiated with the slaves and translated to fair QoS entities. The polling unit, in turn, keeps track of the fractions of these fair QoS treatments that each slave has been given. Simulation results indicate that PFP outperforms the PRR in the scenarios analyzed, and that it performs at least as well as the FEP. The advantage is that it can adapt to different QoS requirements. However, it is considerably complex which makes it harder to implement.

The Demand-Based Bluetooth Scheduling

A flexible polling scheme is proposed in [Rao2001] that initially adopts common polling periods for all slaves, and subsequently increases the polling period for those slaves with less traffic load. Similar to other schemes, the idea here is to poll slaves that probably have to send as little as possible. Ultimately, the goal is to maximize throughput and to reduce the overall Piconet power consumption. This new polling scheme, referred to as Demand-Based Bluetooth Scheduling, is based on a scheduling table. Firstly, bridge nodes and synchronous slaves are scheduled. Secondly, asynchronous dedicated slaves (ADSs) are considered. This scheme has some advantages. The second part (ADSs scheduling scheme) can be used not only in single Piconet scheduling, but also for inter-Piconet scheduling over a scatternet. That means it can cooperate with some scheduling table-based bridge scheduling scheme so as to form a scatternet scheduling scheme. To improve on energy consumption, it allows slaves to be parked. On the other hand, ADSs may increase the access latency of the slaves. It has better throughput, however, when slaves have unequal traffic load.

5.4.2.2 Inter-Piconet Scheduling

We now turn our attention to the issue of inter-Piconet scheduling, also known as scatternet scheduling. As we mentioned earlier, inter-Piconet scheduling addresses the issue of defining appropriate protocols for master of each Piconet so as to negotiate presence of their corresponding bridge nodes. This is necessary to ensure an appropriate scatternet performance, and efficient communication amongst various scatternet devices.

Distributed Scatternet Scheduling Algorithm (DSSA)

Distributed Scatternet Scheduling Algorithm (DSSA) is proposed in [Johansson2001b], which provides a conflict free access to the shared medium. DSSA can be executed in parallel and is adaptive to topological changes, but does nothing as traffic changes. Using graph theory, it has been proved [Johansson2001b] that defining an optimal scheduler for a scatternet is NP-complete. In addition, it has been proved that the bound for DSSA is polynomial in terms of number of nodes.

DSSA is based on the assumption that nodes have distinct identities (IDs) and are aware of the IDs and traffic requirements of their neighbors. In DSSA, each master needs the permission of all its neighbors to schedule its Piconet. Permission is granted to the neighboring master with the highest ID among those neighboring masters that have not yet scheduled by their Piconets. Permissions together with a set of restrictions are passed in messages, specifying which frames cannot be allocated due to previous commitments by this slave with other neighboring masters. After receiving the permission rights from all the slaves, master assigns those timeslots. The algorithm terminates when all masters have scheduled their Piconets.

DSSA is an ideal algorithm as it assumes that nodes are aware of the traffic requirements of their neighbors, which cannot be realized in a real scenario. In addition, given the requirement that the solution is carried out from the highest ID master to lowest ID master, not all master devices are treated equally. As DSSA allocates communication time slots for each pair of nodes for the whole scatternet in advance, it can be classified to have a hard coordination scheme which is suitable for those scatternets whose traffic patterns are known in advance and do not vary much over time. In the case of bursty traffic, DSSA may generate too much overhead as scatternet-wide bandwidth has to be allocated in advance, thus demanding significant computation and signaling overhead. In addition, slots have to be reallocated in response to changes in the traffic intensity and each time a connection is established or released.

Pseudorandom Coordinated Scheduling Scheme (PCSS)

Pseudorandom Coordinated Scheduling Scheme (PCSS) [Racz2001] falls in the category of soft coordination based scheme. Wherein the nodes decide their presence in Piconets based on local information. By nature, soft coordination schemes cannot guarantee conflict-free participation of bridge nodes in different Piconets. However, they will have much lower complexity than a hard coordination scheme. In the PCSS algorithm, coordination is achieved by implicit communication rules without the need of exchanging explicit control information. The low complexity of the algorithm and its conformance to the current Bluetooth specification makes it easy to be incorporated. Every node randomly chooses a communication checkpoint that is computed based on the master's clock and the slave's device address. When both end nodes show up at a checkpoint simultaneously, they start to communicate until one of the nodes leaves for another checkpoint. In order to adapt to various traffic conditions, PCSS adjusts the checking period according to previous link utilization.

The advantage of PCSS is that it achieves coordination among nodes with very little overhead. However, as the density of nodes grows, there will be scheduling conflicts among various checkpoints, resulting in missed communication events between two nodes. In response to changing traffic on a link, PCSS increases or decreases the interval between two successive communication events on that link by a fixed multiple intervals. It is quite rough and inaccurate, however, to adapt to the traffic burstiness. Moreover, it neither changes the duration of communication events nor coordinates with other links.

Locally Coordinated Scheduling (LCS)

While there are conflicts in the PCSS, the Locally Coordinated Scheduling (LCS) scheme coordinates nodes to eliminate all scheduling conflicts. In response to bursty traffic on a link, LCS adjusts both the intervals between communication events and duration of those events, while PCSS changes only the intervals between events. This makes LCS more responsive to bursty traffic as it is based on the concept of scheduled meetings called appointments. It optimizes the overall efficiency of scatternet in terms of throughput, latency and energy, by minimizing missed communication opportunities. It also allows nodes to tradeoff between energy efficiency and latency. However, LCS can only be applied to loop-free scatternet topologies. Simulation results show that LCS achieves better TCP throughput, low packet latency and low node activity time (which corresponds to low energy consumption) for low bandwidth applications. LCS is able to achieve an efficient scatternet wide schedule through the following procedure:

- Computing duration of the next meeting based on queue size and past history of transmissions so that the duration is just large enough to exchange all the backlogged data.
- Computing start time of the next meeting based on whether the data rate observed is increasing, decreasing or stable so that it responds to varying traffic conditions quickly without wasting resources.
- Grouping together meetings with the same traffic characteristics to reduce wasted bandwidth of nodes and end-to-end latency.
- Aligning meetings at various parts of the scatternet in a hierarchical fashion so that the number of parallel communication is high, increasing system-wide throughput significantly.
- Reducing amount of time a node spends transmitting packets while the receiver is not ready, thus conserving energy.

As we can see, the procedure used in LCS is rather complex and its implementation is complicated. In addition, Bluetooth devices may not have the computational capability to implement it efficiently.

Flexible Scatternet wide Scheduling (FSS)

Compared to Demand-Based Bluetooth Scheduling where master devices potentially have a large table, the Flexible Scatternet wide Scheduling (FSS) [Zhang2002] scheme uses a table in a simpler and more efficient manner. FSS consists of two algorithms: a flexible traffic scheduling algorithm executed by each master, and an adaptive switch-table modification algorithm executed by each bridge node. FSS is based on a switch-table concept, which is constructed when the scatternet is formed. Each bridge node uses a switch-table to direct switch between its multiple Piconets. To avoid bridge conflicts, a master polls a bridge node only at those slots when the bridge node is known to be synchronized to the Piconet controlled by the master. Each master, in turn, employs a flexible traffic scheduling algorithm to schedules both dedicated slaves and bridge nodes. Moreover, the switch-table can be dynamically adjusted based on the traffic load so as to improve the system performance. Compared to some static schemes, FSS can significantly improve the system throughput and reduce the packet transmission delay.

Flexible Traffic Scheduling Algorithm: In order to decide the polling frequency, each slave has a polling weight which is represented by (P, R), where P indicates that the slave should be polled every P schedule cycles, and R represents the maximum number of times that the slave can be polled in a cycle. The rules for adjusting the slave's polling weight (P, R) are as follows. If a poll is wasted in the sense that both slots allocated for polling are not used, the value of P associated with the offending slave is increased until it reaches a certain upper threshold; otherwise the polling period is decreased until it reaches one. If the current P value for a slave is already one, the value of R will be adjusted as follows. If a poll is wasted, the value of R is decreased until it reaches one, and in this case, the value of P is increased; otherwise, the value of R is increased until it reaches the upper bound.

Flexible Switch-table Modification Algorithm: Here, the bridge node consistently monitors the outgoing and the incoming queue lengths, where the incoming queue length is obtained by piggybacking this information in packets sent by the masters. Based on this information, the bridge can estimate which path has a heavy traffic load and which path does not. Obviously, the master with the longest queue length should get more time slots as compared to those masters with relatively low queue lengths. In order to avoid bridge conflict, the bridge cannot immediately satisfy the borrower's requirement if there is no idle slot in the switch-table. Instead, the bridge has to find a lender and obtain an acknowledgement from this lender that it will not use its allocated time slot. Only then the bridge node will be able to assign the time slot to the borrower, which may start to use the borrowed time slot. Since the borrowing process involves many messages and the bridge has to wait for its turn to communicate with the masters, the borrowing process should only be started when absolutely necessary.

Credit Based Scheduling (CBS): The Credit Based Scheduling (CBS) [Baatz2001] introduced in this section, the Load Adaptive Algorithm (LAA) and the scheduling algorithm based on the JUMP mode described in the following two sections form a class of scheduling algorithms that are built around the low power modes available in Bluetooth. The CBS algorithm is based on the Bluetooth sniff mode. It defines presence points for each inter-Piconet link at which communication may start. The rationale behind these points is to enable each device to quickly determine whether the peer device is in the same Piconet. If so, the communication may start between the devices. Otherwise, another presence point may be tried without having lost much bandwidth. The length of a particular communication period is not predetermined, as it depends on the current link utilization and the amount of data to be exchanged. Interestingly, the presence points and the dynamic length of communication periods may be mapped directly onto the sniff mode, requiring little or no changes to the current Bluetooth specification. The communication schedule is then determined online for each communication period.

Finally, [Baatz2002] proposes an enhanced adaptive scheduling scheme based on CBS. Here, link level fairness is achieved through a slot accounting scheme that is able to redistribute unused bandwidth following the idea of min-max fairness.

The Load Adaptive Algorithm (LAA): In [Har-Shai2002], the Load Adaptive Algorithm (LAA) is proposed for small-scale scatternets. While CBS uses the sniff mode, LAA uses the hold mode. The primary difference between these two modes is that the duration of the hold period is set every time the slave is placed in the hold mode, whereas the parameters of the sniff mode are set once and can be reused for many intervals. Thus, the hold mode requires repeated negotiations that waste at least a pair of slots, while the sniff mode requires a single negotiation. Therefore, the sniff mode may be more suitable for steady traffic, whereas bursty traffic may be better supported by the hold mode. LAA takes into account a few decision variables and parameters for its functioning, and these are as follows:

- *Idle State (IS)*: The bridge is in IS if either the queue of the current Piconet is empty or it received a NULL (non-data) packet.
- *Max Queue Size (MQS)*: If the queue size is larger than MQS, the bridge node should try to switch Piconets.

- *Time Commitment (TC)*: The bridge node sends this variable before a Piconet switching and indicates the minimum time interval the bridge will spend outside the Piconet. It calculates based on the length of the bridge's queues to the other Piconets which allows the master not to address the bridge throughout this interval, and to readdress it once it expires.
- *Predictability Factor (β)*: The Predictability Factor (β) is used in order to estimate the average packet size of this traffic and to compute the value of TC.
- *Max Time-Share (MTS)*: In cases of heavy traffic, the queue sizes may be huge and therefore the TCs derived from them will also be long. Thus, the maximum time a bridge spends in a Piconet has to be bounded. We refer to this bound as the MTS.

LAA manages the scheduling mechanism of the bridge by determining the duration of the bridge activity in different Piconets, so that the delay incurred by packets requiring inter-Piconet routing can be reduced. The algorithm adapts to varying loads by utilizing information regarding its queues to the different masters, and also by using information transferred by the masters. LAA complies with the Bluetooth specification in the following way. When the bridge switches to another Piconet, it enters hold mode in the first Piconet and sets the hold timeout to TC. Once TC expires, the master polls the bridge every few slots according to its polling scheme. After the bridge returns to the Piconet, the master should then poll it with a higher priority. As the bridge node might not return immediately after TC expires, the value of the link supervision timers should be set to a value that does not create false connection drops.

The JUMP Mode Based Scheduling Algorithm: In [Johansson2001a], a scheduling scheme for scatternets based on the JUMP mode is proposed, and aims at allowing flexible and efficient scatternet operation and overcome the shortcomings of current Bluetooth modes. This scheme includes a set of communication rules that enable efficient scatternet operation by offering a great deal of flexibility for a node to adapt its activity in different Piconets as per the traffic conditions. Using the

JUMP mode, a bridge node divides the time into time windows and then signals about which Piconet it is going to be present for each of these time windows. The time windows are of pseudo random length to eliminate systematic collisions and thereby avoid starvation and live-lock problems without any need for scatternet-wide coordination. Besides enabling scatternet operation, the JUMP mode may also enhance other aspects of Bluetooth such as the low-power operation.

Table 5.4 – Comparison of the various scatternet scheduling algorithms

Characteristics	DSSA	PCSS	LCS	FSS	LAA	CBS	JUMP mode
Dynamic	No	Yes	Yes	Yes	Yes	Yes	Yes
Ideal	Yes	No	No	No	No	No	No
QoS	No	No	No	Yes	No	Yes	No
Scatternet topology	Any	Any	Loop-free	Any	Any	Any	Any
Modifies Bluetooth specifications	No	No	Yes	No	No	No	Yes
Computational complexity	High	Low	High	Quite low	Low	Low	Low
Feasibility	No	No	No	Yes	Yes	Yes	No
Basic technique	Graph theory	Pseudo random technique	Appoint ment	Switch Table	Hold mode	Credit scheme	JUMP mode

Table 5.4 compares various scatternet scheduling algorithms discussed so far. As we can see, they are compared under various criteria taking into consideration various important aspects such as whether the algorithm is dynamic or not, if it is ideal (i.e., the master knows the updated length of all queues), if QoS can support (e.g., if the scheduler can be optimized system wide throughput, end-to-end communication delay, or energy consumption), as long as it is in compliance with current Bluetooth specifications.

5.4.3 Bridge Selection

As we have discussed before, in Bluetooth large ad hoc networks are formed by inter-linking individual Piconets to form a scatternet. Scatternets are formed by sharing one or more slaves (the bridges nodes)

in a time division multiplexed system, wherein the bridges share their active time period between two Piconets. Theoretically, the bridge can be a master in one Piconet and a slave in another Piconet, or a slave in both Piconets. In practice, however, most current research considers bridges in the slave-slave configuration only, as having a bridge to be a master in one Piconet will result in this Piconet being idle every time the bridge is in a part of some other Piconet.

The bridge bears the responsibility of a switch, buffering incoming data packets, then switching to another Piconet and relaying the buffered packets to a new master. This means that they are always transmitting, receiving or switching between Piconets. While this might work for low to even medium traffic conditions, it essentially drains the bridge energy and, at high traffic loads, the bridge may be overwhelmed, causing buffer overflows, packet drops and increased end-to-end delay. A bridge node that is drained of power will die, disrupting inter-Piconet traffic and causing a heavy loss of packets. While this may be inevitable, it must be ensured that the device (and hence the scatternet) is alive for as long as possible. Therefore, in order to achieve energy efficiency in a Bluetooth scatternet, an effective policy for bridge management is needed.

So far, energy efficiency in Bluetooth has been tackled by using the default low power modes with some modifications. In [Lin2002], power is saved by scheduling the occurrence of the SNIFF slots and the length of each occurrence, while [Prabhu2002] discusses power control by using cost metrics associated with routing and switching the roles of the master and the slave. Two mechanisms that focus on sharing the responsibility of being a bridge among devices that are capable of handling such work have been introduced in [Duggirala2003a, Duggirala2003b]. Here, energy usage pattern is spread out and lifetime of the scatternet is increased. The idea is to concentrate on sharing the bridge responsibility and responding to energy changes based on a node's capability, the relative power levels of other prospective bridge slaves and on the traffic conditions. The ultimate goal is to increase the scatternet lifetime by extending bridge's lifetime. As shown in [Duggirala2003a, Duggirala2003b], these policies can be used with any scheduling scheme [Lin2002, Baatz2002, Johansson2001a] or routing mechanism [Prabhu2002]. It is important to note that these policies are executed at the master, and is relevant only for inter-Piconet communication. Each Piconet in the scatternet can run these policies, and to avoid control message overhead, the energy values of every device are piggy-backed together with the existing Bluetooth ACK packet.

5.4.4 Traffic Engineering

If a larger number of connections ought to be supported, it either drastically increases the delay or simply blocks the incoming traffic. These problems are rooted in the master-centric packet-forwarding paradigm, with its inability to serve the additional demands exceeding the 1 Mbps nominal bandwidths provided by Bluetooth.

The bottom line of these problems is the lack of Traffic Engineering techniques in current Bluetooth. Traffic Engineering has been shown to be extremely useful for Internet [Awduche2002], by efficiently transferring information from a source to an arbitrary destination with controlled routing function that steer traffic through the network. A systematic application of Traffic Engineering helps in enhancing the QoS delivered to end-users, and aids in analyzing these results. Traffic Engineering suggests both demand side and supply side policies for minimizing congestion and improving QoS. Demand side policies restrict access to congested resources, dynamically regulates the demand to alleviate the overloaded condition, or control the way the data is routed in the network. Supply side policies augment network capacity to better accommodate the traffic.

Traffic Engineering into Bluetooth has been suggested in [Abhyankar2003] by employing the demand side and supply side policies [Awduche2002] in the form of Pseudo Role Switching (PRS) and Pseudo PaRtitioning (PPR) schemes. PRS would maximize bandwidth utilization and minimize latency within Piconet, while PPR would dynamically partition Piconet as traffic demand exceeds Bluetooth capacity. Preliminary results shows up to 50% reduction in the network overhead and up to 200% increase in the aggregate throughput.

Current Bluetooth specifications do not say anything about slave-to-slave communication. If one of such communication has to be supported, this will take place from source slave to master and from master to destination slave. This would effectively consume double bandwidth and higher delay due to non-optimal communication path. This is first major problem related to inefficient use of the bandwidth. Previous studies [Capone2001, Kalia1999] have indicated the drawbacks of existing scheduling techniques and have suggested several modifications to widen the scope of applications running on Bluetooth devices but did not address the scenarios discussed above.

Maximum throughput that can be obtained theoretically in Bluetooth is 1 Mbps using 5-slot length packet. But, in audio applications, 1-slot length packet is used and 64 kbps bandwidth is supported. This limits the master to support maximum 3 such connections (practically only two such simplex connections). If it happens to be a slave-to-slave request, with the master acting as a relay, only one such audio transmission could be supported within a single Piconet, since the master needs to allocate 2/3 of the total bandwidth for such single connection. In [Lim 2001], it is suggested that a new Piconet be formed for each new connection and can be said to make an effective use of the channel while keeping the delay of all the connections low. However, the increase in the number of Piconet causes noticeable increase in inter-Piconet interference [Cordeiro2003a]. Therefore, formation of a new Piconet for each slave-to-slave transmission is not a good solution. Given this, a PPR scheme was introduced in [Abhyankar2003] in which the Piconet is partitioned dynamically after the master reaches its maximum capacity in order to support a higher traffic rate. Another significant aspect of traffic engineering is minimizing congestion as it helps in a proactive as well as reactive way to improve network performance. PRS and PPR are proactive in nature as they take measure to control congestion before it takes place.

Pseudo Role Switch

PRS would not require any change in FHS as this scheme keeps the Piconet synchronized on the previous Piconet parameters. Demand side traffic engineering is suggested by categorizing the requests based on the type of data being transmitted. For example, *audio* data has critical latency requirements, *telnet* traffic needs quick response time, *ftp* traffic needs reliable communication, etc. In role switching decisions, priority is given to those connections which have stringent QoS requirements while at the same time supporting less constrained communications. So, audio traffic is given priority over telnet, which in turn is given priority over ftp. This scheme should be very useful in numerous situations as follows:

- When master is not involved in any data transfer and receives a connection request from one slave to another slave, it should switch the role to reduce delay and bandwidth consumption.
- A connection with higher priority arrives at master based on the aforementioned categories. Priority should be given to QoS-constrained slave and thus role switching is desirable.
- The existing connections terminate and some ongoing traffic connection still exists between two slaves. Here, one of the slaves involved in the connection should become a master.
- A master device is running out of its battery, which can result in sudden crashing of the Piconet. In this case, role switch is vital for continuing operation of the Piconet.

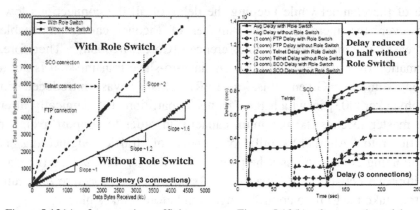

Figure 5.13(a) – 3 connections efficiency   Figure 5.13(b) – 3 connections delay

Note that if the master decides to switch role in response to a new connection request, it has to exchange control information about role switching message, LMP and L2CAP data connections. So, even before data transmission starts, some bandwidth is consumed by exchange of control message. On the other hand, if PRS scheme is not implemented, for every slave-to-slave packet exchange two data transmissions are needed; one from source slave to master and the second from master to destination slave. Figure 5.13(a) shows the graph for data bytes received versus actual information transmitted over the network where ftp connection is followed by telnet connection and telnet is followed by SCO connection. Moreover, the delay characteristics in Figure 5.13(b) reveals that this PRS scheme manages to reduce the delay to almost ½ of its original value.

Pseudo Partitioning

In PPR, the Piconet is partitioned when the need for bandwidth cannot be fulfilled by the current structure. The decision partitions the Piconet in such a way so that devices consuming most of the bandwidth are separated. Also, this type of partitioning should not last forever and rejoining the Piconet should be made possible as soon as the traffic in one of the Piconet ends. Certain threshold value should be maintained to avoid continuous partitioning and rejoining upon every connection arrival and termination. Such a decision can only be taken if the master knows negotiated QoS parameters while establishing all previous connections.

The performance of PPR under overloaded conditions by dynamically generating FTP connection requests has been evaluated with the following conclusions:

- Number of total packets transmitted per data packet received ratio is minimized when both PRS and PPR schemes are in action, as depicted in Figures 5.14(a).
- Increased aggregate throughput. Figure 5.14(b) shows the improved performance of PPR.

5.4.5 QoS and Dynamic Slot Assignment

When we consider the limitations of current Bluetooth, namely, the support of a very limited number of audio connections (e.g., at most one duplex audio connection), no delay or throughput guarantees to data connections, and the lack of end-to-end QoS guarantees, simple QoS primitives ought to be devised in order to support basic application QoS demands.

Therefore, a QoS-driven Enhanced Dynamic Slot Assignment (EDSA) scheme has been proposed in [Cordeiro2004] to address these major shortcomings while keeping the simplicity of Master/Slave in the Bluetooth paradigm. The basic strategy is to combine the QoS-driven Dynamic Slot Assignment (DSA) and the dynamic Piconet partitioning. Here, DSA-only is employed at the Piconet level while dynamic partitioning opens up the scope of DSA to the scatternet level. The basic idea behind DSA is to appropriately manage the polling cycle conducted by the master of the Piconet, given the connection QoS requirements.

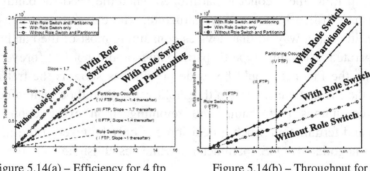

Figure 5.14(a) – Efficiency for 4 ftp connections   Figure 5.14(b) – Throughput for 4 ftp

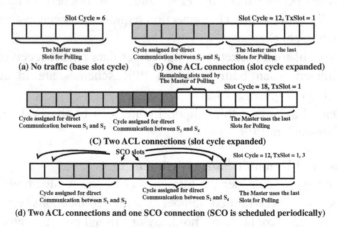

Figure 5.15 – Dynamic assignment of slots and expanding slot cycle

As shown in Figure 5.15, as devices initiate or terminate communication with each other within the Piconet, the Piconet polling cycle is restructured (expanding it with a new connection or shrinking it upon termination), a new transmission schedule is built for each Piconet device, and then the resulting schedule is propagated to the members of the Piconet. This way, slaves know exactly in which slot to transmit and/or listen. Therefore, not only is direct communication between slave devices supported, but also a multicast-like communication by having several destination slaves listen to the same slot is achieved. As detailed in [Cordeiro2004], the Piconet broadcast address is temporarily allocated in DSA so as to implement multicasting. As we have mentioned before, slave-to-slave communication will be present in approximately 75% of all connections, thereby stressing the need for supporting and optimizing such cases.

In order to widen the scope of DSA to scatternets, effectively support application QoS demands, and provide effective scalability, a controlled form of dynamic partitioning has been developed to interoperate with DSA. This new scheme is referred to as Enhanced DSA (EDSA). EDSA dynamically partitions Piconets when application QoS demands cannot be satisfied by the current slot allocation. The partitioning is guided by the connection endpoints as EDSA tries to keep communicating devices within the same Piconet. If this cannot be achieved, EDSA carefully synchronizes slot allocations of neighboring Piconets so that data can be transferred from one Piconet to another over the scatternet, hence providing uninterrupted communication. As shown in [Cordeiro2004], the application of EDSA provides increased application and system performance, effective QoS guarantees, and enhanced scatternet support.

5.4.6 Scatternet Formation

An ad hoc network based on Bluetooth brings with it new challenges. There are specific Bluetooth constraints not present in other wireless networks. As shown in [Miklos2000], the configuration of a scatternet has significant impact on the performance of the network. For example, when a scatternet contains more Piconets, the rate of collisions among packet increases. Therefore, before we can make effective use of Bluetooth ad hoc networking, it is necessary to first devise an efficient protocol to form an appropriate scatternet from isolated Bluetooth devices. In the following we give a brief description into the field of scatternet formation by introducing the most prominent solutions. It is applied in [Miklos2000] heuristics to generate scatternets with some desirable properties. They evaluate these scatternets of different characteristics through simulations. Cross-layer optimization in Bluetooth scatternets is discussed in [Raman2001].

It is introduced in [Aggarwal2000] a scatternet formation algorithm which first partitions the network into independent Piconets, and then elects a "super-master" that knows about all network nodes. However, the resulting network is not a scatternet, because the Piconets are not inter-connected. Here, a separate phase of re-organization is required.

A scatternet formation algorithm denominated as Bluetooth Topology Construction Protocol (BTCP) is described in [Salonidis2001]. BTCP has three phases:

- A coordinator is elected with a complete knowledge of all devices.
- This coordinator determines and tells other masters how a scatternet should be formed.
- The scatternet is formed according to these instructions.

A formation scheme is then presented in [Salonidis2001] for up to 36 devices. Since the topology is decided by a single device (the coordinator), BTCP has more flexibility in constructing the scatternet. However, if the coordinator fails, the formation protocol has to be restarted. BTCP's timeout value for the first phase would affect the probability that a scatternet is formed. In addition, BTCP is not suitable for dynamic environments where devices can join and leave after the scatternet is formed.

In [Law2003], a two-layer scatternet formation protocol is presented. First, it is investigated how these devices can be organized into Piconets. Second, as a subroutine of the formation protocol, a scheme is proposed for the devices to discover each other efficiently. The main idea is to merge pairs of connected components until one component is left. Each component has a leader. In each round, a leader either tries to contact another component or waits to be contacted. The decision of each leader is random and independent.

A distributed Tree Scatternet Formation (TSF) protocol is presented in [Tan2002]. The extensive simulation results indicate relatively short scatternet formation latency. However, TSF is not designed to minimize the number of Piconets. The simulation results suggest that each master usually has fewer than three slaves. While the preceding protocols usually assume all the devices to be within radio range of each other, Bluetree [Zaruba2001] and Bluenet [Wang2002] are scatternet formation protocols for larger-scale Bluetooth networks in which the devices can be out of range with respect to each other. Simulation results of the routing properties of the scatternets have been presented in [Zaruba2001, Wang2002]. However, there are no simulations or theoretical analyses on the performance of the scatternet formation process.

5.5 The IEEE 802.15 Working Group for WPANs

As mentioned earlier, the goal for the 802.15 WG is to provide a framework for the development of short-range, low-power, low-cost devices that wirelessly connect the user within their communication and computational environment. Altogether, the 802.15 WG is formed by seven TGs:

- IEEE 802.15 TG 1 (802.15.1) – The TG 1 was established to support applications which require medium-rate WPANs (such as Bluetooth). These WPANs handles a variety of tasks ranging from cell phones to PDA communications and have a QoS suitable for voice applications. In the end, this TG derived a Wireless Personal Area Network standard based on the Bluetooth v1.1 specifications.

- IEEE 802.15 TG 2 (802.15.2) – Several wireless standards, such a Bluetooth and IEEE 802.11b, and appliances, such as microwaves and cordless phones, operate in the unlicensed 2.4 GHz ISM frequency band. Therefore, to promote better coexistence of IEEE 802 wireless technologies, the TG 2 has developed recommended practices to facilitate collocated operation of WPANs and WLANs.

- IEEE 802.15 TG 3 (802.15.3) – The TG 3 for WPANs has defined standards for high-rate (well over 55 Mbps) WPANs. Besides a high data rate, this standard provides for low power low cost solutions, addressing the needs of portable consumer digital imaging and multimedia applications.

- IEEE 802.15 TG 4 (802.15.4) – The TG 4 has defined a standard having low complexity, cost, and power for a low data-rate (200 Kbps or less) wireless connectivity among fixed, portable, and moving devices. Location awareness is considered as a unique capability of the standard. The TG 4 specifies the physical and MAC layer. Potential applications are sensors, interactive toys, smart badges, remote controls, and home automation.

- IEEE 802.15 TG 5 (802.15.5) – The TG 5 defines mechanisms in the PHY and MAC layers to enable mesh networking [Akyildiz2005]. A mesh network is a PAN that employs one of two connection arrangements: full mesh topology or partial mesh topology. In the full mesh topology, each node is connected directly to each of the others. In the partial mesh topology, some nodes are connected to all the others, but some of the nodes are connected only to those other nodes with which they exchange the most data and more details are covered in Chapter 6.

- IEEE 802.15 TG 6 (802.15.6) – The TG 6 is in the process of developing a body area network (BAN) standard optimized for low power devices and operation on, in or around the human body (but not limited to humans) to serve a variety of applications including medical, consumer electronics/personal entertainment and so on.

- IEEE 802.15 TG 7 (802.15.7) – The TG 7 is in the process of defining a PHY and MAC standard for Visible Light Communications (VLC).

Since 802.15.3 and 802.15.4 are the most well-known representatives of the 802.15 family of standards, in this section we confine our discussion to the 802.15.3 and 802.15.4 standards.

5.5.1 The IEEE 802.15.3

The 802.15.3 Group [IEEE802.15www] has been tasked to develop a MAC layer suitable for multimedia WPAN applications and a PHY capable of data rates in excess of 20 Mbps. The 802.15.3 standard specifies data rates up to 55 Mbps in the 2.4 GHz unlicensed band (basic 803.15.3 standard), and rates up to 5 Gbps in the 60GHz unlicensed band (with 803.15.3c amendment). The technology employs an ad hoc topology not entirely dissimilar to Bluetooth.

5.5.1.1 The 802.15.3 MAC and PHY Layer

The 802.15.3 MAC layer specification is designed from the ground up to support ad hoc networking, multimedia QoS provisioning, and power management. In an ad hoc network, devices can assume either master or slave functionality based on existing network conditions. Devices in an ad hoc network can join or leave an existing network without complicated setup procedures. Figure 5.16 illustrates the MAC superframe structure that consists of a network beacon interval, a contention access period (CAP) and guaranteed time slots (GTS). The boundary between the CAP and GTS periods is dynamically adjustable.

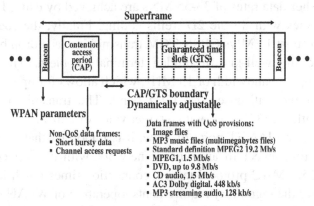

Figure 5.16 – IEEE 802.15.3 MAC superframe

A network beacon is transmitted at the beginning of each superframe, carrying WPAN-specific parameters, including power management, and information for new devices to join the ad hoc network. The CAP period is reserved for transmitting non-QoS frames such as short bursty data or channel access requests made by the devices in the network. The medium access mechanism during the CAP period is CSMA/CA. The remaining duration of the superframe is reserved for GTS to carry data frames with specific QoS provisions. The type of data transmitted in the GTS can range from bulky image or music files to high-quality audio or high-definition video streams. Finally, power management is one of the key features of the 802.15.3 MAC protocol, which is designed to significantly lower the current drain while being connected to a WPAN. In the power saving mode, the QoS provisions are also maintained.

The basic 802.15.3 PHY layer operates in the unlicensed frequency band between 2.4 GHz and 2.4835 GHz, and is designed to achieve data rates of 11-55 Mb/s that could commensurate with the distribution of high-definition video and high-fidelity audio. The 802.15.3 systems employ the same symbol rate, 11 Mbaud, as used in the 802.11b systems. Operating at this symbol rate, five distinct modulation formats are specified, namely, uncoded QPSK modulation at 22 Mb/s and trellis coded QPSK, 16/32/64-QAM at 11, 33, 44, 55 Mb/s, respectively (TCM) [Ungerboeck1987]. The base modulation format is QPSK. Higher data rates of 33-55 Mb/s are achieved by using 16, 32, 64-QAM schemes with 8-state 2D trellis coding. Finally, the specification includes a robust 11 Mb/s QPSK TCM transmission as a drop back mode to alleviate the well-known hidden terminal problem. The 802.15.3 signals occupy a bandwidth of 15 MHz, which allows for up to four fixed channels in the unlicensed 2.4 GHz band. The transmit power level complies with the FCC rules with a target value of 0 dBm.

From a MANET point of view, it is important that devices have the ability to connect to an existing network with a short connection time. 802.15.3 MAC protocol targets connection times much less than 1 s. Reviewing the regulatory requirements, operation of WPAN devices in the 2.4 GHz band is highly advantageous since these devices cannot be used outdoors while operating in the 5 GHz band. Several countries (e.g., Japan) prohibit the use of 5 GHz band for WPAN applications.

5.5.2 The IEEE 802.15.4

IEEE 802.15.4 [IEEE802.15www] defines a specification for low-rate, low-power wireless personal area networks. It is extremely well suited to those home networking applications where the key motivations are reduced installation cost and low power consumption. For example, there is a large market for home automation, security and energy conservation applications, which typically do not require high bandwidth. Application areas include industrial control, agricultural, vehicular and medical sensors and actuators that have relaxed data rate requirements. Inside the home, there are several areas where such technology can be applied effectively: home automation including heating, ventilation, air conditioning, security, lighting, control of windows, curtains, doors, locks; health monitors and diagnostics. These will typically need less than 10 kbps. Maximum acceptable latencies will vary from 10 ms up to 100 ms for home automation.

802.15.4 standard is geared towards those applications which have low bandwidth requirements, very low power consumption and are extremely inexpensive to build and deploy. In 2000, two groups, the Zigbee alliance (a HomeRF spin-off) and the IEEE 802 working group, came together to specify the interfaces and the working of 802.15.4. In this coalition, the IEEE group is largely responsible for defining the MAC and the PHY layers, while the Zigbee alliance is responsible for defining and maintaining higher layers above the MAC. The alliance is also developing application profiles, certification programs, logos and a marketing strategy. The specification is based on the initial work done mostly by Philips and Motorola for Zigbee – previously known as PURLnet, FireFly and HomeRF Lite.

802.15.4 standard – like all other IEEE 802 standards – specifies those layers up to and including portions of the data link layer. The choice of higher-level protocols is left to the application, depending on specific requirements. The important criteria would be energy conservation and the network topology. Therefore, the standard supports networks in both the star and peer-to-peer topology. Multiple address types – both physical (64 bit) and network assigned (8 bit) – are allowed. Network layers are also expected to be self-organizing and self-maintaining to minimize cost. Currently, the PHY and the DLL (Data Link Layer) have been defined. The focus now is on the upper layers, which is largely led by the Zigbee Alliance.

5.5.2.1 The 802.15.4 Data Link Layer

The DLL is split into two sublayers – the MAC and the Logical Link Control (LLC). Figure 5.17 shows the correspondence of the 802.15.4 to the ISO-OSI reference model. The IEEE 802.15.4 MAC provides services to an IEEE 802.2 type I LLC through the Service

Specific Convergence Sub layer (SSCS). A proprietary LLC can access the MAC layer directly without going through the SSCS which ensures compatibility between different LLC sub layers.

The MAC protocol supports association and disassociation, acknowledged frame delivery, channel access mechanism, frame validation, guaranteed time slot management and beacon management. The MAC sub layer provides the data service through the MAC common part sub layer (MAC-SAP), and the management services through the MAC layer management entity (MLME-SAP). These provide the interfaces between the SSCS (or another LLC) and the PHY layer. MAC management service has only 26 primitives as compared to IEEE 802.15.1 which has 131 primitives and 32 events.

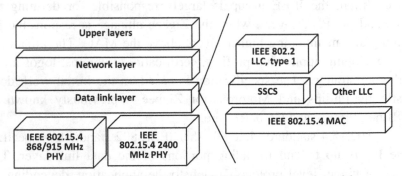

Figure 5.17 – 802.15.4 in the ISO-OSI layered network model

The MAC protocol data unit (MPDU), or the MAC frame, consists of the MAC header (MHR), MAC service data unit (MSDU) and MAC footer (MFR). The MHR consists of a 2 byte frame control field – that specifies the frame type, the address format and controls the acknowledgement, 1 byte sequence number which matches the acknowledgement frame with the previous transmission, and a variable sized address field (0-20 bytes). This allows either only the source address – possibly in a beacon signal – or both source and destination address like in normal data frames or no address at all as in an acknowledgment frame. The payload field is variable in length with the maximum possible size of an MPDU being 127 bytes. The beacon and the data frames originate at the higher layers and actually contain data, while the acknowledgement and the command frame originate in the

MAC layer and are used to simply control the link at a peer-to-peer level. The MFR completes the MPDU and consists of a frame check sequence (FCS) field which is basically a 16-bit CRC code.

By operating in a *superframe* mode, the IEEE 802.15.4 provides dedicated bandwidth and low latencies to certain types of applications. One of the devices – usually one that is less power constrained – acts as the PAN coordinator, transmitting superframe beacons at predetermined intervals that range from 15 ms to 245 ms. The time between the beacons is divided into 16 equal time slots independent of the superframe duration. The device may transmit at any slot, but must complete its transmission before the end of the superframe. Channel access is usually contention based though the PAN may assign time slots to a single device. This is known as a *guaranteed time slot* (GTS) and introduces a contention free period located immediately before the next beacon as in the 802.15.3 MAC. In a beacon enabled superframe network, a slotted CSMA/CA is employed, while in non-beacon networks, an un-slotted or standard CSMA/CA is used.

An important function of MAC is to confirm successful reception of frames. Valid data and command frames are acknowledged; otherwise it is simply ignored. The frame control field indicates whether a particular frame has to be acknowledged. IEEE 802.15.4 provides three levels of security: no security, access control lists and symmetric key security using AES-128. To keep the protocol simple and the cost minimum, key distribution is not specified while could be included in the upper layers.

5.5.2.2 The 802.15.4 PHY Layer

IEEE 802.15.4 offers two PHY layer choices based on the DSSS technique and share the same basic packet structure for low duty cycle low power operation. The difference lies in the frequency band of operation. One specification is for the 2.4 GHz ISM band available worldwide and the other is for the 868/915 MHz for Europe and USA, respectively. These offer an alternative to the growing congestion in the ISM band due to proliferation of devices like microwave ovens, etc. They also differ with respect to the data rates supported. The ISM band PHY layer offers a transmission rate of 250 kbps while 868/915 MHz offers 20 and 40 kbps. The lower rate leads to better sensitivity and larger coverage area, while the higher rate of the 2.4 GHz band can be used to attain lower duty cycle, higher throughput and lower latencies.

The transmission range is dependent on the sensitivity of the receiver which is -85 dB for the 2.4 GHz PHY and -92 dB for the 868/915 MHz PHY. Each device should be able to transmit at least 1 mW while actual transmission power depends on the application. Typical devices (1 mW) are expected to cover a range of 10-20 m, but with good sensitivity and a moderate increase in power, it is possible to cover the home in a star network topology. The 868/915 MHz PHY supports a single channel between 868.0 and 868.6 MHz and 10 channels between 902.0 and 928.0 MHz. Since these are regional in nature, support for all 11 channels need not be provided on the same network. It uses a simple DSSS in which each bit is represented by a 15-chip maximal length sequence. Encoding is done by multiplying m-sequence with +1 or −1, and the resulting sequence is modulated by a carrier signal using BPSK.

The 2.4 GHz PHY supports 16 channels between 2.4 GHz and 2.4835 GHz with 5 MHz channel spacing for easy transmit and receive filter requirements. It employs a 16-ary quasi-orthogonal modulation technique based on DSSS. Binary data is grouped into 4-bit symbols, each specifying one of 16 nearly orthogonal 32-bit chip pseudo noise (PN) sequences for transmission. PN sequences for successive data symbols are concatenated and the aggregate chip is modulated onto the carrier using minimum shift keying (MSK). The use of "nearly orthogonal" symbol set simplifies the implementation, but incurs minor performance degradation (< 0.5 dB). In terms of energy conservation, orthogonal signaling performs better than differential BPSK. However, the 868/915 MHz has a 6-8 dB advantage in terms of receiver sensitivity.

The two PHY layers though different, maintain a common interface to the MAC layer, i.e., they share a single packet structure as shown in Figure 5.18. The packet or PHY protocol data unit (PPDU) consists of the synchronization header, a PHY header for the packet length, and the payload itself which is also referred to as the PHY service data unit (PSDU). The synchronization header is made up of a 32-bit preamble – used for acquisition of symbol and chip timing and possible coarse frequency adjustment and an 8-bit start of packet delimiter, signifying the end of the preamble. Out of the 8 bits in the PHY header, seven are used to specify the length of the PSDU which can range from 0-127 bytes. Channel equalization is not required for either PHY layer because of the small coverage area and the relatively low chip rates. Typical packet sizes for monitoring and control applications are expected to be in the order of 30-60 bytes.

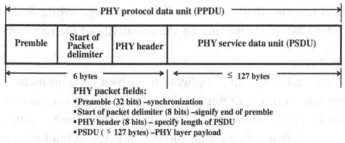

PHY packet fields:
- Preamble (32 bits) –synchronization
- Start of packet delimiter (8 bits) –signify end of premble
- PHY header (8 bits) – specify length of PSDU
- PSDU (≤ 127 bytes) –PHY layer payload

Figure 5.18 – 802.15.4 PHY layer packet structure

Since the IEEE 802.15.4 standard operates in the ISM band, it is important to consider the effects of interference that is bound to occur. Little or no QoS requirements are envisioned by applications of this protocol. Consequently, data unable to go through in the first attempt are retransmitted and higher latencies are tolerable. Too many transmissions also increases the duty cycle and therefore affects the consumption of power. Once again the application areas are such that transmissions will be infrequent as devices are in a passive mode most of the time.

5.6 Comparison between WPAN Systems

To understand the suitability of these systems for WPAN applications, we have identified several criteria keeping in mind the overall goal of forming MANETs using simple, low power, small, cost effective devices as:
- **Range:** The communication range of the device.
- **Data Rate:** The maximum data rate possible in the network.
- **Support for Voice:** Support a protocol or method to allow voice communication.
- **Power Management:** A true method for devices to conserve power

5.6.1 Range

WPAN computing will typically involve communication with devices within a few meters. Ten meters is usually considered sufficient for these devices to collaborate and provide services, like an ad hoc network for meetings in small rooms, study sessions in libraries, or home networking for computers or consumer devices. Bluetooth can support up to 10 meters and 100-meter range if an external power source is utilized. IEEE 802.15.3 can also support a 10 meter range while 802.15.4 can support 10-20 meters depending on the sensitivity of the receiver.

5.6.2 Data Rate

Data rate is an application driven requirement. WPAN computing has a myriad of applications, from simple inventory tracking, to personal information management, ad hoc networking, email, interactive conferencing and web surfing. WPAN technologies cover all kinds of data rates, from a very low data rate to transmit text between two devices to a high data rate for video transmission. It is difficult to place a number on what would be an adequate data rate for a WPAN.

Bluetooth allows for up to eight devices to operate in a single Piconet and transmit data in symmetric (up to 432.6 kbps) or asymmetric (up to 721 kbps and 57.6 kbps) mode. The 802.15.3 is able to provide data rates ranging from 11 Mbps to 55 Mbps. IEEE 802.15.4, on the other hand, seems ideal only for low rate networks providing services of 20-250 kbps (e.g., wireless sensor networks).

5.6.3 Support for Voice

A WPAN technology is most likely to be embedded into existing devices such as mobile phones, PDAs and pagers, and hence voice communication as well as integration with the PSTN is highly desirable. A possible scenario could be using two mobile phones as short wave radios using a WPAN. Bluetooth's voice support is provided by the Telephony Control protocol Specification (TCS) Binary, which is based on ITU-T Recommendation Q.931 for voice. Bluetooth matches standard telephony with a 64 kbps data rate and can support calls for all eight members of a Piconet. It is able to provide voice support without the need of infrastructure such as a Connection Point. In a Bluetooth WPAN, a single Bluetooth enabled voice device (mobile phone) can act as a gateway for all other devices. IEEE 802.15.3 with its GTS can support all kinds of multimedia traffic from simple image files to high definition MPEG-2 at 19.2 Mbps and MP3 streaming audio at 128 kbps. The flexibility of adapting the size of the GTS is certainly proving to be an efficient method of supporting variable QoS requirements. On the other hand, IEEE 802.15.4 was never designed to support voice, though there are mechanisms for time-bounded data services.

5.6.4 Power Management

With battery power being shared by the display, transceiver and processing electronics, a method to manage power is definitely needed in a WPAN device. IEEE 802.15.3, 802.15.4, and, to a less extent, Bluetooth offer power management facilities to prolong battery's life. Bluetooth has a standby and peak power range of less than 1 mA to 60 mA and allows devices to enter low power states without losing connectivity to the WPAN Piconet. It has three low power states – PARK, HOLD, and SNIFF and a normal power state when the device is transmitting. The power savings varies due to the reduced transmit-receive duty cycle.

The IEEE 802.15.3 standard has advanced power management features with a current drain of just 80 mA while actively transmitting and very minimal when in power save mode. It also is able to support QoS functionality, even when it is in a power save mode. It has three modes of power management – the Piconet Synchronized Power Save (PSPS) mode, the Synchronized Power Save (SPS) mode and the Hibernate mode. IEEE 802.15.4 has been designed ground-up for low power operation, in some cases stretching the battery life for several years. The current drains may be as low as 20μA.

5.6.5 Comparison and Summary of Results

Based on the analysis above, it appears that Bluetooth, IEEE 802.15.3 and IEEE 802.15.4 broadly meet the criteria of size, cost, simplicity, and low power consumption. IEEE 802.15.3 definitely has the upper edge since it can offer much higher data rates, good power control, extremely low connection setup times, advanced security features (see

Table 5.5) and a plethora of QoS services for high end multimedia traffic even under low power operation. In the context of WPAN computing today, it is sometimes seen as an excess of everything, whereas Bluetooth may to a large extent cover WPAN computing needs in the short-term future. IEEE 802.15.4, on the other hand, is extremely suitable for very low power applications such as sensor networking and home automation, something that Bluetooth and IEEE 802.15.3 are clearly not meant for. Table 5.5 provides a comparison of the various WPAN systems discussed so far.

Table 5.5 – A comparison of the various WPAN systems

Technology	Bluetooth (802.15.1)	IEEE 802.15.3	IEEE 802.15.4
Operational Spectrum	24 GHz ISM Band	2.402-2.480GHz ISM Band	2.4GHz and 868/915 MHz
Physical Layer Details	FHSS, 1600 hops per second	Uncoded QPSK. Trelis coded QPSK, 16/32/64-QAM scheme	DSSS with BPSK or MSK (O-QPSK)
Channel Access	Master-Slave polling Time Division Duplex (TDD)	CSMA-CA and Guaranteed Time Slots (GTS) in a superframe structure	CSMA-CA and Guaranteed Time Slots (GTS) in a superframe structure
Maximum Data Rate	Up to 1 Mbps	11-55 Mbps	858 MHz
Coverage	< 10m	<10m	<20m
Power Level Issues	10mA – 60 mA	<80mA	Very low. Current drain (20-50 ma)
Interference	Present	Present	Present
Price	Low (<$10)	Medium	Very low
Security	Less secure. Uses the SAFER+ encryption at the baseband layer. Relies on higher layer security	Very high level of security authentication, privacy, encryption, and digital certificate services	Security features in development

5.7 Conclusions and Future Directions

Wireless PANs are experiencing a considerable growth, but clearly not as much as the explosive growth seen in the wireless LANs arena. Obviously, this is largely due to farthest wireless PANs are much more recent than wireless LANs. Nevertheless, the vast availability of Bluetooth devices and the standardization of IEEE of various WPAN systems will take this field to a new level. There are numerous environments where WPANs are very suitable such as in sensor networks (discussed in chapters 10, 11 and 12). In the home and in the office, WPANs will be part of our lives.

But before that can be realized, many technical challenges have to be solved. Interference mitigation with other systems operating in the same frequency band, effective QoS support, decentralized network formation, energy conservation and security are just a few examples. Obviously, many efforts have to be devoted in designing new and exciting applications of this ever expanding technology.

Homework Questions/Simulation Projects

Q. 1. Bluetooth technology has revolutionized the world by providing wireless solution to the short-range connectivity issue. Several Bluetooth devices could be connected to constitute a Piconet. In a situation with many such independent Piconets, each Piconet follows a different frequency hopping sequence.

a. What is the probability that two Piconets use the same hopping frequency at a given time?
b. Does this increase with the number of Piconets?
c. How does the packet size influence the collision probability? Derive the appropriate collision probabilities.
d. Propose an approach to improve collocated operation and derive the new collision probabilities.
e. Can you use reuse factor similar to a cellular structure? If so, what should be the cluster size and the shape?

Q. 2. Describe the main differences between wireless LANs and wireless PANs in terms of range, data rate, power consumption, and suitability for different application scenarios.

Q. 3. Design a problem based on any of the material covered in this chapter (or in references contained therein) and solve it diligently.

References

[Abhyankar2003] S. Abhyankar, R. Toshiwal, Carlos Cordeiro, and D. P. Agrawal, "On the Application of Traffic Engineering over Bluetooth Ad Hoc Networks," in ACM MOBICOM International Workshop on Modeling, Analysis and Simulation of Wireless and Mobile Systems (MSWiM), September 2003.

[Aggarwal2000] A. Aggarwal, M. Kapoor, L. Ramachandran, and A. Sarkar, "Clustering algorithms for wireless ad hoc networks," In Proceedings of the 4th International Workshop on Discrete Algorithms and Methods for Mobile Computing and Communications, pp. 54–63, Boston, MA, August 2000.

[Agrawal2002] D. Agrawal and Q-A. Zeng, Introduction to Wireless and Mobile Systems, Brooks/Cole Publishing, 432 pages, ISBN 0534-40851-6, August 2002.

[Akyildiz2005] I. F. Akyildiz, X. Wang, and W. Wang, "Wireless Mesh Networks: A Survey," Elsevier Computer Networks Journal, March 2005.

[Awduche2002] D Awduche, A Chiu, A Elwalid, I Widjaja, and X Xiao, "Overview and principles of Internet traffic engineering," Internet draft, draft-ieft-tewg-principles-02.txt.

[Baatz2001] S. Baatz, M. Frank, C. Kuhl, P. Martini, and S. Christoph, "Adaptive scatternet support for Bluetooth using sniff mode," In Proceedings of IEEE Local Computer Networks (LCN), Tampa, Florida, November 2001.

[Baatz2002] S. Baatz, M. Frank, C. Kuhl, P. Martini and C. Scholtz, "Bluetooth Scatternets: An enhanced Adaptive Scheduling Scheme," in IEEE Infocom, 2002.

[BlueHocwww] BlueHoc, IBM Bluetooth Simulator, http://oss.software.ibm.com/developerworks/opensource/bluehoc/.
[Bluetooth Central] Bluetooth Central, http://www.bluetoothcentral.com/.
[Bluetoothwww] The Bluetooth SIG, http://www.bluetooth.com/.
[BluetoothSpecwww] The Bluetooth Specifications, http://www.bluetooth.com/.
[Capone2001] A. Capone, M. Gerla, and R. Kapoor, "Efficient polling schemes for bluetooth Piconets," in Proceedings of IEEE International Conference on Communications (ICC), June 2001.
[Chiasserini2003] C-F. Chiasserini and R. Rao, "Coexistence Mechanisms for Interference Mitigation in the 2.4-GHz ISM Band," in IEEE Transactions on Wireless Communications, Vol. 2, No. 5, September 2003.
[Cordeiro2002a] C. Cordeiro and D. P. Agrawal, "Mobile Ad Hoc Networking," Tutorial/Short Course in 20th Brazilian Symposium on Computer Networks, pp. 125-186, May 2002, http://www.ececs.uc.edu/~cordeicm/.
[Cordeiro2002b] C. Cordeiro, S. Das, and D. Agrawal, "COPAS: Dynamic Contention-Balancing to Enhance the Performance of TCP over Multi-hop Wireless Networks," in IEEE IC3N, Miami, USA, October 2002.
[Cordeiro2002c] C. Cordeiro and D. Agrawal, "Employing Dynamic Segmentation for Effective Co-located Coexistence between Bluetooth and IEEE 802.11 WLANs," in IEEE Globecom, November 2002.
[Cordeiro2003a] C. Cordeiro, D. Agrawal and D. Sadok, "Piconet Interference Modeling and Performance Evaluation of Bluetooth MAC Protocol," in IEEE Transactions on Wireless Communications, Vol. 2, No. 6, November 2003.
[Cordeiro2003b] C. Cordeiro, H. Gossain, and D. Agrawal, "Multicast over Wireless Mobile Ad Hoc Networks: Present and Future Directions," in IEEE Network, Special Issue on Multicasting: An Enabling Technology, January/February 2003.
[Cordeiro2004a] C. Cordeiro, S. Abhyankar, and D. Agrawal, "Design and Implementation of QoS-driven Dynamic Slot Assignment and Piconet Partitioning Algorithms over Bluetooth WPANs," in IEEE INFOCOM, March 2004.
[Cordeiro2004b] C. Cordeiro, S. Abhyankar, R. Toshiwal, and D. Agrawal, "BlueStar: Enabling Efficient Integration between Bluetooth WPANs and IEEE 802.11 WLANs," in ACM/Kluwer Mobile Networks and Applications (MONET) Journal, Special Issue on Integration of Heterogeneous Wireless Technologies, Vol. 9, No. 4, August 2004.
[Cordeiro2006] C. Cardeiro, S. Abhyankar, and D. Agrawal, " Scalable and QoS-Aware Dynamic Slot Assignment and Piconet Partitioning to Enhance the Performance of Bluetooth Ad hoc Networks," IEEE Transactions on Mobile Computing, 2006.
[Derfler2000] F. Derfler, "Crossed Signals: 802.11b, Bluetooth, and HomeRF," PC Magazine, ZDNet, 2000.
[Duggirala2003a] R. Duggirala, R. L. Ashok and D. Agrawal, "Energy Efficient Bridge Management Policies for Inter-Piconet Communication in Bluetooth Scatternets," in IEEE Fall VTC, 2003.
[Duggirala2003b] R. Duggirala, R. L. Ashok and D. Agrawal, "A Novel Traffic and Energy Aware Bridge Management Policy for Bluetooth Scatternets," in IEEE MWCN, 2003.
[Ericssonwww] Bluetooth at Ericsson, http://www.ericsson.com/bluetooth/.
[Gossain2002] H. Gossain, C. Cordeiro, and D. Agrawal, "Multicast: Wired to Wireless," IEEE Communications Magazine, June 2002, pp. 116-123.
[Har-Shai2002] L. Har-Shai, R. Kofman, G. Zussman, and A. Segall, "Inter-Piconet scheduling in Bluetooth scatternets," In OPNETWORK Conference, August 2002.
[IEEE802.15.22000www] IEEE 802.15.2 definition of coexistence, http://grouper.ieee.org/groups/802/15/pub/2000/Sep00/99134r2P802-15_TG2-CoexistenceInteroperabilityandOtherTerms.ppt.
[IEEE802.15www] IEEE 802.15 Working Group for WPAN, http://www.ieee802.org/15/.
[IEEE802.15-FAQ2000www] IEEE 802.15 Working Group, "802.15 FAQ," http://grouper.ieee.org/groups/802/15/pub/WPAN-FAQ.htm, April 2000.
[IEEEP802.15-145r12001www] IEEE 802.15.2-01/145r1, SCORT – An Alternative to Bluetooth SCO Link for Operation in an Interference Environment, http://www.ieee802.org/15/pub/TG2-Coexistence-Mechanisms.html.
[IEEEP802.15-300r12001www] IEEE 802.15.2-01300r1, TG2 Mobilian Draft Text, TDMA and MEHTA, http://grouper.ieee.org/groups/802/15/pub/2001/ Jul01/01300r1P802-15_TG2-Mobilian-draft-text.doc.

[IEEEP802.15-316r02001www] IEEE P802.15-01/316r0, Non-Collaborative MAC mechanisms, adaptive packet selection and scheduling, http://grouper.ieee.org/ groups/ 802/15/pub/2001/Jul01/01316r0P802-15_TG2-MAC-Scheduling-Mechanism.doc.

[IEEEP802.15-340r02001www] IEEE P802.15-01/340r0, Alternating Wireless Medium Access techniques, http://grouper.ieee.org/groups/802/15/pub/2001/Jul01/ 01340r0P802-15_TG2-Clause14-1.doc.

[IEEEP802.15-364r2001www] IEEE 802.15.2-01364r, Clause 14.1 Collaborative co-located coexistence mechanism, Deterministic Frequency Nulling, http://grouper.ieee.org/groups/802/15/pub/2001/Jul01/01364r0P802-15_TG2-Clause14p1-Nist.doc.

[IEEEP802.15-366r12001www] IEEE P802.15-TG2_366r1, Clause 14.3 Adaptive Frequency Hopping, http://grouper.ieee.org/groups/802/15/pub/2001/Jul01/ 01366r1P802-15_TG2-Clause-14-3-Adaptive-Frequency-Hopping.doc.

[Infotoothwww] Infotooth, Bluetooth Glossary, http://www.infotooth.com/glossary.htm# authentication.

[IrDAwww] Infrared Data Association. "IrDA Standards," http://www.irda.org/ standards/standards.asp, April 2000.

[Johansson1999] N. Johansson, U. Korner, and P. Johansson, "Performance evaluation of scheduling algorithms for Bluetooth," In Proceedings of IFIP TC 6 Fifth International Conference on Broadband Communications, pp. 139-150, Hong Kong, November 1999.

[Johansson2001a] N. Johansson, F. Alriksson and U. Jonsson, "JUMP-mode a Dynamic window-based Scheduling Framework for Bluetooth Scatternets," in ACM MobiHoc, 2001.

[Johansson2001b] N. Johansson, U. Korner, and L. Tassiulas, "A distributed scheduling algorithm for a Bluetooth scatternet," In Proceedings of the International Teletraffic Congress (ITC-17), pp. 61-72, Brazil, September 2001.

[Kahn1999] J. Kahn, R. Katz, K. Pister, "New Century Challenges: Mobile Networking for Smart Dust," in ACM Mobicom, pages 271-278, 1999.

[Kalia1999] M. Kalia, D. Bansal, R. Shorey, "MAC Scheduling and SAR Policies for Bluetooth: A Master Driven TDD Pico-Cellular Wireless System," 6th IEEE International Workshop on Mobile Multimedia Communications (MoMuC), November 1999.

[Kasten2001] O. Kasten and M. Langheinrich, "First Experience with Bluetooth in the smart-its Distributed Sensor Network," In Proceedings of the Workshop in Ubiquitous Computing and Communications (PACT), October 2001.

[Law2003] C. Law, A. Mehta, K-Y. Siu, "A New Bluetooth Scatternet Formation Protocol," in ACM/Kluwer Journal on Mobile Networks and Applications (MONET), Special Issue on Mobile Ad Hoc Networks, Vol. 8, No. 5, October 2003.

[Lee2002] Y-Z. Lee, R. Kapoor, and M. Gerla, "An Efficient and Fair Polling Scheme for Bluetooth," In IEEE Milcom, 2002.

[Lim2001] Y. Lim, S. Min, and J. Ma, "Performance evaluation of the bluetooth-based public internet access point," in Proceedings of the 15th International Conference on Networking (ICOIN-15), pages 643-648, 2001.

[Lin2002] T-Y. Lin and Y-C. Tseng, "An Adaptive Sniff Scheduling Scheme for Power Saving in Bluetooth," in IEEE Wireless Communications, December 2002.

[Liu1992] Z. Liu, P. Nain and D. Towsley, "On optimal polling policies," In Queuing Systems Theory and Applications, Vol. 11, pp. 59-83, 1992.

[Miklos2000] G. Miklos, A. Racz, Z. Turanyi, A. Valko, and P. Johansson, "Performance aspects of Bluetooth scatternet formation," In Proceedings of The First Annual Workshop on Mobile Ad Hoc Networking and Computing, 2000.

[Misic2003] V. Misic and J. Misic, "Adaptive Inter-Piconet Scheduling in Small Scatternets," in ACM Mobile Computing and Communications Review, Special Issue on Wireless Home Networks, April 2003.

[MobileInfowww] Mobile Info, http://www.mobileinfo.com/bluetooth.

[Muller1999] T. Muller, "Bluetooth Security Architecture," White Paper, July 1999.

[Negus2000] K. Negus, A. Stephens, and J. Lansford, "HomeRF: Wireless networking for the connected home," in the IEEE Personal Communications, February 2000.

[NSwww] Network Simulator (NS) version 2, http://www.isi.edu/nsnam/ns/index.html.

[PCSDATA] http://www.pcsdata.com/CahnersBluetooth.htm.

[Possiowww] Possio, Possion PX20, http://www.possio.com.

[Prabhu2002] B. Prabhu and A. Chockalingam, "A Routing Protocol and Energy Efficient Techniques in Bluetooth Scatternets," IEEE ICC, 2002.

[Racz2001] A. Racz, G. Miklos, F. Kubinszky, and A. Valko, "A pseudo random coordinated scheduling algorithm for Bluetooth scatternets," In Proceedings ACM MobiHoc, pp. 193-203, October 2001.

[Raman2001] B. Raman, P. Bhagwat, and S. Seshan, "Arguments for cross-layer optimizations in Bluetooth scatternets." In Proceedings of Symposium on Applications and the Internet, pp. 176–184, 2001.

[Rao2001] R. Rao, O. Baux, and G. Kesidis, "Demand-based Bluetooth Scheduling," In Proceedings of the Third IEEE Workshop on Wireless Local Area Networks, Boston, Massachusetts, September 2001.

[Rouhana2002] N. Rouhana and E. Horlait, "BWIG: Bluetooth Web Internet Gateway," In Proceedings of IEEE Symposium on Comp. and Communications, July 2002.

[Salonidis2000] T. Salonidis, P. Bhagwat, and L. Tassiulas, "Proximity awareness and fast connection establishment in Bluetooth," in First Annual Workshop on Mobile and Ad Hoc Networking and Computing, pp. 141-42, 2000.

[Salonidis2001] T. Salonidis, P. Bhagwat, L. Tassiulas, R. LaMaire, "Distributed Topology Construction of Bluetooth Personal Area Networks," in IEEE Infocom, 2001.

[Schneier1996] B. Schneier, Applied Cryptography, John Wiley & Sons Inc., 2nd Ed., 758 pages, 1996.

[Spikewww] Spike, http://www.spikebroadband.net/.

[Takagi1991] H. Takagi, Queuing Analysis, North-Holland, 1991.

[Tan2002] G. Tan, "Self-organizing Bluetooth scatternets," Master's thesis, Massachusetts Institute of Technology, January 2002.

[Tanenbaum1996] A. Tanenbaum, "Computer Networks," Prentice Hall, ISBN 0-13-349945-6, 1996.

[UCBTwww] Simulator for Bluetooth/802.15/under ns-2, http://www.cs.uc.edu/~cdmc/ucbt

[Ungerboeck1987] G. Ungerboeck, "Trellis Coded Modulation with Redundant Signal Sets Part 1: Introduction," in IEEE Communications Magazine, Vol. 25, No. 2, February 1987.

[UWBwww] Ultra-Wideband Working Group, "http://www.uwb.org".

[Wang2002] Z. Wang, R. Thomas, and Z. Haas, "Bluenet - a new scatternet formation scheme," in 35th Hawaii International Conference on System Science (HICSS-35), Big Island, Hawaii, pages 7-10, January 2002.

[Wang2005] Q. Wang and D.P. Agrawal, "A Dichotomized Rendezvous Algorithm for Bluetooth Scatternets," International Journal of Ad hoc and Sensor Wireless Networks, Vol 1, Nos. 1-2, pp 65-88.

[Yaiz2002] R. Yaiz and G. Heijenk, "Polling best effort traffic in Bluetooth," Wireless Personal Communications, Vol. 23, No. 1, pp. 195–206, October 2002.

[Zaruba2001] G. Zaruba, S. Basagni, and I. Chlamtac, "Bluetrees - scatternet formation to enable bluetooth-based ad hoc networks," in Proceedings of IEEE International Conference on Communications (ICC), pages 273-277, 2001.

[Zhang2002] W. Zhang and G. Cao, "A flexible scatternet-wide scheduling algorithm for Bluetooth networks," In IEEE IPCCC, Phoenix, Arizona, April 2002.

[Zimmerman1996] T. Zimmerman, "Personal Area Networks: Near-field Intrabody Communication," IBM Systems Journal, Vol 35, No. 3-4, Jan. 1996, pp. 609-617.

Chapter 6

Wireless Mesh Networks

6.1 Introduction

Wireless mesh networks (WMNs) has recently being emerged as a trend for the next generation wireless network with the unique advantages, which are inherited and evolved from many successful wireless network technologies. We briefly introduce key points that enable implementation of a WMN using existing technologies.

These days, people use cell phone in a server-client mode supported by a Base station (BS). Like WLAN access point (AP), the coverage of a single BS is limited and if MH goes out of the service region, the wireless link becomes too weak to maintain. Therefore, a multi-cell network is built with many BSs to cover a wider region whose combined coverage makes ubiquitous access to MHs, thereby supporting MHs' mobility. Learning from cellular system, WMN employs AP known as IGW (Internet Gate Way) to provide network access service for MHs. The IGW does have Internet access that allows useful information for work or leisure; exchange e-mails, talk or chat with friends, or update recent events on Facebook [Facebook] or Linkedin [Linkedin]. WMN is a inherits technology from many existing schemes, thus offers combined merit from these network architectures in supporting next generation wireless communication.

Thus, MHs access the Internet connection by requesting and creating connection to an IGW, which is installed at a fixed location. IGW receives and accepts multiple connection requests from MHs. Different from point-to-point communication, the connections between a serving IGW and multiple MHs are simultaneously active, which forms a star like network topology. This kind of mode is very popular and used in many other applications in our daily life. For example, many computer terminals at home or offices connect to a same server for accessing information. The

IGW can accept multiple connections (within its capacity) and respond to the inquiry from each MH. However, the coverage of a single IGW is very limited. With current technology, the typical transmission of an IGW radio is at the level of hundreds of meters. To have a larger coverage, the transmission power of both IGW and MHs has to be increased greatly, which is not economical and could create strong interference between nearby wireless devices. Thus, to fit in the single cell formed by an IGW, it is used only for the "hot" spots, such as office, coffee shop, airport waiting rooms, and so on, where the clients are always crowed.

To address such a problem, WMN uses relay stations known as Mesh Routers (MRs) in an ad hoc mode. Use of these MRs enable relay of packets originated at MHs by wireless radio links. The relay is achieved similar to the beacon towers used in ancient time for intrusion alarm. Once someone at the border observes enemy's intrusion, a beacon light is fired and the process is continued hop-by-hop to the center at the speed of light. A modern relay station (MR) operates like a beacon tower using modulated wireless signals and is able to carry much rich information. While MR receives the packets from other remote MRs, it either simply amplifies the signal strength or reconstructs the packets with error check, and then forwards to the next MR towards the IGW which is the only unit directly connected to the Internet. To receive and forward packets from its MRs, WMN implements this approach by the ad hoc wireless connectivity so that the packets can be forwarded hop-by-hop fashion to the destination or the IGW over a long distance. To provide the connection service to MHs, MRs also work like APs to aggregate the traffic from MHs, in addition to the relay function.

WMNs have heavily inherited characteristics from the ad hoc networks as an ad hoc wireless network is formed between MRs that primarily act as a backbone. With an increased demand in wireless communication, the information is expected to be delivered from one point to any other random place. Ad hoc network technology enables the MRs to establish a "mesh" like network and in most cases, the only important exception is no mobility of MRs so that they can get power from A/C supply. Each MR connected with ad hoc technology will have multiple neighboring MRs in its communication region. MR may receive packets from these neighboring stations and forward packets to them. As a graph is drawn between the MRs, their links form a mesh. The packets can be delivered along the links from the source MR to the destination MR. To support such a kind of operation, WMN uses key technologies of ad hoc networks such as: self-organization and ad hoc routing. Self-organization is a process that constructs a network without centralized or external control. The MRs in the network are capable of finding its neighboring MRs and form the network automatically. As the MRs may appear or disappear in the network, validation of the links may change dynamically. Thus, it needs to use self-organization to find active and/or disappeared MRs in a very short time and update the database or tables at the related MRs. On the other hand, ad hoc routing is essential for packet delivery in a WMN. Different from conventional relay stations located on a line, multiple paths are present in an ad hoc network between the source and the destination MRs. To choose the optimal path is a challenging problem. WMN can implement routing protocols which have originally been proposed for ad hoc networks.

6.2 Network Architecture

Since WMN has adopted many useful characteristics from other networks, it forms new network architecture, which combines multi-hop and multi-cell fashions with no mobility of cells. Typically, a WMN is comprised by mesh clients (MCs), MRs, and IGWs.

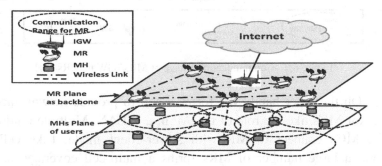

Figure 6.1 – A Generic WMN Architecture

As the backbone plane shown in Figure 6.1, MRs are interconnected with each other with wireless links. MRs have different communication range between MRs and MR to MHs and this is illustrated in Figure 6.2. MRs deliver information from one MR to another with multi-hop transmission using relay technology and has been developed with more advanced feature—ad hoc routing function as indicated in Figure 6.3. MRs can select the next hop MR according to certain measurement with aim to achieve better performance. Since MR usually has multiple neighboring MRs, the ability to select one or more neighboring MRs as the next hop MR will provide more flexibility as compared to a pure relay network. But MRs are usually located at those places where A/C power is readily available and there are no restrictions on the power availability.

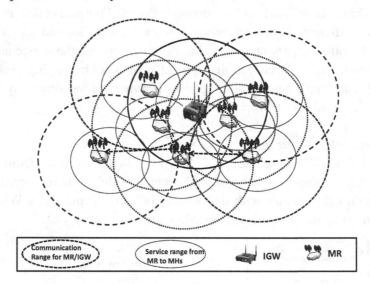

Figure 6.2 –Communication Range of MRs/IGW to MRs-MHs

On the other hand, MR forms a cell to provide network access service for MCs like the role of the AP in a WLAN. Multiple randomly located MCs connect to MR with server-client mode. Like cellular network, a large number of MRs forms a multi-cell coverage, as the circles shown in the user plane in Figure 6.3. The overlapped cell coverage is to ensure that every MC can always find at least one MR for accessing the network. To enable the Internet functionality, there is a special kind of MRs having wired Internet connection and is called as an IGW, which is also shown in Figure 6.1.

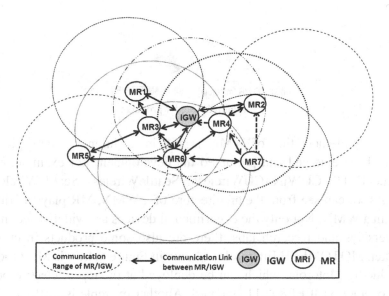

Figure 6.3 – Ad hoc connectivity between MRs and IGW

From the network architecture point of view, a WMN is observed as a two-tier network. Inter-connected MRs constitute a backbone which is formed on an ad hoc basis. This performs ad hoc multi-hop routing; and has a multi-cell network constituted by multiple MRs. WMN has combined these features which has not been provided by other wireless networks.

Although the backbone of a WMN uses ad hoc routing and self-organization, there are some significant differences between ad hoc networks and WMNs. First of all, the MRs are fixed at given locations while wireless devices are mobile in an ad hoc network. The radio used in an ad hoc network may only operate on a single channel. In contrast, WMN's radio could switch and operate on different channels. Ad hoc network is considered as a flat, single-tier network, where each node either forwards the message from neighboring nodes or sends messages to neighboring nodes. WMN is a two-tier network, whose MR does not generate data traffic. MR aggregates the traffic from MCs in its cell and forwards to other neighboring MRs. These differences impose some changes while performing technology migration from basic ad hoc networks.

(a) MOTO MR [motorola]   (b) TRANZEO MR [tranzeo]
Figure 6.4 – MR products

Implementation of WMN is still in its infancy. Many pioneer works have emerged and prompted the development, for example, MIT roofnet [MIT], CUWiN [CUWireless], SeattleWireless [SeattleWireless], etc. As we can see from the architecture of a WMN, MR plays a critical role in a WMN. Not only the experimental devices are widely used in the universities and research labs, there are also some products in current market. MOTOMESH [motorola], shown in Figure 6.4(a), is MR product provide by Motorola, which can be deployed at the top of electric poles and operate on IEEE802.11 standard. Another example is EnRouter500, shown in Figure 6.4(b), manufactured by TRANZEO [tranzeo], which has mesh radios to operate on either 2.4GHz or 5.8GHz bands. In current deployment, MRs are primarily assumed to be located statically. There is also a trend to develop mobile WMN (MWMN), where MRs are also moving in a controlled or uncontrolled way. MWMNs provide more flexibility, but at the cost of increased implementation complexity. A lot of potential applications have been discussed in [meshdynamics].

6.3 Challenging Technologies

There are many challenging technologies in implementing a WMN. Some of them are special underlying issues associated with a WMN.

6.3.1 MR Deployment

There are two assumptions for the MR locations. MRs could be deployed by service providers and located at the selected place. The other assumption is that MRs are deployed by users and the locations are randomly selected. If the location is selected by a service provider, the MR deployment problem can be described like minimizing the number of MRs while satisfying the constraints such as having adequate coverage in a given area and required quality of service. Typically, common MR deployment constraints can be summarized as follows:

- Multiple MRs can adequately cover a given region.
- The overlapped coverage of two neighboring MRs should be enough to support the handoff.
- The MR deployment can support required quality of service.
- The MR deployment shall have certain fault-tolerance.
- The locations of MRs are such that they are able to form a connected network, given constraint on the transmission distance.
- Sometime, geographical location may also be a constraint. It is very possible that MRs can be deployed only at certain locations.

In a realistic WMN, the MR deployment may consider additional geographical issues. In general, the deployment area of the network may not be flat. There may be tall buildings, hills, valley, and other obstacles in the area. Some regions may be prohibitive for the MR installation. These geographical factors can be transformed in to obstacles. On the other hand, the traffic may also depend on some geographical attributes. Some region may have larger traffic demands while others may be light. Those factors are additional considerations as constraints for the deployment problem.

In near future, MRs will be more likely to be deployed by the users. For example, MRs can be deployed by users at home or office to aggregate the traffic from local computers, cell phones or other wireless devices. Under such situations, MR selection issue may disappear. However, it poses a harder problem of IGW deployment.

6.3.2 IGW Deployment

Besides channel assignment, the deployment of IGWs also has significant impact on the capacity and throughput of a WMN. Deployment of IGWs aims to minimize the number of IGWs while satisfying the requirement of quality of service. As IGWs need a wired connection to Internet, the cost will greatly increase if the number of IGWs is large. There are several basic strategies for IGW deployment and main ones are briefly covered.

The location of IGWs can be modeled as a facility location problem. While dealing with random deployed WMN, the IGW deployment will be based on the connectivity graph and form a general facility location problem, which is a NP-hard in a general case. To solve this kind of problem, approximation is desirable to obtain a quick and reasonable solution. If a WMN is deployed with a specified/regular topology, an optimization solution may be obtained in a polynomial time. A method based on facility location has been implemented in [Qiu2004], which takes constraints such as co-channel interference, network capacity, fault-tolerance, traffic demands, and so on. In this scheme, it first prunes the search space of the problem with grouped equivalent classes from a set of available locations. Then, a greedy approach is used to iteratively locate IGWs that maximizes the capacity.

Another scheme is to utilize Cluster formation method that focuses on forming multiple disjoint clusters with given objective and constraints. Each cluster has an IGW as the cluster head and all the MRs in the same cluster are served by the IGW cluster. Such a scheme has been implemented in [He2010]. MR is assumed to have weights representing certain attributes. The objective of the cluster formation is to minimize the number of clusters while satisfying the quality of service constraints. There are two approximation algorithms in cluster formation. One is based on divide-and-conquer method and second is a greedy dominant-independent-set approach.

Another scheme is tree formation which generates multiple trees in the connectivity graph formed by the network. The root of the tree is the selected IGW. The objective is to minimize the number of trees and/or their depth. Such a tree-based IGW selection is used in [He2010]. First, it selects multiple trees from an initial network graph as a linear program. Then, two heuristic algorithms are proposed: degree-based greedy dominating tree set partitioning and width based greedy dominating set partitioning.

6.3.3 Channel Assignment

To enhance the capacity of multi-hop transmission in the backbone of a WMN, MRs are usually equipped multiple radios which are able to switch and operate on multiple channels. The task in arranging the radios and their operating channels are called channel assignment. To establish communication for packet delivery, two neighboring MRs, each of them, is required to have a radio that employs a common channel. The objective of channel assignment is to assign optimal channels for MRs so as to avoid and/or mitigate interference and maximize the number of parallel transmissions. Optimal channel assignment is a NP-hard [Bejerano2004] problem. Many near-optimal solutions have been proposed in the past. In a board sense, these approaches can be divided into three categories as follows:

- Static Channel Assignment: Static channel assignment assigns a channel to be employed by a radio for a long period.
- Dynamical Channel Assignment: Dynamical channel assignment allows each node to potentially access all channels by dynamically switching its interfaces among available channels, which offers flexibility in channel usage according to the node density, traffic, channel conditions, etc.
- Hybrid Channel Assignment: Some of the radios are fixed at certain channels for control purpose and others are dynamically operated.

Static channel assignment allocates the channels over radios of MRs with relatively long duration. It is noted that the allocation has to be carefully scheduled and assigned. While establishing the connection between two neighboring MRs for active communication, they have to employ a common channel on both radios. In a WMN, there are numerous active connections requiring appropriate channels for packet to be delivered over multiple hops. To eliminate interference and ensure the throughput of the network, the channel assignment tries to allocate non-overlapped channels in the same interference region. However, to maintain the network connectivity, the channel assignment needs to keep the network connected with the constraint on the number of radios. Otherwise, the network may be partitioned if some essential links cannot be assigned with appropriate channels, or the performance will be serious degraded due to serious interference.

Dynamic channel assignment requires a radio switch between channels frequently. Channel switching and switching synchronization become two essential factors that affect the network performance. Radio switching from one channel to another costs time, called switching delay. It is negligible in static channel assignment because it is in very small fraction as compared to a long duration that a radio remains on using a switched channel. In contrast, the switching delay becomes a major factor in a dynamic channel assignment as frequently switched radio makes the switching delay to be a relatively large fraction of the whole transmission. Most of the current commercial radios need several milliseconds to hundred milliseconds to switch across channels. Furthermore, switching delay depends on the slower of two scheduled transmitting MRs one has to wait for the completion of both the MRs. Switching delay is a challenging parameter and the essential issue is the switching synchronization which not only needs two transmitting MRs to switch radios to one common channel simultaneously but also requires low overhead and high accuracy.

There are two kinds of switching synchronization methods: out-of-band and in-band. Out-of-band method utilizes external information by each MR for acknowledging the time. For example, GPS (Global Positioning System) technology can be used for MR acknowledgement time. MR equipped with GPS receiver obtains the absolute time information from the satellite so that the radio can be switched to a channel at an agreed time, thereby minimizing the switching synchronization time. However, it is still expensive for every MR to be equipped with a GPS receiver. In-band method synchronizes the switching by exchanging information among MRs, which is cost-efficient. For instant, beacon signal by every MR is taken as an in-band synchronization method. Every MR broadcasts own beacon signal with certain interval and listens to neighboring MRs' beacon signals to have the relative time for channel switching. However, while compared to implementing GPS synchronization, beacon signal suffers multiple negative factors (such as propagation delay of beacon signals) that decrease the accuracy. In static channel assignment, each radio is assigned to a channel for a long duration, hence the number of links of a MR is limited by the number of the radios of that MR. In dynamic channel assignment, the radio switches across channels frequently, a radio can be used for transmissions over multiple neighboring MRs. The ability to switch to any channel helps dynamic channel assignment, and avoids network partitioning which may happen in a static channel assignment.

Hybrid channel assignment approach allows some radios to be fixed while the rest are dynamically assigned. The fixed channel is so called default channel which is used for exchanging control information. It is noted that the radio assigned on the default channel may be used for data transmission in some schemes. The rest of radios of a MR are called "dynamic" which are able to switch to any channel freely. Hybrid channel assignment avoids the low efficiency of pure static assignment by adjusting some of the radios switching across channels on demand. At the same time, it mitigates the requirements on switching delay and synchronization by using the default channels. However, as compared to the previous two approaches, implementation of hybrid channel assignment is more complicated.

6.4 Other Issues:

There are also many other challenging problems inherited from ad hoc networks, cellular networks, relay networks, etc. These issues vary, depending on how they are implemented in a WMN. A thorough analysis is beyond the scope of this chapter.

Homework Questions/Simulation Projects

Q. 1. What are the reasons you cannot effectively utilize routing schemes used for ad hoc networks in wireless mesh networks? Explain clearly.
Q. 2. How do you characterize the traffic in a wireless mesh network? How does that compare with a typical ad hoc network?
Q. 3. In your local area, gather the statistics for cell-phone users for a given service provider. For that traffic pattern and number f users, how many MRs are needed and where are they laced?
Q. 4. For Q. 3, if you have an IGW, where are you going to place that? Can you justify your answer?
Q. 5. In Q. 4, if the traffic is doubled due to some unknown reasons, how many IGWs will be needed and where are you going to place them? Explain clearly.
Q. 6. In Q. 3, if you need to support 20 different users at the same time. How many wireless channels are needed and how are you going to allocate them? Can you justify your answer?

References

[Bejerano2004] Y. Bejerano, "Efficient integration of multihop wireless and wired networks with QoS constraints," *IEEE/ACM Transaction on Networking*, 2004, vol. 12, no. 6, pp. 1064–1078.
[CUWireless] http://www.cuwireless.net/
[He2010] *Bing He,* "Architecture Design and Performance Optimization of Wireless Mesh Networks," *PhD Dissertation*, University of Cincinnati, May 21, 2010.
[Facebook] http://www.facebook.com/
[Linkedin] http://www.linkedin.com/
[meshdynamics] http://www.meshdynamics.com/
[MIT] http://pdos.csail.mit.edu/roofnet/design/
[motorola] http://www.motorola.com/
[Qiu2004] L. Qiu, R. Chandra, K. Jain, and M. Mahdian, "Optimizing the placement of integration points in multi-hop wireless networks," Proceeding of the Twelfth IEEE International Conference on Network Protocols (ICNP), Oct. 2004.
[SeattleWireless] http://map.seattlewireless.net/
[tranzeo] http://www.tranzeo.com/

Chapter 7

Directional Antenna Systems

7.1 Introduction

Researchers have been trying to increase the capacity of ad hoc networks through a variety of innovative strategies. One of the main technological restrictions to the capacity limitations is the omni-directional nature of transmission. Distribution of energy in all directions other than the intended direction of the destination node not only generates unnecessary interference to other neighboring nodes, but also decreases the potential range of transmissions [Liberti1999]. All MAC and routing protocols described in the previous chapters are designed assuming this omni-directional nature of transmission. While this provides for simplicity, it fundamentally limits how high one can push the capacity of the system. A thorough study of the capacity of the ad hoc system is performed in [Gupta2000, Ramanathan2001], where it has been shown that the throughput obtainable by each node is:

$$\Theta(\frac{W}{\sqrt{n\log n}})$$

where W is the data rate and n is the number of nodes in the network. This limitation on capacity exists irrespective of the routing protocol or channel access mechanism. It has been shown that splitting a channel into sub-channels does not impact this value [Gupta2000] in any way.

Directional antenna systems are increasingly being recognized as a powerful way of increasing the capacity, connectivity, and covertness of MANETs. Directional antennas can focus electromagnetic energy in one direction and enhance coverage range for a given power level. They also minimize co-channel interference and reduce noise level in a contention-based access scheme, thereby reducing the collision probability. Further, they provide longer range and/or more stable links due to increased signal strength and reduced multipath components. Increased spatial reuse and longer ranges translate into higher network capacity (more simultaneous transmissions and fewer hops), and longer ranges also provide richer connectivity. On the receiving side, directional antennas enable a node to selectively receive signals only from a certain desired direction [Liberti1999].

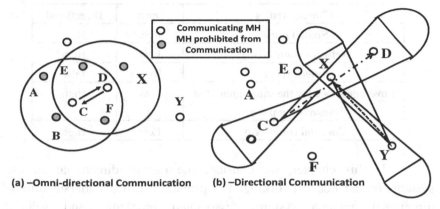

(a) –Omni-directional Communication (b) –Directional Communication

Figure 7.1 – Communication using Omni-directional Antennas and Directional Antennas (from www.crhc.uiuc.edu/~croy/presentation.html)

Figure 7.1 illustrates the increased spatial reuse capability provided with the use of directional antennas when nodes C and D, and X and Y want to simultaneously communicate. If omni-directional antennas are in use as in Figure 7.1(a), only one pair of nodes can communicate as nodes D and X are within the radio range of each other. Although we are confining our discussion here to nodes C, D, X and Y, note that all the nodes within the radio range of these nodes (i.e., nodes A, B, E and F) are also affected when employing omni-directional antennas. In the case of Figure 7.1(a), if we assume that nodes C and D initiated their communication first, all neighbors of C and D, including node X, will stay silent for the duration of their transmission. However, when directional antennas are in place, both the node pairs C-D and X-Y can simultaneously carry out their communication as depicted in Figure 7.1(b). Consequently, the capacity of the network can be considerably increased and the overall interference decreased, as transmissions are now directed towards the intended receiver, thus allowing multiple transmissions in the same neighborhood (which is not possible with omni-directional antennas) to occur in parallel using the same channel.

Table 7.1 briefly compares omni-directional and directional antennas under five self-explaining criteria.

Table 7.1 – Comparison of Omni and Directional Antennas

Characteristics	Omni	Directional
Spatial reuse	Low	High
Network connectivity	Low	High
Interference	High	Low
Coverage range for the same amount of transmitting power	Low	High
Cost and complexity	Low	High

In this chapter, we consider the use of directional antenna systems for ad hoc networking. We provide a broad understanding of directional antenna systems, associated problems, and solution approaches for utilizing these antenna systems. We describe research issues in physical, MAC, neighbor discovery, and routing with directional communications, and survey the state of the art.

7.2 Antenna Concepts

The main function of the antennas used in any communication system is to compensate for the loss of signal strength that occurs when a signal is transmitted from the source to a destination (and in the reverse direction). Most antennas are resonant devices, which operate efficiently over a relatively narrow frequency band. An antenna must be tuned to the same frequency band that the radio system to which it is connected operates in, otherwise reception and/or transmission is impaired.

Until recently, antennas have been the most neglected components in personal communications systems [IECwww, Blostein2003]. Radio antennas have to couple electromagnetic energy from one medium (space) to another (wire, coaxial, waveguide, etc.). The manner in which energy is distributed into and collected from the space has a profound effect on the use of the wireless spectrum. One of the earliest used configurations is a simple dipole antenna whose length depends on the wavelength λ of the transmitted wave and is supposed to be isotropic. These antennas are also termed as omni-directional and their radiation pattern is supposed to be symmetric in all directions (Figure 7.2). On the other hand, more focused directional antennas (also called "yagi") transmit or receive more energy in one direction.

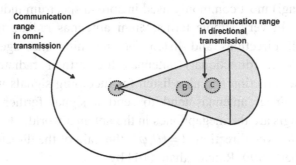

Figure 7.2 – Coverage range of Omni-Directional and Directional Transmissions

7.2.1 Gain

Antenna "gain" is a word that seems to strike fear in the hearts and minds of inexperienced radio users all over the world. It is often used to refer to some sort of mysterious signal amplifier, yet never really understood. However, one antenna with a "higher" gain does not amplify the signal more than another with "less" gain, as most people think. An antenna with greater gain simply focuses the energy of the signal differently.

To get a handle on "gain", let us discuss it in terms of a megaphone. When you want to get your message across a noisy stadium, you have the following two options: 1) you can shout into it as loudly as possible, and 2) you can direct the focused end of the megaphone toward the listener. These two actions can be applied to transmitting a radio signal as well. So, either you can increase the transmit power (to a limit of 1 Watt for spread spectrum radios according to FCC Part 15), or you can "aim" the radiating power from the antenna toward the receiver. Aiming the power is what is meant by "gain". Taking this one step further, if someone in the stadium also had a megaphone and really wanted to hear what you had to say, they could put their megaphone to their ear and aim the open end toward you, thereby focusing in on what is being transmitted from your location. Likewise, a receiving radio gets "gain" by focusing the direction of the "listening" antenna toward the source. In other words, gain is simply how you focus the radiated energy at the transmitter and how you focus the ear of the receiver.

We now discuss how gain applies to the two types of antennas (omni and yagi) most commonly used in spread spectrum industrial radio installations. In very simple terms, omni antennas radiate transmit the signal in all directions and listen for incoming messages from all directions. Yagi (directional) antennas focus their radiated transmit power in one direction and also listen for incoming signals with a more focused ear. Yagi antennas tend to send a signal farther than omni antennas. Yagis are the megaphones in the antenna world.

For a given direction $\vec{d} = (\theta, \varphi)$, the gain of the direction antenna is given [Liberti1999, Ramanathan2001] by

$$G(\vec{d}) = \eta \frac{U(\vec{d})}{U_{avg}}$$

where $U(\vec{d})$ gives the power density in direction $\vec{d}$, U_{avg} is the average power density over all directions, and η is the efficiency of the antenna which accounts for losses. The gain gives the relative power in one direction as compared to an omni-directional antenna, and higher gain means a higher directionality. Gain is generally measured of decibels (dBi), where $G_{dBi} = 10\log_{10}(G_{abs})$.

7.2.2 Radiation Pattern

The antenna pattern describes the relative strength of the radiated field in various directions from the antenna, at a fixed or constant distance. It is used to specify the gain values in all directions of the space. It generally has a main lobe of peak gain and side lobes (smaller gain). Peak gain is the maximum gain taken over all directions. Beam is also used as a synonym for "lobe". A related concept in the antenna system is beam width. A "half power beam width" refers the angular separation between the half power points on the antenna radiation pattern, where the gain is one half of the peak gain. Typically, a more directional antenna has higher gain and lower beam width.

7.2.3 Beam Width

Depending on the radio system in which an antenna is being employed, there can be many definitions of beam width (or simply beam width). A common definition is the half power beam width. Once the peak radiation intensity is found, the points on either side of the peak represent where half the power of the peak intensity are located. The angular distance between the half power points traveling through the peak is the beam width. Half the power is -3dB, so the half power beam width is sometimes referred to as the 3dB beam width.

7.3 Evolution of Directional Antenna Systems

We now give a brief description of different existing directional antenna systems. The growth of directional antenna can be studied at different stages, starting from basic sectorized and diversity antenna systems to more advanced smart antenna systems. The discussion presented here is not intended to cover all aspects of the technology; rather, expose the basics in an informal and intuitive fashion so that it can serve as the basis for understanding its implications at the MAC and routing layers discussed later. Readers wishing to explore more in this field can refer to [Liberti1999].

Present directional antenna systems can be broadly classified into three different categories: sectorized, diversity, and smart. In what follows we describe each of these systems.

7.3.1 Sectorized Antenna Systems

These antenna systems are used extensively in cellular systems where a base station divides the traditional cellular area into independent sectors, and each of these sectors is treated as a sub-cell. By using sectorized antennas, the range of each sector is increased. Also, sectorized antennas increase the possibility of channel reuse and reduce the interference.

7.3.2 Diversity Antenna Systems

Diversity systems maintain multiple antenna elements at the receiving side. These elements are physically spaced to improve reception by minimizing the effect of multipath. Following are two methods generally employed by diversity schemes:

Switched Diversity: In this method, it is assumed that at least one of the antennas is at a different physical location, at a favorable position in a given moment, and the system can continuously switch between these elements to use the element with largest output. Although these systems try to increase the throughput, they do not utilize the gain by multiple antennas as only one of them is used at a given time;

Diversity Combining: These systems use the concept of diversity combining wherein multipath signals received at different antenna elements are mixed, their phase errors are corrected, and their powers are combined to produce the gain, and tackle multipath and fading.

7.3.3 Smart Antenna Systems

A smart antenna system combines an antenna array with digital signal processing capability to receive and transmit in an adaptive and spatially sensitive manner [Sheikh1999]. These systems can automatically change the directionality of its radiation patterns to suite the wireless environment. If properly deployed, smart antenna systems can significantly increase the performance of a wireless ad hoc system.

The concept of smart antenna systems has been around for some time now, but until recent years, cost barriers have prevented their use in commercial products. The advancement in technology and the advent of powerful low-cost digital signal processors, application-specific integrated circuit (ASIC) design, and development of software-based signal processing techniques and algorithms have made these systems practical for not only cellular environment, but also for ad hoc networks. Smart antennas can broadly be classified into two groups, both systems using an array of (omni-directional) antenna elements:

Switched Beam Antenna Systems: A switched beam system consists of a set of predefined beams (see Figure 7.3(a)), out of which the one that best receives the signal from a particular desired user is selected. The beams have a narrow main lobe and small sidelobes, so, signals arriving from directions other than that of the desired main lobe direction are significantly attenuated. This class of smart antenna systems is similar to what is used by existing cellular systems. They employ transmission through directional beams that have finite number of fixed radiation patterns, and provide interference suppression along other beam directions. A variant of switched beam antenna system called steerable antenna systems can also steer the beam to continuously track a transmitter or receiver. Switched beam antenna systems facilitate spatial reuse by concentrating energy in a particular direction only. When a switched beam antenna directs its main lobe with enhanced gain in the direction of the user, it forms lobes, nulls and areas of medium and minimal gain in directions away form the main lobe.

Adaptive Antenna Arrays: These are also called adaptive beamforming antenna systems. This is the most advanced state of the art smart antenna to date and provides highest degree of flexibility in configuring the beam patterns and interference suppression. Adaptive antenna arrays have infinite number of radiation patterns which can be adjusted in real time (see Figure 7.3(b)). They rely on beamforming algorithms [Liberti 1999] to steer the main lobe of the beam in the direction of the desired user and simultaneously place nulls in the direction of the interfering users' signals. By using these signal processing algorithms, these antenna systems can effectively locate and track signals to minimize interference and maximize signal reception quality.

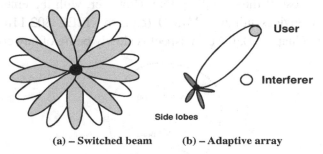

(a) – Switched beam   (b) – Adaptive array

Figure 7.3 – Comparison of switched beam and adaptive array antenna systems
(http://www.antennasonline.com/)

Smart antennas are implemented as an array of omni-directional antenna elements, each of which is fed with the signal, with an appropriate change in its gain and phase. This array of complex quantities constitutes a steering vector, and allows the resultant beam to form the main lobe and nulls in selected directions. With an L-element array, it is possible to specify (L–1) maxima and minima (i.e., nulls) in desired directions, by using constrained optimization techniques when determining the beamforming weights. This flexibility of an L-element array to be able to fix the pattern at (L–1) places is known as the degree of freedom of the array [Godara1997].

Figure 7.4 illustrates transmission ranges of different smart antenna systems. In this figure, the interference rejection capability of the adaptive system provides significantly more coverage than either the sectorized or switched beam systems. The use of multiple antennas at both ends of a communication link provides a significant improvement in link reliability, spectral efficiency, and results into a technology popularly known as multiple input multiple output (MIMO) systems [Liberti1999]. By using multiple antennas at both ends of a link, it is now possible to multiplex the data stream and open up multiple data pipes within the same frequency spectrum to yield a linear bandwidth of the system, with no extra power consumption. In the recent past, the main bottleneck towards commercialization of adaptive antenna array system was the cost since each of the independent antenna beams requires independent digital signal processing controllers, which considerably increase the cost of the overall system. However, with the emergence of commercial devices utilizing MIMO (e.g., the IEEE 802.11n standard discussed in Chapter 4), the cost aspect is slowly being of any concern.

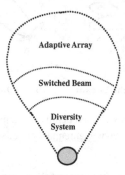

Figure 7.4 – Directional Antenna Coverage Range

As far as the use of directional antennas for ad hoc networks is concerned, most of the research to date has focused on the use of switched beam antenna systems and adaptive antenna arrays. In the context of switched beam antennas, the vast amount of research has concentrated on the networking aspects (read, MAC and routing) when these types of antennas are used. With respect to adaptive antenna arrays, however, the large majority of research is still confined to the physical layer with limited investigation in the MAC and routing layers.

7.4 Advantages of Using Directional Antennas

Directional antenna technology can significantly improve the performance of wireless system. Both infrastructure-based (personal communication systems, cellular, and wireless local loop) and ad hoc networks can benefit from using directional antenna systems. In what follows, we outline some of the advantages of using directional antennas [Sheikh1999].

- **Antenna Gain:** By using multiple antenna beams in ad hoc networks, a node can concentrate its entire transmission energy towards a particular direction, which increases the range of its transmission. This is typically termed as transmission gain. Similarly on the receiving side, a node can selectively receive the packet at a particular antenna beam.
- **Array Gain:** In smart antenna systems, multiple antennas can coherently combine the signal energy. This improves the SNR both at the source and the destination.
- **Diversity Gain:** As discussed above, spatial diversity from multiple antennas can help to combat channel fading. By using diversity, a node can switch between antenna elements to receive the maximum available signal strength.
- **Interference Suppression:** By using the concept of smart antenna system, it is now possible to adaptively combine multiple antennas to selectively cancel or avoid interference and pass the desired signal;
- **Angle Reuse:** By using directional antennas, it is possible to reuse frequency at angles covered by different antenna beams. This is generally termed as space-division multiple accesses (SDMA) and can support more than one user in the same frequency channel. It is to be noted that signal separation of co-channel beams has to be

handled at each node. In MANETs, angle reuse has not been a successful technology because of scattering and mobility which makes signal separation difficult.

- **Spatial Multiplexing:** By using multiple antenna beams at both ends of the wireless link, it is now possible to dramatically increase the bit rates of the wireless link by using a technique termed as spatial multiplexing [Paulraj1994]. In spatial multiplexing, the stream of information is split in N independent streams. These streams are modulated and transmitted one stream per antenna, all in the same radio channel using the required bandwidth to support lower rate sub-streams. If the receiver antenna is well separated, the received sub-streams can be merged to yield the original high bit rate stream. Under favorable channel conditions, spatial multiplexing offers increased spectrum efficiency and does not require prior-knowledge of the channel making it a very robust technique.

7.5 Directional Antennas for Ad Hoc Networks

It is envisioned that different future applications will demand different types of antenna systems so as to meet requirements such as cost, size, energy constraints, performance, and so on. As we have seen so far, the main applications of ad hoc networks can be classified into different categories: military, outdoor, disaster recovery and indoor applications [Ramanathan2001].

For military applications, the nodes (tank, airplanes) are so expensive that the cost of even the most sophisticated antenna may be acceptable. As an added bonus, beamforming antennas can provide a better immunity to jammers and better security provisioning. For fixed outdoor environments, a switched antenna beam can be used to reach different nodes. In this particular application, steerable beam antennas may be too expensive. On top of that, when we consider the use of directional antenna systems for small handheld devices, laptops and PDAs, the size of the antenna becomes another complicating factor. For example, for the 2.4 GHz spectrum band, an 8-element cylindrical array will have a size of approximately 8 cm. However, for the 5 GHz frequency band, the size goes down to 3.3 cm, and for the 24 GHz ISM band the size would be around 0.8 cm.

Therefore, based on the fact that future devices tend to use less crowded and higher frequency bands for communication, the use directional antenna for small devices seems very bright. Also, there is a continuous trend in military to employ directional antenna systems given the tremendous advantages they provide.

7.5.1 Antenna Models

The classification of antenna types discussed earlier has a significant impact on the MAC and routing performance. This is especially true in a a MANET where there is no centralized coordinator. As for the MAC, the antenna type must be taken into account while defining the medium access control scheme, so that the hidden and exposed terminal problems are adequately addressed. As for the routing protocols, depending on the antenna system in use, the routing protocol may have to be redesigned to take into account issues such as the particular direction in which a node can be found, new neighbor discovery mechanisms, possible availability of multiple paths to reach the same destination, the impacts of directional antennas on the route discovery procedure, and many more.

Therefore, it is of paramount importance to study the underlying antenna system in use at the physical layer so as to understand its impact on higher layers. With this in mind, we present two antenna models, namely, the switched beam antenna model and the adaptive antenna arrays model. As we mentioned before, these are the preferred choice for use in MANETs and are the ones upon which most of the existing solutions are based. Obviously, these are just abstract antenna models and the results obtained by the systems based on these models are only as good as these abstractions. Future research needs to investigate more accurate antenna models so as to appropriately design upper layer protocols.

7.5.1.1 Switched Beam Antenna Model

This model possesses two separate modes: Omni and Directional. This may be seen as two separate antennas: an omni-directional and a switched beam antenna which can point towards any specified direction [Choudhury2002]. In principle, both the Omni and Directional modes may be used to transmit as well as to receive signals. However, the Omni mode is used only to receive signals, while the Directional mode is used for transmission as well as reception. In other words, the Omni mode is never used for transmission. This way, both transmitter and receiver take advantage of the increased coverage range provided by beamforming.

In *Omni* mode, a node is capable of receiving signals from all directions with a gain of G^o. While idle (i.e., neither transmitting nor receiving), a node usually stays in *Omni* mode. By employing selection diversity, as soon as a signal is sensed, a node can detect the antenna through which the signal is strongest and go into the *Directional* mode for receiving this particular antenna, going *deaf* in all other directions.

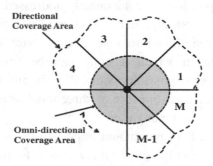

Figure 7.5 – The Antenna Model

In *Directional* mode, a node can point its beam towards a specified direction with gain G^d (with G^d typically greater than G^o), using an array of antennas called array of beams. Due to the higher gain, nodes in *Directional* mode have a greater range in comparison with the *Omni* mode. In addition, the gain is proportional to number of antenna beams given that more energy can be focused on a particular direction, thus resulting in increased coverage range in that particular direction. For example, with the same transmit energy, a 12 antenna array has a higher coverage range than a 6 antenna array, and a 6 antenna array covers, in turn, a larger range than a 4 antenna array. Some proposals take this feature into consideration, while others do not. In order to perform a broadcast with this type of antenna model, a transmitter may need to carry out as many directional transmissions as there are antenna beams so as to cover the whole region around it. This is called *sweeping*, and a negligible delay is generally assumed in beamforming in the various directions.

Figure 7.5 illustrates a switched beam antenna model. Many of existing MAC and routing schemes consider this model [Cordeiro2004, Gossain2004a, Choudhury2002, Korakis2003]. Here, node provides coverage around it by a total of M non-overlapping beams. The beams are numbered from 1 through M starting at the three o'clock position and running counter clockwise. A node can receive and transmit in any of these M antenna beams. Finally, since tracking a user dynamically is a hard task, it is generally assumed that nodes maintain the orientation of their beams regardless of mobility which can be achieved with the aid of a direction-locating device such as a compass [Nasipuri2000a].

7.5.1.2 Adaptive Antenna Array Model

The main distinctive property of the adaptive antenna array and the switched beam antenna models is that the former allows for multiple simultaneous receptions or transmissions (simultaneous transmission and reception is not possible, however), while only a single transmission or reception is possible in the latter.

To clarify distinctive features of the adaptive antennas array model, consider the example of Figure 7.6. Here, nodes are equipped with adaptive antenna arrays, each beam having a beam width of approximately $\pi/2$ radians. In this figure, receiving beams are shown by solid lines while transmitting beams are shown using dashed lines. The figure illustrates that a particular node, say node A, receives from two different nodes, B and C, that lie in different receive beams of A. In this manner, A is able to simultaneously receive from more than one node that has a packet for it. Since transmission and reception are reciprocal processes, a node can also simultaneously transmit to multiple nodes at the same time. Note in Figure 7.6 that even though the areas covered by the transmit beams of nodes B and C overlap, they do not cause collision at node A. This is because the crucial parameter that helps an adaptive antenna array to form a receive beam is the direction of the incident electromagnetic energy [Lal2004]. Therefore, a particular adaptive beam may be seen as the matching of the antenna system to a particular set of directions, i.e., a set of angles for the incoming or outgoing RF. The incident angles for reception are illustrated in the receive beams of node A. Thus, as the incident energy from nodes B and C differ widely in the incident angle at node A, they do not cause interference to each other in reception at A. Interference may be caused when a particular beam has sidelobes in undesirable directions. However, most of existing research in the context of ad hoc networks assumes that perfect switched beams are formed in the desired directions and that sidelobes are negligible.

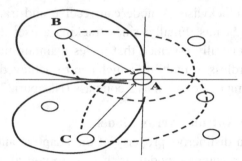

Figure 7.6 – Example of the adaptive antenna array model
[Taken from Lal2004]

7.6 Protocol Issues on the Use of Directional Antennas

In this section, we discuss various protocol issues with directional antennas related to MAC and routing. Here, we note that some of the issues to be discussed next may not be present in all protocols proposed for directional antennas, as they depend highly upon the antenna model under consideration.

7.6.1 Directional Neighborhood

As we know, the neighborhood of a node comprises of all those nodes within its direct communication range. However, in comparison to omni-directional antennas, the notion of a neighbor needs to be reconsidered for directional antennas. To perform a complete "broadcast" in a directional antenna system, a node may have to transmit the broadcast packet in a circular manner as many times as the antenna beams, and this process is called sweeping. Such a scheme emulates a broadcast as performed by omni-directional antenna and theoretically should achieve the same results. However this is not so simple, as there is an obvious delay associated with the sweeping procedure and the need to send the same packet multiple times. As the number of beams increases, so is the sweeping delay. This trade-off and other efficient schemes for broadcasting over directional antenna systems are discussed in a later section.

The notion of directional neighborhood becomes even more subtle if we consider higher gain provided by the directional antennas. We illustrate this point with the aid of Figure 7.7. In Figure 7.7(a), assume nodes A and B are idle listening to the medium omni-directionally with gain G^o. Node C then decides to communicate with node A by increasing its gain to G^d and transmitting its packet. In this scenario, node A is sufficiently close to receive the packet from node C although it is receiving with gain G^o. Node B, however, is not able to receive the packet transmission originated from node C as it finds itself in omni-directional mode and, hence, with receive gain of G^o. Here, we say that C and A are Directional-Omni (DO) neighbors while C and B are not [Choudhury2002]. Consider now the scenario of Figure 7.7(b) where node B tunes towards the direction of node C, thereby increasing its gain to G^d in this particular direction. Node B is now able to receive packets from node C as they are both communicating directionally with gain G^d. Therefore, nodes B and C are only neighbors and we say that C and B are directional-directional (DD) neighbors [Choudhury2002].

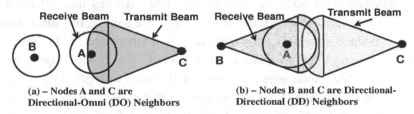

(a) – Nodes A and C are Directional-Omni (DO) Neighbors

(b) – Nodes B and C are Directional-Directional (DD) Neighbors

Figure 7.7 – Directional neighborhood in directional antenna systems
[Taken from http://www.crhc.uiuc.edu/~croy/presentation.html]

As we can see from this example, the notion of neighbors in directional antennas needs to be drastically reconsidered. Along with it, broadcasting is another issue deserving more attention. In later sections, we present proposals which aim at mitigating or overcoming neighboring issues arisen with the introduction of directional antenna systems.

As we studied in Chapter 4, the hidden terminal problem is one of the main sources of performance degradation of wireless MAC protocols. To overcome this problem, RTS/CTS handshake mechanism has been used in protocols such as IEEE 802.11 and MACA [Karn1990] have usd as a way to reserve the channel before data communication and is considered next.

7.6.2 New Types of Hidden Terminal Problems

An implicit assumption in the RTS/CTS handshake is that the underlying antenna transmits omni-directionally, which allows all nodes who receive the handshake packets to accordingly set their NAVs for the forthcoming transmission duration (see Chapter 4). As we present below, directional antennas bring along new instances of the hidden terminal problem requiring new innovative solutions [Choudhury2002, Korakis2003, Takai2002]. We discuss the Directional NAV scheme as a proposed solution to help mitigate the hidden terminal problems.

7.6.2.1 Asymmetry in Gain

The first type of hidden terminal problem that may arise as the result of employing directional antennas is due to asymmetry in gain. To exemplify this, consider Figure 7.8(a) where all nodes are initially in omni-directional mode and hence with gain G^o. Assume node B wants to communicate with node C. To this end, node B goes into directional mode towards node C and sends it a RTS (for the sake of this discussion, assume node B somehow knows the direction where node C is located). Upon receiving the RTS from node B, node C which was originally in omni-directional mode goes into directional mode (with gain G^d) towards node B and sends back CTS. Here, assume that node A (which is still in omni-directional mode with gain G^o) is far enough from node C so that it does not hear node C's CTS.

Once the RTS/CTS between nodes B and C is completed, node B initiates DATA transmission to node C, where both of these nodes now point their transmission and reception beams (with gain G^d) to each other. This is illustrated in Figure 7.8(b), where we highlight that node A is not able to sense the ongoing communication between nodes B and C as it is still in omni-directional mode with gain G^o. Now assume that while the communication between nodes B and C is in progress, node A receives a packet to be sent to node B. Node A then performs a directional carrier sensing of the channel towards node B and concludes the channel to be idle. It then sends its RTS towards node B as depicted in Figure 7.8(b). However, as node C is receiving a DATA packet (with gain G^d) with its beam pointed towards node B, there is a high chance that the RTS sent by node A (sent with gain G^d) will interfere with the DATA reception at node C as these two nodes are now DD neighbors.

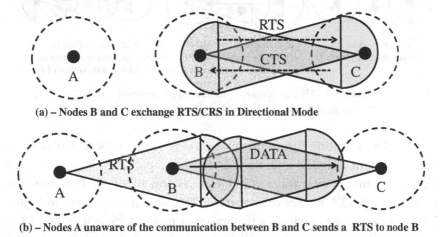

(a) – Nodes B and C exchange RTS/CRS in Directional Mode

(b) – Nodes A unaware of the communication between B and C sends a RTS to node B

Figure 7.8 – Hidden terminal problem due to asymmetry in gain
[Taken from http://www.crhc.uiuc.edu/~croy/presentation.html]

7.6.2.2 Unheard RTS/CTS

Another type of terminal problem that may arise in ad hoc networks employing directional antenna systems is due to unheard RTS/CTS. To illustrate this problem, consider the scenario depicted in Figure 7.9(a) where node A is currently communicating with node D (hence beamformed in the direction of node D) while nodes B and C are currently idle in omni-directional mode. Now, assume that while the communication between nodes A and D is in progress, node B has a packet to send to node C. In this case, nodes B and C will exchange the RTS/CTS followed by node B's transmission of the DATA packet to node C as illustrated in Figure 7.9(b). The crucial remark to make here is that while node A is beamformed towards node D, it is said to be locked in node D's direction and becomes deaf towards all other directions. If applied to the scenario of Figure 7.9(b), this means that node A is unable to hear the RTS/CTS handshake between nodes B and C, and hence is completely unaware of this ongoing communication.

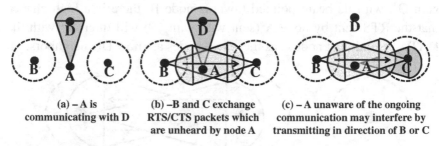

(a) – A is
communicating with D

(b) –B and C exchange
RTS/CTS packets which
are unheard by node A

(c) – A unaware of the ongoing
communication may interfere by
transmitting in direction of B or C

Figure 7.9 – Hidden terminal problem due to unheard RTS/CTS
[Taken from http://www.crhc.uiuc.edu/~croy/presentation.html]

The resulting effect of this deafness scenario is that if node A ends its communication with node D and has a packet to be sent to node C, it will transmit the packet as it does not know that node C is currently receiving a DATA packet from node B. This scenario is shown in Figure 7.9(c). In this case, node A's transmission may interfere with node C's reception, hence causing a collision. Obviously, this type of hidden terminal problem will not take place in case of omni-directional transmissions as node B would be aware of the ongoing communication between nodes A and D, and hence would not send its RTS to node C.

7.6.2.3 The Directional NAV (DNAV)

A discussion about the hidden terminal problem in directional antennas can be found in [Choudhury2002, Korakis2003, Takai2002] where a Directional NAV (DNAV) scheme [Ko2000] is proposed. DNAV is an extension to the NAV concept used in IEEE 802.11 for directional antennas. Essentially, DNAV is a table that keeps track for each direction of the time during which a node must not initiate a transmission through this direction. With this scheme, nodes continuously update the DNAV table upon overhearing a packet transmission in order to keep it from transmitting through one particular direction and hence garble the ongoing transmission. As we discuss later, DNAV is widely used in existing MAC protocols for directional antennas as it is a logical extension to the NAV mechanism found in IEEE 802.11.

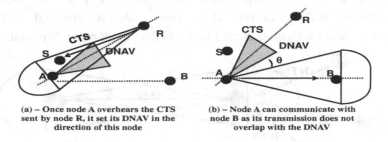

(a) – Once node A overhears the CTS sent by node R, it set its DNAV in the direction of this node

(b) – Node A can communicate with node B as its transmission does not overlap with the DNAV

Figure 7.10 – Extension to the NAV scheme in directional antennas
[Taken from http://www.crhc.uiuc.edu/~croy/presentation.html]

Figure 7.10 depicts an example of the DNAV concept in directional antennas. When node S sends an RTS to node R, that sends CTS back. With DNAV, nodes overhearing either the RTS or the CTS set their DNAV in the corresponding Direction of Arrival (DoA). In Figure 7.10(a), node A overhears the CTS sent by node R and in response sets its DNAV in node R's direction. This will prevent node A from initiating any transmission towards node R which would, in this case, cause a collision with the DATA packet sent by node S. If node A wants to initiate a transmission, it is only possible if the direction of the transmission does not overlap with the DNAV. In Figure 7.10(b), node A is allowed to transmit to node B as the angular separation between the direction of transmission and the DNAV is larger than a threshold θ.

7.6.3 Deafness

Deafness is defined as the phenomenon when a node X is unable to communicate with a neighbor node A, as A is presently tuned to some other directional antenna beam. In the specific case of widely used IEEE 802.11 MAC protocol, at each unsuccessful attempt of node X to communicate with node A, the backoff interval is doubled, hence considerably degrading network performance. Deafness is a serious issue in directional antennas as it may considerably impact performance, not only at the MAC layer but also at upper layers [Gossain2004b]. Deafness may also occur if A's NAV in the direction towards node X (i.e., the DNAV) is set, and hence node A becomes unable to reply back with a CTS. Figure 7.11 illustrates a deafness scenario where nodes X, Y and Z keep on trying to communicate with node A (e.g., by sending consecutive RTS packets) while this node is engaged with a communication with node B, and hence is deaf to all other directions.

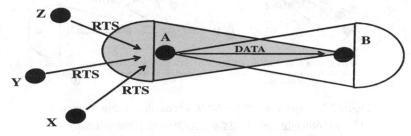

Figure 7.11 – Node A is deaf to the packets sent by nodes X, Y and Z
[Taken from http://www.crhc.uiuc.edu/~croy/presentation.html]

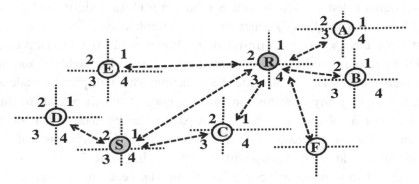

Figure 7.12 – Deafness due to the persistent hearing of DATA packets

Therefore, unless node A informs all its neighbors in advance that it is going to start communicating with node B, its neighbors might unsuccessfully try to contact it. On the other hand, a large overhead may be generated in informing a node's neighbors about a forthcoming communication. Clearly, this raises a tradeoff issue between deafness and overhead, and is discussed in detail in later sections. Another type of deafness is due to persistent hearing of DATA packets. As a matter of fact, this problem may occur in almost all MAC protocols proposed for directional antennas, and is explained through Figure 7.12 where each node is assumed to have a total of four antennas beams.

When a node sends RTS/CTS (either directional or omni), all neighbors who receive it set their NAV (or DNAV) accordingly. Whenever the source node starts transmitting the DATA packet, neighboring nodes which are idle and overhear this DATA transmission will move to the directional mode so as to receive the DATA packet, hence becoming deaf to all other directions. For example, assume in Figure 7.12 that a data communication is to be carried out between nodes S and R. Clearly, node C will detect the forthcoming data communication due to the RTS/CTS handshake between S and R. As a consequence, node C will move to directional mode (i.e., tune it's receive beam) towards node S whenever the DATA transmission originating at node S starts. Therefore, if in the meantime node F tries to send an RTS to C (received through beam 4), node C will not reply as it is currently beamformed (i.e., tuned) in the direction of node S's DATA transmission. Finally, note that this deafness scenario is different from the one explained through Figure 7.11. Here, the deaf node in question (i.e., node C in Figure 7.12) is not involved in the actual communication (which takes place between nodes S and R) while in the scenario of Figure 7.11 the deaf node (i.e., node A) is the actual source of the DATA transmission. As deafness is very detrimental to the performance, recent MAC protocols for directional antennas have attempted to properly handle it. We come back to this issue later in this chapter.

7.7 Broadcasting

As we have seen so far, broadcasting is a widely employed mechanism in ad hoc networks. Routing protocols including DSR, AODV, ZRP, LAR, and so on (all discussed in Chapter 2), use variants of a network-wide broadcasting to establish and maintain routes. Ultimately, these protocols use simple flooding for broadcasting. Here, every node forwards each broadcast packet it receives exactly once. As pointed out in [Ni1999], simple flooding causes redundancy and increases the level of contention and collisions in a network. The omni-directional nature of transmissions makes things even worse as all hosts receiving the broadcast packet will retransmit it, generating unnecessary interference to other nodes and considerably reducing network capacity. On the other hand, with directional transmission both transmission range and spatial reuse can be substantially enhanced by having nodes concentrate transmitted energy only towards their destination's direction, thereby achieving higher signal to noise ratio.

A simple solution to broadcasting with directional antennas (here we confine our discussion to switched beam antenna systems) is to sequentially sweep across all the pre-defined beams of the antenna system. Due to the increased range of a directional beam, a greater number of nodes are covered in a single broadcast sweep as compared to an omni-directional broadcast. However, broadcast by sweeping incurs a sweeping delay [Choudhury2003]. Therefore, it is of paramount importance to investigate the issue of broadcasting over directional antennas such that efficient schemes can be designed to take the antenna system characteristics into consideration, and hence reduce redundancy and the sweeping delay. In this section we study proposed schemes for broadcasting over directional antennas. To set the foundation for our discussion, initially we discuss simple flooding broadcast over a directional and an omni-directional antenna model. Next, we delve into the proposed schemes and show how they take advantage of directional communication. Finally, we note that all the schemes discussed here have their transmission rules governed by the basic IEEE 802.11.

7.7.1 Broadcasting Protocols

As we described in Chapter 2, a variety of broadcasting schemes have been proposed for omni-directional antennas. For directional antennas, however, there is not much done [Hu2003, Joshi2004]. In this section, we discuss broadcasting techniques over directional antennas and show how these protocols overcome new challenges arising in this type of directional environments. Essentially, the existing schemes can be classified according to considered antenna model. They can be either adaptive antenna array based or switched beam antenna based.

7.7.1.1 Adaptive Antenna Array Based

In this section we discuss the proposed broadcasting protocols for adaptive antenna array based systems. As we discussed earlier, this type of antenna system allows multiple simultaneous transmission of a packet by a node. As a result, the schemes presented in this subsection do not consider the effect of sweeping delay which is only present in switched beam antenna systems.

7.7.1.1.1 Simple Enhanced Directional Flooding (SEDF)

In SEDF [Hu2003], whenever a node receives a packet to be forwarded, it starts a delay timer. If the same packet is received again before the expiration of this timer, the node makes a note of all the beams where that packet arrived at, and sets them to passive mode. Upon expiration of the delay timer, the node will forward the packet in only those beams/directions other than those in which the packet arrived (i.e., which have been marked as passive).

7.7.1.1.2 Single Relay Broadcast (SRB)

In any relay based scheme, whenever a node receives a broadcast packet it chooses a subset of its neighbors to forward the packet. Only members of this subset are allowed to forward the packet. Here, it is the responsibility of the broadcasting node to explicitly designate the broadcast relay nodes within a broadcast packet header.

In the particular case of the SRB scheme [Hu2003], each node designates one and only one relay node in each direction on the basis of the received signal strength of hello packets through this particular direction. For the purpose of maintaining one-hop neighbor information, every node periodically transmits hello packets. The node whose hello packet is received with the weakest signal is selected to be the relay in that direction. Before forwarding, a node waits for a random delay and does not designate any relay node in directions from which the packet arrives. It should be noted that a node may also discard a packet even if it has been designated as a relay, which happens in case it has already seen the packet before. Finally, it is important to note that SRB requires one-hop neighbor information to be maintained.

7.7.1.2 Switched Beam Antenna Based

We now turn our attention to broadcasting schemes designed for switched beam antenna systems. Contrary to adaptive antenna array based schemes, this class of broadcasting protocols is harder to design as the effect of sweeping becomes a major concern. To mitigate the sweeping delay, most of the schemes presented in this section require a node to transmit in fewer necessary beams instead of performing a complete sweep. Usually, this is done by estimating which of a node's beams have neighbors and which do not.

7.7.1.2.1 Protocol Design Considerations

All the schemes discussed next make use of some basic design considerations. In this section we discuss these mechanisms which aim at enhancing the protocols performance as well as overcoming the new challenges presented in switched beam antenna systems.

One-Hop Neighbor Awareness

A periodic exchange of hello packets amongst the nodes is assumed by these schemes. A node at any time is aware of which antenna beam its one-hop neighbor lies. Usually, this exchange comes at no additional cost as most directional MAC/Routing protocols (discussed later in this chapter) need one-hop neighbor awareness to operate. To accomplish this, a node, say S has to resort to a circular directional transmission of the broadcast packet through all its antenna sectors. It should be noted that while S is engaged in this circular sweep, it remains deaf to any incoming packet. Hence, it has been showed in [Gossain2004a] that a sender node S needs to inform its neighbors the additional time they should wait before initiating a transmission towards it. To this end, the sender node S includes in the broadcast packet the value $(K-c-1)$ where c is an integer (initially equal to zero) that keeps track of how many sectors the broadcast packet has been already sent, and K is the number of idle antenna beams at S. Now, if T_{bc} is the time the receiver takes to completely receive the broadcast packet, receiver node R waits for an additional time equals to $(K-c-1)*T_{bc}$ before initiating any transmission in the direction from which it received the packet from node S.

Novel Optimized Deferring while Sweeping

In the schemes that follow, the IEEE 802.11 MAC is followed before transmitting in the first beam of a particular sweep. For subsequent beams of the same sweep, only carrier sensing is done before transmission. However, if a beam has been marked as busy (i.e., the DNAV is set in this direction), that beam is ignored and the next free beam is picked. It should be noted that these schemes do not wait for the beam to become free. It is argued that deferring in every beam could prove to be disastrous, leading to extremely high sweeping delays.

Random Delay Timer

Similar to broadcasting with omnidirectional antennas discussed in Chapter 3, here a node also starts a random delay timer (RDT) before forwarding a packet so as to avoid any global synchronization.

7.7.1.2.2 New Enhanced Directional Flooding (NEDF)

In NEDF [Joshi2004], whenever the MAC layer receives a packet to be broadcast, it marks the beams where it will not retransmit. First of all, similar to SEDF, the NEDF protocol does not rebroadcast in the antenna beams where it received the broadcast packet. In addition, using the neighbor table, it marks as passive those beams in which there are no neighbors. Obviously, it will also not rebroadcast in beams which have been marked as busy. Next, amongst the resulting selected beams, the goal should be to transmit first in regions with maximum uncovered nodes so as to reduce the chances of collisions. Also, in regions where there is a high probability that nodes have already received the broadcast packet, re-broadcasting should be delayed. Keeping these objectives in mind, NEDF defines an order in which the transmission is carried out. Initially, NEDF chooses the beams which are vertically opposite to the beams where the node received the broadcast packet. Next, the beams which are adjacent to these vertically opposite beams are chosen. This continues until all the selected beams are covered.

NEDF uses the received power of an incoming packet to decide the order of transmitting in the vertically opposite beam. The beam vertically opposite to the one where the lowest power packet has been received is selected first. Clearly, the idea here is that this packet came from the farthest node. Therefore, transmitting first in this beam guarantees maximum additional coverage, and if a uniform distribution of nodes is assumed, this translates into covering the maximum number of uncovered nodes. For example, in Figure 7.13 node (a) receives broadcast packets from nodes (b) and (c) through beams 5 and 4, respectively, before its RDT expires. It shall hence rebroadcast only in beams 1, 0, 2, 7, and 3 in that order. Here we note that beam 6 is ignored as it has no neighbors in that direction. Finally, beam 1 is chosen over beam 0 as node (b) is farther from node (a) than node (c).

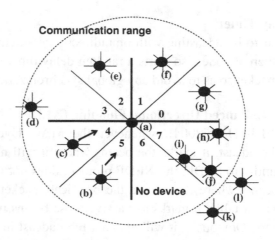

Figure 7.13 – The NEDF scheme [Taken from Joshi2004]

7.7.1.2.3 Probabilistic Relay Broadcasting (PRB)

An inherent flaw in the SRB scheme (discussed earlier) is that it uses a single relay node in each direction. This can lead to a partition in the network. For example, in Figure 7.13, node (h) is the obvious choice to serve as a relay for node (a) through its beam 7 (as it is the farthest). Hence, node (i) shall not forward the broadcast it hears from node (a). As a consequence, if nodes (j), (k) and (l) are outside the radio range of node (h), they will never receive the broadcast packet, thereby resulting in a partition. Furthermore, it may also happen that the relay is engaged in a conversation with another node and hence may be deaf [Gossain2004a] to the broadcast packet. This too can cause a partition in the network. Therefore, although a single relay node translates into very few packet transmissions, it may also result in higher latencies and poor connectivity, especially under conditions of heavy traffic.

The PRB [Joshi2004] protocol aims at overcoming the drawbacks found in SRB. In PRB, each node is required to record the received power of the hello packet from the farthest node (weakest signal) in each direction. Let us denote this power as P_f. Upon receiving a broadcast packet and after the expiration of RDT, the node forwards the packet on all the beams except the ones on which it received the packet. For each beam, it includes P_f of the corresponding beam in the packet header. Whenever a node receives this packet, it retrieves its received power, says P_r, and calculates the ratio of P_f / P_r.

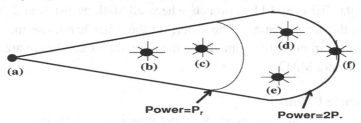

Figure 7.14 – The PRB scheme [Taken from Joshi2004]

In PRB, this is the probability with which it will re-broadcast. In addition, the order of re-broadcasting will be similar to the one proposed in NEDF – vertically opposite beams followed by their adjacent beams. Similarly, neighbor-less and busy sectors will be ignored. Thus, in PRB, nodes which are very close to the broadcast originator have very little probability to rebroadcast, and very close nodes could be eliminated from forwarding. Therefore, in each sector only nodes receive the packet at a power less than or equal to $2 * P_f$ will retransmit. Note that the farthest node in each sector has probability 1 to rebroadcast. Figure 7.14 illustrates this idea where nodes (b) and (c) do not forward at all while nodes (d), (e) and (f) forward with probability P_f/P_r.

7.8 Medium Access Control

As we observed above, in directional antennas, new types of hidden node problems arise. In addition, issues such as the node deafness problem and the determination of neighbors' locations have to be properly handled [Korakis2003]. Therefore, the development of an adequate MAC protocol, overcoming these limitations, is of paramount importance. In addition, due to higher antenna gain, directional antennas have a greater transmission range than omni-directional antennas. This enables far-away nodes to communicate over a single hop, and results in increased throughput and reduced delay. An efficient MAC protocol for directional antennas should, therefore, attempt to take advantage of both these benefits of directionality: spatial reuse and higher transmission range.

We categorize existing directional MAC protocols into single channel and multi-channel. The former type of MAC protocols often follow the IEEE 802.11 approach where all stations listen and receive through the same channel. The latter, on the other hand, use more than one channel in order to organize the medium access and hence are called multi-channel MAC.

7.8.1 Single Channel

In this section we present the most prominent single channel MAC protocols proposed for use over directional antennas. As we shall see below, the design of the vast majority of protocols in this category has been inspired by the IEEE 802.11 MAC, as the functioning of the IEEE 802.11 MAC is very well understood.

7.8.1.1 The Directional MAC (DMAC) Protocol

The Directional MAC (DMAC) protocol [Choudhury2002] has been derived from IEEE 802.11 and is similar to the schemes introduced in [Takai2002, Ko2000]. DMAC attempts to achieve spatial reuse of the channel and take advantage of the higher transmission range by using DO links. In essence, DMAC is similar to the IEEE 802.11 but on a per-antenna basis. The DMAC protocol assumes nodes know their neighbors' location, that is, they are aware through which antenna beam a given neighbor can be reached. Similar to IEEE 802.11, channel reservation in DMAC is performed using a RTS/CTS handshake both being transmitted directionally. An idle node listens to the channel in omni mode, i.e., omni-directionally. Whenever a node receives a signal from a particular direction, it locks onto that signal directionally and receives it. Please note that collisions may happen during signal reception while the node finds itself in omni mode. Only when a node is beamformed in a specific direction, it can avoid interference in the other remaining directions. The RTS transmission in DMAC is as follows. Before sending a packet, the transmitter node S performs a directional physical carrier sensing towards its intended receiver R. If the channel is sensed idle, DMAC checks its DNAV table to find out whether it must defer transmitting in the direction of node R. An example of DMAC can be seen in Figure 7.15.

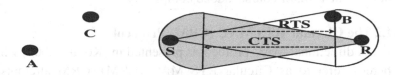

Figure 7.15 – The DMAC protocol

The DNAV maintains a virtual carrier sense for every DoA (i.e., for every antenna beam) in which it has overheard a RTS or CTS packet. If node S finds it is safe to transmit, then similar to IEEE 802.11 it enters the backoff phase and transmits the packet in the direction of node R when the backoff counter counts down to zero. If idle, the receiver node R remains in omni mode listening to the channel omni-directionally. When node R receives the RTS from S, it is able to detect the DoA of the RTS and locks in the corresponding direction. Upon complete reception of the RTS packet, node R beamforms in the direction of node S and sends the CTS packet directionally towards S, provided its DNAV indicates to be free. Similar to IEEE 802.11, the CTS is transmitted after SIFS duration after reception of the RTS. The complete RTS/CTS handshake between nodes S and R is shown in Figure 7.15.

Note that the nodes other than S and R (for example, node B in Figure 7.15) which receive either the RTS or CTS packet, update their DNAV in the captured DoA with the duration field specified in the RTS or CTS packet. This prevents node B from transmitting any signal in the direction which may interfere with the ongoing transmission between nodes S and R. On the other hand, node A in Figure 7.15 could initiate communication with node C, as none of these two nodes have received the RTS or CTS from nodes S and R.

The DMAC protocol has its drawbacks. First of all, both types of new hidden terminal problem (i.e., due to asymmetry in gain and unheard RTS/CTS) are present. In addition, deafness scenarios are also possible. Finally, the built-in assumption in DMAC that neighbors' locations are known in advance may not be practical in reality. Although simple, all these issues may considerably limit the applicability of DMAC. In later sections, we discuss protocols which attempt to overcome some of these limitations but with, of course, added complexity.

7.8.1.2 The Circular RTS MAC (CRM) Protocol

A directional MAC protocol is presented in [Korakis2003] which we hereby refer to as Circular RTS MAC (CRM). CRM attempts to overcome some of the limitations found in DMAC. Contrary to DMAC, CRM does not depend on the availability of neighbors' location information. To accomplish the same effect, CRM employs a circular directional transmission of the RTS packet, that is, a node S with an RTS to be sent to node R directionally transmits the same through all of its antenna beams in a sequential and circular way. This way, node R will eventually receive the RTS packet coming from node S.

Based on the circular directional transmission of the RTS packet, CRM may also decrease the occurrence of node deafness as it informs all nodes within the transmitter's directional radio range about the forthcoming transmission. This way, nodes overhearing the RTS defer their transmission in the direction of the transmitter, hence minimizing deafness scenarios. In addition, CRM includes extra information in the RTS and CTS packets so as to enable other nodes to determine whether they need to defer in the direction of the transmitter or receiver, thus also minimizing the hidden terminal problem.

In CRM, upon receipt of an RTS packet, the receiver node R delays the transmission of its CTS for a period of $T_{CRM} = K *$ RTS_Tx_Time + SIFS, where K is the number of antenna beams on which the sender node S will transmit the circular directional RTS (i.e., K = the total number of beams minus one), RTS_Tx_Time is the time required for the transmission of a single RTS, and SIFS is as defined in

IEEE 802.11. Therefore, CTS is only transmitted after the sender node has swept through all of its beams. The CRM protocol does not completely overcome the limitations of DMAC, and it introduces new shortcomings. First of all, CRM only prevents node deafness in the neighborhood of the transmitter node. As we have seen earlier, CRM employs a circular directional RTS transmission coupled with a single directional CTS transmission by the receiver. As a result, CRM is only able to cope up with node deafness at the sender neighborhood, while deafness may still occur in the neighborhood of the receiver.

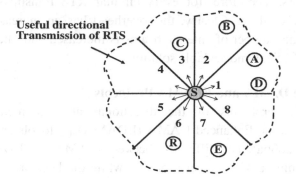

Figure 7.16 – The CRM protocol

A more serious problem with CRM is in the design of its RTS/CTS handshake mechanism. In CRM, a sender node S initiates the circular directional transmission of its RTS although it is not at all sure whether its intended receiver node R has correctly received its RTS or not. To illustrate this case, consider the example in Figure 7.16 where nodes are equipped with an eight-beam antenna array. Further, consider that the sender node S, initiates transmission of a circular RTS through antenna beam 1 and its intended destination node R is located at the antenna beam 6. As node S circularly transmits the RTS packets, nodes in the corresponding directions update their DNAV for the duration contained in the RTS packet. Now, assume that when node S transmits its RTS through antenna beam six towards node R, node A also sends a RTS to node R thus causing a collision. In this case, node R will not respond to node S's RTS. The side effect of this is that nodes in the neighborhood of node S and which correctly received its circular RTS will not be able to initiate any transmission either towards node S or node R, since their DNAV is set towards both nodes S and R. Clearly, this degrades the network capacity as neighbors of node of S could eventually initiate a transmission as the previous node R's RTS transmission failed.

Another limitation in CRM can also be seen through the example in Figure 7.16. Here, we see that the sender node S transmits its circular directional RTS through four "empty" sectors. That is, out of the eight sectors covered by the eight antenna beams of node S, four of them have no neighbors. Therefore, for every circular RTS transmission node S wastes four of them. Clearly, this overhead has an increasingly larger impact as the number of antenna beams is increased and, therefore, may not be an efficient and scalable solution.

7.8.1.3 The DAMA and EDAMA Protocols

Two protocols of the directional antenna medium access (DAMA) and the Enhanced DAMA (EDAMA) protocols are introduced in [Gossain2006a]. The difference between DAMA and EDAMA is that the first is only a MAC layer protocol, while the later proposes a cross-layer design which integrates the MAC and routing functionality and is shown to improve on the performance of DAMA.

DAMA

The DAMA protocol addresses the hidden node problem and node deafness by employing a novel scheme of selective circular directional transmission of RTS and CTS, where these packets are transmitted only through the antennas to neighbors. To make the protocol implementation simple, DAMA design has been inspired by the IEEE 802.11 MAC just like DMAC and CRM.

The DAMA protocol aims to effectively overcome the limitations found in both DMAC and CRM by utilizing a new combination of adaptive mechanisms. To take advantage of the increased gain obtained by directional antennas, all transmissions in DAMA are directional. In addition, DAMA does not rely on prior availability of neighbors' location such as in DMAC, while it learns its neighbors with time as communication between nodes takes place. To this end, DAMA employs a self-learning algorithm to determine the presence or absence of nodes in given directions.

To prevent node deafness and the new types of hidden node problems, DAMA employs circular directional transmission of both RTS and CTS. However, contrary to CRM, DAMA only transmits the RTS and CTS packets through the antenna beams with neighbors. Initially, DAMA performs similar to CRM by sweeping through all antenna beams. However, as responses are received, it collects and caches neighboring information. With this information, it is possible to selectively transmit the RTS/CTS packets through only those sections where neighbors can be found, hence cutting down on the protocol overhead.

Another important aspect in the design of DAMA is that the first RTS sent is always transmitted in the sector where its intended neighbor is located, and the circular directional RTS and CTS procedure is only initiated once the RTS/CTS handshake is successfully completed. This is done to overcome one of the major limitations in CRM (discussed earlier) where it initiates the circular directional transmission of RTS packet (thus reserving the channel by having neighbor nodes set their DNAV) before the sender node knows if any of its RTS has or will ever be correctly received by its intended destination node.

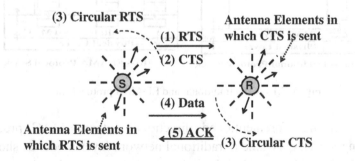

Figure 7.17 – RTS/CTS/DATA/ACK packet exchange in DAMA

In DAMA, upon reception of an RTS packet sent by node S – shown by step (1) in Figure 7.17 – the receiver node R proceeds similar to IEEE 802.11. That is, it waits for a period of time equal to SIFS and sends back CTS as shown by step (2). Only after the RTS/CTS handshake is completed and the channel is reserved in their direction, will both sender and receiver nodes simultaneously initiate the circular directional transmission of their RTS and CTS packets, respectively, to inform their neighboring nodes. This simultaneous transmission of RTS and CTS is observed to save time and effectively takes care of the hidden node problem and deaf nodes at both the neighborhood of the sender and receiver. Figure 7.17 illustrates the simultaneous circular transmissions through step (3), where we note that nodes S and R do not send their RTS and CTS through all sectors, but only through those where neighbors can be found. Finally, the sender S and receiver R synchronize back again in order to carry out the DATA/ACK transmission as depicted in steps (4) and (5) in Figure 7.17.

EDAMA

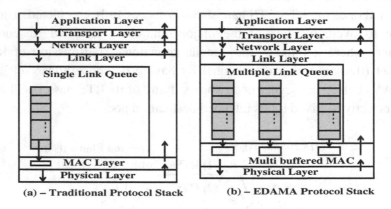

(a) – Traditional Protocol Stack   (b) – EDAMA Protocol Stack

Figure 7.18 – The traditional and EDAMA protocol stacks

All the aforementioned directional antenna MAC protocols, including DAMA, assume a traditional network layer model as shown in Figure 7.18(a). Here, the link layer has a single queue of packets waiting to be handed over to the MAC layer which is, in turn, a single buffered entity. Whenever the network layer has a packet to send, it determines the next hop for the packet and places it in the link layer queue. In case MAC is in idle state, it signals for a packet from the link queue and subsequently buffers it. It then determines the antenna beam required to transmit the packet and enters into send state. The MAC will only request another packet from the link layer queue when it has successfully transmitted or given up (e.g., the next hop is unreachable) on the packet it is currently handling.

In existing directional MAC protocols, in the event that the packet to be transmitted is for a beam whose DNAV was set, the MAC waits for the medium to become idle. While doing so, it could so happen that other packets in the link layer queue could be transmitted over beams which are not busy at that time. In such a scenario, waiting for the medium to become idle reduces the overall throughput of the system. This is called self-induced blocking phenomenon which is a consequence of using a single MAC buffer for all antenna beams. EDAMA proposes overcoming this problem by employing a cross-layer design approach wherein the network layer is aware of different antenna beams at the MAC layer. The MAC, in turn, has separate buffers for each the antenna beams. Accordingly, the link layer follows this approach by maintaining separate queues for each beam. The modified protocol stack of EDAMA is shown in Figure 7.18(b). As we can see, in EDAMA, the MAC layer has multiple buffers for each corresponding antenna beam, where each of them corresponds to a specific queue in link layer.

In order to place the packet in the correct link layer queue, the routing algorithm needs to determine the antenna beam which the MAC will use for transmission of this packet. This is done in EDAMA by augmenting the routing table with an additional entry called Antenna Beam, which corresponds to the antenna beam the MAC uses to reach the corresponding next hop. Whenever the MAC receives a packet from a node, it informs the network layer the antenna beam through which it received the packet. The network layer, in turn, updates the beam entry for that destination in the routing table. As time progresses, the network layer will eventually learn about the antenna beams used to reach each of its neighbors. In case the network layer's beam entry field is empty for a given next hop, a simple broadcast is done when this entry is needed (broadcast packets are kept in a special dedicated queue as they are to be transmitted through all antenna beams).

7.8.1.4 The Multi-Hop RTS MAC (MMAC) Protocol

All the protocols we have seen so far are able to support DO links only. This is because the receiver node always finds itself in idle mode before it receives any packet. As we know, in idle mode a node listens to the channel omni-directionally with gain G^o. Therefore, when a sender node S directionally transmits a packet (with gain G^d) to a receiver node R (who listens to the medium omni-directionally and hence with gain G^o), a DO link is established. Now assume the case similar to the one explained in Figure 7.7(b), wherein nodes A (the sender) and B (the receiver) would not be able to establish a communication unless they are DD neighbors. That is, unless node B tunes itself directionally towards node A (hence with gain G^d) before node A sends it a packet, nodes A and B will not be able to communicate directly. Here, nodes A and B are neighbors only when they beamform towards each other. If this is possible, the MAC protocol is able to fully exploit the higher transmission range offered by directional antennas by supporting DD links. Clearly, the problem here is how to inform node B in advance to beamform towards node A, as node A will be sending it a packet.

With this in mind, the Multi-hop RTS MAC (MMAC) protocol has been proposed in [Choudhury2002] which builds on the basic DMAC. The idea of MMAC is to propagate the RTS packet through multi-hopping until it reaches the intended destination which, in turn, beamforms towards the transmitting node and sends back CTS. This way, MMAC is able to establish links between distant nodes, and then transmit CTS, DATA and ACK over a single hop.

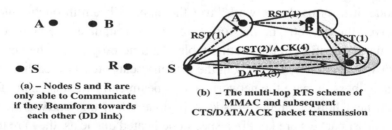

(a) – Nodes S and R are only able to Communicate if they Beamform towards each other (DD link)

(b) – The multi-hop RTS scheme of MMAC and subsequent CTS/DATA/ACK packet transmission

Figure 7.19 – The MMAC protocol
[Taken from http://www.crhc.uiuc.edu/~croy/presentation.html]

To better understand MMAC, let us consider the topology in Figure 7.19(a) where node S has a packet to send to node R, and these two nodes are not neighbors unless they beamform toward each other. So that node S can inform node R that it has a packet destined to it, MMAC utilizes a form of multi-hop RTS packet transmission where this packet traverses multiple intermediate nodes, until it reaches the destination. For the specific case of the example of Figure 7.19(a), the multi-hop RTS packet transmitted by node S would traverse nodes A and B before it reaches node R as depicted in Figure 7.19(b). It is assumed in MMAC that the multi-hop route information is provided to the MAC layer by upper layer module, which should be capable of selecting an appropriate DO-neighbor route to the intended DD-neighbor node (i.e., node R in the example of Figure 7.19).

In addition, this upper layer module specifies the MAC layer what is called a transceiver profile which contains, among other things, the necessary beamforming information required to propagate the multi-hop RTS packet through intermediate nodes and also to enable the destination node to beamform towards the source node (hence forming the DD link). In the example of Figure 7.19(b), the transceiver profile provided by node S allows all nodes in the route to the intended destination node R to beamform towards each other. That is, it allows node S to beamform towards node A, node A to beamform towards node B, node B to beamform towards node R and, finally, node R to beamform towards node S. This way, RTS/CTS/DATA/ACK transmission can be carried out between nodes S and R who are now DD neighbors. Figure 7.19(b) indicates the steps in the packet exchange performed in MMAC.

Clearly, we can identify some disadvantages with MMAC. First of all, it relies too much on an upper layer module whose functioning is not clear. It is not discussed in MMAC how transceiver profiles can be constructed. Secondly, MMAC assumes that all nodes in the route from the source to the destination are available at the time of RTS transmission. As discussed earlier, deafness and hidden node scenarios are commonplace in directional antennas, and this is more serious in the case of MMAC which is built on top of DMAC. Thirdly, MMAC brings some routing functionality to the MAC protocol and this by itself incurs significant overhead at the MAC layer. The complexity of MMAC may prohibit its applicability. Fourth, assume that there is a route of length n hops from the source node S to the destination node D. Since this is an ad hoc network where nodes are mobile, a situation might occur where node R moves out, but this will only be figured out once the multi-hop RTS packet has traversed n-1 hops; hence generating too much unnecessary overhead. As a matter of fact, mobility may completely diminish any of the advantages introduced by MMAC. Obviously, other issues such as the estimation of the transmission time all the way from source to destination, the management of MAC layer timers, retransmission, and so on, have to properly addressed.

7.8.2 Multi-Channel

As we have seen before, single-channel MAC protocols often have to put extra effort if the goal is to effectively overcome the issues arising in directional antennas such as deafness and the hidden terminal problem. To a certain extent, this extra effort is largely due to the availability of a single channel. Therefore, multi-channel MAC protocols try to remove some of the MAC layer complexity by employing multiple channels and, for example, using one of the channels to convey important control information. Obviously, multi-channel protocols also have disadvantages such as higher physical layer complexity, bandwidth allocation requirements and, sometimes, throughput limitations. This section presents some of the prominent multi-channel MAC protocols proposed for use over directional antennas.

7.8.2.1 The Simple Tone Sense (STS) Protocol

The Simple Tone Sense (STS) [Yum1992] protocol is based on the concept of busy tones introduced by the Busy Tone Multiple Access (BTMA) protocol [Tobagi1975]. In BTMA, a station broadcasts a busy tone signal whenever it is receiving a packet. This way, all the nodes within the transmission range of the receiving station will sense the signal and remain silent for the duration of the busy tone, thus avoiding collisions. Clearly, BTMA employs at least two channels: one for the busy tone and one for data and/or control information.

The STS protocol reuses the idea of busy tones and applies it to directional antennas. In STS, each node is assigned a tone, which is simply a sinusoidal wave at a particular frequency. In addition, the STS protocol employs algorithms to guarantee that this tone frequency is unique in the neighborhood of any given node. Hence, whenever a node starts receiving a packet destined to itself through a particular direction, it transmits its tone in that direction. A neighbor node receiving the tone through antenna beam B, assumes that one of its neighbors is receiving a packet towards the direction of beam B and hence does not transmit using this beam. Obviously, antenna beams other than B not sensing busy tones can be used for transmission.

The STS protocol has some drawbacks. First of all, assigning tones to nodes is a hard task. In addition, these tones have to be unique in a node's neighborhood which further complicates the matter. Second, the STS protocol assumes that the direction and angles of the beams can be arbitrarily chosen. This is necessary for the protocol operation as it requires nodes to be evenly distributed among the antennas beams and is clearly a major limitation.

7.8.2.2 The DBTMA/DA Protocol

The Dual Busy Tone Multiple Access for Directional Antennas (DBTMA/DA) [Huang2002] is an extension to directional antennas of the DBTMA protocol [Deng1998] protocol designed for omni-directional antennas. The DBTMA is, in turn, based on the BTMA protocol and employs two busy tones instead of one as in BTMA.

The DBTMA protocol employs the exchange of RTS/CTS frames to turn on a pair of transmit and receive busy tones, which jointly reserve the data channel. More specifically, DBTMA divides a single channel into two sub-channels, namely, a data channel for data frames and a control channel for control frames. This control channel is further split into two busy tones operating in separate frequencies: transmit busy tone (BTt) and receive busy tone (BTr). Whenever a node starts transmitting or receiving data, it turns on the BTt or BTr, respectively, which can be heard by all nodes within its transmission range. By employing the dual busy tones, DBTMA can reserve the channel in both directions. DBTMA operates as follows. Whenever a sender node has data to be sent, it first senses the channel for BTr to make sure that the intended receiver is not currently receiving from another hidden node. The sender then transmits an RTS frame to the intended receiver if BTr is idle. Upon receipt of the RTS packet, the receiver senses for BTt to ensure that the data it is expected to receive will not collide with any other ongoing neighboring data transmission. If BTt is idle, it replies with a CTS frame and turns on BTr until the data packet is completely received. Upon receipt of the CTS packet the sender node will transmit the data packet and turn on BTt for the duration of this data transmission.

The DBTMA/DA protocol applies the concept of DBTMA to directional antennas and improves channel capacity by transmitting the RTS/CTS frames, data frame, and the dual busy tones directionally. Clearly, the major enhancement of DBTMA/DA over DBTMA is the possibility of transmitting the busy tones directionally and, hence, being able to support multiple simultaneous transmissions. It is observed in [Huang 2002] that the directional transmission of busy tones often gives the best results.

7.8.3 Other Protocols

The study of MAC protocols for directional antennas is gaining considerable attention from both industry and academia. As a result, there have been many other proposed protocols in the literature.

In [Nasipuri2000a] a protocol similar to IEEE 802.11 is introduced but adapted to directional antennas. As opposed to the schemes we discussed so far, it is assumed that several directional antennas can be used simultaneously. Based on this, the protocol in [Nasipuri2000a] is able to support omni-directional reception which is not possible in all previous schemes. This is done by employing all directional antennas together, at the same time. Clearly, this antenna model simplifies many of the problems.

In any case, the protocol described in [Nasipuri2000a] works as follows. The sender node transmits an omni-directional RTS and the receiver sends back an omni-directional CTS. Upon reception of the RTS/CTS, the receiver/sender determines the antenna beam from which these packets were received by using the DoA information. All neighboring nodes overhearing either the RTS or CTS packet defer their transmissions so as not to cause a collision. Finally, both the DATA and ACK transmissions are carried out directionally.

One of the benefits of this protocol is that it reduces interference by transmitting the DATA and ACK packets directionally. It is able to reduce interference further as the range of all directional transmissions is identical. On the other hand, it has a major drawback as it is not able to take advantage of the increased spatial reuse provided by directional antennas. In this scheme, nodes have to defer their transmissions in all directions whenever they overhear either a RTS or CTS packet, hence making a poor reuse of the channel.

The use of TDMA with directional antennas is suggested within the context of the Receiver Oriented Multiple Access (ROMA) protocol [Bao2002]. ROMA is a distributed channel access scheduling protocol where each node uses multiple beams, and can participate in multiple transmissions simultaneously. ROMA follows a different approach than the random access protocols we have discussed so far which employ either on-demand handshakes (e.g., RTS/CTS packet transmission) or signal scanning (e.g., busy tones). ROMA is a scheduled access scheme based on a link activation scheme that prearranges or negotiates a set of timetables for individual nodes or links, such that the communicating nodes are coupled with each other accordingly, and the transmissions from the nodes or over the links are collision-free in the time and frequency.

ROMA employs a neighbor-aware contention resolution algorithm to derive channel access schedules for a node. As per this algorithm, each entity among a group of contending entities knows its direct and indirect contenders to a shared resource. Contention to the shared resource is resolved for each context (e.g., a time slot) according to the priorities assigned to the entities based on the context number and their respective identifiers. The entities with the highest priorities among their contenders are elected to access the common resource without conflicts. As the channel is time-slotted in ROMA (after TDMA), the contention context is identified by the time slot number.

7.9 Routing

Most of the efforts on the study of directional antennas for ad hoc networks have concentrated on the MAC layer. For multihop directional communication, we can clearly identify a tradeoff between the antenna beamwidth and the delay. Small beamwidth translates into large transmission ranges, and hence fewer hops and low latency at the routing layer for data packets sent from source to destination. On the other hand, small beamwidth results in a high sweeping delay and high overhead. Therefore, the design of a directional antenna solution has to be carefully planned, as there are a series of tradeoffs between number of antenna beams, MAC and routing characteristics. MAC and routing protocols which are designed without the knowledge of radio hardware are bound to have a poor performance.

7.9.1 Protocols

We now briefly discuss existing work on routing for directional antennas. Additional information can be found in [Gossain2006b, Choudhury2003].

Scheme 1

The mechanism introduced in [Nasipuri2000b] is applicable to on demand routing protocols running over directional antennas. As we know, on demand protocols use flooding of route requests packets in order to discover the intended destination. Therefore, the scheme in [Nasipuri2000b] proposes to reduce the number of packets transmitted during route discovery. It does so by requiring nodes to record the last known direction to other nodes. Therefore, whenever a node S has a route request for node R, it checks if it knows the last direction it used to communicate with R. If so, it will forward the route request packet only through the direction towards node S. Otherwise, it forwards the route request by sweeping.

A major drawback with this scheme is that it does not cope up well with mobile scenarios that are very common in ad hoc networks. Therefore, this scheme is only suitable for quasi-static networks. Another limitation is that it assumes that nodes can keep their orientation at all times. That is, even when nodes are not mobile, they can rotate around themselves which will change the orientation. This will, as a result, lead this scheme to erroneous conclusions about the direction of destination nodes.

Scheme 2

Contrary to the approach taken by scheme 1 which concentrates on reducing the overhead on the route discovery procedure, this mechanism focuses on reducing the overhead in route maintenance. More specifically, this scheme takes advantage of the increased range provided by directional antennas to bridge possible network partitions. As nodes are mobile in ad hoc networks, it might so happen that the network becomes either permanently or intermittently partitioned. With the possibility of transmitting over longer distances, directional antennas can help bridge the partitions which would not be otherwise possible with omni-directional antennas.

With this is mind, this scheme suggests a modified version of the DSR on demand routing protocol which employs directional transmissions only when necessary and for selected packets. This is done as follows. Whenever a node S has a packet to be sent to node R, it checks its route cache to determine weather it has a route to R. If so, it uses this route. Otherwise, it broadcasts a route request packet. The first route request packet is transmitted with the required power to cover the same area equivalent to an omni-directional transmission, hence conserving energy and reducing interference. If no reply is received within the timeout interval, the node S sets a partition bridging flag in the route request and resends it with higher power in directional mode, thus covering a longer distance. All intermediate nodes receiving this route request update a passive acknowledgment table and perform a similar procedure. This way, possible network partitions can be overcome.

Scheme 3

This scheme evaluates the performance of the DSR routing protocol over directional antennas. As DSR was designed based on the omni-directional antenna assumption, several performance degrading issues are then identified and enhancements are proposed to adapt DSR to directional antenna environments.

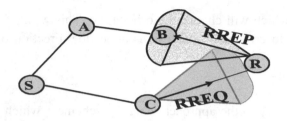

Figure 7.20 – Route reply is sent over a sub-optimal route

In DSR, whenever a source node S wants to find a route to an unknown destination D it floods the network with route request packets. As far as the MAC layer is concerned, this route request flooding is mapped to a MAC layer broadcast which is often performed by sweeping (improved broadcasting schemes were discussed in Section 7.7). As sweeping incurs delay, nodes receive route request packets at different times. In addition, if nodes are currently receiving a route request packet form one direction (hence deaf to all other directions) they may miss route request packets coming from other directions. An example of this issue is depicted in Figure 7.20 where node S floods a route request for node R. Here, we see that node R receives the first RREQ through node B and, as per DSR, send back a RREP right away. While it sends the RREP to node B, node R misses the RREQ coming from node C which is shorter (hence better as per our routing metric) in terms of number of hops. Therefore, the route reply is sent over a sub-optimal route. This scenario may be quite common in current generation of on demand routing protocols which respond to a RREQ as soon as it is received. While this may be a good idea if we consider omni-directional antennas, with directional antennas this is no longer the case (Figure 7.20).

Due to this problem, this scheme proposes a delayed route reply mechanism. Here, whenever the destination node receives a route request, it waits for a certain amount of time before responding with a route reply. The idea is that during this time span, other route requests coming through better routes may arrive, and hence the route reply could be sent over the shortest route of all. In the example of Figure 7.22, once node R receives the RREQ coming from node B, it delays the transmission of the RREP. In the meantime, node R would receive the

RREQ from node C and realize that the route through C is shorter than the route through B. Once the timer expires at node R, it sends back a RREP through node C which is the shortest of all. There is one drawback with this delayed route reply mechanism. While this scheme works well if the destination is the only node responding to route requests to it, it may not work so well if intermediate nodes are allowed to send route replies on behalf of the destination. In this case, the same issue will be present.

7.10 Conclusions and Future Directions

With current technological advancements, there is no doubt that directional antennas will become an integral part of future MANETs as means to considerably enhance their capacity. However, a directional antenna creates many difficulties with regard to protocol design as it impacts almost all layers of the protocol stack. There are many other issues still open and need further investigation. First of all, the antenna models being used for the design of upper layer protocols (e.g., MAC and routing) are too ideal and will most likely not conform to real world scenarios. Therefore, more realistic antenna models are needed in order to prove (or disprove) the effectiveness of proposed protocols and also serve as a tool to tune them. Coupled with this issue is the fundamental need to better understand the physical layer which is often neglected by many protocol designers. Secondly, power control is also an important aspect, as it may also create problems in a heterogeneous environment characterized by a combination of legacy omni-directional nodes and nodes having different number of antenna beams. For example, most existing protocols assume links to be bidirectional which would not be true in this heterogeneous environment. Thirdly, it is necessary to analyze in detail the effect of using directional antennas on the transport layer. More specifically, preliminary analysis reveals that TCP can suffer considerable performance degradation if special care is not taken. In mobile scenarios, TCP retransmission timers go off, considerably degrading its performance (more details in Chapter 7).

Another challenge is to analyze the capacity improvement of directional antenna systems. Although we know that network capacity can be considerably augmented, it would be very interesting to know the capacity limitations of ad hoc networks with directional antennas. Finally, further investigations are needed to explore the issue of multi-packet transmission and reception [Lal2004]. That is, how to design efficient PHY and MAC protocols which allow multiple receptions and transmissions simultaneously, hence increasing the spectral efficiency.

Homework Questions/Simulation Projects

Q. 1. There are 900 PDAs distributed uniformly in a University campus in the form of a Rectangular grid and constitute an ad hoc network. An omni-directional antenna covers one node in both X and Y directions. It was decided to replace the antennas with smart MIMOs.
  a. What is the number of nodes covered by each beam of the MIMO if directional antennas with 4-, 6-, and 16-beams are used with the same energy level per beam?
  b. Can you characterize the interference levels?
  c. What will be the average number of hops a message has to travel if an arbitrary source-destination pair is selected?
  d. What will be the impact on the performance if PDAs are randomly distributed?
Q. 2. What is the approximate range improvement that the use of directional antennas can provide over omni-directional antennas? Assume that all the power is to be directed within a beamwidth of b. Let d be the diagonal range and P the total transmit power.
Q. 3. Design a problem based on any of the material covered in this chapter (or in references contained therein) and solve it diligently.

References

[Bao2002] L. Bao and J.J. Garcia-Luna-Aceves, "Transmission Scheduling in Ad Hoc Networks with Directional Antennas," in ACM Mobicom, Sept. 2002.
[Blostein2003] S. Blostein and H. Leib, "Multiple Antenna Systems: Their Role and Impact in Future Wireless Access," IEEE Communications Magazine, July 2003.
[Choudhury2002] R. Choudhury and N. Vaidya, "Using Directional Antennas for Medium Access Control in Ad hoc Networks," in ACM Mobicom, September 2002.
[Choudhury2003] R. Choudhury and N. Vaidya, "Impact of Directional Antennas on Ad Hoc Routing," in Proc. of Personal Wireless Comm. (PWC) Conf., Sept. 2003.
[Cordeiro2004] C. Cordeiro, H. Gossain, and D. Agrawal, "Cross-Layer Directional Antenna MAC and Routing Protocols for Wireless Ad Hoc Networks," University of Cincinnati Intellectual Property Office No. 104-006, 2004 (filed for U.S. patent).
[Deng1998] J. Deng and Z. Haas, "Dual Busy Tone Multiple Access (DBTMA): A New Medium Access Control for Packet Radio Networks," in ICUPC, 1998.
[Godara1997] L. Godara, "Application of Antenna Arrays to Mobile Communications, Part I: Performance Improvement, Feasibility and System Considerations," in Proc. of the IEEE, July 1997.
[Gossain2004a] H. Gossain, C. Cordeiro, and D. Agrawal, "MDA: A Novel MAC Protocol for Directional Antennas over Wireless Ad Hoc Networks," University of Cincinnati Intellectual Property Office No. 104-014, 2004 (filed for U.S. patent).
[Gossain2004b] H. Gossain, C. Cordeiro, and D. Agrawal, "The Deafness Problems and Solutions in Wireless Ad Hoc Networks using Directional Antennas," in *IEEE Workshop on Wireless Ad Hoc and Sensor Networks*, in conjunction with *IEEE Globecom*, November 2004.

[Gossain2006a] H. Gossain, C. Cordeiro, T. Joshi, and D. Agrawal, "Cross-Layer Directional Antenna MAC and Routing Protocols for Wireless Ad Hoc Networks," Wiley *Wireless Communications and Mobile Computing (WCMC) Journal*, Special Issue on Ad Hoc Wireless Networks, to Appear, 2006.
[Gossain2006b] H. Gossain, T. Joshi, C. Cordeiro, and D. Agrawal, "DRP: An Efficient Directional Routing Protocol for Mobile Ad Hoc Networks," *in IEEE Transactions on Parallel and Distributed Systems*, to Appear, 2006.
[Gupta2000] P. Gupta and P. Kumar, "The Capacity of Wireless Networks," IEEE Trans. on Information Theory, Vol. IT-46, No. 2, March 2000.
[Hu2003] C. Hu, Y. Hong, and J. Hou, "On Mitigating the Broadcast Storm Problem with Directional Antennas," in IEEE ICC, 2003.
[Huang2002] Z. Huang, C.-C. Shen, C. Srisathapornphat, and C. Jaikaeo, "A Busy-Tone Based Directional MAC Protocol for Ad Hoc Networks," in IEEE Milcom, 2002.
[IECwww] International Engineering Consortium (IEC), "Smart Antenna Systems," Web ProForum Tutorials, http://www.iec.org.
[Joshi2004] T. Joshi, S. Vogety, C. Cordeiro, H. Gossain, and D. Agrawal, "Broadcasting over Switched Beam Antennas for Ad Hoc Networks," IEEE Int. Conf. on Networks, Nov. 16-19, 2004.
[Karn1990] P. Karn, "MACA - A new channel access method for packet radio," in Proc. of ARRL/CRRL Amateur Radio 9th Computer Networking Conf., Sept. 1990.
[Ko2000] Y.-B. Ko, V. Shankarkumar, and N. Vaidya, "Medium access control protocols using directional antennas in ad hoc networks," in IEEE Infocom, 2000.
[Korakis2003] T. Korakis, G. Jakllari, and L. Tassiulas, "A MAC protocol for full exploitation of Directional Antennas in Ad-hoc Wireless Networks," in ACM Mobihoc, June 2003.
[Lal2004] D. Lal, V. Jain, Q-A. Zeng, D. Agrawal, "Performance Evaluation of Medium Access Control for Multiple-Beam Antenna Nodes in a Wireless LAN", IEEE Trans. on Parallel and Distributed Systems, Vol. 15, No. 12, Dec. 2004.
[Liberti1999] J. Liberti and T. Rappaport, *Smart Antennas for Wireless Communications: IS-95 and Third Generation CDMA Applications*, Prentice Hall, Upper Saddle River, NJ, 1999.
[Nasipuri2000a] A. Nasipuri, S. Ye, J. You, and R. Hiromoto, "A MAC Protocol for Mobile Ad Hoc Networks using Directional Antennas," in Proc. of IEEE WCNC, Sept. 2000.
[Nasipuri2000b] A. Nasipuri, J. Mandava, H. Manchala, and R. Hiromoto, "On-Demand Routing Using Directional Antennas in Mobile Ad Hoc Networks," in Proc. of the IEEE Int. Conf. on Computer Comm. and Networks (ICCCN), Oct. 2000.
[Ni1999] S-Y. Ni, Y-C. Tseng, Y-S. Chen, and J-P. Sheu, "The Broadcast Storm Problem in a Mobile Ad Hoc Network," in Proc. of ACM/IEEE Mobicom, 1999.
[Paulraj1994] A. Paulraj and T. Kailath, "Increasing Capacity in Wireless Broadcast System Using Distributed Transmission/Directional Reception (DTDR)," U.S. Patent No. 5,345,599, Sept. 1994.
[Ramanathan2001] R. Ramanathan, "On the Performance of Ad Hoc Networks with Beamforming Antennas," In ACM MobiHoc, 2001.
[Sheikh1999] K. Sheikh, D. Gesbert, D. Gore, and A. Paulraj, "Smart antennas for broadband wireless access networks," IEEE Comm. Magazine, Nov. 1999.
[Takai2002] M. Takai, J. Martin, A. Ren, and R. Bagrodia "Directional Virtual Carrier Sensing for Directional Antennas in Mobile Ad Hoc Networks," in ACM MobiHoc, June 2002.
[Tobagi1975] F. Tobagi and L. Kleinrock, "Packet switching in radio channels: Part II – The hidden terminal problem in carrier sense multiple access and the busy tone solution," in IEEE Trans. on Comm., Vol. 23, Dec. 1975.
[Yum1992] T. Yum and K. Hung, "Design Algorithms for Multihop Packet Radio Networks with Multiple Directional Antennas," in IEEE Trans. on Comm., Vol. 40, Nov. 1992.

Chapter 8

Cognitive Radio and Networks

8.1 Introduction

The proliferation of wireless services and devices for uses such as mobile communications, public safety, Wi-Fi, and TV broadcast serves as the most indisputable example of how much the modern society has become dependent on radio spectrum. While land and energy constituted the most precious creation of resources during the agricultural and industrial eras respectively, radio spectrum has become the most valuable resource of the modern era [Calabrese2003]. Notably, the unlicensed bands (e.g., ISM and UNII) play a key role in this wireless ecosystem given that many of the significant revolutions in radio spectrum usage have originated in these bands, and which resulted in a plethora of new applications including last mile broadband wireless access, health care, wireless PANs and LANs, ad hoc and sensor networks, and cordless phones. This explosive success of unlicensed operations and the many advancements in technology that resulted from it, led regulatory bodies (e.g., the US Federal Communications Commission (FCC) through its Spectrum Policy Task Force (SPTF) [FCC2002]) to analyze the way spectra is currently used and, if appropriate, make recommendations on how to improve the usage of radio resources.

As indicated by the SPTF and also by other reports [Kolodzy2003], the use of radio resource spectrum has experienced significant fluctuations. For example, based on the measurements carried out in [McHenry2005] for frequency bands below 3GHz and conducted from January 2004 to August 2005, only about 5.2% of the spectrum is actually in use in the US on an average in any given location and at any given time. Interestingly enough, these measurements reveal that heavy spectrum utilization often takes place in unlicensed bands while licensed bands often experience low (e.g., TV bands) or medium (e.g., some cellular bands) utilization.

Therefore, as a result of such spectrum measurement campaigns coupled with recent advancements in radio technology, worldwide spectrum regulation is going through fundamental changes. It has been realized that not only spectrum usage is very low in certain licensed bands, but also that the scarcity of radio resources is becoming a crisis, hindering the development of many wireless applications, including broadband access (not only in urban/suburban areas, but especially in rural/remote areas), public safety, health care, business, and leisure.

Cognitive Radios (CRs) and Cognitive Networks (CNs) [Mitola1999, Mitola2000, Haykin2005] are seen as the solution to the current low usage of the radio spectrum. It is the key technology that will enable flexible, efficient and reliable spectrum use by adapting the radio's operating characteristics to the environmental real-time conditions. The study of cognitive radios and networks is relatively new and there are many questions and aspects to be tackled before these devices can seamlessly and on an interference-free basis reuse spectrum licensed to PU(s). Therefore, in this chapter we provide an overview of CRs and CNs, including the most important concepts and technologies that are poised to make it a future reality.

8.2 Cognitive Radio and Networks

CRs have the potential to utilize the large amount of unused spectrum (also known as "white spaces") in an intelligent way, while not interfering with other devices in frequency bands already licensed for specific uses. CRs are enabled by the rapid and significant advancements in radio technologies (e.g., software-defined radios, frequency agility, beamforming, power control, etc.), and can be characterized by the utilization of disruptive techniques such as wideband spectrum sensing, real-time spectrum allocation and acquisition, and real-time measurement dissemination (also refer to the DARPA neXt Generation (XG) program RFCs [DARPA-XG] for an overview of issues in and the potential of CRs). CNs are formed by collection of CRs, possibly employing the same communication protocols. Typically, two types of users can be found in an area covered by a CN:

Primary Users (PUs): These incumbent wireless devices have priority access to the wireless medium, subject to certain QoS constraints which must be guaranteed [Ghosh2009]. For example, in television (TV) bands, these devices would be TV sets. Primary users generally represent the license-holders of the spectrum band of interest, and in most practical cases are not CRs.

Secondary Users (SUs): Wireless CR devices that are allowed to access the spectrum which is allocated to PUs, and therefore known as SUs of the wireless spectrum. These wireless devices employ CR techniques to communicate among themselves, while ensuring the communication of the PUs is kept at an "acceptable" level.

By taking advantage of CR techniques, SUs are able to reuse the spectrum that is allocated to PUs. Among the most popular techniques that support the concept of CR include transmit power control, dynamic frequency selection, beamforming, spectrum sensing, geolocation, etc. Later in this chapter, we will investigate CR technologies that take advantage of several of these schemes.

8.3 Spectrum Access Models

To date, there are two models with which spectrum access can be done as dictated by regulatory bodies: spectrum property rights and the spectrum commons approach. The spectrum property rights approach suggests that spectrum can be treated like land, and private ownership of spectrum is viable. This approach entails the holder of spectrum (i.e., PUs) to have exclusive use of the spectrum portion they possess, without the potential of harmful interference from other parties. Alternatively, they would be able to trade their spectrum in a secondary market. The use of spectrum would be flexible, in that the authorized party could use the spectrum portion for any purpose. The US FCC has chosen a partial implementation of this approach by employing spectrum auctions as a means of licensing. But this approach has efficiency issues as observed for TV bands.

The spectrum commons approach allows the bands to be open to technologies as long as they follow the rules of access in the specific band; as smart technologies evolve, communicating devices will become able to avoid interference through mutual cooperation and coexistence, and the spectrum will become abundant. This phenomenon has been seen in 2.4 GHz ISM band as well as 5 GHz UNI I and UNI II bands, with technologies such as Wi-Fi and Bluetooth. The rules for access are governed by regulatory bodies such as the FCC, and device manufacturers have to comply with rules and regulations for operating in that band. Thus, even the commons regime is a form of lightly controlled shared access [Ileri2008].

Over time, as CRs and CNs evolve, a third model of spectrum access or a mixture of existing models could rise. A model wherein both PUs and SUs are allowed to use the same spectrum band, whereby SUs would be allowed to coexist with other SUs, but are required to avoid harmful interference to PUs. This is the approach being taken by most CR and CN technologies under development nowadays, and by the one that is assumed in this chapter when CR and CN technologies are described.

Table 8.1 depicts a taxonomy of the different modes of dynamic spectrum access depending upon the spectrum access model, i.e., licensed (spectrum property rights) or license-free (spectrum commons approach). From the perspective of SUs, access to the spectrum can be done in an underlay or overlay approach. In underlay access, the SU employs a combination of low power transmission and wider bandwidth to generate a signal that appears as noise to the PU. Underlay access typically does not require elaborate CR techniques. On the other hand, overlay access generally uses higher transmission power and narrower transmission bandwidth, which then requires the use of elaborate CR techniques to avoid interference to PUs [Ghosh2010a, Ghosh2010b].

Under both overlay and underlay, users may coordinate access to the spectrum through exchange of real-time spectrum usage information. Spectrum usage information can be in the form of characteristics such as frequency band in use, expected duration and location of spectrum usage, transmit power employed, among others. This is known as coordinated spectrum access to offer better spectrum utilization and less interference

Table 8.1 – Taxonomy of dynamic spectrum access modes

License regime of target spectrum	Spectrum Reuse mode by CR/SU			
	Underlay		Overlay	
	Uncoordinated	Coordinated	Uncoordinated	Coordinated
Licensed (i.e., with PUs)	PUs and SUs do not exchange information. Wide bandwidth and low transmit power (e.g., ultra-wideband). Better spectrum utilization due to underlay mode.	PUs and SUs exchange information. Flexible bandwidth vs. transmit power allocation. Better spectrum utilization due to underlay mode.	PUs and SUs do not exchange information. Requires the use of CR techniques such as spectrum sensing and geolocation to allow higher transmit power. Strict limitations on CR parameters including transmit power and antenna height. Spectrum underutilization.	PUs and SUs exchange information. In some cases, may require use of CR techniques to allow higher transmit power. More flexible CR operation. Better spectrum utilization.
License-free or unlicensed (i.e., no PUs)	No exchange of information amongst SUs. Flexible bandwidth vs. transmit power allocation. More interference and spectrum underutilization.	Secondary users exchange information. Flexible bandwidth vs. transmit power allocation. Flexible bandwidth vs. transmit power allocation. Less interference and better spectrum utilization.	No exchange of information amongst SUs. Flexible bandwidth vs. transmit power allocation. More interference and spectrum underutilization.	Secondary users exchange information. Flexible bandwidth vs. transmit power allocation. Flexible bandwidth vs. transmit power allocation. Less interference and better spectrum utilization.

due to the cooperation between radios through the exchange of real-time spectrum usage information. In contrast, uncoordinated spectrum access tends to lead to less efficient spectrum utilization and higher levels of interference since radios do not exchange real-time spectrum usage information. However, uncoordinated spectrum access is conducive to more robust CRs/CNs implementations which do not rely on external spectrum information for its operation. A mixture of coordinated and uncoordinated access can also be envisioned.

Under both overlay and underlay access, users may coordinate access to the spectrum through the exchange of real-time spectrum usage information. Spectrum usage information can be in the form of characteristics such as frequency band in use, the expected duration and location of spectrum usage, the transmit power employed for transmissions, among others. This is known as coordinated spectrum access, which can offer better spectrum utilization and less interference due to the cooperation between radios through the exchange of real-time spectrum usage information. In contrast, uncoordinated spectrum access tends to lead to less efficient spectrum utilization and higher levels of interference since radios do not exchange real-time spectrum usage information. However, uncoordinated spectrum access is conducive to more robust CRs/CNs implementations which do not rely on external spectrum information for its operation. A mixture of coordinated and uncoordinated access can also be envisioned.

8.4 Cognitive Radio Technologies and Challenges

CR technologies implemented within SU devices can have varying degrees of sophistication depending upon the requirements of PU detection, the need for coexistence with other SUs, the application requirements, and so on. To facilitate the understanding of CR techniques, it is useful to classify them broadly into three categories [Cordeiro2006, Challapali2006]:

Dynamic Spectrum Access (DSA): Involves techniques for quick and robust detection of PUs' presence to avoid interference since PUs have preemptive rights to access the spectrum.

Dynamic Spectrum Sharing (DSS): involves techniques that aware of other CNs of likely similar access rights and include mechanisms to coexist with those networks.

Table 8.2 – CR technologies and challenges

Layer #		DSA	DSS	DSM
Application layer		QoS requirements; user utility/policy; security	QoS requirements; user utility/policy; security	QoS requirements; multi-layer management
Transport layer		Spectrum handoff; delay, jitter, loss	Inter-system and inter-flow communication; delay, jitter, loss; spectrum handoff	Delay, jitter, loss; multi-flow management
Network layer		Spectrum aware routing; policy; delay tolerant networking; topology management for distributed sensing	Load (node and radio resource) balancing; interference-aware routing	Multi-channel assignment; multi-channel and multi-path routing
Data link layer	LLC sub-layer	Radio environment characterization; policy; power control; centralized (e.g., broker-based) or distributed (e.g., opportunistic); interference temperature; dynamic frequency selection	Policy; power control; interference aware dynamic frequency selection; channel assignment; centralized (e.g., broker-based) or distributed (e.g., opportunistic)	Multi-channel assignment
	MAC sub-layer	Coordination of quiet periods; directionality; spectrum sensing management	Space, time and code division multiple access; fairness; directionality; (non-) coordinated resource sharing	(Non-) contiguous multi-channel operation; inter-channel synchronization; multi-channel access; real-time and dynamic resource allocation
Physical layer		Spectrum sensing algorithms; low SNR signal detection; wideband or narrowband sensing; beacon detection; multiple antenna beamforming and beamnulling	Adaptive modulation and coding; waveform shaping; spreading; multiple antenna beamforming and beamnulling	Wideband antennas; ADC; programmable filters; multicarrier modulation; multiple antennas; bandwidth (RF, baseband) scalability

Dynamic Spectrum Multi-channel operation (DSM): involves techniques that enable spectrum-agility and provide seamless operation over multiple channels, possibly simultaneously.

On the basis of this categorization, Table 8.2 identifies the most prominent CR schemes and challenges investigated under each category. Table 8.2 also describes where each category lies with respect to the network protocol architecture model described in Chapter 1. In the next few sections, we discuss a few of these techniques in detail.

8.5 The IEEE 802.22 Standard

Over the last couple of years, the CR techniques depicted in Table 8.2 have been the subject of intense research; some more than others. As a result of the strong technical progress made so far, numerous activities were initiated to standardize CR techniques and thus enable worldwide adoption of consumer grade devices with CR capabilities.

The first of these activities, and arguably the most advanced of them all, was the IEEE 802.22 working group (WG) [IEEE 802.22-2010]. This WG has been chartered with the development of an air interface (i.e., PHY and MAC) based on CRs for unlicensed operation in the TV broadcast bands. For being the most mature CR-based air interface under development, in this section we give an overview of the IEEE 802.22 air interface, which aims at providing broadband access using TV broadcast bands while ensuring that no harmful interference is caused to PUs of this band (e.g., TV receivers).

8.5.1 TV Band and PUs

The first commercial CR-based device is expected to operate in the TV bands, as this is the first spectrum band targeted by regulatory bodies [FCC2003]. In the United States, TV stations operate from channels 2 to 69 in VHF and UHF portions of the radio spectrum. All channels are 6 MHz wide and span from 54–72 MHz, 76–88 MHz, 174–216 MHz, and 470–806 MHz. In other parts of the world, channels are 7 MHz or 8 MHz wide.

In addition to the TV service, also called primary service, other services such as wireless microphones are also allowed to operate on vacant TV channels on a non-interfering basis (please refer to Part 74 of the FCC rules), and so are private land and commercial mobile radio services (PLMRS/CMRS) including public safety (please refer to Part 90 of the FCC rules). Therefore, from the perspective of secondary CR-based devices, the PUs of the TV bands include TV signals, wireless microphones and PLMRS/CMRS services. As we will see later, contrary to detection of TV transmission, detection of wireless microphone operation is much harder as they transmit at a much lower power level (typically 50 mW for a 100 m coverage range) and occupy much smaller bandwidths (200 KHz).

8.5.2 Applications of IEEE 802.22

The most prominent target application of 802.22 systems is wireless broadband access in rural and remote areas with performance comparable to those of existing fixed broadband access technologies (e.g., DSL and cable modems) serving urban and suburban areas [ITU2005]. While availability of broadband access may not be so critical in urban and perhaps suburban areas, whose costs remain high, this certainly is not the case in rural and remote areas where about half of the United States and most of the developing nations (other countries located in South America, Africa, and Asia) populations are concentrated. Therefore, this has triggered the academia and industry to stimulate the development of CR-based technologies that could increase availability of broadband access in these underserved markets.

8.5.3 Reference Architecture

The 802.22 system specifies point-to-multipoint wireless air interface whereby a base station (BS) manages its own cell and all associated CPEs, as depicted in Figure 8.1. The BS controls the medium access and transmits to the various CPEs, which respond back to the BS in the upstream direction. To ensure the protection of primary services, the 802.22 system follows a strict masters/slave relationship, with BS as the master and the CPEs slaves. CPE is allowed to transmit after authorization from a BS controlling the radio frequency (RF) (e.g., modulation, coding, and frequencies of operation) used by the CPEs. In addition to the traditional role of a BS, which is to regulate data transmission in a cell, an 802.22 BS manages a unique feature of distributed sensing. This ensures proper PU protection and is managed by the BS, which instructs the various CPEs to perform distributed measurement activities. Spectrum sensing is covered later in this chapter.

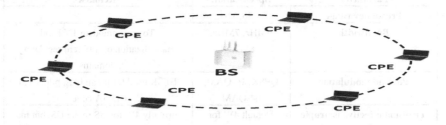

Figure 8.1 – An IEEE 802.22 network

A distinctive feature of an 802.22 wireless network as compared to other IEEE 802 standards is the BS coverage range, which can go up to 100 Km if power is not an issue (current specified coverage range is 33 Km at 4 Watts CPE EIRP (effective isotropic radiated power)). This is primarily due to higher transmit power allowed for this usage, and the favorable propagation characteristics of TV frequency bands in comparison with other unlicensed spectrum in the 2.4/5GHz bands.

8.5.4 The IEEE 802.22 Physical (PHY) Layer

The 802.22 PHY layer is specifically designed to support a system which uses vacant TV channels to provide wireless communication access over distances of up to 100Km. The PHY specification is based on Orthogonal Frequency Division Multiple Access (OFDMA) for both the upstream (US) and downstream (DS) access, and its key parameters are summarized in Table 8.3. The PHY modes for a reference 6MHz TV channel are given in Table 8.4.

8.5.4.1 Preamble, Control Header and MAP definition

Here we define the key PHY features that support basic superframe and frame structures described in the section 8.5.5. In 802.22, allocation of resources in OFDMA frame can be made in terms of sub-channels and symbols. A sub-channel is defined as a set of 28 contiguous

OFDM sub-carriers (24 data and 4 pilot), and there are 60 sub-channels per symbol.

Table 8.3 – IEEE 802.22 System Parameters

Parameter	Specification	Remark
Frequency range	54-862MHz	
Bandwidth	6MHz, 7MHz, 8MHz	To accommodate TV band channelization of different regulatory domains
Payload modulation	QPSK, 16-QAM, 64-QAM	BPSK used for preambles, pilots and CDMA codes
Transmit effective isotropic radiated power (EIRP)	Default 4W for CPEs	Currently 4W for BS in the US, but may vary in other regulatory domains
Multiple access	OFDMA	
FFT size	2048	
Cyclic prefix modes	1/4, 1/8, 1/16, 1/32	
Duplexing	TDD	

In the first frame of the superframe, the first symbol is the superframe preamble, followed by a frame preamble. The third symbol is the SCH, and the fourth symbol contains the FCH and, if needed, DS-MAP, US-MAP, DCD and UCD follow. Due to presence of SCH in the first frame of the superframe, the first frame payload will contain two fewer symbols than the remaining 15 frames to keep the frame length to 10 ms. The other 15 frames of the superframe contain a frame preamble, the FCH and the DS-MAP, US-MAP, DCD, and UCD messages, and the data bursts.

In each frame, transmit-receive turnaround (TTG) gap is inserted between the DS and US to allow the CPE to switch between the receive mode and transmit mode. To allow BS to switch between its receiving and transmit modes, a gap of receive-transmit turnaround (RTG) is inserted at the end of each frame. The value of the TTG and RTG changes based on the cyclic prefix and channel bandwidth under consideration, and this is indicated in Table 8.5. Note that the calculations in Table 8.5 only include the number of symbols required for the FCH, DS-MAP, US-MAP, CDC, and UCD symbols.

Table 8.4 – PHY modes (the data rates are calculated based on T_{CP} to T_{FFT} ratio of 1/16)

PHY Mode	Modulation	Coding rate	Data rate (Mbps)	Spectral efficiency (6MHz channel)
1	BPSK	Uncoded	4.54	0.76
2	QPSK	1/2	1.51	0.25
3	QPSK	1/2	4.54	0.76
4	QPSK	2/3	6.05	1.01
5	QPSK	3/4	6.81	1.13
6	QPSK	5/6	7.56	1.26
7	16-QAM	1/2	9.08	1.51
8	16-QAM	2/3	12.10	2.02
9	16-QAM	3/4	13.61	2.27
10	16-QAM	5/6	15.13	2.52
11	64-QAM	1/2	13.61	2.27
12	64-QAM	2/3	18.15	3.03
13	64-QAM	3/4	20.42	3.40
14	64-QAM	5/6	22.69	3.78

Table 8.5 – Symbol and TTG/RTG values per frame

Cyclic prefix	Number of symbols per frame			TTG			RTG		
Bandwidth (MHz)	6	7	8	6	7	8	6	7	8
1/4	26	30	34				83.33μsec	190μsec	270μsec
1/8	29	33	38		210		46μsec	286μsec	214μsec
1/16	30	35	41		μsec		270μsec	270μsec	32μsec
1/32	31	37	42				242μsec	22μsec	88μsec

Preamble definition

To facilitate burst detection, synchronization and channel estimation, two types of frequency domain sequences are defined:

Short Training Sequence (STS): This sequence is formed by inserting a non-zero binary value on every 4th sub-carrier. In time domain, this results in 4 repetitions of a 512-sample sequence in each OFDM symbol.

Long Training Sequence (LTS): This sequence is formed by inserting a non-zero binary value on every 2nd sub-carrier. In the time domain, this results in 2 repetitions of a 1024-sample sequence in each OFDM symbol.

The STS is used to form a superframe and coexistence beacon protocol (CBP) preamble, while the LTS is used to form the frame preamble. The sequences use binary (+1, -1) values in the frequency domain and are generated in an algorithmic way from m-sequences to ensure low peak-to-average-power-ratio (PAPR).

The superframe preamble is used by the receiver for frequency and time synchronization. The superframe preamble is 1 OFDM symbol in duration and consists of 4 repetitions of STS in the time domain preceded by a cyclic prefix of length ¼ (T_{CP}= ¼ T_{FFT}). The frame preamble is 1 OFDM symbol in duration and consists of 2 repetitions of the LTS in the time domain preceded by a cyclic prefix of length ¼ (T_{CP}= ¼ T_{FFT}). Finally, CBP preamble is 1 OFDM symbol in duration and consists of 5 repetitions of the STS in the time domain. The CBP preamble is designed to have low cross-correlation with the superframe preamble and is differentiated from a superframe preamble.

Control Header and MAP Definition

The IEEE 802.22 standard makes use of control headers and MAPs as they are key components for the entire system operation. There are two control headers (SCH and FCH) and four MAPs (DS-MAP, US-MAP, DCD and UCD). SCH is transmitted using the PHY mode 1 (see Table 8.4) and T_{CP} = ¼ T_{FFT}. It is transmitted over all data sub-carriers and is encoded by a rate-1/2 convolutional coder after interleaving, and is mapped using QPSK constellation resulting in 336 QPSK symbols. In order to improve the robustness and make better utilization of the available sub-carriers, spreading by a factor of 4 is applied to the output of the mapper, resulting in a maximum length of 42 bytes.

FCH is transmitted as a part of the DS protocol data unit (PDU) in the DS sub-frame and uses the basic data rate. The length of FCH is 4 bytes and it also carries the length (in bytes) for the DS-MAP if it exists, or the length of the US-MAP. The FCH is sent in the first two sub-channels immediately following the preamble symbol. In order to increase the robustness of the FCH, the encoded and mapped FCH data may be re-transmitted, which is indicated through the SCH. The receiver can combine corresponding symbols from two or three OFDM slots and decode the FCH data to determine the lengths of the fields in the frame.

MAPs are transmitted using the base data rate mode in the logical sub-channel immediately following the FCH logical sub-channels. The length of the DS-MAP PDU (if present) or the US-MAP PDU is indicated by the FCH. If the DS-MAP is present, the length of the US-MAP, DCD and UCD are specified in the DS-MAP and follow the DS-MAP in that order. The number of sub-channels required to transmit these fields is determined by their lengths and could possibly exceed the number of sub-channels allocated per symbol. In this case, the transmission of these PDUs will continue as the next symbol, starting with the first logical sub-channel. In most situations, no more than 2 symbols will be required to transmit the FCH and MAPs. Any unused sub-channels in the second symbol can be used for data transmissions.

8.5.4.2 CBP packet format

CBP preamble (1 symbol, 4 repetitions)	CBP data 1 (1 symbol, data + pilots)	Optional CBP data 2 (1 symbol, data + pilots)

Figure 8.2 – CBP packet format

The CBP protocol is described in detail in Section 8.5.5, and its packet format is depicted in Figure 8.2. The first symbol of the CBP packet is the preamble, followed by the CBP data payload 1 and an optional data payload 2. The length field in the first symbol enables a receiver to determine the presence or absence of the second data symbol. The CBP data symbols consist of the data and pilot sub-carriers. From the 1680, 426 sub-carriers are designated as pilot sub-carriers and the remaining 1254 are designated as data sub-carriers. Figure 8.3 shows a simplified block diagram of CBP data encoder and mapper. The CBP payload is divided into blocks of 418 bits before encoding and mapping.

Each block of 418 bits is first encoded using a rate-1/2 convolutional code. The encoded bits are then mapped using QPSK constellation, which results in 418 symbols. Each of these QPSK symbols is transmitted on three sub-carriers in order to provide additional frequency diversity.

A simple receiver can combine the pilot symbols with the preamble symbols and then perform interpolation to derive channel estimates. These channel estimates can then be used to equalize the CBP data symbols. The receiver can also use maximal ratio combining to despread the data symbols, and can perform the decoding using a Viterbi algorithm.

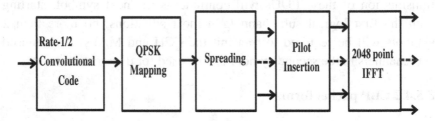

Figure 8.3 – CBP data encoding

8.5.4.3 Channel coding and modulation schemes

Figure 8.4 describes the channel coding process used in 802.22. Channel coding includes data scrambling, convolutional coding or advanced coding, puncturing, bit interleaving and constellation mapping.

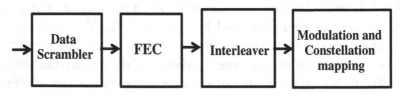

Figure 8.4 – Channel Coding

The frame payload data is first processed by the data scrambler using a pseudo-random binary sequence generator with generator polynomial $(1 + X^{14} + X^{15})$. The preamble and the control header fields of the frame are not scrambled. The forward error correction (FEC) scheme follows the data scrambler. The mandatory burst data coding scheme in 802.22 is half rate binary convolutional encoder. Duo-binary convolutional turbo code, low density parity check codes (LDPC) and shortened block turbo codes (SBTC) are optional advanced coding schemes.

For the interleaving stage, the same interleaver is used for sub-carrier. The interleaving algorithm used in 802.22 is described by the block size K and three integer parameters {p, q, j}. The global equation of the algorithm depends on the interleaving pattern of the previous iteration (j-1), the position index of samples (k) and two integer parameters (p, q). Here p indicates the interleaving size as a multiple of interleaving block size K. Finally, the output of bit interleaver is entered serially into modulation and constellation mapper. The input data to the mapper is first divided into groups of N_{CBPC} (2 for QPSK, 4 for 16-QAM, and 6 for 64-QAM) bits and then converted into complex numbers representing QPSK, 16-QAM or 64-QAM constellation points which is mapped according to Gray-coding constellation mapping. As for the pilot subcarriers, they are modulated according to the mapping. The pilot subcarriers are modulated according to the BPSK modulation using a modulation-dependent normalization factor equal to 1.

8.5.5 The IEEE 802.22 Medium Acess Control (MAC) Layer

Previous works [Cordeiro2005, Cordeiro2007] have provided an overview of the 802.22 MAC based on early drafts of the standard. However, only recently the 802.22 working group has released a draft that is in the final stages to become a standard [IEEE 802.22-2010]. Although most of the basic features have not changed, we provide an up to date overview of 802.22 MAC layer in this section with emphasis on the CR features that are key to support required incumbent protection, self-coexistence amongst networks and QoS.

The 802.22 standard defines a connection oriented and centralized MAC layer, which borrowed some of its basic features, including resource allocation and QoS supporting features, from the IEEE 802.16 standard for Wireless Metropolitan Area Networks [IEEE802.16]. As in 802.16, the medium access within a cell in controlled by a Base Station (BS), which uses TDM (Time Division Multiplex) in the downstream (DS) direction and allocates resources using a DAMA (Dynamic Assigned Multiple Access) approach in the upstream (US) direction.

8.5.5.1 Superframe and Frame Details

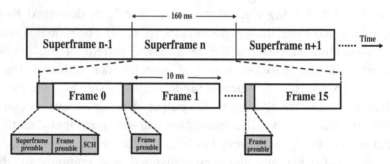

Figure 8.5 – MAC Superframe Structure

The 802.22 MAC uses a synchronous timing structure where frames are grouped into superframe, which was introduced to allow for better incumbent protection and self-coexistence. The superframe structure depicted in Figure 8.5 consists of 16 frames with fixed duration of 10 msec each. The BS starts with superframe preamble within the superframe, followed by the frame preamble and the Superframe Control Header (SCH). The superframe preamble is used for time synchronization, while the frame preamble is for channel estimation, allowing robust decoding of SCH.

The SCH carries the BS MAC address along with the schedule of quiet periods for sensing, as well as other information about the cell. The SCH is transmitted at a very robust rate to allow for successful decoding over long distances, which is important to ensure neighboring networks discover each other and avoid harmful interference. After the SCH the BS transmits the Frame Control Header (FCH), which is followed by the messages within the first frame. The remaining 15 frames within the superframe start with the frame preamble followed by the FCH and subsequent data messages.

In the US sub-frame, the BS can allocate resources for contention based access before the data bursts, which can be used for ranging, bandwidth (BW) requests and Urgent Coexistence Situation (UCS) notification. The UCS window is another new feature in the 802.22 MAC, which can be used by CPEs to transmit an indication that an incumbent has been detected on the channel. Furthermore, the BS may also reserve up to 5 symbols at the end of the frame as SCW which is used for execution of the Coexistence Beacon Protocol (CBP), which involves transmission of coexistence beacons (or CBP packets) carrying information about the cell and specific coexistence mechanisms. The SCW and CBP packets are new CR features that allow for over-the-air coordination among neighboring 802.22 cells to facilitate incumbent protection and spectrum sharing mechanisms. They are described in more detail in the following sections.

The two-dimensional (time-frequency) MAC frame structure is shown in Figure 8.6. Interesting points to notice in this figure are the different ways that MAC data can be mapped into the two-dimensional OFDM structure used at the PHY layer. In DS sub-frame, the MAC data is first layered vertically by sub-channels and then advanced into the time domain (symbol) horizontally. This structure reduces the delay for the DS data.

Two options are possible in the US sub-frame. First, the US MAC data can be mapped horizontally symbol by symbol in a logical sub-channel, and advanced to the next sub-channel only when full capacity is filled in the previous one. This option with long US bursts allows the CPEs to maximize power per sub-carrier, which can contribute to increase in the communication range. The second option is to use US bursts with maximum length of 7 symbols. If this option is used, the width of the last US bursts may be between 7 and 13 symbols. This optional allocation can provide better delay performance in the US at the cost of losing in range, which may be a good trade-off for real-time multimedia applications.

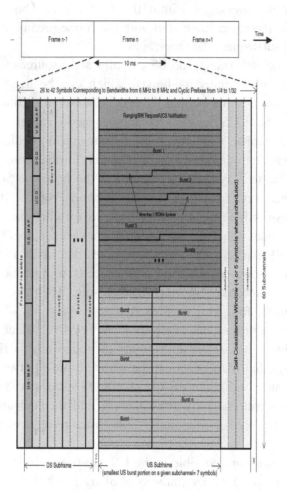

Figure 8.6 – Time-frequency illustration of the Frame Structure
[IEEE 802.22-2010]

8.5.5.2 Incumbent detection and notification support

One of the design challenges for the 802.22 MAC layer is to allow for reliable protection of incumbents, while maintaining QoS for the users. This problem was addressed with incumbent detection through spectrum sensing and multi-channel operation. In this section we describe how the 802.22 MAC layer supports incumbent detection and notification, which may generate trigger events for frequency agility operations. A summary of the sensing algorithm adapted in the standard are discussed in section 8.5.7.

Two important capabilities have been introduced in the MAC layer to support reliable incumbent detection, namely, network quiet periods and channel measurement management.

Network quiet periods: In order to avoid interference with spectrum sensing, which has to meet very low Incumbent Detection Threshold (see Table 8.6), the BS can schedule network-wide quiet periods (QPs) during which all transmissions are suspended, and hence sensing can be performed more reliably. Without QPs for sensing, the network may face a high false alarm rate, especially in areas where the coverage area of multiple networks overlaps.

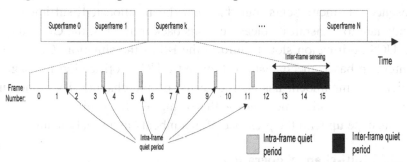

Figure 8.7 – Intra-frame and inter-frame quiet periods

Two types of QPs can be scheduled: intra-frame and inter-frame QPs (see Figure 8.7). Intra-frame QPs, as the name suggests, are of short duration QP (less than a frame) and are useful for regular sensing of in-band channels[1] without impacting the QoS for network users. However, the BS can also schedule longer inter-frame QPs across multiple frames in case more time is needed for sensing. Inter-frame QPs should be used on an on-demand basis, since it impact the QoS of the SUs. Overall, the BS can limit the number and duration of QPs to the minimum necessary to meet the sensing requirements in terms of probability of detection and probability of false alarm. The BS can schedule QPs by using the QP

---

[1] In-band channels are the operating channel (N) and its first adjacent channels (N-1 and N+1).

scheduling fields in the SCH or it can use a specific management message, called Channel Quiet Request (CHQ-REQ) to stop traffic at anytime within its cell.

Channel measurement management: In case an incumbent is detected by the BS, the BS can take different actions to avoid interference; but, when a CPE detects an incumbent it has to report to the BS. For that, the MAC layer includes channel measurement request and report messages, which allows the BS to take full control of the incumbent detection and notification process within its cell. The BS can also use management frames to request CPEs to perform other type of measurements, such as detection of other nearby networks and other performance related measurements. The BS is also responsible for allocating US resources for the CPEs to transmit their measurement reports once the sensing is completed. In case a CPE has detected the presence of incumbents but has not been allocated sufficient US bandwidth to transmit its measurements back to the BS, the CPE can use the UCS notification slots to inform the BS of the situation. CPEs use a contention based mechanism to transmit a UCS notification message to indicate an incumbent has been detected on the channel. By combining the UCS notification messages with sensing results received from CPEs, BS can cope up with the presence of incumbents in a timely manner.

8.5.5.3 Multi-channel operation

Regulatory rules require that SUs have to vacate the channel shortly after an incumbent is detected (Channel Move Time in Table 8.6). This requirement imposes an important challenge for 802.22 networks, namely, how to maintain network connectivity and meet application QoS requirements in an incumbent detection situation.

The 802.22 specification tackles this issue by using the concept of backup channels. During normal operation, the BS proactively maintains a list of backup channels. In case an incumbent is detected on in-band channels, the BS is responsible for triggering a switch to a backup channel within the required moving time, which should occur seamlessly in order to maintain QoS guarantees for users. Obviously, the backup channel must also be clear of incumbents in order to be used right away. Therefore, incumbent detection must also be done in out-of-band channels, i.e., channels that may be used as backup. However, sensing backup channels may be done during the CPEs' idle time and does not require QPs in the operating channel.

In order to achieve seamless transition to the backup channel without service interruption, two channel management modes are specified in the 802.22 standard. In the implicit mode, the BS may use the action fields in the DCD broadcast message to signal the transition to all its CPEs. In the explicit mode, the BS may use a specific management message to schedule a channel switch event for a time in the future. The explicit approach provides the flexibility to allow channel management for a single or a group of CPEs, since the channel management message can be sent as a broadcast, unicast or a multicast frame. In this way, a BS operating with multiple co-located 802.22 radios in different channels could direct a CPE to another channel and continue operation in its current channel. This feature is especially useful in case individual CPEs detect low power incumbents (e.g., wireless microphones) in areas with limited channel availability.

8.5.5.4 Synchronization

Synchronization is a key factor for successful operation of CR networks, and it is not only needed for communication purposes between BS and CPEs but also for the incumbent's protection. The BS and CPEs in a cell must be synchronized to ensure no transmissions occur during the QP for sensing. Also, neighboring networks sharing the same channel (N) or operating on first and second adjacent channels (N-2, N-1, N+1, N+2) must synchronize their QPs to avoid interference with incumbent sensing and reduce the false alarms rate. Although in-band channels include only up to first adjacent channel (N, N+1 and N-1), synchronization of QPs up to second adjacent channel (N+2 and N-2) is needed to avoid interference when sensing the first adjacent channels.

In order to facilitate synchronization of QPs, all the BSs are required to be equipped with a satellite based positioning system (e.g., GPS), which is used to derive the timing information for the superframes. Therefore, by specifying a common reference time, the standard ensures all the superframes will be synchronized and the next step is to synchronize the intra-frame and inter-frame QPs. The inter-frame QPs are expected to be used for sensing of in-band channels, since its limited duration (less than a frame) has only minimal impact on QoS. To facilitate synchronization, such QPs can only be scheduled at the end of each frame. Therefore, neighboring BSs need only to advertise the frames in which intra-frame QPs are scheduled and the QPs duration. This information is broadcasted in the SCH at the beginning of every superframe and it is also included in the CBP packets that are transmitted by CPEs. The intra-frame QPs schedule can apply across multiple superframes – which allows upcoming neighboring BSs to discover existing cells operating in the area and align their intra-frame QPs accordingly. Also, once a BS discovers the QPs scheduled by other BSs using the same channel or adjacent channels, it has to align and adjust its intra-frame QP duration to the largest amongst the overlapping BSs.

8.5.5.5 Self-Coexistence

The unlicensed spectrum used by CRs and CNs underscores the importance of self-coexistence[2] mechanisms to ensure efficient and fair spectrum utilization. Self-coexistence also plays a key role in protecting the incumbents and provides required coordination for reliable spectrum sensing by synchronization of QPs. Proper coordination of sensing schedules also minimizes the number of required QPs and leaves more time for data communication.

The self-coexistence problem is approached in 802.22 with the joint application of neighboring network discovery and coordination, the coexistence beacon protocol and resource sharing mechanisms.

Neighboring network discovery and coordination: The non-deterministic availability of spectrum coupled with the ability to switch channels makes the CR operating spectrum environment dynamic. This requires sensing not only for incumbent detection, but also for other neighboring systems. Network discovery is part of the initialization

---

[2] Self-coexistence refers to coexistence amongst 802.22 systems.

procedures for both BSs and CPEs, but it is also continuously done as part of normal network operation. Networks can be discovered through the SCH transmitted by the BSs or by CBP packets, which are transmitted during the SCW window by CPEs or BSs. CPEs that discover other neighboring networks send this information back to its BS. Upon discovery of new neighboring networks, the BS must consider whether QP synchronization is required.

Coexistence Beacon Protocol: The CBP protocol plays a key role in enabling efficient discovery and coordination of neighboring networks. BSs and CPEs can discover other networks by detecting CBP packets transmitted during SCWs. The CBP packets carry information about the cell (e.g., BS MAC address, schedule of QPs, backup channels) and BSs should open regular SCWs for transmission of CBPs. During normal operation, the CPE should listen to its BS's SCH to identify any changes relevant to its cell, and since superframes are synchronized across cells, CPEs may not be able to detect neighboring BSs SCH transmissions. The CBP packets provide additional support for network discovery, since they are transmitted during the SCWs, and the BS can explicitly request CPEs to listen to the channel during the SCW in order to detect CBP packets from other neighboring cells. The CBP packets may also carry information required for executing several resource sharing mechanisms and serve as an underlying protocol for inter-network communication. As can be seen in Figure 8.6, the SCW impose extra overhead (4 or 5 symbols), and BSs must take into account the overhead and timing constraints for inter-network communication when scheduling SCWs.

Resource sharing mechanisms: After network discovery, neighboring networks may have to consider how to share the available spectrum. Consider an example scenario where BS A and CPE A1 and part of the same network and are operating on a given channel N. When a new BS B and CPE B1 start operation, they will first scan the available channels and CPE B1 will eventually detect BS A's SCH or CBP packet transmitted by A1. At this point, BS B must execute the first coexistence mechanism, which is called spectrum etiquette. The main idea behind spectrum etiquette is to avoid operating in co-channel with other existing networks. Therefore, BS B will first search for an available channel which is not used by BS A. Only if no other channel is available, BS B can share the same channel with BS A. Although this is not a desirable situation, it might happen in areas with only a few channels available, and the standard provides the required support for coexistence in such cases. BSs may also engage in a negotiation process to share the channel on a frame by frame basis. Such negotiation may be based on spectrum renting/offering or spectrum contention mechanisms. In both cases, the negotiation between BSs is carried out through CBP packet exchanges. It should be noted that such negotiation process may become a bottleneck, as the number of neighboring BSs on the same channels increase.

Another problem that should be considered in coexistence scenarios where multiple networks share the same channel is the potential collision of SCH transmissions from neighboring cells. Synchronization of the superframes across cells results in simultaneous transmissions of SCHs by all BSs in the beginning of every superframe, which may lead to collisions at CPEs in the overlapping areas. Due to the robust transmission mode, a CPE may be able to decode the SCH from its BS, even in presence of other SCH transmissions. However, depending on the distance between interfering BSs, SCH collisions might occur, which could prevent operation at some specific locations.

This is still an open issue, but its practical impact depends on the availability of incumbent free channels. Neighboring BSs always try to find empty channels, and co-channel operation with other BSs is pursued only as the last resort to maintain connectivity.

8.5.6 Spectrum Management Model

The cognitive radio protocols described in the previous sections provide all flexibility to operators to detect and share the available spectrum, as well as comply with the requirements of the DSA model for secondary operation in TV bands. In this section, we describe the architecture for managing CR capabilities and protocols.

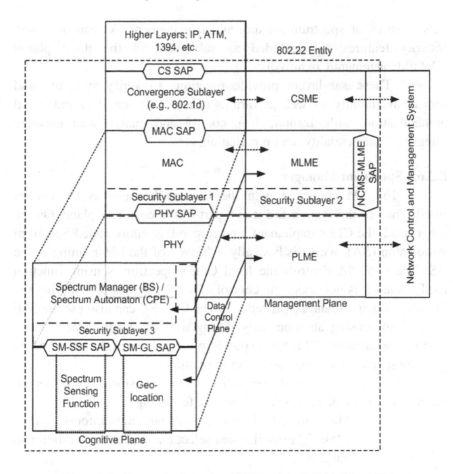

Figure 8.8 – Protocol reference architecture for a 802.22 BS or CPE
[Taken from Mody2008]

As described in [Cavalcanti2008], the 802.22 standard has adopted a spectrum management model where each BS has a central entity, called Spectrum Manager (SM), that is responsible to make key decisions and triggers proper events/actions to ensure protection of incumbents and efficient spectrum utilization while complying with regulatory policies. The SM shown in Figure 8.8 is a part the 802.22 protocol reference architecture [Mody2008], which also includes a cognitive plane and security features. The data and management planes are separated from the cognitive plane, which was added to support the new features of spectrum sensing and management. As can be noted, security features are included as sub-layers in the three planes (Data/Management/Cognitive).

These sub-layers provide functions to verify spectrum and service availability, as well as various forms of device, data and signal authentication, authorization, data, control and management message integrity, confidentiality, non-repudiation, etc.

8.5.6.1 Spectrum Manager

The SM colocated with the BS can be seen as the central intelligence of the system and it is a part of the cognitive plane. On the other hand, the CPEs implement a corresponding entity called Spectrum Automaton (SA), which is basically a "slave" of the SM running at the BS. The CPE SA controls the local CPE's spectrum sensing function (SSF) when it is not under the control of a BS, such as done during the initialization and scanning procedures. The CPE SA can also use the SSF to perform sensing autonomously during the CPE's idle time. During normal operation the CPE SA responds to requests from the BS's SM, to perform spectrum sensing and report the results.

Within a cell, the SM centralizes all the decisions with respect to spectrum management, which includes the following:

- o   Maintain up to date spectrum availability information.
- o   Classify, prioritize and select channels for operation and backup.
- o   Association control.
- o   Trigger frequency agility related actions (i.e. channel switch).
- o   Manage mechanisms for self-coexistence (interference-free scheduling, renting/offering and spectrum contention).

Although some of the decisions are not specified in the standard (i.e., they are left to provide flexibility for manufacturers), any SM implementations must comply with the regulatory rules for the operating domain (e.g., rules established by the FCC in the case of the US). Moreover, the specific regulatory rules may differ from one domain to another, which requires a flexible standard in order to achieve worldwide success. Most of the regulatory rules deal with incumbent protection, and two main capabilities are mandated in the standard to support potentially different incumbent protection requirements, such as spectrum sensing function and incumbent database support and are described next.

8.5.6.2 Spectrum sensing function

Every CPE is required to implement a spectrum sensing function (SFF), which is defined in the standard as a "black box" that performs spectrum sensing. The standard specifies inputs and outputs for the SSF, as well as the performance requirements for the sensing algorithm implemented (e.g., probability of detection, incumbent detection threshold and probability of false alarm). The specific sensing algorithm is implementation dependent which have been proposed and evaluated during the standardization work. Details on spectrum sensing algorithms are described in Section 8.5.7.

8.5.6.3 Incumbent database support

Although spectrum sensing can provide required protection to incumbents, some regulatory domains may require access to incumbent databases that store information about the presence of incumbents in a given area which may reduce the overhead on the system due to spectrum sensing, especially for TV signals. For instance, having a TV station occupying a channel can be stored in a database and then a BS may decide to reduce the frequency where sensed by its associated CPEs or it may even decide not to perform sensing at all on that channel.

There are also issues related to incumbent databases, such as accuracy of the database information and maintenance costs. In addition, incumbent databases may not be efficient for more dynamic incumbents, such as wireless microphones, and the cost of maintaining a database for a large number of low power devices (i.e., wireless microphones) may be prohibitive. In fact, one of the main open questions in enabling commercial CR applications is related to maintenance of such databases. Regulatory bodies could maintain a database infrastructure, but that would come at a certain cost. A group of services provide and incumbents could be another option to maintaining such a database. At the present time, it is not clear yet how the implementation of

346 AD HOC & SENSOR NETWORKS

incumbent databases will play along with the business models of operators.

Nevertheless, to allow for aggregation of information from multiple available sources and to support potential regulatory policies, the 802.22 standard requires BSs to be able to connect to external databases. Access to incumbent databases is implemented through higher layer protocols and it is outside the scope of the standard. However, in order to allow for modular and interoperable implementation, 802.22 defines interfaces and primitives to connect to the databases. If an incumbent database is not available or accessing a database is not required by regulation, the SM is still able to make decisions based on spectrum sensing information. This is an important flexibility, especially in less populated and rural areas, due to the low cost in implementing spectrum sensing which allows operation with minimal infrastructure costs and without dependency on third parties to maintain an incumbent database.

8.5.7 Spectrum Sensing

Spectrum sensing is a key CR technology that will likely be part of any commercial grade device with CR capabilities, particularly when considering uncoordinated spectrum access. Spectrum sensing is probably the area in CR that has received most attention by researchers in the past few years. Sensing algorithms can be classified as blind and nonblind. In the former, no assumption is made on signals of the incumbents, while in the latter the characteristics of the incumbent signal are used to detect it. To be able to evaluate and compare spectrum sensing algorithms, we need to specify the requirements that these algorithms must meet. Table 8.6 depicts various parameters that have been defined to ensure incumbent protection. These parameters are discussed throughout this chapter.

8.5.7.1 Power detector

The power detector is the simplest detector that can be constructed in practice [Kay1998, Shellhammer2006a, Shellhammer2006b]. This detector uses very limited a-priori information regarding the incumbent signal. The detection is based on only the signal power. Though this detector is unlikely to be used by itself in practical systems, it does give a lower bound on sensing performance, since other complex detectors will outperform the power detector. The use of more complex detectors should be weighed if higher complexity and sophisticated techniques result in significant performance gain. The power and energy detectors are very similar. The power detector estimates the signal power, while the test statistic for the energy detector is an estimate of the signal energy during the sensing time. Since we prefer test statistics where mean value does not change with the sampling duration, we choose to use the power detector as just a matter of convenience.

Table 8.6 – Incumbent Protection Parameters [IEEE 802.22-2005]

Parameter	Wireless microphones	TV services
Incumbent detection threshold (IDT)	-107 dBm (over 200 KHz)	-116 dBm (over 6 MHz)
Probability of detection (PD)	90%	90%
Probability of false alarm (PFA)	10%	10%
Channel detection time (CDT)	≤ 2 sec	≤ 2 sec
Channel move time (CMT)	2 sec	2 sec
Channel closing transmission time (CCTT)	100 ms	100 ms

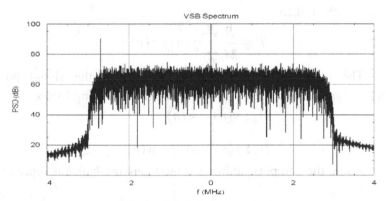

Figure 8.9 – DTV spectrum

To illustrate the power detector, let us consider a digital TV (DTV) signal according to the Vestigial Sideband (VSB) standard used in the USA. The DTV signal has a bandwidth of 6 MHz as illustrated in

Figure 8.9. The frequency ranges from −3MHz to +3MHz for the signal at the baseband. It is assumed that the signal is sampled at the Nyquist rate, resulting in a complex signal with 6MHz sampling. If the sampling rate is much higher than the Nyquist rate, then the assumption about the independence of the noise samples is no longer valid.

There are two hypotheses in this problem. The null hypothesis is that the channel is vacant which is represented as H_0. The alternative hypothesis is that the channel is occupied by the DTV transmitter represented as H_1. The test statistic is an estimate of the received signal power, which is given by:

$$T = \frac{1}{T_S} \sum_{n=1}^{M} y(n) y^*(n),$$

where T_S represents the sensing time, which is defined as the product of number of samples, M, with the sampling rate T_S; $y(n)$ is the sample at the nth time instant; and $y^*(n)$ represents the complex conjugate. The sampling time is given by:

$$T_S = \frac{M}{f_S}$$

The sampling is performed at bandwidth B such that $f_S = B$. If the test statistical, then:

$$T = \frac{B}{M} \sum_{n=1}^{M} y(n) y^*(n).$$

The mean value of the test statistics is the signal power. Observed signals under the hypotheses take the following forms:

$$H_0 : y(n) = w(n)$$

$$H_1 : y(n) = x(n) + w(n),$$

where the complex additive white Gaussian Noise is represented as:

$$w(n) = w_R(n) + jw_I(n).$$

The real and imaginary parts of the noise are independent and identically distributed Normal (Gaussian) random variables:

$$f_{WR}(w) = f_{WI}(w) = N(0, N/2).$$

Here, the noise spectral density is N. We chose to use N instead of N_0, since the receiver is likely to have a nonzero noise figure. It is not necessary to consider multipath fading for the power detector explicitly, since the power detector collects all the energy of the signal and the time variation of the total energy over a TV channel bandwidth (6, 7 or 8 MHz) due to multipath is insignificant. We have to explicitly consider multipath fading in more complex detectors. Shadow fading is addressed by allowing the signal level to have a lognormal distribution. The test statistic for the power detector is an estimate of the signal power.

Probability of False Alarm for the Power Detector

The threshold in the power detector is based on the probability density function of the test statistic under hypothesis H_0. As we can see from Table 8.6, the probability of false alarm (P_{FA}) should be around 0.1. Here, we describe how to select the detector threshold to meet the probability of false alarm requirements.

The probability density function for the test statistic under H_0 is given by:

$$f_T(t) = N\left(NB, \frac{(NB)^2}{M}\right).$$

If we define a new threshold, γ_0, as:

$$\gamma_0 = \frac{\sqrt{M}}{NB}(\gamma - NB),$$

then, we have that $T > \gamma$ if and only if $S > \gamma_0$. Hence, the probability of false alarm can also be written as: $P_{FA} = P(S > \gamma_0)$. Since S is a zero-mean Gaussian random variable with unit variance, the probability of S exceeding a threshold is given by the Q function [Kay1998]:

$$Q(x) = \frac{1}{\sqrt{2\pi}} \int_x^\infty \exp\left(\frac{-y^2}{2}\right) dy.$$

The CBP protocol is described in detail in Section 8.5.5, and its packet format is depicted in Figure 8.2. The first symbol of the CBP packet is the preamble, followed by the CBP data payload 1 and an optional data payload 2. The length field in the first symbol enables a receiver to determine the presence or absence of the second data symbol.

The CBP data symbols consist of the data and pilot sub-carriers. From the 1680, 426 sub-carriers are designated as pilot sub-carriers and the remaining 1254 are designated as data sub-carriers. Figure 8.3 shows a simplified block diagram of CBP data encoder and mapper. The CBP payload is divided into blocks of 418 bits before encoding and mapping. Each block of 418 bits is first encoded using a rate-1/2 convolutional code. The encoded bits are then mapped using QPSK constellation, which results in 418 symbols. Each of these QPSK symbols is transmitted on three sub-carriers in order to provide additional frequency diversity.

The threshold γ_0 is then given by $\gamma_0 = Q^{-1}(P_{FA})$. Therefore, the value of the detector threshold for test statistic T is:

$$\gamma = NB\left[1 + \frac{Q^{-1}(P_{FA})}{\sqrt{M}}\right].$$

Probability of Misdetection for the Power Detector

Having set the detection threshold based on the required P_{FA}, we can calculate the probability of misdetection (P_{MD}). The probability of misdetection is simply the probability that the test statistic will not exceed the threshold under hypothesis H_1:

$$P_{MD} = P(T < \gamma) \quad under\, H_1.$$

Under hypothesis H_1, the probability density function of the test statistic is given as:

$$f_T(t) = N\left[P + NB, \frac{(P+NB)^2}{M}\right].$$

To determine the probability of misdetection, it is convenient to transform the test statistic into a zero-mean Gaussian random variable with unit variance as was done previously. Therefore, similar to what was done for hypothesis H_0, P_{MD} can be expressed as:

$$P_{MD} = P(S < -\gamma_0) = P(S > \gamma_0) = Q(\gamma_0).$$

Therefore, the probability of misdetection can be written as:

$$\tilde{x}(n) = \sum_{k=0}^{K} f(k)x(n-k), n = 0,1,\ldots$$

$$P_{MD} = Q(\gamma_0) = Q\left\{\frac{\sqrt{M}}{(P+NB)}[(P+NB)-\gamma]\right\}.$$

To illustrate these theoretical results, specific DTV detection problem have been applied in [Shellhammer2006c]. To specify the detection threshold, we must specify the noise power spectral density, the bandwidth, the probability of detection, and the number of samples observed to generate T. As in [ATSC2004], we assume the receiver has an 11 dB noise figure that includes the noise figure of the RF front end and any other cable and coupling losses. Hence, the noise power spectral density is given by:

$$N = N_0 + 11 = -174 + 11 = -163 \, dBm / Hz.$$

Assuming a signal bandwidth of B = 6 MHz, the total noise power is:

$$NB = -174 + 11 + 10\log(6x10^6) = -95.2 \, dBm.$$

8.5.7.2 Eigenvalue Sensing Technique

Another blind sensing algorithm is the eigenvalue sensing technique [IEEE 802.22-2010]. Let y(t) be the continuous time received signal of interest with central frequency f_c, bandwidth W, that is sampled at a rate $f_S(>> W)$. Let $T_S = 1/f_S$ be the sampling period. The received discrete signal is then $x(n) = y(nT_S)$. There are two hypotheses:

H_0: signal does not exist.

H_1: signal exists.

The received signal samples under the two hypotheses are therefore as follows:

$$H_0 : x(n) = \eta(n)$$

$$H_1 : x(n) = s(n) + \eta(n),$$

where s(n) is the transmitted signal passed through a wireless channel (including fading and multipath effect), and $\eta(n)$ is the white noise samples. Note that s(n) can be a superposition of multiple signals. The received signal is generally passed through a filter. Let f(k), k = 0, 1, ...,K be the filter. After filtering, the received signal becomes:

$$\tilde{x}(n) = \sum_{k=0}^{K} f(k)x(n-k), n = 0,1,...$$

Let:

$$\tilde{s}(n) = \sum_{k=0}^{K} f(k)s(n-k), n = 0,1,...$$

$$\tilde{\eta}(n) = \sum_{k=0}^{K} f(k)\eta(n-k), n = 0,1,...$$

Then: $H_0 : \tilde{x}(n) = \tilde{\eta}(n)$

$$H_1 : \tilde{x}(n) = \tilde{s}(n) + \tilde{\eta}(n)$$

Note that here the noise samples $\tilde{\eta}(n)$ are correlated. If the sampling rate f_S is larger than the channel bandwidth W, we can down-sample the signal. Let $M \geq 1$ be the down-sampling factor. If the signal to be detected has a much narrower bandwidth than W, it is better to choose $M \geq 1$. For notation simplicity, we still use $\tilde{x}(n)$ to denote the received signal samples after down-sampling; that is, $\tilde{x}(n) \triangleq \tilde{x}(Mn)$ (mathematically, the symbol $\triangleq$ means "is defined as" or "is equivalent to"). Then, choose a smoothing factor $L > 1$ and define:

$$x(n) = [\tilde{x}(n) \quad \tilde{x}(n-1) \quad ... \quad \tilde{x}(n-L+1)]^T, n = 0,1,...,N_S -1$$

Based on this, we now define a Lx[K + 1 + (L − 1)M] matrix as:

$$H = \begin{bmatrix} f(0) & \cdots & \cdots & f(K) & 0 & \cdots & 0 \\ 0 & \cdots & f(0) & \cdots & f(K) & \cdots & 0 \\ \cdots & \cdots & \cdots & \cdots & \cdots & \cdots & \cdots \\ 0 & \cdots & \cdots & \cdots & f(0) & \cdots & f(K) \end{bmatrix}$$

Let $G = HH^H$. Decompose the matrix into $G = QQ^H$, where Q is a LxL Hermitian matrix. The matrix G is not related to signal and noise and can be computed offline. If an analog filter or both analog and digital filters are used, the matrix G should be revised to include the effects of all the filters. In general, G is found to be the covariance matrix of the received signal, when the input signal is white noise only. The matrix G and Q are computed only once and only Q is used in detection.

We can now employ these relations for the purpose of signal detection. For example, the algorithm for the Maximum-Minimum Eigenvalue (MME) detection is as follows:

Step 1: Sample and filter the received signal as described previously.

Step 2: Choose a smoothing factor L (e.g., L=10) and compute the threshold γ to meet the requirement for the probability of false alarm.

Step 3: Compute the sample covariance matrix:

$$R(N_S) = \frac{1}{N_S} \sum_{n=0}^{N_S-1} x(n)x^H(n).$$

Step 4: Transform the sample covariance matrix to obtain:

$$\tilde{R}(N_S) = Q^{-1}R(N_S)Q^{-H}.$$

Step 5: Compute the maximum eigenvalue and the minimum eigenvalue of the matrix $\tilde{R}(N_S)$ and denote them as λ_{max} and λ_{min}, respectively.

Step 6: Determine the presence of the signal based on the eigenvalues and the threshold: if $\lambda_{max}/\lambda_{min} > \gamma$, the signal exists; otherwise, the signal does not exist.

Similar to the MME detector, other eigenvalue detectors can be developed following the same principle.

8.5.7.3 FFT-based pilot sensing technique

The FFT-based pilot sensing technique described here is a nonblind sensing technique [IEEE 802.22-2010]. It is a very effective mechanism for detecting signals that employ pilots. As an example, consider the ATSC system which is used for DTV transmission in the US. The ATSC signal has a pilot at the lower band edge in a known location relative to the signal. It is assumed that the signal to be sensed is a bandpass at a low intermediate frequency (IF) of 5.38 MHz with the nominal pilot location at 2.69 MHz and is sampled at 21.52 MHz. However, the basic steps of the sensing algorithm can be implemented with suitable modifications for any IF and sampling rate. The steps of this method are:

The signal is demodulated to the baseband by the nominal frequency offset of $f_c = 2.69$ MHz. Hence, if x(t) is the real band-pass signal at low IF, $y(t)=x(t)e^{-j2\pi t f_c}$ is a complex demodulated signal at baseband.

The signal y(t) is filtered with a low-pass filter of bandwidth, say, 40 kHz (±20 kHz). The filter bandwidth should be large enough to accommodate any unknown frequency offsets.

The output of the low-pass filter is down-sampled from 21.52 MHz to 53.8 kHz, to form the signal z(t).

FFT is applied to the down-sampled signal z(t) and the length of the FFT will vary depending on the sensing period. For example, a 1 ms sensing window allows a 32-point FFT while a 5 ms window allows a 256-point FFT.

Then, the FFT output is squared and the maximum value and location are determined.

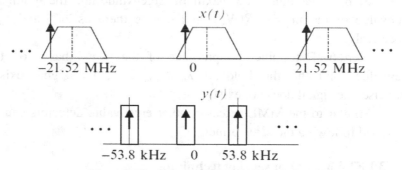

Figure 8.10 – Frequency Domain Description

Steps 1 and 3 are shown in Figure 8.10. Signal detection can then be done either by setting a threshold on the maximum value or by observing the location of the peak over successive intervals. Instead of the FFT, other well-known spectrum estimation methods, such as the Welch periodigram, can be used in Step 5. The basic method just described can be adapted to a variety of scenarios: i) multiple short sensing windows (e.g., 1-5 ms) with sensing dwells at every 10 ms. The squared 256-point FFT output from each sensing window can be averaged to form a composite statistic and the location information from measurements can be used to derive a detection metric; ii) a single long sensing window (e.g., 10-20 ms) with sensing dwells at every 40 ms. In this case, a 512-point FFT or periodigram can be used to obtain better detection performance. The parameters of the sensor can be chosen depending on the desired sensing time, complexity, probability of missed detection, and probability of false alarm. Detection based on location is robust against noise uncertainty, since the position of the pilot can be pinpointed with accuracy. Various schemes can be developed for combining both pilot-energy and pilot-location sensing.

8.5.7.4 Cyclostationary sensing technique

It has been recognized that many random time series encountered in the field of signal processing are appropriately modeled as cyclostationary than stationary, due to the underlying periodicities of these signals [Kay1998, IEEE 802.22-2010]. Thus, cyclostationarity provides a way to separate desired signals from noise, since random signals such as white Gaussian noise are not cyclostationary.

Consider the ATSC system, which is used for DTV transmission in the US. ATSC signals are VSB modulated [ATSC2004]. A constant value of 1.25 is added to the eight-level pulse amplitude modulated (8-PAM) signal before VSB modulation. Therefore, there is a strong pilot tone on the power spectrum density of the ATSC DTV signal. Let s(t) be this pilot tone signal, which is sinusoidal in time domain, and further assume that this strong pilot tone is located at frequency f_0; that is,

$$s(t) = \sqrt{2P} \cos(2\pi f_0 t + \theta) \otimes h(t),$$

where P and θ are the power and the initial phase of the sinusoidal function, respectively. The function h(t) is the channel impulse response and $\otimes$ is the convolution operator. The received signal must contain $x(t) = s(t)e^{-j2\pi vt} + w(t)$, the signal:

where w(t) is the additive white Gaussian noise and v is the amount of frequency offset in units of Hertz. We assume that w(t) is zero-mean with autocorrelation function $Rw(\tau) = E[w(t)w*(t-\tau)]$ $= \sigma^2 \delta(\tau)$. The cyclic spectrum of the received signal must contain the cyclic spectrum of x(t), which is given by:

$$S_x^\alpha(f) = \begin{cases} \dfrac{P}{2}[\delta(f - f_0 - v) + \delta(f + f_0 + v)]\,|\,H(f)\,|^2 + \sigma^2, \, for \, \alpha = 0 \\[2mm] \dfrac{P}{2}\delta(f)H(f - f_0 - v)H*(f + f_0 + v), \, for \, \alpha = \pm 2(f_0 + v) \\[2mm] 0, otherwise \end{cases}$$

where H(f) is the frequency response of the channel. The parameter α is the cyclic frequency. From this equation, ideally, the noise does not contribute to the cyclic spectrum of x(t) when cyclic frequencies $\alpha = \pm(2f_0 + v)$. Thus, performing spectrum sensing by detecting the peaks on the cyclic spectrum of the signal is generally better than using the power spectral density.

8.6 Other Activities

There are many other ongoing industry activities that involve CR technologies. In this section we introduce some of them.

8.6.1 IEEE 802.22.1 and IEEE 802.22.2

Within the IEEE 802.22 working group there are two task groups that are not as popular as its parent working group: the IEEE 802.22.1 and IEEE 802.22.2 task groups. The IEEE 802.22.1 task group was created to define a standard for CR devices to provide enhanced protection to low-powered licensed devices, such as wireless microphones, which also operate in the TV bands. The IEEE 802.22.2 task group is defining a technical guidance to installers, deployers, and operators of IEEE 802.22-compliant systems to help assure that such systems are correctly installed and deployed.

8.6.2 IEEE SCC 41

The IEEE Standards Coordinating Committee (SCC) 41 [SCC41] is also developing standards in the area of dynamic spectrum access networks. Among other things, SCC 41 intends to provide coordination and information exchange among standards developing activities in the IEEE. The focus of SCC 41 is on improved use of the spectrum, which includes the definition of new techniques and methods of dynamic spectrum access for managing interference, coordination of wireless technologies and network management, and information sharing.

8.6.3 IEEE 802.11af

The IEEE 802.11af task group is the most recent industry activity to be formed around CRs. As described in Chapter 4, the 802.11af task group works under the 802.11 working group on WLANs. The 802.11af task group is defining modifications to both the 802.11 PHY and MAC layers to meet the legal requirements for channel access and coexistence in the TV White Space. The purpose of this amendment is to allow 802.11 wireless networks to be used in the TV white space.

8.7 Conclusions and Future Directions

Better spectrum utilization through the use of sophisticated communication and computing technologies is undoubtedly the next major wave of innovation that will ignite a revolution in wireless applications and usages. Cognitive radios and cognitive networks are at the center of this innovation. In recent years, a plethora of activities have started activities in this area. Simultaneously, academia has been extremely active in developing new CR-based technologies that will enable future, smarter networks.

The main obstacle to the market success of CRs and CNs remains the business model. As an example, operators may face problems in developing a business model wherein the network sometimes works and sometimes does not work due to varying spectrum availability – and charging users for such a service does not seem trivial. While this may not be an issue for data services, this may be a challenge for networks intending to carry video or voice services. Therefore, more research and development is needed in this area. Unquestionably, however, the future is bright in the area of cognitive radio and network technologies.

Homework Questions/Simulation Projects

Q. 1. Explain how the superframe structure and the SCH can be used to support reliable incumbent detection and to meet the incumbent protection requirements described in Table 8.6.

Q. 2. Explain the main reasons for supporting the two different formats of upstream data burst allocations illustrated in the 802.22 frame structure.

Q. 3. What features in the 802.22 frame structure and MAC protocol enable over the air signaling between different 802.22 cells, and in which scenarios these features are most useful?

Q. 4. Explain why and in which situations neighboring 802.22 BSs need to synchronize their quiet periods for spectrum sensing.

Reference

[ATSC2004] ATSC A74 Recommended Practice Guideline Document entitled: "ATSC Recommended Practice: Receiver Performance guidelines," Sections 4.5.2 & 4.5.3, ATSC A74, June 2004.

[Calabrese2003] M. Calabrese and J. Snider, "Up in the Air," The Atlantic Monthly, New America Foundation, September 2003.

[Cavalcanti2008] D. Cavalcanti and M. Ghosh, "Cognitive Radio Networks: Enabling New Wireless Broadband Opportunities," in the Third International Conference on CrownCom, May 2008.

[Challapali2006] K. Challapali, C. Cordeiro and D. Birru, "Evolution of Spectrum-Agile Cognitive Radios: First Wireless Internet Standard and Beyond," in WICON, August 2006.

[Cordeiro2005] C. Cordeiro, K. Challapali, D. Birru, and S. Shankar, "IEEE 802.22: The First Worldwide Wireless Standard based on Cognitive Radios," in IEEE DySPAN, November 2005.

[Cordeiro2006] C. Cordeiro, K. Challapali and M. Ghosh, "Cognitive PHY and MAC Layers for Dynamic Spectrum Access and Sharing of TV Bands", in ACM TAPAS, August 2006.

[Cordeiro2007] C. Cordeiro, M. Ghosh, D. Cavalcanti and K. Challapali, "Spectrum sensing for dynamic spectrum access of TV bands" CROWNCOM 2007, August 2007 (Invited).

[DARPA-XG] DARPA Next Generation (XG) Program, http://www.darpa.mil/ato/programs/xg/.

[FCC2002] Federal Communications Commission (FCC), "Spectrum Policy Task Force," ET Docket no. 02-135, November 15, 2002.

[FCC2003] FEDERAL COMMUNICATIONS COMMISSION (2003a) Notice for Proposed Rulemaking (NPRM 03 322): Facilitating Opportunities for Flexible, Efficient, and Reliable Spectrum Use Employing Cognitive Radio Technologies. ET Docket No. 03 108, December 2003.

[Ghosh2009] C. Ghosh, S. Chen, D. P. Agrawal, and A. M. Wyglinski, "Priority-based Spectrum Allocation for Cognitive Radio Networks Employing NC-OFDM Transmission," MILCOM 2009, Boston, MA, October 18-21, 2009.

[Ghosh2010a] C. Ghosh, S. Pagadarai, D. P. Agrawal, and A. M. Wyglinski, "A Framework for Statistical Wireless Spectrum Occupancy Modeling," IEEE Transactions on Wireless Communications, vol. 9, no. 1, Jan. 2010, pp. 38-44.

[Ghosh2010b] C. Ghosh, S. Roy, M. B. Rao, and D. P. Agrawal, "Spectrum Occupancy Validation and Modeling Using Real-time Measurements," CoRoNet workshop, Mobicom, Sept. 20-24, 2010.

[Haykin2005] S. Haykin, "Cognitive Radio: Brain-Empowered Wireless Communications," in IEEE J-SAC, vol. 23, no. 2, February 2005.

[IEEE802.16] IEEE 802.16 Working Group on Broadband Wireless Standards. http://wirelessman.org.

[IEEE 802.22-2005] Functional Requirements for IEEE 802.22, 802.22-05/0007r46, September 2005.

[IEEE 802.22-2010] IEEE 802.22 Draft standard, "IEEE P802.22™/D4.0 Draft Standard for Wireless Regional Area Networks," http://www.ieee802.org/22/, Draft 4.0, July 2010.

[Ileri2008] O. Ileri and N. B. Mandayam, "Dynamic Spectrum Access Models: Toward an Engineering Perspective in the Spectrum Debate," in IEEE Communications Magazine, Jan. 2008.

[ITU2005] International Telecommunication Union (ITU) New Broadband Statistics, http://www.itu.int/osg/spu/newslog/ITUs+New+Broadband+Statistics+For+1+Janua ry+2005.aspx, January 1, 2005.

[Kay1998] S. Kay, Fundamentals of Statistical Signal Processing: Detection Theory, Prentice Hall, 1998.

[Kolodzy2003] P. Kolodzy, "Spectrum Policy Task Force: Findings and Recommendations," in International Symposium on Advanced Radio Technologies (ISART), March 2003.

[McHenry2005] M. McHenry, "Report on Spectrum Occupancy Measurements," Shared Spectrum Company, http://www.sharedspectrum.com/?section=nsf_summary.

[Mitola1999] J. Mitola et al., "Cognitive Radios: Making Software Radios more Personal," IEEE Personal Communications, vol. 6, no. 4, August 1999.

[Mitola2000] J. Mitola, "Cognitive radio: An integrated agent architecture for software defined radio," PhD Dissertation, Royal Institute of Technology (KTH), Stockholm, Sweden, 2000.

[Mody2008] A. Mody et al, "Protocol Reference Model Enhancements in 802.22", document number 22-08-0121-06, July 2008.

[SCC41] IEEE Standards Coordinating Committee 41, http://www.scc41.org/

[Shellhammer2006a] S. Shellhammer, "Performance of the Power Detector," IEEE 802.22-06/0075r0, May 2006.

[Shellhammer2006b] S. Shellhammer, S. Shankar, R. Tandra and J Tomcik, "Performance of Power Detector Sensors of DTV Signals in IEEE 802.22 WRANs," ACM TAPAS, August 2006.

[Shellhammer2006c] S. Shellhammer, V. Tawil, G. Chouinard, M. Muterspaugh, and M. Ghosh, "Spectrum Sensing Simulation Model," IEEE 802.22-06/0028r6, June 2006.

Chapter 9

TCP over Ad Hoc Networks

9.1 Introduction

Over the past few years, MANETs have emerged as a promising approach for future mobile IP applications. This scheme can operate independently from existing underlying infrastructure and allows simple and fast implementation. Obviously, the more this technology evolves, the higher will be the probability of being an integral part of the global Internet [Perkins2002]. Therefore, considerable research efforts have been put on the investigation of the Transmission Control Protocol (TCP) [Comer1995] performance over MANETs.

As we know, TCP is the prevalent transport layer protocol in the IP world today and is employed by a vast majority of applications, especially those requiring reliability. More specifically, TCP is the most widely used transport protocol for data services like file transfer, email, web browsing, and so on. Therefore, it is essential to look at the TCP performance over MANETs. TCP is already fine-tuned to work well in wired environments; however in a MANET environment, its performance is highly degraded due to the high error rates and longer delays of wireless mediums, as well as mobility. In recent years, several improvements for TCP have been proposed for cellular [Balakrishnan1997, Zhang2001] wireless networks, but still much work has to be done for the MANETs paradigm.

Unlike cellular networks, where only the last hop is based on a wireless medium, MANETs are composed exclusively of wireless links, where multi-hop connections are often in place. Besides, in an ad hoc scenario, all nodes can move freely and unpredictably, which makes the clock based TCP congestion control quite hard. In particular, as the errors in wireless networks occur not only due to congestion but also due to medium constraints and mobility, TCP needs to distinguish nature of the error so that it can take the most appropriate action. Factors such as path asymmetry and congestion window size also impact the performance. These and few other issues are addressed in this chapter.

Although there are a number of differences between cellular and MANETs, some of the ideas developed for the former can be used in the latter as well. As a matter of fact, many of the proposed TCP solutions for MANETs is a mixture of old concepts used for cellular networks. Nevertheless, we cover some solutions specifically tailored for MANETs, while many issues are still open.

9.2 TCP Protocol Overview

The TCP protocol is defined in the Request For Comment (RFC) standards document number 793 [Postel1981] by the Internet Engineering Task Force (IETF) [IETFwww]. The original specification written in 1981 was based on earlier research and experimentation in the original ARPANET. It is important to remember that most applications on the Internet make use of TCP, relying upon its mechanisms that ensure safe delivery of data across an unreliable IP layer below. In this section we explore the fundamental concepts behind TCP and how it is used to transport data between two endpoints. More specifically, we focus on those aspects of TCP that are of importance to understand its performance over MANETs. TCP adds a great deal of functionality to the IP service it is layered over:

- **Streams:** TCP data is organized as a stream of bytes, much like a file. The datagram nature of the network is concealed. A mechanism (the *Urgent Pointer*) exists to let out-of-band data be specially flagged.
- **Reliable delivery:** Sequence numbers are used to coordinate which data has been transmitted and received. TCP will arrange for retransmission if it determines that data has been lost;
- **Network adaptation:** TCP will dynamically learn the delay characteristics of a network and adjust its operation to maximize throughput without overloading the network.
- **Flow control:** TCP manages data buffers, and coordinates traffic so that its buffers will never overflow. Fast senders will be stopped periodically to keep up with slower receivers.

9.2.1 Designed and Fine-Tuned to Wired Networks

The design of TCP has been heavily influenced by what is commonly known as the "end-to-end argument" [Clark1988]. As it applies to the wired Internet, the system gets unnecessarily complicated by putting excessive intelligence in physical and link layers to handle error control, encryption or flow control. While these functions need to be done at the endpoints anyway, the result is the provision of minimal functionality on a hop-by-hop basis and maximal control between end-to-end communicating systems.

TCP performance is often dependent on the flow control and the congestion control. Flow control determines the rate at which data is transmitted between a sender and a receiver. Congestion control defines the methods for implicitly interpreting signals from the network in order for a sender to adjust its rate of transmission. Ultimately, intermediate devices, such as IP routers, would only be able to control congestion. A recent study on congestion control examines the current state of activity [Kristoff2000].

Timeouts and retransmissions handle error control in TCP. The nature of TCP and underlying packet switched network provide formidable challenges for managers, designers and researchers. It is important to incorporate link layer acknowledgements and error detection/correction functionality in TCP for wireless networks. Furthermore, when we consider MANETs, mobility comes into picture. Therefore, higher error rates, longer delays, and mobility makes MANET environments extremely challenging to the implementation of TCP as it tears down most the assumptions over which TCP was designed.

9.2.2 TCP Basics

TCP is often described as a byte stream, connection-oriented, full-duplex, reliable delivery transport layer protocol. In this subsection, we discuss the meaning for each of these descriptive terms.

9.2.2.1 Byte Stream Delivery

TCP interfaces between the application layer above and the network layer below. When an application sends data to TCP, it does so in 8-bit byte streams. It is then up to the sending TCP to segment or delineate the byte stream in order to transmit data in manageable pieces to the receiver. It is this lack of "record boundaries" which gives it the name "byte stream delivery service".

9.2.2.2 Connection-Oriented

Before two communicating TCP entities (the sender and the receiver) can exchange data, they must first agree upon the willingness to communicate. Analogous to a telephone call, a connection must first be made before two parties exchange information.

9.2.2.3 Full-Duplex

No matter what a particular application may be, TCP almost always operates in full-duplex mode. It is sometimes useful to think of a TCP session as two independent byte streams, traveling in opposite directions. No TCP mechanism exists to associate data in the forward and reverse byte streams, and only during connection start and close sequences can TCP exhibit asymmetric behavior (i.e., data transfer in the forward direction but not in the reverse, or vice versa).

9.2.2.4 Reliability

A number of mechanisms help providing the reliability TCP guarantees. Each of these is described briefly below:

- **Checksums**: All TCP segments carry a checksum, which is used by the receiver to detect errors with either the TCP header or data.
- **Duplicate data detection**: It is possible for packets to be duplicated in packet switched network; therefore TCP keeps track of bytes received in order to discard duplicate copies of data that has already been received.
- **Retransmissions**: In order to guarantee delivery of data, TCP must implement retransmission schemes for data that may be lost or damaged. The use of positive acknowledgements by the receiver to the sender confirms successful reception of data. The lack of positive acknowledgements, coupled with a timeout period (see timers below) calls for a retransmission.
- **Sequencing**: In packet switched networks, it is possible for packets to be delivered out of order. It is TCP's job to properly sequence

segments it receives so that it can deliver the byte stream data to an application in order.

- **Timers**: TCP maintains various static and dynamic timers on data sent. The sending TCP waits for the receiver to reply with an acknowledgement within a bounded length of time. If the timer expires before receiving an acknowledgement, the sender can retransmit the segment.

9.2.3 TCP Header Format

As we know, the combination of TCP header and TCP in one packet is called a TCP segment. Figure 9.1 depicts the format of all valid TCP segments. The size of the header without options is 20 bytes. Below we briefly define each field of the TCP header.

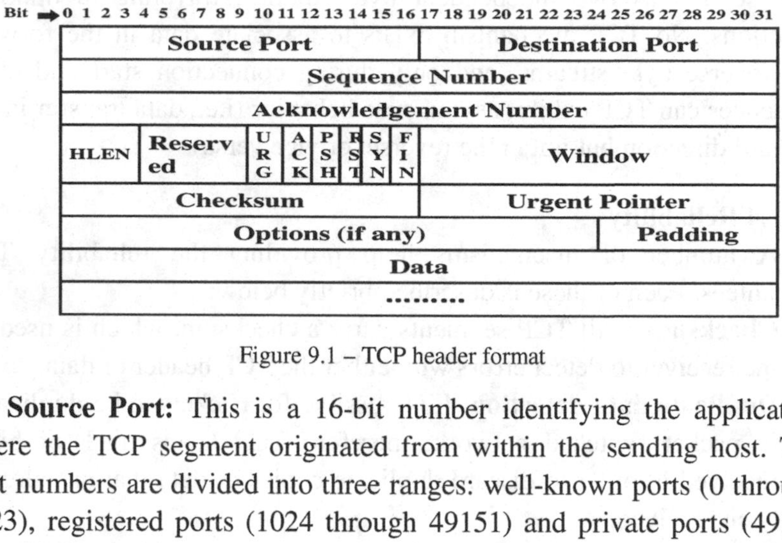

Bit → 0 1 2 3 4 5 6 7 8 9 10 11 12 13 14 15 16 17 18 19 20 21 22 23 24 25 26 27 28 29 30 31

Source Port							Destination Port	
Sequence Number								
Acknowledgement Number								
HLEN	Reserved	URG	ACK	PSH	RST	SYN	FIN	Window
Checksum							Urgent Pointer	
Options (if any)								Padding
Data								

Figure 9.1 – TCP header format

Source Port: This is a 16-bit number identifying the application where the TCP segment originated from within the sending host. The port numbers are divided into three ranges: well-known ports (0 through 1023), registered ports (1024 through 49151) and private ports (49152 through 65535). Port assignments are used by TCP as an interface to the application layer. For example, the TELNET server is always assigned to the well-known port 23 by default on TCP hosts. A pair of IP addresses (source and destination) plus a complete pair of TCP ports (source and destination) defines a single TCP connection that is globally unique.

- **Destination Port:** A 16-bit number identifying the application TCP segment is destined for on a receiving host. Destination ports use the same port number assignments as those set aside for source ports.

- **Sequence Number:** Within the entire byte stream of the TCP connection, a 32-bit number, identifying the current position of the first data byte in the segment. After reaching 2^{32} -1, this number will wrap around to 0.
- **Acknowledgement Number:** This is a 32-bit number identifying the next data byte the receiver expects from the sender. Therefore, this number will be one greater than the most recently received data byte. This field is used only when the ACK control bit is turned on.
- **Header Length:** A 4-bit field that specifies the total TCP header length in 32-bit words (or in multiples of 4 bytes). Without options, a TCP header is always 20 bytes in length. On the other hand, the largest a TCP header is 60 bytes. Clearly, this field is required because the size of the options field(s) cannot be determined in advance. Note that this field is called "data offset" in the official TCP standard, but header length is more commonly used.
- **Reserved:** A 6-bit field currently unused and reserved for future use.
- **Control Bits:**
  - *Urgent Pointer (URG)* – If this bit field is set, the receiving TCP should interpret the urgent pointer field (see below);
  - *Acknowledgement (ACK)* – If this bit is set, the acknowledgment field is valid.
  - *Push Function (PSH)* – If this bit is set, the receiver should deliver this segment to the receiving application as soon as possible. An example of its use may be to send a Control-C request to an application, which can jump ahead of queued data.
  - *Reset Connection (RST)* – If this bit is present, it signals the receiver that the sender is aborting the connection and all the associated queued data and allocated buffers can be freely relinquished.
  - *Synchronize (SYN)* – When present, this bit field signifies that the sender is attempting to "synchronize" sequence numbers. This bit is used during the initial stages of connection establishment between a sender and a receiver.
  - *No More Data from Sender (FIN)* – If set, this bit field tells the receiver that the sender has reached the end of its byte stream for the current TCP connection.

- **Window:** This is a 16-bit integer used by TCP for flow control in the form of a data transmission window size. This number tells the sender how much data the receiver is willing to accept. The maximum value for this field would limit the window size to 65,535 bytes. However, a "window scale" option can be used to make use of even larger windows.

- **Checksum:** A TCP sender computes the checksum value based on the contents of the TCP header and data fields. This 16-bit value will be compared with the value the receiver generates using the same computation. If the values match, the receiver can be very confident that the segment arrived intact.

- **Urgent Pointer:** In certain circumstances, it may be necessary for a TCP sender to notify the receiver of urgent data that should be processed by the receiving application as soon as possible. This 16-bit field tells the receiver when the last byte of urgent data in the segment ends.

- **Options:** In order to provide additional functionality, several optional parameters may be used between a TCP sender and a receiver. Depending on the option(s) used, the length of this field varies in size, but it cannot be larger than 40 bytes due to the maximum size of the header length field (4 bits). The most common option is the maximum segment size (MSS) option where a TCP receiver tells the TCP sender the maximum segment size it is willing to accept. Other options are often used for various flow control and congestion control techniques.

- **Padding:** Because options may vary in size, it may be necessary to "pad" the TCP header with zeroes so that the segment ends on a 32-bit word boundary as defined by the standard.

- **Data:** Although not used in some circumstances (e.g., acknowledgement segments with no data in the reverse direction), this variable length field carries the application data from TCP sender to receiver. This field coupled with the TCP header fields constitutes a TCP segment.

9.2.4 Congestion Control

TCP congestion control and Internet traffic management issues in general is an active area of research and experimentation. Although this section is a very brief summary of the standard congestion control algorithms widely used in TCP implementations today, it covers the main points necessary to understand behavior of the TCP over MANETs. These algorithms are defined in [Jacobson1988] and [Jacobson1990a], and the most update version of the TCP congestion control algorithm can be found in [Allman1999].

9.2.4.1 Slow Start

Slow Start, a requirement for TCP software implementation, is a mechanism used by the sender to control the transmission rate, otherwise known as sender-based flow control. The rate of acknowledgements returned by the receiver determines the rate at which the sender can transmit data. Whenever a TCP connection starts, the Slow Start algorithm at the sender initializes a *congestion window* (CWND) to one segment. As the connection is carried out and acknowledgements are returned by the receiver, the congestion window increases by one segment for each acknowledgement returned. Thus, the sender can transmit the minimum of the congestion window and the advertised window (contained in the header of the acknowledgment packet) of the receiver, which is simply called the transmission window and is increased exponentially.

9.2.4.2 Congestion Avoidance

During the initial data transfer phase of a TCP connection, the Slow Start algorithm is used. However, there may be a point during Slow Start that the network is forced to drop one or more packets due to overload or congestion. If this happens, Congestion Avoidance is used to reduce the transmission rate. However, Slow Start is used in conjunction with Congestion Avoidance in order to get the data transfer going again so it does not slow down and stay slow. In the Congestion Avoidance algorithm, the expiration of a timer called *retransmission timeout* (RTO) or the reception of duplicate ACKs implicitly signal the sender that a network congestion situation is occurring. The sender immediately sets its transmission window to one half of the current window size (the minimum of the congestion window and the receiver's advertised window size). If congestion is indicated by a timeout, the congestion window is reset to one segment, which automatically puts the sender into Slow Start mode. There are many variants such as TCP Tahoe, TCP Reno that provide enhancement in TCP performance of wired networks.

As data is received during Congestion Avoidance, the congestion window is increased. However, Slow Start is only used up to the halfway point where congestion originally occurred. This halfway point was recorded earlier as the new transmission window. After this halfway point, the congestion window is increased by one segment for all segments in the transmission window that are acknowledged. This mechanism will force the sender to slowly grow its transmission rate, as it will approach the point where congestion had previously been detected.

9.2.4.3 Fast Retransmit

When a duplicate ACK is received, the sender does not know if this is because a TCP segment was lost or simply because a segment was delayed and received out of order at the receiver. If the receiver can re-order segments, it should not be long before the receiver sends the latest expected acknowledgement. Typically, no more than one or two duplicate ACKs should be received when a simple out of order conditions takes place. If, however, more than two duplicate ACKs are received by the sender, it is a strong indication that at least one segment has been lost. Here, the TCP sender will assume enough time has lapsed for all segments to be properly re-ordered by the fact that the receiver had enough time to send three duplicate ACKs.

Thus, whenever three or more duplicate ACKs are received, the sender does not even wait for the RTO to expire and retransmits the segment (as indicated by the position of the duplicate ACK in the byte stream). This process is called the Fast Retransmit algorithm and was first defined in [Jacobson1990a]. Immediately following Fast Retransmit, there is the Fast Recovery algorithm described next.

9.2.4.4 Fast Recovery

Since the Fast Retransmit algorithm is used when duplicate ACKs are being received, the TCP sender has implicit knowledge that there is data still flowing to the receiver. Why? The reason is because duplicate ACKs can only be generated when a segment is received. This is a strong indication that serious network congestion may not exist and that the lost segment was a rare event. Therefore, instead of reducing the flow of data abruptly by going all the way into Slow Start, the sender only enters Congestion Avoidance mode. Rather than start at a window of one segment as in Slow Start mode, the sender resumes transmission with a larger window, incrementing as if in Congestion Avoidance mode. This allows for higher throughput under the condition of only moderate congestion [Lin1998].

9.2.5 Round-Trip Time Estimation

When a host transmits a TCP packet to its peer, it must wait a period of time for an acknowledgment. If the reply does not come within the expected period, the packet is assumed to have been lost and the data is retransmitted. The obvious question - How long do we wait? - lacks a simple answer. Over an Ethernet, no more than a few microseconds should be needed for a reply. If the traffic must flow over the wide-area Internet, a second or two might be reasonable during peak utilization times. If we are talking to an instrument package on a satellite hurtling toward Mars, minutes might be required before a reply. There is no one answer to the question - How long?

As a result, all modern TCP implementations seek to answer this question by monitoring the normal exchange of data packets and developing an estimate of how long is "too long". This process is called Round-Trip Time (RTT) estimation which computes one of the most important performance parameters in a TCP exchange, especially when you consider that on an indefinitely large transfer all TCP implementations eventually drop packets and retransmit them, no matter how good the quality of the link is. If the RTT estimate is too low, packets are retransmitted unnecessarily; if too high, the connection can sit idle while the host waits to timeout.

If we bring this discussion to MANETs, things become quite complicated. Once a TCP connection is established over multiple wireless hops in an MANET and packets start flowing, TCP initiates the process of RTT estimation. As some nodes may be mobile, some station on the route from the source to the destination may move causing a link breakage. Very often when this happens, the RTO of TCP (set based on the RTT estimation) goes off and triggers TCP's congestion control algorithms. Clearly, this is not what TCP is supposed to do as there is no congestion in place. Rather, the expiration of the RTO was caused by mobility and not by congestion. As we discuss later, most of the current research on TCP over ad hoc concentrates on addressing this issue.

9.3 TCP and MANETs

MANETs pose some tough challenges to TCP primarily because TCP has not been designed to operate in such a highly dynamic environment [Oliveira2002]. In general, we can identify three main different types of challenges posed to TCP over MANETs.

- First, as the topology changes, the path is interrupted and TCP goes into exponentially increasing time-outs repeatedly with severe performance degrade. Few efficient retransmission strategies have been proposed to overcome such problems (e.g., [Chandran2001, Holland1999a]) and are discussed later in this chapter;
- The second problem has to do with the fact that the TCP performance in ad hoc multi-hop environment depends critically on the congestion window in use. If the window grows too large, there are too many packets (and ACKs) on the path, all competing for the same wireless shared medium. Congestion builds up and causes "wastage" of the broadcast medium with consequent throughput degradation [Fu2005, Oliveira2005];
- The third problem is significant TCP unfairness in multihop MANETs which has been recently revealed and reported through both simulations and testbed measurements [Cordeiro2002].

In reality, even though TCP has evolved significantly over the years toward a robust and reliable service protocol, the focus has been primarily on wired networks. In wired scenario, the additive-increase/multiplicative-decrease strategy coupled with the fast recovery and fast retransmits mechanisms [Allman1999], inherent in most of current TCP versions; provide an effective congestion control solution. The key idea of TCP is to probe the network in order to determine the availability of resources. It injects packets at an increasing rate into the network until a packet loss is detected, whereby it infers the network is facing congestion. Then, the TCP sender shrinks its CWND, retransmits the lost packet and resumes transmission at a lower increasing rate. If the losses persist (no timely ACK received), at every retransmission the sender doubles (up to 64s) its wait timer (i.e., the RTO) so that it can wait longer for the ACK of the current packet being transmitted. This is the *exponential backoff* strategy depicted in Figure 9.2. More details on this mechanism can be found in [Stevens1994].

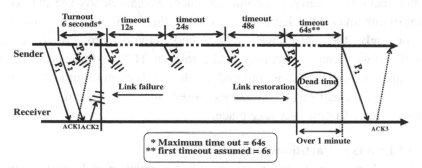

Figure 9.2 – Drawback of the TCP exponential backoff algorithm in MANETs

Clearly, the aforementioned mechanisms work quite well in a wired network where the Bit Error Rate (BER) is typically very low allowing any lost packet to be treated as an indication of network congestion. In a wireless mobile MANET, however, two more factors can induce packet losses in addition to network congestion: non-negligible *wireless medium losses* (high BER) and frequent *connectivity disruptions* caused by node mobility. As a result, lost packets no longer serve as an indication of congestion for TCP. In order for TCP to perform efficiently in wireless networks, it needs to be aware of what caused which loss.

MANETs have a high level of BER as all links involved are wireless and hence suffer from fading, multipath effects, and interference [Rappaport1999]. In addition, since nodes can move freely and unpredictably, there is a high probability of the peers in the ongoing connections to get abruptly disconnected by a temporary or permanent lack of a route between them, and this can last for a significant period of time. Such a discontinuity (whether temporary or permanent) is typically referred to as *partition* and is generally very degrading for TCP performance, as explained later.

Another important characteristic of TCP is its dependence on reception of timely ACKs from the receiver to the sender so as to increase its data transfer rate. Thus, in case there is an asymmetrical path in place, with lower ACK rate as compared to the data rate, TCP can be prevented from sending at a higher rate. If we apply this fact to MANETs where path asymmetry is a common place, we can clearly see that TCP algorithms may also have to deal with this issue. Finally, and not surprisingly, the lower layers (MAC and network) protocol strategies used in this scenario also play a key role on TCP performance, which demands special handling tailored to the ad hoc environment. In the following, we address all the issues aforementioned in detail in order to clarify why and how they take place.

9.3.1 Effects of Partitions on TCP

A network partition occurs when a given mobile node moves away, or is interrupted by the medium, thereby splitting its adjacent nodes into two isolated parts of the network that are called *partitions*. Here, the term *partition* is defined within the context of a connection which is interrupted due to a link breakage, and not necessarily because there are no alternate paths available through some other nodes.

Figures 9.3 and 9.4 depict a scenario in which a partition takes place and interrupts an ongoing connection between nodes 3 and 9. Note that in Figure 9.3 there is a direct link between nodes 6 and 7, while this is not the case in Figure 9.4. We consider these two cases to illustrate the different between a short-term (Figure 9.3) and a long-term (Figure 9.4) partition. Both figures show the case when node 5 moves away from node 3 causing a link breakage and consequently disrupts the connection between nodes 3 and 9. In the case of Figure 9.3, the routing protocol in use will find an alternate route from node 3 to node 9 through node 6, possibly, in one attempt. The resulting connection topology is shown in Figure 9.5. Therefore, we refer to this type of partition as being short-term.

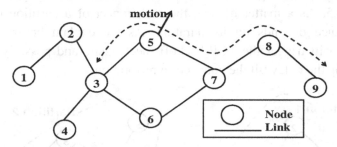

Figure 9.3 – Node 5 moves away from Node 3 (short-term partition)

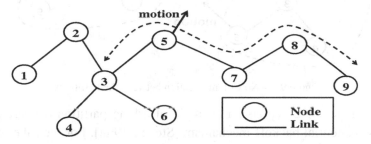

Figure 9.4 – Node 5 moves away from Node 3 (long-term partition)

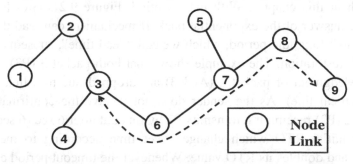

Figure 9.5 – The Routing Protocol reestablishes the path through Node 6

Another possibility is as shown in Figure 9.4 where there is no link between nodes 6 and 7. In this case, upon movement of node 5 and consequent link breakage between nodes 3 and 5, not only is the connection between nodes 3 and 9 disrupted, but a long-term network partition takes place as can be seen from Figure 9.6. In this case, the connection can only be reestablished whenever the topology rejoins. For example, in Figure 9.6 node 6 moves towards node 7 rejoining the network and allowing the connection to be reestablished as presented in Figure 9.5. As a matter of fact, the real impact of a partition on TCP performance depends on its duration. As we explain below, a long partition will trigger the TCP backoff mechanism, and possibly end up increasing the delay till the connection restoration.

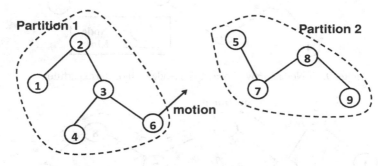

Figure 9.6 – No Communication between the Partitions

Figure 9.2 depicts an example of a long partition triggering the TCP exponential backoff mechanism [Stevens1994]. For the sake of this discussion, we use the term *packet* and *segment* interchangeably throughout this chapter. With that in mind, Figure 9.2 shows how the delayed answer of the exponential backoff mechanism can lead the TCP sender to a long idle period, which we call "dead time", subsequently to the link restoration. The example shows that both packet 3 (P3) and the acknowledgment of packet 2 (ACK3) are dropped due to a link failure (left vertical line). As the sender does not receive the confirmation of packet 2 (P2) receipt, it retransmits P2 by timeout after 6 sec (6 sec is the typical initial RTO which changes over time according to measured RTTs), and doubles its RTO value. Whenever the timeout period expires, the TCP sender retransmits P2 and doubles the RTO up to the limit of 64sec that refers to the maximum timeout allowed (note that after 12 unsuccessful attempts TCP would give up). The example shows that shortly after triggering its timer for 64 sec, the link is recovered.

However, it is too late for TCP and it will stay over one minute frozen which is indeed a dead time for the assumed starving connection. In terms of percentage, we have roughly 61% (100 sec/163 sec) of interruption due to link failure and the remaining 39% (63 sec/163 sec) is completely caused by TCP, which is certainly too much to be acceptable.

It is appropriate to mention that the infrequent (not necessary low) and transient packet losses caused by mechanisms such as fading or short-term partitions, can also lead TCP to take inappropriate actions. That is, even though such losses are improbably to cause a long period of disconnection, they do can undesirably lead TCP to invoke its exponential backoff algorithm. Consequently, TCP prevents the applications from using precious bandwidth by decreasing its transmission rate when, in fact, it should not. More details about this issue can be found in [Caceres1994] where the focus is on cellular networks which can face a similar problem. Therefore, it is clear that the standard TCP needs to be adapted to work satisfactorily in the new MANET paradigm. An effective TCP algorithm must be capable of distinguishing the origin of a packet loss so as to take the most appropriate action. In other words, its error-detection mechanism needs to detect the nature of the error so that its error recovery mechanism can be tailored for each specific case [Tsaoussidis2002].

9.3.2 Impact of Lower Layers on TCP

Given the fact that TCP is a reliable protocol, providing end-to-end guarantees over a variety of local and unreliable protocols running in the lower layers (notably, MAC and routing protocol layers), it is no surprise that its performance depends strongly on such protocols. What is not so clear, however, is how either TCP or lower protocols can be fine tuned to avoid as much as possible any undesired interferences between them. In this subsection we give an appropriate insight into the issues resulted from TCP interaction with lower layers.

9.3.2.1 MAC Layer Impact

Like the transport layer, the wireless MAC layers almost invariably also rely on error control mechanisms in order to improve the transmission efficiency. However, while the former deals with end-to-end recovery, the latter concentrates on link (one hop) recovery. Hence, unless a well defined synchronism between these both protocols is put in place, negative interactions can substantially deteriorate end-to-end throughput provided by TCP.

As we have studied in Chapter 4, the IEEE 802.11 DCF is nowadays the standard MAC layer protocol adopted for MANETs (obviously, other WLAN/WPAN systems could also be used). This MAC protocol, which defines both physical and link layer mechanisms, is intended for providing an efficient shared broadcast channel through which the involved mobile nodes can communicate. The main novelty of this protocol refers to the inclusion of acknowledgment for data frames (link layer's ACKs) in addition to RTS/CTS control frames to make it possible for link layer retransmissions. This makes it possible to recover a packet loss at the link level instead of waiting for TCP to detect the loss only at the destination when it has already taken too long time. Another mechanism introduced by the IEEE 802.11 MAC is the virtual carrier sense used to track medium activity.

In IEEE 802.11, RTS/CTS handshake is only employed when the DATA packet size exceeds some predefined threshold. However, given the fact that most existing commercial implementations of IEEE 802.11 use RTS/CTS handshake before any DATA/ACK transmission (obviously, broadcast packets are not preceded by RTS/CTS), for the sake of our discussion here we assume RTS/CTS handshake is always employed before any DATA/ACK exchange. As we saw in Chapter 4, each of these frames carries the remaining duration of time for the transmission completion, so that other nodes in the vicinity can hear it and postpone their transmissions accordingly. In fact, every node maintains an information parameter called NAV which is updated according to other nodes' transmission schedules.

In order to provide fair access to the medium at the end of every such sequence, the nodes must await an IFS interval and then contend for the medium again. The contention is carried out by means of a binary exponential backoff mechanism which imposes a further random interval, aiming to avoid collisions to allow all nodes the same probability of gaining access to the medium. At every unsuccessful attempt, this random interval tends to become higher (its range of randomness is doubled at every attempt) and after some number of attempts (typically, seven times) the MAC layer gives up and drops the data, which is reported as a route failure to the network layer.

Therefore, the MAC protocol is certainly very robust in dealing with the possibility of collisions in the wireless shared medium. By using short RTS/CTS control frames to reserve the medium, bandwidth wastage is considerably minimized in case of collisions as these frames are much smaller than DATA frames. As for the virtual carrier sense mechanism, it prevents the so-called hidden node problem (discussed in Chapter 4). In such cases, under IEEE 802.11, the first node to succeed reserves the medium and the other (or others) becomes aware and defers its transmission. However, although this MAC protocol has been designed to be fair in the sense that all neighboring stations would have equal chances of accessing the wireless medium, in fact it may result in strong unfairness for TCP traffic as some stations will have larger backoff values while others have smaller ones.

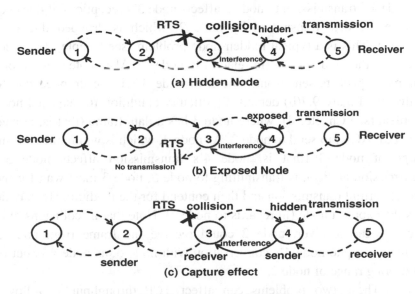

Figure 9.7 – Issues at the MAC layer

IEEE 802.11 relies on the assumption that every node can reach each other or at least sense any transmission into the medium, which is not always true in an ad hoc scenario. Consequently, the hidden node and exposed node problems can arise in some conditions inducing what is termed as the *capture problem* [Cordeiro2002], which impairs not only TCP performance but also results in unfairness amongst simultaneous TCP connections. We explain in the following, by means of examples, how these problems can take place.

In Figure 9.7, we consider a linear topology in which each node can only communicate with its adjacent neighbors. In addition, consider that in Figures 9.7(a) and 9.7(b) there exist a single TCP connection running between nodes 1 and 5. Further, suppose node 1 starts transmitting to node 5 as illustrated in Figure 9.7(a). Once the first few packets reach the destination node 5, there will be a condition in which node 2 wishes to communicate with node 3, while node 4 is transmitting to node 5. As node 2 cannot hear the transmission from node 4, it senses the medium idle (both physically and virtually using NAV) and so attempts its transmission by sending a RTS toward node 3.

Even so, since node 3 is within node 4's interference range (that is, node 4's transmission to node 5 affects node 3's reception), it does not receive the data transmitted by node 2, which is dropped due to a collision. This is a typical hidden node problem (see Chapter 4), where node 4 is the hidden node (in relation to node 2). Also, note that not one but many packets sent from node 2 to node 3 will be dropped due to collisions. Figure 9.7(b) depicts a particular condition for exposed node problem (see Chapter 4), where node 3 has a data frame (that is, related to a TCP ACK) to send to node 2. As node 4, which is within the sensing range of node 3 (that is, node 4's transmission affects node 3's transmission ability), is transmitting to node 5, node 3 must wait for the end of current transmission and then contend for the medium. Here, node 4 is the exposed node (in regards to node 3). Note that as collisions only occur at the receiver, node 2 could receive the frame from node 3 correctly despite node 4's conversation with node 5, as node 4 is out of interfering range of node 2.

These two problems can affect TCP throughput as follows. When a hidden node condition as illustrated in Figure 9.7(a) takes place, the MAC layer of node 2 invokes its backoff mechanism which attempts retransmission (locally) of the lost frame up to a maximum number of times (typically seven). In case it does not succeed (e.g., due to high traffic between nodes 4 and 5), node 2 will drop the packet and send a route error packet back to node 1. As a consequence, the routing protocol in node 1 will attempt to find a new route to the destination which will, by itself, delay the forwarding of TCP packets. Usually in these circumstances, the TCP sender will time out, reduce its transmission rate, and further delays the retransmission. Likewise, under the exposed node depicted in Figure 9.7(b), as long as the traffic between nodes 4 and 5 is high enough to delay the pending TCP ACK beyond the TCP timeout interval, the TCP sender at node 1 will also timeout.

If we look carefully at the example in Figure 9.7, we will notice that the same exposed and hidden terminal problems takes place in the reverse traffic (i.e., ACK packet flow) from the TCP receiver at node 5 to the TCP sender at node 1. In other words, the TCP backward ACKs need to compete for access to the wireless shared medium with its own TCP forward data packets. However, in terms of link layer frames, there are many more data frames in the forward direction (i.e., TCP data packets) than TCP ACK packets, as ACKs packet sizes are much smaller and also due to the possibility of the receiver sending cumulative ACKs through a single ACK packet. Therefore, the wireless medium will be highly loaded with by TCP data frames and, as a result, a significant amount of ACK packets will be lost.

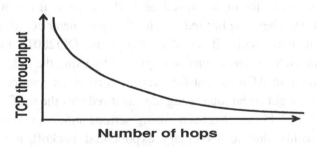

Figure 9.8 – TCP Throughput is inversely proportional to the Number of hops

Due to the above facts, both hidden and exposed node problems can cause a considerable lack of ACKs at the TCP sender, which is characterized as an asymmetrical path problem for TCP as discussed later in this section. Such a problem will either trigger a TCP retransmission by timeout when the lost ACK regards a confirmation of a successfully received data packet, or impair the TCP fast retransmit mechanism for which three duplicate ACKs are required in order to trigger a fast retransmission without the need to wait for a timeout event. Hence, in general, the larger the number of nodes a TCP connection needs to span, the lower is end-to-end throughput as there will be more medium contention taking place in several regions of the network. In this way, it has been shown [Holland1999a] that the TCP throughput over IEEE 802.11 decreases sharply and exponentially as a function of number of hops, and is shown in Figure 9.8.

Similar problems have been evaluated in [Xu2001a] where it has been shown that using smaller values for both the segment packet size and the maximum window size in TCP setup can mitigate such problems to a certain extent. The idea behind using smaller values for these parameters is to prevent TCP from sending too much data packets before receiving an ACK (i.e., to reduce the number of outstanding packets). As a result, the probability of collisions is decreased and the local MAC retransmission scheme has a better chance to succeed within its seven times retry limit.

In addition, it has been found in [Xu2001a] that a maximum window size of four segments should be enough to provide maximum stable throughput. Nevertheless, further analysis is needed in order to validate such results for higher speed networks, such as IEEE 802.11a/g (see Chapter 4), where the limited size for that parameter could represent a throughput bottleneck. It is also shown in [Xu2001] that TCP throughput in such a scenario can be improved by using the *delayed ACK* option in which an ACK is sent for every two received data packets. In principle, this might be an interesting idea as it reduces the traffic load.

IEEE 802.11 has also been raising serious *unfairness* concerns in MANETs mainly due to its binary exponential backoff mechanism, which leads to what is known as capture conditions [Cordeiro2002]. In fact, such a phenomenon is also related to the hidden node or exposed node conditions. For simplicity, here we explain only the situation in which the hidden node problem induces the capture effect (the capture effect caused by the exposed node is similar). Consider Figure 9.7(c) where there are two independent connections, one between nodes 2 and 3 (connection 2-3) and another between nodes 4 and 5 (connection 4-5). Assuming that connection 2-3 experiences collision due to the hidden node problem caused by the active connection 4-5 (as explained earlier), node 2 will back off and retransmit the lost frame. As we know, at every retransmission the binary exponential backoff mechanism imposes an increasingly (although random) backoff interval. Implicitly, this is actually decreasing the possibility of success for the connection 2-3 to send a packet as connection 4-5 will "dominate" the medium access once it has lower backoff value for most of the time. Besides, if the MAC retransmission scheme fails, TCP will eventually time out and will also invoke its exponential backoff mechanism, further increasing the delay for the next attempt. In consequence, the connection 2-3 will hardly obtain access to the medium while connection 4-5 will capture it. Note that the MAC protocol is designed in such a way that if the connection between nodes 4 and 5 has a large data to transfer, it will fragment and transmit it in smaller data frames with higher priority over all the other nodes, which is done by using a short IFS between the transfer of each fragment. Clearly, this behavior also contributes to the unfairness. Finally, the burstiness in TCP is another component which forces a connection to continue using the medium.

Contrary to mobility related TCP issues, the capture problem is mostly present when network nodes are static or possess small mobility since nodes stay longer within radio range of each other [Cordeiro2002], while in high mobility networks nodes are often moving out of range of each other and hence rarely have the chance to capture or to be captured. The capture problem is severe enough that nodes cannot access the medium for some amount of time they generate route error packets, even though the network is completely static. For TCP traffic, this causes retransmission timers to go off and throughput to degrade drastically.

Capture conditions are very likely in current generation routing protocols as the same route is used for forward and reverse traffic given a <source, destination> pair [Perkins2001, Johnson2001]. For TCP, this implies that data packets in the forwarding direction and ACK packets in the reverse direction compete to access the same shared medium, frequently causing ACK packets to be unable to reach the source and, thus, TCP executing its congestion control algorithms. It has been shown that TCP data packets often capture the medium preventing ACK packets from reaching their destination [Xu2001a, Xu2001b]. This problem is worsened by the presence of multiple TCP flows.

Fairness problems due to capture conditions have been investigated in [Gerla1999], and it was found that it can be mitigated by properly adjusting some MAC layer timers. Specifically, it has been shown that, as far as IEEE 802.11 is concerned, better fairness can be achieved by increasing the IFS interval (called *yield time*). However, it comes at the cost of degrading the aggregate throughput, which is somewhat expected as it makes medium idle for a longer time. Therefore, it is clear that alternative solutions for inherently unfair behavior detected in this environment need to be explored. Furthermore, no solution appears to be effective enough by simply configuring either TCP or MAC parameters. Rather, hidden and exposed node problems have to be addressed to have a deeper robust approach that would provide not only a fairer but also throughput effective MAC protocol. Moreover, the solutions presented so far do not address the scenario where multiple TCP connections are simultaneously competing for the medium access. That is, solving the problem from the point of view of a single connection clearly is the exception to the rule. As we discuss later, a general solution to the fairness problem has been proposed [Cordeiro2002] regardless of the number of active TCP connections in the network.

Apart from what has been explained here with regards to the interactions between the MAC and TCP layers, there are many other issues that are potential sources of complications. For example, if the nodes have different interfering (and sensing) and communication ranges, then the exposed node problem gets exacerbated. Likewise, either hidden or exposed node problems would be quite difficult to be controlled if the nodes had different battery power levels, which may be likely the case in an actual network [Poojary2001]. Additionally, the inherently node mobility can give rise to synchronization issues which would compromise the effectiveness of the reservation scheme provided by the MAC protocol.

9.3.2.2 Network Layer Impact

As MANETs consist of a highly dynamic environment where frequent route changes are expected, routing strategies play a key role on TCP performance as well. Unlike the MAC layer, for which the IEEE 802.11 protocol has been widely used as a testbed, the network layer has been a subject of most research efforts on mobile MANETs area towards standardization. As we have seen in Chapter 2, there have been a lot of proposed routing schemes and, typically, each of them have different effects on the TCP performance. To understand the network layer effect on TCP, in this subsection we consider two of the major routing protocols proposed for MANETs (both of them covered in Chapter 2) and show how a network layer protocol can affect the performance of an upper layer protocol.

DSR

As we know, the DSR protocol operates on an on-demand basis in which a node wishing to find a new route broadcasts a RREQ packet. Then, the destination node, or any other node which knows a route to the destination, responds back with a RREP packet. This packet informs the sender node the exact path to be followed by the data packets, which are sent with a list of nodes through which they must go. In addition, each node keeps a cache of routes it has learned or overheard. As a result, intermediate nodes do not need to keep an up-to-date table of routes, thereby avoiding periodic route advertisements that cause considerable overhead. The problem with this approach concerns the high probability of stale routes in environments where high mobility as well as medium constraints may be normally present. That can happen, for instance, when a RREP message is in its way back to the sender but the replied route is no longer valid due to either an involved node that has moved away or a link that has somehow been interrupted. The problem is exacerbated by the fact that other nodes can overhear the invalid route reply and populate their buffers with stale route information. Therefore, unless stale routes can be detected and recovered in a fast way, TCP can be led to backoff state which considerably degrades its performance.

This problem has been studied in [Holland1999] and shown that it can be mitigated by either manipulating TCP to tolerate such a delay or by making the delay shorter so that the TCP can deal with them smoothly. In these studies, it has been observed that by disallowing route replies from caches can improve route accuracy at the expense of the routing performance in terms of overhead, since every new route discovery implies flooding the network. On the other hand, such an additional overhead is outweighed by the accuracy in the route determination, mainly for high mobility conditions, resulting in an enhanced TCP performance.

TORA

As we saw in Chapter 2, TORA is also an on-demand based protocol but has pro-active features as well. TORA has been designed to be highly dynamic by establishing routes quickly and concentrating control messages within a small set of nodes close to the place where the topological change has occurred. It is accomplished by maintaining multiple routes between any possible peers. In consequence, most topological changes should entail no reaction at all concerning route discovery, as it only reacts when all routes to a specific destination are lost. As we have seen before, TORA makes use of directed acyclic graphs, where every node has a path to a given destination. In other words, all neighbors of a given sender have an alternative path to a given destination, which define multiple potential paths for every peer. The directed acyclic graph is established initially by each node advertising a query packet and receiving update packets when it first tries to discover a route. A new query is only necessary when no more routes are available for a given sender. This can happen as the invalid routes, caused by partition, are removed from the nodes by having the affected node send a clear packet.

From the TCP viewpoint, this protocol can also suffer from stale route problem similar to the DSR protocol. Nevertheless, as route discovery procedures are confined to situations less probable (no available path), such a drawback can be considered not too harmful to TCP. On the other hand, multiple path routing can indeed cause significant performance degradation. The problem occurs mainly because TORA does not prioritize shorter paths, which can yield considerable amount of out-of-sequence packets for the TCP receiver, triggering retransmission of packets. A typical situation could be to send an earlier packet through a longer path and then, due an instantaneous route problem, a new packet being forced through a shorter enough path to arrive first at the destination. Therefore, it should be interesting to have a self-adaptive mechanism to avoid this possibility. Later in this chapter, we discuss this further.

In conclusion, the characteristics presented here in regards with DSR and TORA reveal that the design of a routing protocol should take into consideration its impacts on the upper layer, especially on the widely used TCP protocol. Once more, solutions can be placed in both layers and cooperation between them (cross-layer design) may be extremely advantageous [Cordeiro2002].

9.3.2.3 Path Asymmetry Impact

As we know, TCP relies on time sensitive feedback information to perform its flow control and asymmetrical paths can seriously compromise its performance. In other words, if TCP does not receive timely ACKs, it cannot expand its CWND to make full use of the available channel capacity, thereby wasting the bandwidth. Hence, in case the *forward path* characteristics are considerably different from the ones of the *backward path*, TCP will quite likely face performance problems. In MANETs where the topology as well as the environment conditions can change quite frequently and unpredictably, asymmetry can occur by different reasons, including lower layer strategies. Based on the work presented in [Balakrishnan2001], asymmetry in a TCP-based wireless mobile MANET can be categorized into the following classes:

- **Loss rate asymmetry:** This sort of asymmetry takes place when the backward path is significantly more error prone than the forward path. In ad hoc environment, this can be a serious factor as all links involved are wireless having high error rate which depends on local constraints that can vary from place to place and due to mobility patterns as well;
- **Bandwidth asymmetry:** This is the classical asymmetry found in satellite networks in which forward and backward data follow to be distinct paths with different speeds. In MANETs this can happen as well, since all nodes need not have the same interface speed. So,

even if a common path is used in both directions of a given flow, not necessarily they have the same bandwidth. Besides, as the routing protocols can assign different paths for forward and backward traffics [Cordeiro2002], asymmetry can definitely occur in MANETs;

- **Media access asymmetry:** This type of asymmetry may occur due to characteristics of the wireless shared medium. As explained before, in this kind of network, TCP ACKs may have to contend for the medium along with TCP data, and this may cause excessive delay as well as drops of TCP ACK packets;

- **Route asymmetry:** Unlike the previous three forms of asymmetry where forward and backward path can be the same, route asymmetry implies in distinct paths in both directions [Cordeiro2002]. Route asymmetry is associated with the possibility of different transmission ranges for the nodes. In fact, the transmission range of each node depends on its instantaneous battery power level that, in most cases, is likely to vary over time. The inconvenience with different transmission ranges is that it can lead to conditions in which the forward data follow a considerably shorter path than the backward data (TCP ACK) due to lack of power in one (or more) of the nodes in the backward path. Whenever a given node has low power to transmit, instead of directly communicating with the destination it has to communicate in a multi-hop fashion. However, as we discussed earlier, multi-hop paths are prone to have low throughput (see Figure 9.8). Consequently, TCP ACKs may face considerable disruptions. Finally, mobility and variations in the battery power level make the problem even worse as they may cause frequent route changes.

All the above forms of asymmetry can lead to lack of ACKs for the sender node, thereby impairing the forward throughput. The problem might be exacerbated when the ACKs arrive bunched up at the sender, causing bursty traffic in the forward path, which is known as *ACK compression* phenomenon. Additionally, asymmetry condition can lead to inaccuracy in RTT estimation. As stated in [Balakrishnan2001], potential solutions should either mitigate the low ACK flow problem or deal with it in another effective manner. Some proposed solutions are discussed next.

Based on the fact that TCP header field are frequently left unchanged in a stream of packets, *TCP header compression* [Jacobson1990b] has been proposed to reduce the size of ACK packets in the backward path. However, this option alone can be ineffective [Balakrishnan2001, as the IEEE 802.11 presents considerable overhead which impairs the enhancement provided by such a compression. Furthermore, despite being compressed, the packets still have to interact with other backward traffics.

ACK filtering is another approach that attempts to minimize the amount of ACK in the backward path. This scheme takes advantage of the fact that ACK packets are cumulative. Thus, when a new ACK is to be enqueued, the receiver node first verifies whether there are outstanding ACKs to be sent. If so, then only the latest ACK is sent while the others are discarded. A drawback with this scheme is that state information needs to be maintained for the connections with enqueued packets. *ACK congestion control* and *ACK-first scheduling* schemes [Balakrishnan2001] extend congestion control mechanism for ACK packets and give priority for them over data packets, respectively. The former relies on a random early detection (RED) mechanism [REDwww] for detecting congestion in the backward path, which is signaled back to the sender that, in turn, slows down its transmission rate. The latter assumes that ACKs must be scheduled with high priority so that the sender does not starve.

The schemes presented above attempt to avoid lack of ACKs at the sender, but they do not guarantee that the sender will not starve of ACK. Thus, in order to mitigate the effects of reduced ACK feedbacks even when some of the above schemes are in place, [Balakrishnan2001] proposed the *TCP sender adaptation* and *ACK reconstruction* techniques. The former relies on the idea that the sender node should avoid slowing down in its CWND by considering the amount of data acknowledged by each ACK, instead of the number of ACKs themselves, to trigger CWND increases. In addition, the sender should estimate the admitted rate of the connection so that it never sends excessive bursts into the network. On the other hand, *ACK reconstruction* attempts to avoid standard TCP senders from being affected by reduced ACK frequency. This technique encompasses a soft-state agent (at the sender) called *ACK reconstructor* which receives the spaced ACKs from the receiver and paces its relaying to the sender in a regulated rate. This rate is based on the amount of data acknowledged by the received ACK and also on the actual rate of the backward path. As a result, the CWND growing at the sender is controlled by the actual rate in the backward path and no burst behavior arises.

As mentioned above, asymmetry can also induce inaccuracy on RTT estimation. As we have seen before, TCP estimates its RTT based on measurements of the delay suffered by a packet in both direction of the flow. Therefore, it can so happen that such estimation does not suit the actual forward path necessity, which might lead TCP to retransmit either prematurely or too late. In [Parsa1999], *TCP Santa Cruz* is proposed as a solution for decoupling CWND growth from the number of ACKs received. This scheme relies on measurements of relative delay that one packet experiences in relation to the previous packet, rather than on measurements of absolute delay for sampled packets as standard TCP does. This way, it pursues not only to provide enhanced RTT estimation but also to be resilient to ACK losses. Thus, under realistic assumption that MANETs will experience asymmetrical paths, the above mechanisms as well as other related ones [Oliveira2005] should be carefully investigated in the context of this challenging environment.

9.4 Solutions for TCP over Ad Hoc

We now present the most prominent schemes that have been specifically proposed to overcome TCP performance problems in MANETs. Here, we classify the proposed solutions into *Mobility-related* and *Fairness-related* based on the key TCP issue they aim to overcome. The mobility-related approaches address the TCP problems resulting from node mobility which may mistakenly trigger TCP congestion control mechanisms. On the other hand, fairness-related solutions tackle the serious unfairness conditions raised when TCP is run over MANETs.

9.4.1 Mobility-Related

Notably, most of the solutions in this category have limitations which can compromise their widespread deployment. Nevertheless, it is of paramount importance to understand them as they may serve as the basis for future research. In the following we discuss the details of each one of them.

9.4.1.1 TCP-Feedback

As the name suggests, TCP-Feedback (TCP-F) [CRVP97] is a feedback-based scheme in which the TCP sender can effectively distinguish between route failure and network congestion by receiving Route Failure Notification (RFN) messages from intermediate nodes. The idea is to push the TCP into a "snooze state" whenever such messages are received. In this state, TCP stops sending packets and freezes all its variables such as timers and CWND size, which makes sense once there is no available route to the destination. Upon receipt of a Route Re-establishment Notification (RRN) message from the routing protocol, indicating that there is again an available path to the destination, the sender leaves the frozen state and resumes transmission using the same variables values prior to the interruption. In addition, a *route failure timer* is employed to prevent infinite wait for RRN messages, and is started whenever a RFN is received. Upon expiration of this timer, the frozen timers of TCP are reset hence allowing the TCP congestion control to be invoked normally.

Results from TCP-F shows gains over standard TCP in conditions where the route reestablishment delay are high, which are due to a fewer number of involved retransmission. Nevertheless, the simulation scenario employed to evaluate TCP-F has been quite simplified and so the results might not be a true representative. For example, the RFN and RRN messages employed in TCP-F are to be carried by the routing protocol, but no such protocol has been considered.

9.4.1.2 The ELFN Approach

The Explicit Link Failure Notification (ELFN) [Holland1999] is a cross-layer proposal in which TCP also interacts with the routing protocol in order to detect route failure and take appropriate actions. Here, ELFN messages are sent back to the TCP sender from the node detecting the failure. Such messages are carried by the routing protocol that needs to be adapted for this purpose. In fact, the DSR's route error message has been modified to carry a payload similar to the "host unreachable" message of the Internet Control Message Protocol (ICMP) [Tanenbaum1996]. Basically, the ELFN messages contain sender and receiver addresses and ports, as well as the TCP sequence number. This way, the modified TCP is able to distinguish losses caused by congestion from the ones due to mobility. In ELFN, whenever the TCP sender receives an ELFN message it enters a "stand-by" mode in which its timers are disabled and probe packets are sent regularly towards the destination in order to detect route restoration. Upon receiving an ACK packet, the sender leaves the "stand-by" mode and resumes transmission using its previous timer values and state variables.

This scheme was only evaluated for the DSR routing protocol where the stale route problem was found to be crucial for the performance of ELFN. Additionally, the interval between transmission of probe packets and the choice of type of packet to be sent as a probe have also been evaluated. In essence, it has been suggested that a varying interval based on RTTs values could perform better than the fixed probe interval. In general, the ELFN approach provides meaningful enhancements over the standard TCP, but further evaluation may be needed. For instance, different routing protocols should be studied, and the performance of ELFN under congestion conditions should be considered. Last, but not the least, more appropriate values for the probe interval should be determined.

9.4.1.3 Fixed RTO

The fixed RTO scheme [Dyer2001] relies on the idea that routing error recovery should be accomplished in a fast fashion by the routing algorithm. As a result, any disconnection should be treated as a transitory period which does not justify the regular exponential backoff mechanism of TCP being invoked, as this can cause unnecessarily long recovery delays. Thus, it disables such a mechanism whenever two successive retransmissions due to timeout occur, assuming that it actually indicates route failure. By doing so, it allows the TCP sender to retransmit at regular intervals instead of at increasingly exponential ones. In fact, the TCP sender doubles the RTO once and if the missing packet does not arrive before the second RTO expires, the packet is retransmitted again and again but the RTO is no longer increased. It remains fixed until the route is recovered and the retransmitted packet is acknowledged.

The fixed RTO approach has been evaluated in [Dyer2001] considering different routing protocols along with TCP selective and delayed acknowledgements options. Sizeable enhancements have been accomplished with on-demand routing protocols, but only marginal improvements have been noticed when using the TCP options. Nevertheless, this proposal is limited to wireless networks only, which makes it somewhat discouraging as interoperation with wired networks is a mandatory requirement in the vast majority of applications.

9.4.1.4 The ATCP Protocol

Different from previously discussed approaches, the Ad hoc TCP (ATCP) protocol [Liu2001] does not impose changes to the standard TCP itself. Rather, it implements an intermediate layer between the network and the transport layers in order to provide an enhanced performance to TCP and still maintain interoperation with non-ATCP nodes.

More specifically, ATCP relies on the ICMP protocol and on the Explicit Congestion Notification (ECN) [ECNwww] scheme to detect/distinguish network partition and congestion, respectively. This way, the intermediate layer keeps track of the packets to and from the transport layer so that the TCP congestion control is not invoked when it is not really needed, which is done as follows. Whenever three duplicate ACKs are detected, indicating a lossy channel, ATCP puts TCP in "persistent mode" and quickly retransmits the lost packet from the TCP buffer; after receiving the next ACK, the normal state is resumed. In case an ICMP "Destination Unreachable" message arrives, pointing out a network partition, ATCP also puts the TCP in "persistent mode" which only ends when the connection is reestablished. Finally, when network congestion is detected by the receipt of an ECN message, the ATCP does nothing but forwards the packet to TCP so that it can invoke its normal congestion control mechanism.

ATCP was implemented in a testbed and evaluated under different constraints such as congestion, lossy scenario, partition, and packet reordering. In all cases, the transfer time of a given file by ATCP yielded better performance as compared to TCP. However, the scenario employed has been somewhat special as neither wireless links nor ad hoc routing protocols have been considered. In fact, such experiments relied on simple Ethernet networks connected in series, in which each node had two Ethernet cards. Moreover, some assumptions such as ECN-capable nodes as well as the sender node being always reachable might be somehow hard to be met in reality. In case if the latter is not fulfilled, for example, the ICMP message might not even reach the sender which would retransmit continuously instead of entering the "persist mode". Also, the deployment of the ECN scheme is known to raise many security concerns [ECNwww], and it might compromise the viability of ATCP.

9.4.1.5 TCP-DOOR

Due to its dynamic environment, mobility in MANETs is extremely frequent. Therefore, a natural effect of mobility is that the packet usually arrive out-of-order (OOO) at the destination. If the OOO delivery event is appropriately monitored, it might be just enough to detect link failure inside the network and, hence, be able to effectively distinguish between mobility and congestion. The TCP-DOOR (Detection of Out-of-Order and Response) [Wang2002] protocol focuses on the idea that OOO delivery of packets can happen frequently in MANETs as a result of nodes mobility. TCP-DOOR imposes changes to TCP code but does not require intermediate nodes to cooperate, which represents its main differentiation from all previously described proposals. In this way, TCP-DOOR detects OOO events and responds accordingly as explained below.

Based on the fact that not only data packets but also ACK packets can experience OOO deliveries, TCP-DOOR implements a detection of such deliveries at both entities: TCP sender and TCP receiver. For this, additional ordering information is used in both types of packets (data and ACK) which are conveyed as TCP options, where one extra byte is required for ACKs and two extra bytes are required for data. Thus, for every packet sent the sender increments its *own stream sequence number* inside the two-byte option regardless whether it is a retransmission or not (standard TCP does not increment sequence number of retransmitted packets). This allows the receiver to precisely detect OOO delivery of data packets and notify the sender via a specific bit into the return ACK packet. Additionally, because all ACKs associated with a given missing data packet have identical contents, the receiver increments its *own ACK stream sequence number* inside the one-byte option for every retransmitted ACK, so that the sender can distinguish the exact order of every (retransmitted or not) packet sent. Therefore, the explained mechanism provides the sender with reliable information about the order of the packet stream in both directions, allowing the TCP sender to act accordingly.

After detecting OOO events, the TCP sender can respond with two mechanisms: *temporarily disabling congestion control* and *instant recovery during congestion avoidance*. In the former, the TCP sender keeps its state variables constant for a while after the OOO detection. The rationale behind this is that such condition might be short term (route change) not justifying the invocation of the congestion avoidance mechanism. In the latter, whenever an OOO condition is detected the TCP sender checks if the congestion control mechanism has been invoked in the recent past. If so, the connection state prior to the congestion control invocation is restored, as such an invocation may have been caused by temporary disruption instead of by congestion itself.

Different scenarios combining all the aforementioned mechanisms have been simulated in [Wang2002]. The effects of the route cache property of the DSR routing protocol on TCP-DOOR performance have also been considered. The results indicate that only sender detection mechanism (ACK OOO detection) should suffice. Both responses mechanisms are important and *instant recovery during congestion avoidance* performs better than *temporarily disabling congestion control*. In addition, the DSR route cache impaired the performance improvement mainly due to stale caches. In general, TCP-DOOR improves TCP performance by an average of 50%, while other protocols such as ATCP report from 200% to 300% improvement (contrary to other solutions, however, TCP-DOOR confines the changes to the TCP protocol only).

On the other hand, the assumption in TCP-DOOR that OOO packets are the exclusive result of route disturbance deserves much more careful analysis. Multipath routing algorithms (e.g., TORA) can induce OOO packets that are not necessarily related to route failures. Besides, as we have seen before, diverse factors can cause path asymmetry inducing OOO events as well. Obviously, the independence from intermediate nodes makes TCP-DOOR quite attractive which calls for further developments towards a more general approach.

9.4.1.6 Discussion

The main drawbacks of the proposed schemes are as follows. The approaches that rely on feedback information from inside the network (TCP-F, ELFN-based, ATCP) may fail in situations where TCP sender is unable to receive data from the next hop node (e.g., due to mobility). In such cases, the TCP sender would retransmit continuously instead of entering a frozen state. Furthermore, the usage of explicit notification by the intermediate nodes, such as ECN, raises many security concerns. The fixed RTO scheme, on the other hand, does not seem to be appropriate for possible future interoperation with wired networks. Finally, the assumption in TCP-DOOR that OOO packets are exclusive results of route disturbance may not be true in a quite a few scenarios.

In fact, the main concern addressed by the approaches presented so far is how to avoid the TCP exponential backoff mechanism when losses take place by factors other than congestion. However, as discussed in the previous section, other factors such as path asymmetry and fine tuning with lower layers, among others, should also be considered by an effective design. Moreover, challenging approaches should also address prominent issues such as power management, interoperation with wired networks (e.g., Internet), security, and so on. Thus, it is noticeable that the proposed approaches are somewhat limited, highlighting the necessity for further investigation in this area.

9.4.2 Fairness-Related

Compared to the number of mobility-related studies, little has been done to address the serious unfairness conditions raised by TCP over MANETs. In this section we present some important solutions in this category. Other related studies can be found in [Bottigliengo2004].

9.4.2.1 COPAS

As we have seen before, the problem of capture is severe due to the interplay of the MAC layer and TCP backoff policies and results in a single node within its radio range being able to access the medium at all times, while others in its neighborhood starve. To the same extent as mobility related issues, the capture problem drastically affects TCP performance and is stressed in wireless MAC protocols that employ exponential backoff schemes such as IEEE 802.11 and FAMA (Floor Acquisition Multiple Access) [Fullmer1995] as their backoff mechanisms always favor the last successful station. A protocol called COPAS (COntention-based PAth Selection) has been proposed in [Cordeiro2002] to address TCP performance drop due to the capture problem and resulting unfairness. COPAS implements two novel routing techniques in order to contention-balance the network, namely, the use of disjoint forward (for TCP data) and reverse (for TCP ACK) paths to reduce the conflicts between TCP packets traveling in opposite directions, as well as a dynamic contention-balancing technique that continuously monitors network contention and selects routes with minimum MAC layer contention.

COPAS works as follows. In on-demand protocols, a route discovery process is initiated when a route to a destination is needed and none is available. The source floods the network with a RREQ packet to discover a route to the destination. When the destination receives the RREQ, it responds with a unicast RREP packet back to the source. In COPAS, upon receipt of a non-duplicate RREQ packet, to this packet nodes append a weighted average of the number of times it has backed off in a "recent past" due to activity in the medium. The RREQ packet is then re-broadcasted. By keeping track of the recent average number of times backed off, COPAS determines how busy the wireless shared medium is. More times a node backs off, means that more busy is the medium around it. This provides precise information on the contention experienced and the destination waits for an appropriate amount of time to learn all possible routes. The destination node accepts duplicate RREQ received from different previous nodes. When the RREQ collection timer expires, COPAS employs two selection criteria in order to choose exactly two routes: path disjointness, and least contented routes.

Disjoint path routing has been explored before in both DSR [Nasipuri2001] and AODV [Marina2001] routing protocols. COPAS uses similar techniques to choose all possible node-disjoint routes (between source and destination) at the destination and selects two least contented routes based on the information collected by the arriving RREQ packets. Least contented routes are computed by evaluating the sum of the contentions experienced by the RREQ packets on each node-disjoint route and then minimizing the sum over all disjoint routes available. Ties are resolved by favoring lower route lengths (in hops) and then by the arrival order of the RREQ packets. In the absence of disjoint paths, COPAS behaves similarly to existing routing protocols with the difference that it can take advantage of network contention information.

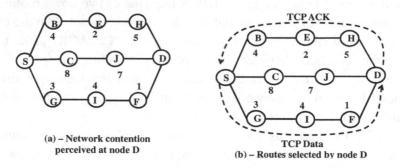

(a) – Network contention
perceived at node D

TCP ACK

TCP Data
(b) – Routes selected by node D

Figure 9.9 – Route establishment in COPAS [Taken from Cordeiro2002]

The destination responds with at most two RREPs along the chosen paths as the destination also sets a direction flag in the packet header to indicate to the source node which path is to be used as forward (for TCP data packets) and reverse (for TCP ACK packets) traffic. This direction information is also kept in a node's routing table.

To illustrate this, consider the scenario of Figure 9.9(a) wherein the source node S sends a RREQ packet towards the destination node D.

In this case and with the contention values as depicted in the Figure 9.9(a), the destination first applies the disjointness path rule and finds out routes i = <S-B-E-H-D>, j = < S-C-J-D >, and k = < S-G-I-F-D> to be disjoint. Next, it applies the minimum contention sum rule and end up selecting routes i and k to be used as reverse (for TCP ACK) and forward (for TCP data) paths respectively, as showed in Figure 9.9(b).

Employing disjoint forward and reverse paths is also desirable for robustness reasons. Capture conditions can be so severe that links appear to be broken even when there is no mobility. Therefore, to guarantee continuous operation even in link breakage situations COPAS makes use of previously established forward and reverse routes. In a capture scenario, it is usually the MAC layer which reports to the network layer the link breakage since it is in this layer where the capture problem is rooted. When a route is disconnected, the immediate upstream node of the broken link sends a RERR message to the source of the route notifying the route invalidation. Nodes along the path to the source remove the route entry upon receiving this message and relay it to the source.

In traditional on-demand routing protocols, the source reconstructs a new route by flooding a RREQ when informed of a route disconnection. In COPAS – in addition to flooding a RREQ to reconstruct the broken route, TCP packets are redirected using the second alternate path when available, hence providing uninterrupted communication. It is up to TCP to recover from potential lost packets due to link breakage, while COPAS attempts to minimize the route disruption by rerouting data packets. In this case, COPAS behaves similar to existing approaches. COPAS also includes provisions for dynamic contention-balancing. Traffic pattern across the network changes a lot with time and space. Therefore, routes that were optimal during the initial route construction process may no longer be good paths as contention might have increased with the new traffic pattern. Therefore, COPAS implements a scheme to dynamically monitor and change routes between any <source, destination> pair that have their contention increased noticeably.

Recent research either evaluates a single TCP session [Holland1999, Wang2002], or when multiple TCP sessions are considered the network is fully mobile [Dyer2001], or the connections mostly cover one hop employing unrealistic topologies such as ring and string [Gerla1999, Xu2002]. However, in [Cordeiro2002] simulations are performed where it is considered multiple TCP connections under several scenarios, and where the network comprised of only static hosts. This is the worst case scenario where capture conditions are mostly severe since nodes remain within radio range of each other continuously, and where multiple TCP flows compete to have access to the shared medium. Nevertheless, COPAS could cooperate with any of the other proposed schemes in [Chandran2001, Dyer2001] as it tackles TCP degradation in static to low mobility network while they cope up with mobility related issues.

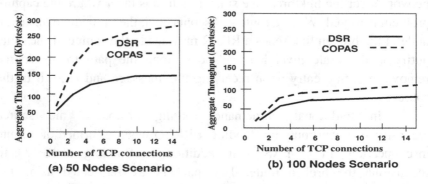

Figure 9.10 – Average aggregate throughput [Taken from Cordeiro2002]

COPAS has been evaluated and compared with the DSR routing protocol. Figures 9.10(a) and 9.10(b) show some simulation results of COPAS applied to scenario of 50 and 100 nodes, respectively, where TCP connections range from 1 to 15. As we can see, for the 50-nodes scenario, COPAS is shown to drastically improve TCP throughput by up to 90%, whereas in the 100-nodes scenario, COPAS still achieves a considerable improvement but it is not as sizeable as in the 50-nodes scenario due to the large number of routes from any given source and destination which reduces conflicts among TCP connections. It has also been observed that nodes running COPAS experience much less medium contention due to the dynamic contention-balancing mechanism, while keeping the routing overhead low. However, there still issues that need to be addressed including how the protocol can handle unidirectional links, and the interrelationship between TCP and UDP traffic.

9.4.2.2 Neighborhood RED

In [Xu2003], a scheme called Neighborhood RED (NRED) is proposed, where it is claimed that two unique features of ad hoc wireless networks are the key to understand unfair TCP behaviors. One is the spatial reuse constraint, and the other is the location dependency. The former implies that space is also a kind of shared resource. TCP flows, which do not even traverse common nodes, may still compete for "shared space" and hence interfere with each other. The later, location dependency, triggers many of the problems we have mentioned discussed so far (channel capture, hidden and exposed terminals, and so on), which are often recognized as the primary reasons for TCP unfairness. Clearly, TCP flows with different relative positions in the bottleneck may get different perception of the bottleneck situation in terms of packet delay and packet loss rate. Since obtaining correct feedback information of the bottleneck is critical to the fairness of TCP congestion control, limited information of the bottleneck situation causes significant unfairness.

If we view a node and its interfering neighbors to form a neighborhood (the neighborhood of a node X is formed by all nodes within communication range of X), the local queues at these nodes can be considered to form a distributed queue for this neighborhood (for instance, the neighborhood of node A and its distributed queue in Figure 9.11). Obviously, this distributed queue is not a FIFO queue. Flows sharing this queue have different and dynamic priorities determined by the topology and traffic patterns due to channel capture, hidden and exposed terminal situations, and so on. Therefore, they get different feedback in terms of packet loss rate and packet delay when congestion happens. The uneven feedback makes TCP congestion control diverge from the fair share. Similar situations may occur in wired networks when a buffer is full and drop tail queue management scheme is used. In these wired networks, the RED scheme has been shown to improve TCP fairness under such situations by keeping the queue size relatively small and dropping or marking packets roughly proportional to the bandwidth share of each flow through the gateway.

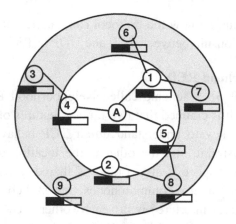

Figure 9.11 – Node A's neighborhood and its distributed queue

The idea of the NRED scheme is to extend the original RED mechanism to operate on the distributed neighborhood queue. Similar to RED, each node employing NRED keeps estimating the size of its neighborhood queue. Once the queue size exceeds a certain threshold, a drop probability is computed by using the algorithm from the original RED scheme. Since a neighborhood queue is the aggregate of local queues at neighboring nodes, this drop probability is then propagated to neighboring nodes for cooperative packet drops. Each neighbor node computes its local drop probability based on its channel bandwidth usage and drops packets accordingly. The overall drop probability will realize the calculated drop probability on the whole neighborhood queue. Thus, the NRED scheme is basically a distributed RED suitable for ad hoc wireless networks.

In [Xu2003], NRED is mostly evaluated under relatively long-lived TCP flows such as FTP connections transferring a medium or large size file. This is because TCP unfairness issues are more serious to such TCP traffic. If a TCP connection finishes its transfer in seconds, NRED may not have enough time to detect the network congestion and perform proper actions. Therefore, NRED may not be suitable for all types of TCP connections, even though short-lived TCP flows may not hurt other flows too much as they tend to quickly end.

The main achievement of NRED is the ability to detect early congestion and drop packets proportionally to a flow's channel bandwidth utilization. By doing this, the NRED scheme is able to improve the TCP fairness. Finally, one major contribution of NRED is the design of a network layer solution that does not require any MAC modification.

9.5 Conclusions and Future Directions

According to what has been discussed in the preceding sections, it is clear that several issues remain to be addressed toward a complete, robust and efficient approach for TCP over MANETs. In reality, new paradigms may have to be devised and we summarize the main points to be addressed as well as some potential future directions in this area. The error-detection strategies can be classified as *network detection* and *end node detection*. For example, TCP-Feedback, ELFN-based, ATCP and NRED approaches rely on network signaling for detecting and tackling path anomalies. On the other hand, DOOR, Fixed RTO and COPAS schemes count on end node detection. Each approach has its advantages and disadvantages, and there should be efforts to combine advantages of each one and a hybrid approach could be based on tradeoffs.

In addition, we have seen that the interactions between TCP and MAC protocols could be improved by using either using smaller values for the maximum TCP window size or larger MAC IFS intervals, respectively. In addition, it might be useful to investigate the possibility of increasing the maximum number of possible retransmissions at the MAC layer as an attempt to increase the probability of success of the local retransmission scheme. However, an increase may trigger TCP timeouts and cause longer delays. In fact, approaches to improve MAC layer performance under TCP represent a very broad open area.

In terms of routing protocols effects on TCP, we saw that route caching and multipath routing strategies can significantly degrade TCP performance. To date, evaluations of schemes based on route caching have revealed that the overhead of discovering new routes at every enquiry is outweighed by the accuracy of the new routes achieved. Therefore, it seems that such an approach (as originally proposed) is not viable from TCP point of view.

With regards to multipath routing strategies, further evaluation in improving TCP support is needed. As we have discussed in previous chapters, multipath is interesting as it makes it possible for efficient bandwidth utilization by using several routes in parallel. Additionally, it may be quite useful in providing routing protocol robustness by establishing redundant paths between the sender and the receiver. Thus, improvements on either network or transport layers in order to avoid TCP disruption due to out-of-order packets are desirable. In principle, it should be possible to have a self-adaptive mechanism as to avoid out-of-order packets effects at the receiver. For instance, the TCP receiver could delay the response for an out-of-order packet hoping that an in-order packet arrives, similarly to what is suggested in [Wang2002].

As for path asymmetry, its impact on TCP depends primarily on the routing protocol strategy in place. As this issue has not been appropriately investigated in MANETs, quantitative analysis need to be carried out in order to understand to what extent it affects TCP. This will help us determine what specific mechanisms need to be considered.

Power management is a very important topic within MANETs as they are supposed to be composed mostly of battery powered devices. Thus, power aware approaches should be followed as little has been done with respect to TCP. For example, TCP retransmits continuously regardless weather the destination is reachable or not, and this may cause considerable energy wastage. At first, TCP could be enhanced by stopping transmissions when no connectivity to the next hop is available. This could be possible, for instance, by providing TCP with MAC layer information similar to [Cordeiro2002]. Selective acknowledgement, like TCP SACK [Mathias1996], might also be useful in avoiding unnecessary retransmissions upon a link recovery. Last, but not the least, some options available in recent TCP versions could also be adopted for MANETs in order to meet this requirement [Agrawal2001].

Interoperation between wireless mobile MANETs and wired networks is another subject that has not been adequately addressed from TCP perspective. However, there is a high probability that MANETs will need to communicate with the fixed world (e.g., Internet) to take advantage of the vast number of existing services. Consequently, effective approaches have to take this issue into consideration as well.

In this case, other concepts such as base stations, Mobile IP, interoperation of routing protocols, amongst others, will have to be dealt with for this challenging environment. In particular, the multi-hop scheme in the wireless side poses new tough challenges for TCP congestion control which will need to distinguish between losses in both sides of the base station so as to take proper actions. In addition, *scalability* issues will also have to be considered.

Security considerations have become nowadays a hot issue in wireless environments as wireless mediums are much more susceptible to malicious attackers than the wired ones. Thus, many studies have been carried out on this area as it will be discussed in Chapter10. In this case, however, approaches that rely on explicit feedback from intermediate nodes can face problems as no direct access to the IP header is allowed for such nodes. Therefore, alternative ways of detecting congestion in the network may need to be designed for MANETs. In this sense, the congestion control mechanism of TCP Vegas appears to be more promising as it probes the network to detect congestion and adjusts its CWND accordingly.

We have now reached a point where we have covered most of the issues related to MANETs up to the transport layer of the network protocol stack. Therefore, in the next chapter we delve into an exciting new application area of Wireless Sensor Networks.

Homework Questions/Simulation Projects

Q. 1. TCP has become standard transport protocol for computer communication. This allows slow start increase of transmission rate when doing cold start and then adjust rate when a threshold is crossed. Why do you have several variations of TCP and what are their relative advantages and disadvantages? Are any of these variations suited for wireless MANETs? How does the hidden terminal problem affect TCP over multihop MANETs?

What will be the impact on the performance when the congestion window is changed by:

a) Linear decrease, and

b) Geometric decrease.

Do you have a feel for relative traffic conditions and the size of the network under which these schemes may be useful?

Q. 2. Design a problem based on any of the material covered in this chapter (or in references contained therein) and solve it diligently.

References

[Agrawal2001] S. Agrawal and S. Singh, "An Experimental Study of TCP's Energy Consumption over a Wireless Link," In Proceedings of the 4th European Personal Mobile Communication, February 2001.

[Allman1999] M. Allman, V. Paxson, and W. Stevens, "TCP Congestion Control," IETF RFC 2581, April 1999.

[Balakrishnan1997] H. Balakrishnan, V. Padmanabhan, S. Seshan, and R. Katz, "A Comparison of Mechanisms for Improving TCP Performance over Wireless Links," In IEEE/ACM Transactions on Networking, December 1997.

[Balakrishnan2001] H. Balakrishnan and V. N. Padmanabhan, "How Network Asymmetry affects TCP," In IEEE Communication Magazine, pp. 2-9, April 2001.

[Bottigliengo2004] M. Bottigliengo, C. Casetti, C.-F. Chiasserini, M. Meo, "Short-term Fairness for TCP Flows in 802.11b WLANs," In Proceedings of the IEEE Infocom, 2004

[Caceres1994] E. Caceres and L. Iftode, "Effects of Mobility on Reliable Transport Protocols," In Proceedings of the 4th International Conference on Distributed Computer Systems, June 1994.

[Chandran2001] K. Chandran, S. Raghunathan, S. Venkatesan, and R. Prakash, "A feedback-based scheme for improving TCP performance in ad hoc wireless networks," IEEE Personal Communications Magazine, Vol. 8, No. 1, February 2001.

[Clark1988] D. Clark, "The Design Philosophy of the DARPA Internet Protocols," In Proceedings SIGCOMM Computer Communications Review, Vol. 18, No. 4, August 1988.

[Comer1995] D. Comer, Internetworking with TCP/IP, Volume I: Principles, Protocols and Architecture, Prentice Hall, ISBN: 0-13-216987-8, 1995.

[Cordeiro2002] C. Cordeiro, S. Das, and D. Agrawal, "COPAS: Dynamic Contention-Balancing to Enhance the Performance of TCP over Multi-hop Wireless Networks," in IEEE IC3N, Miami, USA, October 2002.

[CRVP97]

[Dyer2001] T. Dyer and R. Boppana, "A comparison of TCP performance over three routing protocols for mobile MANETs," Proceedings of ACM MobiHoc, October 2001.

[ECNwww] ECN (Explicit Congestion Notification) in TCP/IP, http://www.icir.org/floyd/ecn.html.

[Fu2005] Z. Fu, H. Luo, P. Zerfos, S. Lu, L. Zhang, and M. Gerla, "The Impact of Multihop Wireless Channel on TCP Performance," in IEEE Transactions on Mobile Computing, Vol. 4, No. 2, March/April 2005.

[Fullmer1995] C. Fullmer and J.J. Garcia-Luna-Aceves, "Floor Acquisition Multiple Access (FAMA) for packet radio networks," Computer Communication Review, October 1995.

[Gerla1999] M. Gerla, K. Tang, and R. Bagrodia, "TCP Performance in Wireless Multi-hop Networks," in IEEE WMCSA, February 1999.

[Holland1999] G. Holland and N. Vaidya, "Analysis of TCP performance over mobile MANETs," Proceedings of ACM Mobicom, August 1999.

[IETFwww] The Internet Engineering Task Force (IETF), http://www.ietf.org.

[Jacobson 1990b] V. Jacobson, "Compressing TCP/IP Header for Low-Speed Serial Links," IETF RFC 1144, February 1990.

[Jacobson1988] V. Jacobson, "Congestion Avoidance and Control," Computer Communications Review, Vol. 18, No. 4, pp. 314-329, August 1988.

[Jacobson1990a] V. Jacobson, "Modified TCP Congestion Control Avoidance Algorithm," end-2-end-interest mailing list, April 30, 1990.

[Johnson2001] D. Johnson, D. Maltz, Y.-C. Hu, and J. Jetcheva, "The dynamic source routing protocol for mobile MANETs (DSR)," IETF Internet-Draft, November 2001 (work in progress).

[Kristoff2000] J. Kristoff, TCP Congestion Control, March 2000.

[Lin1998] D. Lin and H. Kung, "TCP Fast Recovery Strategies: Analysis and Improvements," in IEEE Infocom, 1998.

[Liu2001] J. Liu and S. Singh, "ATCP: TCP for Mobile MANETs," IEEE Journal on Selected Areas in Communications, Vol. 19, No. 7, July 2001.

[Marina2001] M. Marina and S. Das, "On-demand Multipath Distance Vector Routing in MANETs," in Proceedings of the International Conference for Network Protocols (ICNP), November 2001.

[Mathias1996] M. Mathias, J. Mahdavi, S. Floyd, and A. Romanow, "TCP Selective Acknowledgement Options," IETF RFC 2018, October 1996.

[MICS] Mobile Information and Communications Systems (MICS) Project, http://www.terminodes.com.

[Nasipuri2001] A. Nasipuri, R. Casteneda, and S. Das, "Performance of Multipath Routing for On-Demand Protocols in Mobile MANETs," in ACM/Kluwer Mobile Networks and Applications (MONET) Journal, Vol. 6, No. 4, 2001.

[Oliveira2002] R. Oliveira and T. Braun, "TCP in wireless mobile MANETs," Technical Report IAM-02-003, University of Berne, Switzerland, July 2002.

[Oliveira2005] R. Oliveira and T. Braun, "A Dynamic Adaptive Acknowledgment Strategy for TCP over Multihop Wireless Networks," IEEE Infocom, Miami, USA, 2005.

[Parsa1999] C. Parsa and G. Aceves, "Improving TCP Congestion Control over Internets with Heterogeneous Transmission media," IEEE Intl. Conference on Network Protocols (ICNP), October 1999.

[Perkins2001] C. Perkins, E. Royer, and S. Das, "Ad Hoc on Demand Distance Vector Routing (AODV)," IETF Internet Draft, March 2001 (work in Progress).

[Perkins2002] C. Perkins, J. Malinen, R. Wakikawa, A. Nilsson, and A. Tuominen, "Internet Connectivity for Mobile MANETs," in Wireless Communications and Mobile Computing, 2002.

[Poojary2001] N. Poojary, S. Krishnamurthy and S. Dao, "Medium Access Control in a Network of Ad Hoc Mobile Nodes with Heterogeneous Power Capabilities," In Proceedings of ICC, June 2001.

[Postel1981] J. Postel, "Transmission Control Protocol," IETF RFC 793, September 1981.

[Rappaport1999] T. Rappaport, Wireless Communications Principles and Practices, Prentice Hall PTR, July 1999.

[REDwww] RED (Random Early Detection), http://www.icir.org/floyd/red.html.

[Stevens1994] W. Stevens. TCP/IP Illustrated Volume 1. Addison Wesley, 1994.

[Tanenbaum1996] A. Tanenbaum, Computer Networks, Prentice Hall, ISBN 0-13-349945-6, 1996.

[Tsaoussidis2002] V. Tsaoussidis and I. Matta, "Open Issues on TCP for Mobile Computing," In the Journal of Wireless Communications and Mobile Computing, Wiley Academic Publishers, Issue 2, Vol. 2, February 2002.

[Wang2002] F. Wang and Y. Zhang, "Improving TCP Performance over Mobile Ad-Hoc Networks with Out-of-Order Detection and Response," In Proceeding of ACM Mobihoc, June 2002.

[Xu2001a] S. Xu, T. Saadawi and M. Lee, "On TCP over Wireless Multi-hop Networks," In Proceedings of IEEE MILCOM, October 2001.

[Xu2001b] S. Xu and T. Saadawi, "Evaluation for TCP with Delayed ACK Option in Wireless multi-hop Networks," In Proceeding of Semiannual Vehicular Technology Conference (VTC Fall), October 2001.

[Xu2002] S. Xu and T. Saadawi, "Revealing the Problems with 802.11 MAC Protocol in Multi-hop Wireless MANETs," Journal of Computer Networks, Vol. 38, No. 4, March 2002.

[Xu2003] K. Xu, M. Gerla, L. Qi, and Y. Shu, "Enhancing TCP Fairness in Ad Hoc Wireless Networks Using Neighborhood RED," in ACM Mobicom, September 2003.

[Zhang2001] C. Zhang and V. Tsaoussidis, "TCP Real: Improving Real-time Capabilities of TCP over Heterogeneous Networks," In Proceedings of the 11th IEEE/ACM NOSSDAV, New York, 2001.

Chapter 10

Applications of Sensor Networks

10.1 Introduction

Wireless Sensor Networks (WSNs) were primarily introduced for defense application and one simple application is illustrated in Figure 10.1. The objective here is to monitor enemy's activities without any human intervention and using a large number of wireless sensor nodes (SNs). A low-flying airplane, an unmanned aerial or ground vehicle or a powerful laptop can act as a Base Station (BS) or a sink to collect information from all SNs. The BS broadcasts a query message to all SNs simultaneously as illustrated in Figure 10.2 while the response could include useful surrounding information and efforts were encouraged by the US Defense Advanced Research Projects Agency (DARPA).

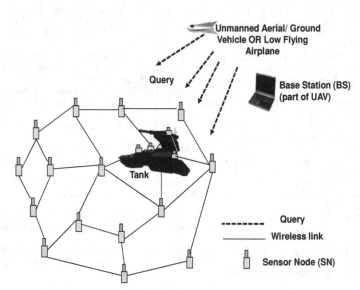

Figure 10.1 – Deployment of Sensors and Network Formation

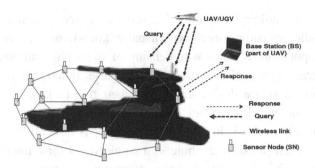

Figure 10.2 – WSN and Query Injection/Response

WSNs can be considered as a special case of ad hoc networks with reduced or no mobility, and are known as "data centric". This means, unlike traditional ad hoc networks where data is requested from a specific node or location, data is requested based on sensed attributes such as, "which area has temperature the maximum/minimum or over 100°C or 212°F". The use of a particular type of query might depend on the application requirements. Sometimes, the query may ask for multiple parameters such as temperature, pressure, humidity, etc., and may be required to sense and transmit the values only once, or over a period of time, or use past history to gain statistical information. Based on these, the query can be divided into three categories:

1. One time queries.
2. Persistent queries.
3. Historical queries.

In one time queries, the information about the sensed value is needed only one time, possibly the snapshot of the current values, while persistent queries implies data over an extended period of time, preferably at a regular interval. The historical query incorporates the data collected over a specified period of time.

In recent years, advances in miniaturization, low-power circuit design, improved low cost, and small-size batteries have made a new technological vision possible for numerous civilian applications [Akyildiz2002, Estrin1999, Jain2005, Kahn1999]. The size of a sensor is expected be a few cubic millimeters while the target price range may be less than one US dollar. The National Science Foundation (NSF) considers a wireless sensor network (WSN) as "the next generation of the information technology revolution" [nsf109]. A SN typically combines wireless radio transmitter-receiver (usually known as transceiver) and limited computation facilities with sensing of some physical phenomenon using different types of transducers. All these functional components together form a single device called a *sensor* due to lack of any better terminology or word capturing or encompassing the concept and is illustrated in Figure 10.3. In other words, sensor is basically a device that converts a sensed physical attribute into an electrical form understandable by a user. These could include temperature, pressure, velocity, acceleration, stress and strain, fatigue, tilt, light intensity, sound, humidity, gas-sensors, biological, pollution, impurity level detection, nuclear radiation, civil structural sensors, blood pressure, sugar level, white cell count, and many others. As there are so many physical quantities that could be monitored, different types of transducers are needed and we first discuss possible use of sensors in various application areas.

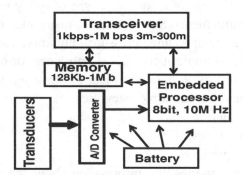

Figure 10.3 – Functional Components of a Wireless Sensor Node

10.2 Applications of WSNs

Sensors are being extensively used for a long period of time such as constantly monitoring automobile or airplane both from the inside and the outside using hundreds or even thousands of sensors over strategic locations. These wired sensors are large (and expensive) and need a continuous power supply so as to transfer their data to an appropriate location or a controller. The strategic position should be planned carefully as failure of a single sensor might bring down the whole system. Therefore, some degree of fault tolerance is desirable, especially where the sensors may be permanently embedded or terrain could be inaccessible. Undoubtedly, WSNs have been conceived with military applications in mind, including battlefield surveillance and tracking of enemy activities. However, recent surge in civilian security and surveillance applications could considerably outnumber the military ones.

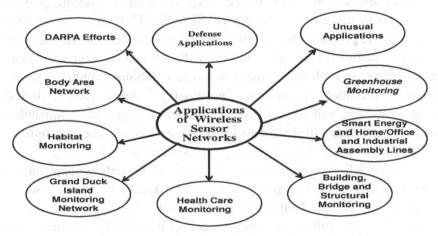

Figure 10.4 – Applications of Wireless Sensor Networks

Judging by the interest shown by military, academia, and the media, innumerable applications do exist for WSNs. Examples include weather monitoring, security and tactical surveillance, distributed computing, fault detection and diagnosis in machinery, large bridges and tall structures, detecting ambient conditions such as temperature, movement, sound, light, radiation, stress, vibration, smoke, gases, impurities, or the presence of certain biological and chemical agents. One most quoted example of a WSN is the SN nodes being deployed from a low-flying airplane or unmanned ground or aerial vehicle in the enemy's territory to sense and monitor activities on the ground and is illustrated in Figure 10.1. Besides this, there has been increased emphasis on monitoring current physical condition of soldiers in the battlefield using SNs attached to the body and acting as a Body Area Network (BAN). Under the civil category, envisioned applications can be classified into environment observation and forecast system, habitat and human rehabilitation monitoring, large structures and other commercial applications and one classification is shown in Figure 10.4. In the following sections we briefly describe each of these categories and their underlying characteristics.

10.3 DARPA Efforts towards Wireless Sensor Networks

The DARPA has identified networked micro sensors technology as a key application for the future. There are many interesting projects and experiments going under the DARPA SensIT (Sensor Information Technology) program [SensITwww] which aims to develop the software. On the battlefield of the future, a networked system of smart, inexpensive and plentiful micro sensors, combining multiple sensor types, embedded processors, positioning ability and wireless communication, will pervade the environment and provide commanders and soldiers alike with heightened situation awareness. These techniques could be very useful in battle damage assessment as well.

Vehicle type identification is important for defense applications and an experiment was performed for two weeks by placing sensor boards in the Marine Corps Air Ground Combat Center in Twenty-nine Palms, CA for collecting acoustic data [Durate2004]. To detect the presence of a vehicle, the sensor board is equipped with acoustic, seismic and passive Infra-Red sensors under program control and local processing is done to obtain local classification and storage. Distributed information from sensors is fused together by the Cluster Head (CH) of a group of sensors called a cluster. Testing runs are performed by driving four different types of vehicles in the sensor field approximately covering an area of 900x300 m^2, with adjacent sensors separated by 20-40m. A number of runs have been performed to determine false alarm rate and sensors energy consumption. A database has been created and can be downloaded for further experimentation.

A recent DARPA program has also supported in assisting recovery of rare and endangered plant species [Biagioni2002] by a comprehensive environmental measurement using a sensor network. Each sensor unit contains a computer, a wireless transceiver, environmental sensors like thermostat for temperature, photo resistor for light sensing. Flexible piezoelectric strip for sensing the wind, relative humidity sensors, and some units with a high-resolution digital camera are also employed. Both geometric routing and multi-path on demand routing are used to enhance the communication reliability. Rechargeable batteries are employed in conjunction with thermoelectric unit for auto-charging. The weather data is collected and stored every ten minutes while high-resolution images are taken once an hour. Some of the sensors are also linked to the Internet. These could be very useful in space explorations as well.

A recent work supported by the National Science Foundation [Xing2005] explores the use of highway sensors networks for safety so that a warning signal can be generated to alert the driver of a possible danger in the forward direction. Highway sensors are placed at fixed locations on the highway for data collection from the cars, measuring the distance between adjacent cars and forwarding the event information (accident or seriousness of an event such as traffic jam), occurrence of fog intensity and duration. Another set of sensors are placed on each auto for receiving signals from the highway sensors and providing location information through GPS capability. Requirements for both maximum and average storage have also been given scenario and known traffic conditions. A similar arrangement can be used to detect congested parts of a city, especially downtown or main shopping areas.

10.4 Body Area Network

Specialized sensors and transducers are being developed to measure human body characterizing parameters in a non-invasive way, so that human conditions could be predicted efficiently and accurately. There has been increased interest in the biomedical area and numerous proposals have recently been introduced [Roy2003]. The use of a micro sensor array for artificial retina, glucose level monitoring, organ monitors, cancer detectors and general health monitoring have been suggested in [Schwiebert2001]. A detailed design of a wearable sensor vest that measures, records and transmits physical characteristics such as heart rate, temperature and movement, has been discussed in [Knight2005]. This could also be very useful in assisted living of elderly and handicapped people or keeping track of endangered species.

Efforts have also been made to detect human daily life pattern by measuring physiological, behavioral and environmental parameters [Chen2004] using sensors like accelerometers, audio sensors and electrical signals and gathering data from sensors, refining them by segmentation etc., integrating and finally interpreting for an event detection. A wearable computing network has also been suggested [Kimel2005] to remotely monitor the progress of a physical therapy done at home and an initial prototype has been developed using electroluminescent strips indicating the range of human body's motion. An indoor/outdoor wearable navigation system has been suggested [Ran2004] for blind and visually impaired people through vocal interfaces about surrounding environment and changing the mode from indoor to outdoor and vice-versa using simple vocal command. A differential GPS receiver has been used to provide accurate location information outdoor while ultrasound position devices are used for indoor coverage. Limitations of the current prototype and a summary of future work have also been included.

A wearable sensor network has been demonstrated [Harada2004] to find the environmental information and to control home electric appliances by having a small Bluetooth-based network or a large scatternet network. The data throughput and communication delay have also been measured and battery life is also observed. A thin multi-resolution flat sensors that adopt resolution based on the regions of interest or the information contents have been proposed [Christensen2005], which can be placed at a soldier's helmet to provide survey of the entire scene simultaneously. A recent work [sensorprod] allows development of new vests that could optimally distribute soldiers' load, thereby alleviating discomfort and reducing the fatigue. A recent U.S. patent no. 6833274 at the Johns Hopkins University describes a Cortisol Sensor [jhuapl] that can detect level of hormone, indicating stress and fatigue levels of soldiers in a combat situation.

10.5 Habitat Monitoring

In [Wang2003a], methods for habitat monitoring are discussed such as target classification by maximum cross-correlation between measured acoustic signal and reference signal, localization using time difference of arrival (TDOA) based beam forming, and data reduction using zero-crossing rate technique. A prototype test bed consisting of iPAQs (i.e., a type of handheld device) has been built to evaluate the performance of these target classification and localization methods. As expected, energy efficiency is one of the design goals at every level: hardware, local processing (compressing, filtering, etc.), MAC and topology control, data aggregation, data-centric routing and storage. Preprocessing is proposed in [Wang2003b] for habitat monitoring applications, a tiered network is solely used for communication. The proposed 2-tier network architecture consists of micro nodes and macro nodes, wherein the micro nodes perform local filtering and data to significantly reduce the amount of data transmitted to macro nodes. A preliminary experiment shows that the data reduction and event filtering using cross-zero rate are effective.

In Habitat Monitoring involving sensing and bio-complexity mapping researchers have proposed [Cerpa2001] a tiered architecture for such applications and a model optimizing energy efficiency while monitoring moving phenomenon. Sensors can also be used to detect and monitor changes in the surrounding area such as inside a house, a forest, or an ocean.

10.6 The Grand Duck Island Monitoring Network

Researchers from the University of California at Berkeley (UCB) and Intel Research Laboratory deployed in August 2002 a mote-based tiered sensor network in Great Duck Island (GDI), Maine, aimed at monitoring the behavior of storm petrel [Mainwaring2002].

10.6.1 Architecture

The overall system architecture is depicted in Figure 10.5. A total of 32 motes have been placed in the area to be sensed grouped into sensor patches to transmit sensed data to a gateway which is responsible for forwarding the information from the sensor patch to a remote base station through a local transit network. The base station then provides data logging and replicates the data every 15 minutes to a database in Berkeley over a satellite link.

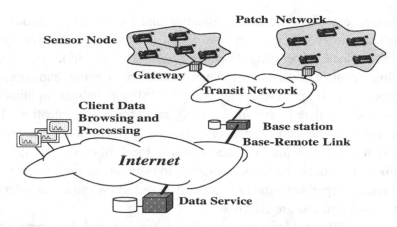

Figure 10.5 – Island Monitoring System Architecture [Mainwaring2002]

Users can interact with the sensors in two ways. Remote users can access the replica database server in Berkeley, while local users make use of a small PDA-size device to perform things like adjusting the sampling rates and managing power and other parameters.

10.6.2 A Remote Ecological Micro-Sensor Network

PODS [Biagioni2002] is a research project undertaken at the University of Hawaii that has built a wireless network of environmental sensors to investigate why endangered species of plants will grow in one area but not in the neighboring areas. They deployed camouflaged sensor nodes, (called Pods), in the Hawaii Volcanoes National Park. The Pods consist of a computer, radio transceiver and environmental sensors, sometimes including a high resolution digital camera, relaying sensed data via wireless link back to the Internet. Bluetooth and 802.11b are chosen as the MAC layer, while data packets are delivered through the IP. Energy efficiency in PODS is identified as one of the design goals and an ad hoc routing protocols called Multi-Path On-demand Routing (MOR) has been developed. Two types of sensor data are collected. Weather data are collected every ten minutes and image data are collected once per hour. Users employ the Internet to access the data from a server in University of Hawaii at Manoa.

The placement strategy for the sensor nodes is then investigated in [Biagioni2003]. Topologies of 1-dimensional and 2-dimensional regions such as triangle tile, square tile, hexagon tile, ring, star, and linear are discussed. The sensor placement strategy evaluation is based on three goals: resilience to single point of failure, the area of interest has to be covered by at lease one sensor, and minimum number of nodes. Finally, it is found that the choice of placement depends on distance between SNs and the sensing radius.

10.6.3 Environmental Monitoring

The use of sensors in monitoring the landfill and the air quality have been suggested [Agrawal2004]. Household solid waste and non-hazardous industrial waste such as construction debris and sewer sludge are being disposed off by using over 6000 landfills in USA and associated organic components undergo biological and chemical reaction such as fermentation, biodegradation and oxidation-reduction. This causes harmful gases like methane, carbon monoxide (CO), carbon dioxide, nitrogen, sulfide compounds and ammonia to be produced and migration of gases in the landfill causes physical reactions which eventually lead to ozone gas that is known to be a primary air pollutant and an irritant to our respiratory systems.

The current method of monitoring landfill employs periodic drilling of collection well, collecting gas samples in airtight bags and analyze off-site, making the process very time consuming. So, the idea is to interface gas sensors with custom-made devices and wireless radio and transmit sensed data for further analysis and appropriate collective action. Deployment of a large number of sensors allows real-time monitoring of gases being emitted by the waste material or from industrial spills and allows cost-effective remote control. Among the hazardous air pollutants, CO is monitored continuously at very few selected places with high traffic volumes due to installation and maintenance cost. The idea is to place a large number of sensors throughout the area of interest and appropriate type of sensors can be placed according to the type of pollutant anticipated in a given area [Agrawal2004]. A large volume of raw data from sensors can be collected and processed. Efficient retrieval of information using appropriate queries is also discussed to have distributed decision making [Biswas2005]. Effective power aware ways of gathering information from sensors are also covered using clustering approach and use of sub-optimal paths for long periodic queries and shortest paths for time-critical queries are also discussed. Selection and placement of sensors, their calibration and interfacing with the wireless network and data integration with geographical information systems, are also considered. The generic scheme can be easily used and adopted for other applications like smoke and gas detector, coastal observing system, etc.

10.6.4 Environment Observation and Forecasting System

The Environment Observation and Forecasting System (EOFS) is a distributed system that spans large geographic areas and monitors, models and forecasts physical processes such as environmental pollution, flooding, among others. Usually, it consists of three components: sensor stations, a distribution network, and a centralized processing.

CORIE [CORIEwww] is a prototype of EOFS for the Columbia River (Oregon, USA) which integrates a real-time 13-sensors placed on piers with one mobile sensor drifting off-shore for data management. The stationary stations are powered by a power grid, while the mobile station uses solar panel to harness solar energy. Sensor data are transmitted via wireless links toward on-shore master stations which, in turn, forward the data to a centralized server where a computationally intensive physical environment model is used to guide vessel transportation and forecasting.

Practical difficulties arise from issues such as power supply and antenna affixation for the off-shore sensor nodes and frequently obscured line-of-sight due to height of surface waves. The Automated Local Evaluation in Real-Time (ALERT) [Alertwww] is probably the first well-known wireless sensor network being deployed across most of the western United States and is heavily used for flood alarming in California and Arizona. It was developed by the National Weather Service in the 1970's providing important real-time rainfall and water level information to evaluate the possibility of potential flooding. ALERT sensor sites are usually equipped with meteorological/hydrological sensors, such as water level sensors, temperature sensors, and wind sensors. Data are transmitted via line-of-sight radio communication from the sensor site to the base station, where a Flood Forecast Model is adopted to process the data and issue automatic warnings. Web-based queries are also available.

10.6.5 Drinking Water Quality

A sensor based monitoring system has recently been proposed [Ailamaki2003], with the emphasis on placement and utilization of in situ sensing technologies and doing spatial-temporal data mining for water-quality monitoring and modeling. The main objective is to develop data-mining techniques to water-quality databases and use them for interpreting and using environmental data. This also helps in controlling addition of chlorine to the treated water before releasing to the distribution system. Detailed implementation of a bio-sensor for incoming wastewater treatment has been discussed in [Melidis2005]. A pilot-scale and full scale system has also been described.

10.6.6 Disaster Relief Management

Novel sensor network architecture has been proposed in [Cayirci2004] that could be useful for major disasters including earthquakes, storms, floods, fires and terrorist attacks. The SNs are deployed randomly at homes, offices and other places prior to the disaster and data collecting nodes communicate with database server for a given sub area which are in-turn linked to a central database for continuous update. Under normal operating conditions, the database servers from different sub areas are connected by a backbone and any disruption due to disaster could force them to be connected either via a satellite or a low-flying aerial vehicle. Steps are also discussed for establishing a route, disseminating a task and getting sensed data using a selected route. Hello messages are used to determine if a SN is alive or dead and signal strength indicates relative distance and direction of reporting SN.

Based on the statistical data from 1999 Izmit earthquake, various performance curves are obtained to indicate required average number of active SNs to detect a disaster, probability of the disaster to be within the sensing range of at least one SN for percentage of SNs failed, total number of transmitted packets, and the number of SNs failed due to energy depletion. This could also be very helpful in seismic monitoring, glacier movements, and underwater sensing.

10.6.7 Soil Moisture Monitoring

A soil moisture monitoring scheme using sensors, have been developed over a one hectare outdoor area [Oliver2005] and various performance parameters have been measured from an actual system. A custom made moisture sensor is interfaced with Mica 2 Mote wireless board. In place of monitoring of the moisture level, a rain gauge is used to wake up the SNs from the sleep mode (2-hour after 1 mm of rains) and such reactive triggering eliminates the need for clock synchronization, clock-drift, time stamping and time setting, thereby drastically saving the energy consumed by WSN. This also helps in achieving robustness and longevity.

A base node linked to GSM gateway, collects information for SNs and SMAC with 4-way handshake of (RTS-CTS-DATA-ACK) the MAC protocol. 100% message delivery success has been observed in the laboratory tests while the field trails show dependence of delivery rate whether it is rainy or dry. End-to-end message delivery in field trails is also given for 10 minutes of rain followed by 2 hours of dry period. The longest interval with no data is observed to be 12 hours (6-readings) for dry rate and 3-5 hours (21 readings) for rain rate. For improving robustness, the researchers also suggest the use of multi-path routing between critical pairs or simply doubling the numbers of SNs and plan to undertake in their future work. The sensors are observed to be $\pm 1\%$ accurate when calibrated for specific soil and $\pm 3\%$ without calibration. The Berkley Mote hardware is seen to draw 5-20 milliamps during active period and 5 microamps during sleep. With 100% duty cycle of motes, the alkaline batteries falls below 2.7 V after 18 hours while 100 hours for NiMH batteries. The researchers are experimenting to generalize the event conditions.

10.7 Health Care Monitoring

Applications in this category include telemonitoring of human physiological data, tracking and monitoring of doctors and patients inside a hospital, drug administrator in hospitals, and so on [Akyildiz2002]. The use of fixed sensors has also been explored and preliminary results provided in monitoring health of cattle [Mayer2004] by checking food and water availability. This is done by measuring intra-rumenal movement of cows by accelerometer and characterizing the feeding cycle. Micomotes have been used to set up an experimental system. A considerable difference is also observed on the feed cycle and the amount of water at different places. The health of a cow is to be predicted by observing when drinking occurs, how much water is digested, what kind of mixing occurs in the rumen and the amount of heat generated when fermentation of digested feed occurs, and how cold the optimum distribution of water within a paddock.

10.8 Building, Bridge, and Structural Monitoring

Several recent projects have explored the use of sensors in monitoring the health of buildings, bridges and highways. A Bluetooth based scatternet has been proposed [Mehta2004] to monitor stress, vibration, temperature, humidity etc. in civil infrastructures and rational for using. Simulation results are given to justify effectiveness of their solution by having a set of rectangular Bluetooth equipped sensor grids to model a portion of bridge span.

Fiber optic based sensors have been proposed for monitoring crack openings due to strain and corrosion of the reinforcement in concrete bridge decks and structures [Casas2003]. Possible use of different types of interferometer sensors for gradual structural degradation has been explored and their relative advantages and disadvantages are discussed. Impact of temperature on the accuracy of strain monitoring sensors, have also been pointed out and the use of a new family of inclinometers has been suggested to overcome the temperature sensitivity. Corrosion of steel bars is measured by using special super glue and angular strain sensors.

Feasibility of monitoring various risks for buildings using wireless acceleration sensors and consider microphones interfaced to Mica 2 Motes have been tested [Kuratawww]. The idea is to check a building for degraded structural performance, fatigue damage, gas leaks, intrusion, fires, etc. for appropriate actions such as structural control, maintenance, evacuation advice, alarms and warning, fire fighting and rescue operation, necessary security measures, etc. The acceleration and strain at different parts of house beam and columns, temperature/light and sound in each room can be used to detect earthquake/wind, fires and intrusions. Test results for damage detection in buildings are also given. Heterogeneous sensors such as for motion, temperature, acoustics, vibration can also be placed around the house or an office building to reduce energy dissipation as well as securing the area such as home land security [Heinzelman2004], breakage of glass enclosures, surveillance of shopping mall unattended parking structure, or indicating open spaces in a large parking lot.

The use of sensors for controlling civil and mechanical systems, has been explored in [Tomizaka2004] by having acquisition, monitoring and recording of data, detecting failure, monitoring health, estimating the state and making control decisions. Such control systems and smart dampers have been installed in more than 40 buildings and in the construction of numerous bridge towers or large-span structures. Tiny hair-like sensors is being used to design acoustic sensors and such "magnetic hairs" can be fabricated easily in the form of pressure detector arrays. A wearable non-invasive ring sensor designed at MIT can detect human circulatory signals and pulses. Smart sensors have also been proposed to do complex aerospace structural health monitoring by generating diagnostic signals, measuring physical parameters, collaborating and interpreting the data into useful information. A new class of self-sensing composite materials, capable of monitoring temperature flow pattern, are also been developed.

10.9 Smart Energy and Home/Office and Industrial Assembly Lines Applications

Societal-scale sensor networks can greatly improve the efficiency of energy-provision chain, which consists of three components: the energy-generation, distribution, and consumption infrastructure. It has been reported that 1% load reduction due to demand response can lead to a 10% reduction in wholesale prices, while a 5% load response can cut the wholesale price in half. In the wake of recent energy regulation in California, it is proposed [Smartwww] that a gradual roll-out plan to make energy-supply chain part of an integrated network of monitoring, information processing, controlling, and actuating devices, in a hope to spread the consumption of energy over time, will reduce peak demand. Obviously, this is a complex and long-term project.

Similar approach can be easily extended to provide efficient control in an industrial assembly line by sensing the product on the conveyor belt, ensuring alignment and caution about preventive maintenance from one step to another [globalspec]. The Boeing Company has started using a 1.6" per minute moving assembly line in congregating 777 planes more efficiently by quickly fixing root cause of the problems [sensorsmag].

10.10 Greenhouse Monitoring

Nowadays, we witness more and more electronic appliances in an average household. Therefore, great commercial opportunities exist for home automation and smart home/office environment cooling, heating and humidity control. An example application in this category is described in [Srivastava2001], where a "Smart Kindergarten" consisting of a sensor-based wireless network for early childhood education is discussed. It is envisioned that this interaction-based instruction method will soon replace the traditional stimulus-responses based methods. In addition, emission of CO_2 from coal power plant can be minimized if efficient ways of distributing energy can be devised [Atkinson2008]. Another alternative is to use sensor controlled solar panels to optimize capturing of sun power.

Grape Networks [westernfarmpress] announced it will be deploying soil moisture sensors in addition to those used to measure light, temperature and humidity. The data collected includes the exact location of the sensor modules, and critical vineyard and wine grape information on powdery mildew and degree days in the vineyard. The sensor modules are mobile and are buried next to the wine-producing grapes. The vineyard operations manager can view the data on any Web-enabled cellular phone or PC, and can also set the threshold values for alerts over the Internet or via e-mail or SMS. The soil moisture is obtained by capacitance sensors and is not affected in extreme dry conditions. The sensor modules can accommodate up to three soil moisture sensors at various depths from 6 inches to 14 feet. A complete system of 15 modules and up to 75 sensors costs approximately the same price as two high-end weather stations. A similar project has been followed in Australia [crcv] and other parts of the world to enhance production of grapes by analyzing variation in yields and assessing response to appropriate management.

A recent work [Wang2008] describe how carbon-mono-oxide (CO), a colorless, odorless and poisonous gas emitted by incomplete combustion of gases in automobiles. This is done by installing CO sensors at various interactions around the University of Cincinnati campus and transmitting sensed data in a multi-hop fashion every 3-minute to a BS located in a close by building. As the actual sensor reading is a function of temperature and humidity, these values are also transmitted and CO sensor is also calibrated in a laboratory environment. Software interface was also developed to have these sensors reading at a web site. Some details are included in a recent article [Wang2010]. Such sensors could be very useful in monitoring coal mines when two main causes of accidents are the presence of CO and water. Similar schemes could also be used in monitoring different types of pollution by having an appropriate type of transducers.

10.11 Unusual Applications of WSNs

There are many unusual applications of sensors and few of them we are familiar with, are described here. The first one is to use sensors in determining instantaneous position of dancers on the stage and their movements and translate them in to corresponding music [Mostafa2008]. A joint work at the University of Cincinnati between Computer Science group and the College of Conservatory Music has presented four different performances [Agrawal2009] during last two years and continuing the efforts in enhancing the quality of the musical events [Helmut2010]. Another recent work is a follow up of this in automatically converting sign language in to text or audio form [Agrawal2010] by detecting hand and fingers movements by sensors and using RFIDs in locating the position of fingers. Another interesting ongoing work at UC deals with detecting the boundary of forest wildfire [Kelkar2009] by deploying sensors using UAVs such that the boundary can be detected as soon as possible. The basic idea is to minimize the number of sensors to be deployed to detect the boundary while been able to send information to close by Base Station. The work is in progress in minimizing sensor requiring by deploying sensors in some controlled way and avoiding wastage of sensors inside the event. Such continuous monitoring could also help indentify spreading direction and movement of the fire. Such event boundary detection could be useful in disaster like recent Louisiana Gulf oil spill by BP undersea well [saveusenergyjobs].

10.12 Conclusions and Future Directions

Sensor networks are perhaps one of the fastest growing areas in the broad wireless ad hoc networking field. As we could see various applications throughout this chapter, the use of sensor networks is flourishing at a rapid pace and one need to use one's own imagination. But, there are many challenges to be addressed such as:

- Energy Conservation - Nodes are battery powered with limited resources while still having to perform basic functions such as sensing, transmission and routing. How can battery power be best utilized?
- Sensing - Many new sensor transducers are being developed to convert physical quantity to equivalent electrical signal and many new development is anticipated;
- Communication - Sensor networks are very bandwidth-limited. How to optimize the use of the scarce resources in an energy efficient way? Along the same lines, how can sensor nodes minimize the amount of communication while still achieving the overall application goal;
- Computation - Here, there are many open issues in what regards signal processing algorithms and network protocols; and
- Applications - Many new applications are being developed and people are using their imagination to explore potential use. It is not inappropriate to say that sky is the limit, even though a killer application is yet to be identified.

The challenges are many and most important ones are discussed in the next two chapters. While we have partial answers or roadmaps to some of the above questions, there is still much more work that needs to be done.

Homework Questions/Simulation Projects

Q. 1. Think about an application of sensors and identify associated design problems. Can you think of ways to address them?

Q. 2. Simulate the problem discussed in Q. 2, and verify the correctness and accuracy of your design strategy? What are the performance parameters you can think about and how can you improve them?

Q. 3. Enumerate at least 5 different ways of conserving energy in wireless sensor networks. How do these energy conserving strategies affect the design of network algorithms and protocols?

Q. 4. Can you rank the importance of different performance parameters? How can you assign weights to them and how can you optimize the overall performance?

Q. 5. How do you compare mobile over static WSNs? Can you envision potential application areas of each category? Do some web search and justify your answer.

References

[Agrawal2004] D.P. Agrawal, M. Lu, T.C. Keener, M. Dong, and V. Kumar, "exploiting the use of wireless sensor networks for environmental monitoring," *Journal of the Environmental Management,* August 2004, pp 35-41.

[Agrawal2009] Dharma P. Agrawal, Jung Hyun Jun, Ahmad Mostafa, Margaret M. Helmuth, "Wireless Sensor Based Music Generation," *Invention Disclosure entitled,* UC 109-103 dated May 21, 2009.

[Agrawal2010] Dharma P. Agrawal, "Wireless Sensor Based Sign Language Translation," UC Invention Disclosure 110-003 dated January 28, 2010, *Patent Filing EFS ID: 7963193,* Application Number: 12831230, Confirmation Number: 2028, Receipt Date: 06-JUL-2010, Time Stamp: 22:45:25.

[Ailamaki2003] A. Ailamaki, C. Faloutsos, P.S. Fiscbeck, M.J. Small, and J. VanBriesen," An environmental sensor network to determine drinking water quality and security," *Proceedings SIGMOD Record,* Vol. 32, No. 4, December 2003, pp 47-52.

[Akyildiz2002] I. Akyildiz, W. Su, Y. Sankarasubramaniam, and E. Cayirci, "A survey on sensor networks," *IEEE Communications Magazine,* vol. 40, no. 8, August 2002.

[ALERTwww] http://www.alertsystems.org.

[Atkinson2008] R. Atkinson, and D. Castro, "Digital Quality of Life – Understanding the Personal & Social Benefits of the Information Technology Revolution," *The Information Technology and Innovation Foundation,* Washington DC, October, 2008.

[Biagioni2002] E. Biagioni and K. Bridges, "The application of remote sensor technology to assist the recovery of rare and endangered species," In *Special issue on Distributed Sensor Networks for the International Journal of High Performance Computing Applications,* Vol. 16, No. 3, August 2002. pp. 315-324.

[Biagioni2003] E. Biagioni and G. Sasaki, "Wireless sensor placement for reliable and efficient data collection," In *Proceedings of the Hawaii International Conference on Systems Sciences,* January 2003.

[Biswas2005] R. Biswas, N. Jain, N. Nandiraju, and D. Agrawal, "Communication Architecture for processing Spatio-temporal continuous Queries in Sensor Networks," *Annals of Telecomm, 2005.*

[Casas2003] J.R. Casas and P.J.S. Cruz, "Fibre Optic sensors for bridge monitoring," journal of bridge monitoring, *ASCE*, Nov.-Dec. 2003, pp 362-373.

[Cayirci2004] E. Cayirci and T. Coplu, "SENDROM:sensor networks for disaster relief operations management," Proceedings *Third Annual Mediterranean ad hoc Networking Workshop*, June 27-30, 2004, Bodrum, Turkey, pp 535-546.

[Cerpa2001] A. Cerpa, J. Elson, D. Estrin, L. Girod, M. Hamilton, and J. Zhao, "Habitat monitoring: Application driver for wireless communications technology," In *Proceedings of the ACM SIGCOMM Workshop on Data Communications in Latin America and the Caribbean*, April 2001.

[Chen2004] W.P. Chen, J.C. Hou, and L. Sha, "Dynamic clustering for acoustic target tracking in wireless sensor networks," *IEEE Transactions on Mobile Computing*, Vol. 3, No. 3, July-Sept. 2004, pp 258-271.

[CORIEwww] CORIE, "A pilot environmental observation and forecasting system (EOFS) for the Columbia River," http://www.ccalmr.ogi.edu/CORIE/.

[crcv] http://www.crcv.com.au/research/programs/one/finalreport.pdf.

[Durate2004] M.F. Durate and Y.H. Hu, "Vehicle classification in Distributed Sensor Networks," *Journal of Parallel and Distributed Computing*, 2004, pp 826-838.

[Estrin1999] D. Estrin, R. Govindan, J. Heidemann, and S. Kumar, "New Century Challenges: Scalable Coordination in Sensor Networks," *ACM Mobicom*, 1999.

[globalspec] http://www.globalspec.com/reference/7587/Torque-Sensor-Application-Assembly-Line-Automation.

[Harada2004] T. Harada, T. Nagai, T. Mori, and T. Sato, "Realization of Bluetooth-equipped module for wireless sensor network," *Proceedings Symposium of Networking and Sensing Technologies*, Tokyo, June 22-23, 2004.

[Heinzelman2004] Wendi B. Heinzelman, Amy L. Murphy, Hervaldo S. Carvalho, and Mark A. Perillo, "Middleware to Support Sensor Network Applications," *IEEE Network*, Vol. 18, no. 1, January 2004, pp. 6-14.

[Helmut2010] Mara Helmuth, Jung Hyun Jun, Talmai Oliveira, Kazuaki Shiota, Amitabh Mishra, Ahmad Mostafa, and Dharma Agrawal, "Wireless Sensor Networks and Computer Music, Dance and Installation Implementations," *2010 International Computer Music Conference ICMC2010*, New York, NY, June 1 & 2; Stony Brook University, June 3, 4, 5, 2010.

[Jain2005] N. Jain and D.P. Agrawal, "Current Trends in Wireless Sensor Network Design," International Journal of Distributed Sensor Networks, Vol. 1, No. 1, 2005, pp 101-122.

[jhuapl] http://www.jhuapl.edu/ott/technologies/technology/articles/P01841.asp.

[Kahn1999] J.M. Kahn, "New Century Challenges: Mobile Networking for Smart Dust," *ACM Mobicom*, 1999.

[Kelkar2009] Harshvardhan Kelkar, "Boundary Marking of Phenomenon using Wireless Sensor Networks," *MS Thesis*, University of Cincinnati, November 16, 2009.

[Kimel2005] J.C. Kimel, "Thera-Network: A Wearable computing network to motivate exercise in patients undergoing physical therapy," *ICDCS 2005*, pp 491-495

[Knight2005] J.F. Knight, A. Schwirtz, F. Psomadelis, C. Baber, H.W. Bristow, and T.N. Arvanitis," *Pervasive Ubiquitous Computing*, 2005, pp 6-19.

[Kuratawww] N. Kurata, B.F. Spencer, and M.R. Sandoval, "applications of wireless sensor networks. Mote for building risk monitoring," www.unl.im.dendai.ac.jp

[Mainwaring2002] A. Mainwaring, J. Polastre, R. Szewczyk, D. Culler, and J. Anderson, "Wireless Sensor Networks for Habitat Monitoring," in *the First ACM Workshop on Wireless Sensor Networks and Applications*, September 2002.

[Mayer2004] K. Mayer, K, Ellis and K. Taylor, "Cattle health monitoring using wireless sensor networks," *Proceedings 2nd IASTED International Conference on Communication and Computer Networks*, Cambridge, Massachusetts, Nov 8-10, 2004.

[Mehta2004] V. Mehta and M.E. Zarki, "A Bluetooth based sensor network for civil infrastructure health monitoring," *Wireless Networks*, Vol. No. , 2004 pp 401-412.

[Melidis2005] P. Melidis and A. Aivasidis, "Biosensor for toxic detection and processs control in Nitrification plants," *Journal of Environmental Engineering*, April 2005, pp 658-663.

[Mostafa2008] Ahmad Mostafa, Jung Hyan Jun, Dharma P. Agrawal, and Mara Helmuth, "Dancing with the Motes," *Fifth IEEE International Conference on Mobile Ad-hoc and Sensor Systems, (IEEE MASS 2008),* September 29-October 2, 2008, Atlanta, Georgia, pp. 538-540.

[nsf109] http://www.nsf.gov/about/congress/109/alb_homelandsec060805.jsp.

[Oliver2005] R.C. Oliver, M. Kranz, K. Smettem, and K. Mayer, "A reactive soil moisture Sensor Network: Design and Field evaluation," *International Journal of Distributed Sensor Networks*, Vol. 1, 2005, pp 149-162.

[Ran2004] L. Ran, S. Helal, and S. Moore, "Drishti: An integrated indoor/oudoor blind navigation system and service," *PERCOM 2004*.

[Roy2003] L.H. Roy, and D.P. Agrawal, "Wearable Networks: Present and Future," *IEEE Computer*, Vol. 36, No. 11, Nov. 2003, pp 31-39.

[saveusenergyjobs] http://www.saveusenergyjobs.com/category/blog/.

[Schwiebert2001] L. Schwiebert, S. K. S. Gupta, and J. Weinmann, "Research challenges in Wireless Networks of biomedical sensors," *Proceedings Mobicom 2001*.

[SensITwww] DARPA SensIT Program, http://www.darpa.mil/ito/ research/sensit.

[sensormag] http://www.sensorsmag.com/automotive/news/sensor-enables-boeings-777-moving-assembly-line-2154.

[sensorpod] www.sensorprod.com.

[Smartwww] http://www.citris.berkeley.edu/smartenergy/smartenergy.html.

[Srivastava2001] M. Srivastava, R. Muntz, and M. Potkonjak, "Smart kindergarten: sensor-based wireless networks for smart developmental problem-solving environments," In *Mobile Computing and Networking*, 2001.

[Tomizuka2004] M. Tomizuka, "Sensor and control technologies, the engineering of modern civil and mechanical systems," *Proceedings of the 3rd International Conference on Earthquake Engineering*, Oct. 19-20, 2004.

[Wang2003a] H. Wang, J. Elson, L. Girod, D. Estrin, and K. Yao, "Target Classification and localization in habitat monitoring," In *Proceedings of the IEEE ICASSP*, Hong Kong, April 2003.

[Wang2003b] H. Wang, D. Estrin, and L. Girod, "Preprocessing in a tiered sensor network for habitat monitoring," In *EURASIP Journal on Applied Signal Processing*, No. 4, March 2003.

[Wang2008] *Demin Wang*, "Wireless Sensor Networks: Deployment Alternatives and Analytical Modeling," *PhD Dissertation*, University of Cincinnati, Department of Computer Science, November 17, 2008.

[Wang2010] Demin Wang, Dharma P. Agrawal, Wassana Toruksa, Chaichana Chaiwatpongsakorn, Mingming Lu and Tim C. Keener, "Monitoring Ambient Air Quality with Carbon Monoxide Sensor based Wireless Network," *Communications of the ACM*, vol. 53, no. 5, May 2010, pp. 138-141.

[westernfarmpress] http://westernfarmpress.com/grapes/111407-monitoring-system/.

[Xing2005] K. Xing, M. Ding, X. Chen and S. Rotenstreich, "Safety warning based on highway sensor networks," *IEEE WCNC*, Vol. 4, March 13-17, 2005, pp 2355-2361.

Chapter 11

Sensor Networks Design Considerations

11.1 Introduction

Various possible applications of wireless sensor networks (WSNs) have been discussed in the previous Chapter. Now, we need to learn about various characteristics and requirements for different areas. Sensors have been used in many areas for a long time and household thermometer is a well-known example. But, miniaturization of devices, having small local memory, processing of raw data using a processor and inclusion of wireless transceiver have made a wireless sensor node (SN) versatile and useful in an unattended monitoring of a given area. Various functional components of a SN have been briefly described in Chapter 10. We need to study various underlying issues and associated parameters.

(a) – Mica Motes 2   *(b) – The Mica Sensor Board*
Figure 11.1 – Mica SN [microstrain]

In this respect, a comprehensive SN has been developed by University of California at Berkeley and marketed by Crossbow and the product Mica Mote [MICA] is shown in Figure 11.1(a). It uses an Atmel Atmega 103 microcontroller running at 4 MHz, with a radio operating at 916 MHz band which provides bidirectional communication at 40 kbps with a pair of AA batteries. The Mica Board, shown in Figure 11.1(b), is

Table 11.1 Parameters of different Sensors Units [Hill2004]

Sensor Node	CPU	Power	Memory	I/O	Trans Rate	Radio	Trans Range
Special-purpose Sensor Nodes							
Spec 2003	4-8 MHz	3mW (peak) 3µW (idle)	3K RAM	8-bit on chip ADC	50-100 Kbps		50 feet (indoors)
Generic Sensor Node							
Rene 1999	ATMEL 8535	60 mW (active) 0.036 mW (sleep)	8K Flash 32 K EEProm	Large expansion connector	10 Kbps	916 MHz	Up to 100 feet (use External antenna)
Mica-2 2001	ATMEGA 128	60 mW (active) 0.036 mW (sleep)	128K Flash 4K RAM	Large expansion connector	76 Kbps	433 MHz	100 meters
Telos 2004 (Moteiv Inc.)	Motorola HC508	32 mW (active) 0.001 mW (sleep)	4K RAM	USB and Ethernet	250 Kbps	2.4 GHz IEEE 802.15.4	50 meters (indoors) 125 meters (outdoors)
Mica-Z 2004	ATMEGA 128		4K RAM 128K Flash	Expansion connector	250 Kbps	2.4 GHz IEEE 802.15.4	30 meters (indoors) 100 meters (outdoors)
iB5324 (Millennial Net Inc.)					250 Kbps	2.4 GHz	30 meters
Rockwell WINS	Intel StrongArm SA 1100		1MB SRAM 4MB Flash			900 MHz	
High-Bandwidth Sensor Nodes							
BT Node 2001	ATMEL Mega 128L 7.328 MHz	50mW (sleep) 285 mW (active)	128KB Flash 4KB EEPROM 4K SRAM	8channel 10 bit A/D, 2 UARTS, Expandable connectors	Bluetooth	433-915 MHz	
Imote 1.0 2003	ARM 7TDMI 12-48MHz	1 mW (idle) 120 mW (active)	64KB SRAM 512K Flash	UARTS, USB, GPIO, I²C, SPI	Bluetooth 1.1		
Gateway Nodes							
Stargate 2003	Intel PXA 255		64 K SRM	2PCMCIA/CF, com ports, Ethernet, USB			
Inrsync Cerfcube 2003	Intel PXA 255		32K Flash 64K SRAM	Single CF card, General I/O		Serial connection to sensor network	
PC104 nodes	X86 processor		32K Flash 64K SRAM	PCI Bus			

stacked to the processor board via the 51 pin extension connector. It includes temperature, photo resistor, barometer, humidity, and thermopile sensors. To conserve energy, later designs include an A/D Converter and an 8x8 power switch on the sensor board, bypassing of the DC booster, etc. To protect sensors from variable weather condition, mica mote is packaged in an acrylic enclosure, which does not obstruct sensing functionality or radio communication. MICA 2 motes have three modes based RF frequency band, MPR400 using 915 MHz, MPR 410 employing 433 MHz and MPR420 based on 315 MHz. Three miniaturized sizes (1/4) of Mote are also available as MICA2DOT. Another version of MICA 2 using a combination of ultrasound and RF signal has been jointly developed by Crossbow and MIT and has a flexibility to configure either as a listener or a beacon transmitter. More details on Mica Motes and version 2 can be obtained from [www.xbow.com]. Details of different types of sensor transducers could be obtained from many web sites [microstrain]. A comparison between different sensor boards is shown in Table 11.1.

This table gives the power consumed in actual commercial units. Specific details of energy consumption in functional units of MICA2 under different operating conditions are summarized in Table 11.2. It is clear that more power is consumed in data communication as compared to local computation as arithmetic operations are done at much higher speed than bit transfer using transceiver. As all WS units get power from the battery, it is critical to conserve energy as much as possible.

Table 11.2 Energy consumption in MICA2 motes sensor unit [memsic]

Component	Current	Power
A/D Converter	7.5 mA	3.1µW
Radio Idle	16 mA	40nJ
Radio Receiver	16 mA	40nJ
Radio Transmit	21 mA	50 nJ
Computation only	10 mA	10nJ

11.2 Empirical Energy Consumption

Minimizing the energy consumption of WSs is critical yet a challenge for the design of WSNs. The energy consumption in WSN involves three different components: Sensing Unit (Sensing transducer and A/D Converter), Communication Unit (transmission and receiver radio), and Computing/Processing Unit. In order to conserve energy, we may make some SNs go to sleep mode and need to consider energy consumed in that state.

Sensing transducer is responsible for capturing the physical parameters of the environment. Its basic function is to do physical signal sampling and convert into electrical signals. The energy consumption of this part depends on the hardware and the application and sensing energy is only a small fraction of the total energy consumed.

A/D Converter: Based on a recent work [moteiv], an AD Converter for sensor consumes only $3.1\ \mu W$, in 31 *pJ/8-bit* sample at 1Volt supply. The standby power consumption at 1V supply is $41pW$. Assuming the D/C is not noise limited, the lower bound on energy per sample for the successive approximation architecture is roughly computed as: $E_{min}=C_{total}V_{ref}^{2}$, where C_{total} is the total capacitance of the array including the bottom plate parasites, and V_{ref} is a common mode input voltage the comparator works under.

Transmission Energy: Based on [Heinzelman2000a] the transmission energy transmits a k-bit message to distance d can be computed as: $E_{Tx}(k,d)=E_{Tx\text{-}elec}(k)+E_{Tx\text{-}amp}(k,d)=E_{elec}*k+\varepsilon*k*d^{2}$, where $E_{Tx\text{-}elec}$ is the transmission electronics energy consumption, $E_{Tx\text{-}amp}$ is the transmit amplifier energy consumption. Their model assumes $E_{Tx\ elec}=E_{Rx\ elec}=E_{elec}=50nJ/bit$, and $\varepsilon_{amp}=100pJ/bit/m^{2}$

Receiver Energy: To receive a k bit message, the energy consumed is [Heinzelman2000a] $E_{Rx(k)}=E_{Rx\text{-}elec}(k)=E_{elec}*k$

Computation: The computing unit associated with a WS is a microcontroller/ processor with memory which can control and operate the sensing, computing and communication unit. The energy consumption of this unit has mainly two parts: switching energy and leakage energy. Switching energy is expressed as [Shih2001a] $E_{switch}=C_{total}V_{dd}^{2}$, where C_{total} is the total capacitance switched by the computation and V_{dd} is the supply voltage. Dynamic voltage scaling (DVS) scheme is used to adaptively adjust operating voltage and frequency meet the dynamically changing workload without degrading performance thus saving energy [Ilyas2005].

Leakage energy is the energy consumed when no computation work is done. It can be expressed as: $E_{leakage,up}=(V_{dd}t)I_{0}e^{\frac{V_{dd}}{n'V_{T}}}$, where V_T is the thermal voltage, n' and I_0 are the parameters of processor and can get from experiment. For StrongARM SA-1100, $n'=21.26$ and $I_0=1.196mA$ [Shih2001a]. It is widely accepted that the energy consumed in one bit of data can be used to perform a large number of arithmetic operations in the sensor processor (power consumed in 1-bit transfer to 100m equals 3000 instructions [Pottie2000]).

Residual Energy: SN usually gets power from an attached battery. Energy is consumed by different functional units of a SN. The remaining energy in SN is usually known as residual energy and reflects how long the SN can perform all its functions correctly.

Sleeping: To save the energy, sensors can be put into sleep-wake up cycles. When a sensor is in sleep slate, it can turn off some units to save energy. There are different types of sleep modes and some useful states are discussed in [Shih2001b].

11.3 Sensing and Communication Range

A WSN consists of a large number of SNs and exploring their best possible use is a challenging problem. As the main objective of a SN is to monitor some physical quantity in a given area, the sensors need to be deployed with adequate density so that sensing of the complete area can be done, without leaving any void or un-sensed area. In other words, given the sensing range of a SN_1, the adjacent sensor SN_2 ought to be located close by so that there is no overlapping nor any area should be left uncovered as illustrated by final position of SN_2 in Figure 11.2. As the sensor data is to be sent a central location of BS, it is important to adjust transmitting power of the transceiver so that SN_1 can transfer data to SN_2 and vice versa. That means the communication distance r_c ought to be at least twice that of r_s given by: $r_c \geq 2 . r_s$.

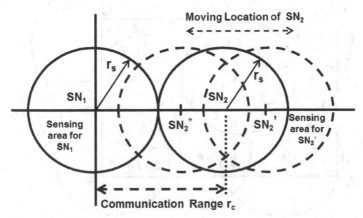

Figure 11.2 – Sensing Range r_s of a SN, location of an adjacent SN, and desired Communication Range r_c

This implies that the wireless communication coverage of a sensor must be at least twice the sensing distance [Zhang2005]. Transmission between adjacent SNs is feasible only when there is at least one SN within the communication range of each SN. Therefore, not just the sensing coverage, but the communication connectivity is at least equally important so that sensor data could be received by other SNs. It may be noted that the data from a single SN is not adequate to make any useful decision [Dasgupta2003] and data need to be collected from a set of SNs in arriving at an intelligent interpretation.

If the area to be covered is large, adequate number of SNs, each with circular sensing range, should be needed to be placed throughout the area so that no corner is left out. So, the real question is how far away the SNs can be located. As illustrated in Figure 11.3, enough sensors need to be deployed, with some overlapping area. Different areas are marked as 1, 2, or 3, depending on a given small sub-area can be represented by the reading from 1, 2, or 3 sensors. Figure 11.3 is just an example of random deployment and not necessarily an optimal placement of sensors.

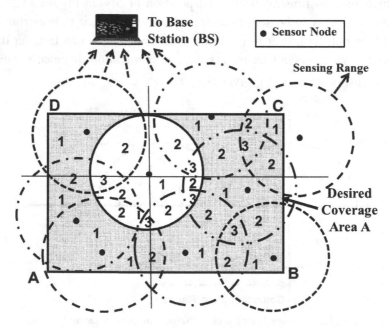

Figure 11.3 – Coverage of a given area *A* with multiple SNs

If N SNs are put in an area A=LxL, then the SNs density can be given by $\lambda_s = N/A$. The SNs are uniformly distributed with the SN density of λ_s. Then [Liu2004]:

Disk shape covered by a single sensor $\alpha = \pi r_s^2$
Probability that point p is located at an arbitrary sensor $= \pi r_s^2/A = \alpha/A$.
Probability that point p is not located at an arbitrary sensor $= 1 - \alpha/A$.
Probability that point p is not located at any N sensors $= (1 - \alpha/A)^N$.

$$= (1 - \frac{\alpha\lambda_s}{N})^N$$

When $A \Rightarrow \infty$, *this probability* $= e^{-\alpha\lambda} = e^{-\lambda_s \pi r_s^2}$.
The fraction of Area being covered $f_a = (1 - e^{-\lambda_s \pi r_s^2})$.
So, required density $\lambda_s = -\frac{ln(1 - f_a)}{\pi r_s^2}$.

The sensing model is known as Boolean sensing model as any event occurring outside the sensing radius r_s *is assumed to be unknown.* The probability that monitored area is covered by at least 1 SN is given by:

$$f_{1-cov\ ered} = 1 - f(0) = 1 - e^{-\lambda S},$$

Where $f(0)$ is coverage by zero sensor. In many situations, an event need to be sensed by at least k close-by sensors for a cooperative decision (such as relative location using triangulation), then concurrent sensing by k SNs can be expressed as:

$$f_{k-cov\ ered} = 1 - \sum_{m=0}^{k-1} f(m) = 1 - \sum_{m=0}^{k-1} \frac{(\lambda S)^m}{m!} e^{-\lambda S}.$$

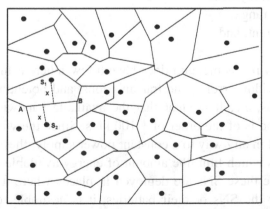

Figure 11.4 – Voronoi Diagram for Randomly Placed SNs

Another way to determine the area to be covered by each SN is to form a Voronoi diagram [Aurenhammer1991] and one such example is shown in Figure 11.4. The basic idea is to partition the area in to a set of convex polygons such that all polygons edges are equidistant from neighboring sensors. For example, a line AB is drawn such that it is at equal distance x from two adjacent sensors S_1 and S_2 [Nayak2010]. This process is repeated till all such equidistance lines are drawn for all adjacent pairs of SNs. Then, it is easier to let each sensor sense the area covered by its surrounding polygon and the maximum distance to be covered by a SN in a polygon will govern minimum required sensing range. Similarly, minimum wireless transmission range can be determined by finding the maximum distance between any pair of adjacent SNs. This kind of scheme has also been used in checking the coverage problem in a WSN [Megerian2005]. The process of drawing Voronoi diagram is very complex and many heuristics have been given [Shah2004].

11.4 Design Issues

The advancement in technology has made it possible to have a network of 100s or even thousands of extremely small, low powered devices equipped with programmable computing, multiple parameter sensing and wireless communication capability, enhancing the reliability, accuracy of data and the coverage area. In short, some of the advantages of WSN over wired ones are:

- Ease of deployment,
- Extended range,
- Fault tolerant, and
- Mobility.

The wireless medium does have a few inherent limitations such as low bandwidth, error prone transmissions, and potential collisions in channel access, etc. It is clear that the available bandwidth for sensor data is low and is of the order of 1-100 kb/s. Since the wireless SNs are not connected in any way to a constant power supply, they derive energy from batteries which limit the amount of energy available to the SNs. In addition, since these SNs are deployed in places where it is difficult to replace either the SNs or their batteries, it is desirable to increase the longevity of the network and, preferably, all the SNs should die together so that new SNs could be replenished simultaneously in the whole area. Finding individual dead SNs and then replacing those SNs selectively would require dynamic deployment and eliminates major advantages of these networks. Thus, the protocols designed for these networks must strategically distribute the dissipation of energy, which also enhances the average life of the overall system. In addition, as we mentioned before, environments in which these SNs are expected to operate and respond are very dynamic in nature, with fast changing physical parameters. An ideal WSN should have the following additional features:

- *Attribute-based addressing.* This is typically employed in WSNs where addresses are composed of a group of attribute-value pairs which specify certain physical parameters to be sensed. For example, an attribute address may be (temperature > 35°C, location = "NorthwestQuadrant"). So, all SNs located in "Northwest Quadrant" which sense a temperature greater than 35°C should respond.

- *Location awareness* is another important issue. Since most data collection is based on location, it is desirable that the SNs know their position, at least relative position in the area in place of absolute longitude and latitude values.

- Another important requirement in some cases is that the SNs should react immediately to *drastic changes* in their environment, for example, in a time-critical application. The end user should be made aware of any drastic deviation in the situation with minimum delay, while making efficient use of the limited wireless channel bandwidth and SN energy.

- *Query Handling* is another important feature. Users should be able to request data from the WSN through the BS (also known as a sink) or through any of the SNs, whichever is closer. So, there should be a reliable mechanism to transmit the query to appropriate SNs which can respond to the query. The answer should then be re-routed back to the BS as quickly as possible.

Traditional routing protocols defined for MANETs (discussed in previous chapters) are not well suited for WSNs due to the following reasons:

- WSNs are "data centric", where data is requested based on particular criteria such as "which area has temperature 35°C";

- In traditional wired and wireless networks, each SN is given a unique identification (e.g., an IP address) used for routing. This cannot be effectively used in WSNs because, being data centric, routing to and from specific SNs in these networks is not required;
- Adjacent SNs may have similar data. So, rather than sending data individually from each SN, it is desirable to aggregate similar data before sending it to the BS.
- The requirements of the network change with the application and hence, it is application-specific. For example, in some applications, the SNs are fixed and not mobile, while data is needed based only on some specified attributes (viz., attribute is fixed in this network).

In WSNs where efficient usage of energy is very critical, longer latency for non-critical data is preferable for longer SN lifetime. However, queries for time critical data should not be delayed and should be handled immediately. Some protocols try to use the energy of the network very efficiently by reducing unnecessary data transmission for non-critical data but transmitting time-critical data immediately, even if we have to keep the sensors on at all times. Periodic data is transmitted at longer intervals so that historical queries can also be answered. All other data is retrieved from the system on-demand. As we can see, WSNs cover a very broad area and impacts the design of every layer in the network protocol stack. Many of the things we have discussed earlier on ad hoc networking may need to be revisited as the applications (and hence the requirements) have changed, which impact the appropriate solutions in support of these applications. Therefore, in this chapter we discuss WSN with respect to every layer of the protocol stack and associated issues.

11.4.1 Challenges

Despite their innumerable applications, WSN have several restrictions, e.g., limited energy supply, limited computing power, and limited bandwidth of the wireless links connecting SNs. One of the main design goals of WSNs is to prolong the lifetime of the network and prevent connectivity degradation by employing aggressive energy management techniques. As we have seen before, existing routing protocols designed for other wireless networks and traditional networks cannot be used directly in WSNs for the following reasons:

- SNs should be self-organizing as the ad hoc deployment of these SNs requires the system to form connections and cope with the resultant nodal distribution. Coupled with the fact that the operation of the WSNs is un-attended, the network organization and configuration should be performed automatically and more often due to SNs failure;
- In most application scenarios, SNs are stationary. SNs in other traditional wireless networks are free to move, which results in unpredictable and frequent topological changes. However, in some applications, some SNs may be allowed to move and change their location (although with very low mobility);
- WSNs are application specific, i.e., design requirements of a WSN change with application. For example, the challenging problem of low-latency precision tactical surveillance is different from that required for a periodic weather-monitoring task;
- Data collected by many nearby sensors is based on common phenomena, thus there is a high probability that the data has redundancy. Therefore, data aggregation and in-network processing are desirable to yield energy-efficient data delivery before being sent to the destinations;
- In traditional networks, data is requested from a specific SN. WSNs are data centric i.e., data is requested based on certain attributes, i.e., attribute-based addressing. An attribute-based address is composed of a set of attribute-value pair query. For example, if the query is something like temperature $> 35°C$, then only those devices sensing temperature $> 35°C$ need to respond and report their readings. Other sensors can remain in the sleep state. Once an event of interest is detected, the system should be able to configure itself so as to obtain very high quality results;
- WSNs have relatively large number of SNs, which may be on the order of thousands of SNs. Therefore, SNs need not have a unique ID as the overhead of ID maintenance is high. In data-centric WSNs, the data is more important than knowing the IDs of which SNs sent the data;

- Position awareness of SNs is important since data collection is based on the location. Currently, it is not feasible to use GPS hardware for this purpose. Methods based on triangulation [Bulusu2000], for example, allow SNs to approximate their position using radio strength from a few known points. Algorithms based on triangulation can work quite well under conditions where only very few SNs know their positions a priori, e.g., using GPS hardware [Bulusu2000].

Routing protocol design for WSNs is heavily influenced by many challenging factors, which must be overcome before efficient communication can be achieved. These challenges can be summarized as follows:

- **Ad hoc deployment** – SNs are randomly deployed which requires that the system be able to cope up with the resultant distribution and form connections between the SNs. In addition, the system should be adaptive to changes in network connectivity as a result of SN failure.
- **Computational capabilities** – SNs have limited computing power and therefore may not be able to run sophisticated network protocols leading to light weighted and simple versions of routing protocols.
- **Energy consumption without losing accuracy** – SNs can use up their limited energy supply carrying out computations and transmitting information in a wireless environment. As such, energy-conserving forms of communication and computation are crucial as the SN lifetime shows a strong dependence on the battery lifetime. In a multi-hop WSN, SNs play a dual role as data sender and data router. Therefore, malfunctioning of some SNs due to power failure can cause significant topological changes and might require rerouting of packets and reorganization of the network.
- **Scalability** – The number of SNs deployed in the sensing area may be in the order of hundreds, thousands, or even more. Any routing scheme must be scalable enough to respond to events and capable of operating with such large number of SNs. Most of the SNs can remain in the sleep state until an event occurs, with data from only a few remaining SNs providing a coarse quality of information.
- **Communication range** – The bandwidth of the wireless links connecting SNs is often limited, hence constraining inter-sensor communication. Moreover, limitations on energy forces SNs to have

short transmission ranges. Therefore, it is likely that a path from a source to a destination consists of multiple wireless hops.

- **Fault tolerance** – Some SNs may fail or be blocked due to lack of power, physical damage, or environmental interference. If many SNs fail, MAC and routing protocols must accommodate formation of new links and routes to the data collection BSs. This may require actively adjusting transmit powers and signaling rates on the existing links to reduce energy consumption, or rerouting packets through regions of the network where more energy is available. Therefore, multiple levels of redundancy may be needed in a fault-tolerant WSN.

- **Connectivity** – High SN density in WSNs precludes them from being completely isolated from each other. Therefore, SNs are expected to be highly connected. This, however, may not prevent the network topology from varying and the network size from shrinking due to SNs failures. In addition, connectivity depends on possible random distribution of SNs.

- **Transmission media** – In a multi-hop WSN, communicating SNs are linked by a wireless medium. Therefore, the traditional problems associated with a wireless channel (e.g., fading, high error rate) also affect the operation of the WSN. In general, bandwidth requirements of sensor applications will be low, in the order of 1-100 kb/s. As we have seen in earlier chapters and in the previous section, the design of the MAC protocol is also critical in terms of conserving energy in WSNs.

- **QoS** – In some applications (e.g., some military applications), the data should be delivered within a certain period of time from the moment it is sensed, otherwise the data will be useless. Therefore, bounded latency for data delivery is another condition for time-constrained applications.

- **Control Overhead** – When the number of retransmissions in wireless medium increases due to collisions, the latency and energy consumption also increases. Hence, control packet overhead increases with the SN density. As a result, tradeoffs between energy conservation, self-configuration, and latency may exist.

- **Security** – Security is an important issue which does not mean physical security, but it implies that both authentication and encryption should be feasible. But, with limited resources, implementation of any complex algorithm needs to be avoided. Thus, a tradeoff exists between the security level and energy consumption in a WSN.

11.5 Localization Scheme

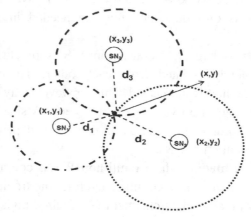

Figure 11.5 – Trilateration of SN *(x,y)* using three reference SNs *(x₁,y₁)*,
(x₂,y₂), and *(x₃,y₃)*

As indicated earlier, it is important to have location information of each SN so that it could be used for making important decision. One approach is to equip each SN with GPS capability which utilizes information from at least 3 satellites to determine its own location in the 3-dmensional space. But GPS devices are expensive and it is not practical to incorporate that with each SN. An alternative is to select some SNs as anchor or reference SNs and determine relative locations of all SNs. This can be done in many different ways. One simpler way is to use trilateration which determines the location by measuring distance between reference points. The distance is computed using RSSI (Received Signal Strength Indicator) as the signal strength is reversely proportional to the square of distance. This is illustrated in Figure 11.5, with location of SN *(x,y)* determined using three reference SNs at (x_1,y_1), (x_2,y_2), and (x_3,y_3) at respective distance of d_1, d_2, and d_3. Such location estimate is inaccurate due to multi-path fading, background interference, and irregular signal propagation. An alternative is to do triangulation which finds the location by measuring angles from known points and such Angle of Arrival estimates relative angles between neighbors. A more complex system is to do multilateration that finds the location by measuring time difference of signal from reference points. So, taking few reference points such as corners A, B, C, and D of a rectangular area of Figure 11.3 allows location information of all sensors in a given area.

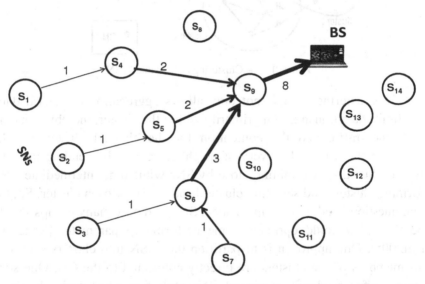

Figure 11.6 – Energy hole due to message concentration near BS

Energy Hole Problem – As the packets are to be accumulated at the BS, the number of packets may grow as they move closer to the BS. This is illustrated as the number of packets by 1, 2, 3 and 8 in Figure 11.6 and is known as the "energy hole" problem [Olariu2006]. To address this, either the use of larger density near the BS (viz. Gaussian distribution [Wang2008]) or data-aggregation schemes have been suggested [Banerjee2007].

11.6 Clustering of SNs

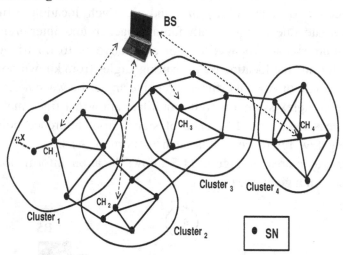

Figure 11.7 – Clustering of SNs in a WSN

Clustering of SNs not only allows aggregation of sensed data, but limits data transmission primarily within the cluster, thereby reducing both the traffic and the contention for the channel clustering. The sequence starts with discovery of neighboring SNs by sending periodic Beacon Signals, determining close by SNs with some intermediate SNs, forming clusters and selecting cluster head (CH) for each cluster. So, the real question is how to group adjacent SNs, and how many groups should be there that could optimize some performance parameter [Son2005, Liu2005]. One approach is to partition the WSN into clusters such that all members of the clusters are directly connected to the CH. One such example for randomly deployed SNs with four clusters is shown in Figure 11.7. SNs in a cluster can transmit directly to their respective CH without any intermediate SN. This minimizes the energy consumed within individual clusters. But, then CHs also need to transmit information among them and the energy consumed in wireless transmission is proportional to the square of the distance between the SNs acting as CHs. Therefore, it may desirable to partition a WSN into clusters with distance d (=2 in Figure 11.7 when SN x is added to CH_1), which can be defined as all SNs in a cluster reachable by a path length to CH less than or equal to d.

Determining optimal value of d that minimizes overall energy consumption, is a very complex problem and also need to take into account the amount of data to be transferred within each cluster and between clusters, frequency of transmissions, maximum allowable latency, local computation to be done and maintaining partial data base information. It is also unrealistic to assume that each SN has the information about the whole WSN connectivity as hundreds or thousands of WSs are expected to be deployed. Such centralized solutions, even through optimal, are fairly expensive in terms of collecting WSN-wide information. Therefore, if one can obtain close to optimal solution, by using appropriate heuristics in a distributed way, it should be more than acceptable.

The next question is how to select the CH of a cluster. A simplest scheme is to select WS with the lowest ID in the cluster by choosing largest neighborhood connectivity in selecting the CH. Another option is to use the SN with highest degree (largest number of neighbors) in the cluster as indicated by CH_1, CH_2, CH_3, and CH_4 in Figure 11.7. As the CH does aggregation of data received from its cluster members, it is usually trusted with the transmission schedule to its members and the BS. As CH needs to perform a lot more than others, it may run out of energy at a much faster rate. A dynamically changing CH has also been suggested [Heinzelman2000a] so as to distribute energy consumption as evenly as possible.

As indicated earlier, the SNs within each cluster can be assigned different time slot in a time-multiplex mode by the CH so that there is no collision between them. This necessitates adjacent clusters to use either a different channel of FDMA or a different code, if CDMA is used. An alternative strategy is to let each SN use one specific channel and there will not be any conflict if no other SN is using that channel with 2-hop distance of the transmission range. This leads to a 2-hop graph coloring problem [Chowdhury2005] which also helps in clustering the SNs of a WSN. Such as approach is feasible as programmable software defined wireless radios have now become a reality. It may be noted that the field may be covered by more than one SN. In that case, some of the SNs can be allowed to go to sleep mode as long as the area is covered by at least one active SN. Defining the sleep-awake cycles may be easier if a WSN is first divided into d-clusters, and compute the coverage of each cluster and then define the sleep cycle.

11.6.1 Architecture of WSNs

Due to the principle differences in application scenarios and underlying communication technology, the architecture of WSNs will be drastically different both with respect to a single WS and the network as a whole. The typical hardware platform of a wireless SN will consist of:

- A simple embedded microcontrollers, such as the Atmel or the Texas Instruments MSP 430. A decisive characteristic here is, apart from the critical power consumption, an answer to the important question whether and how these microcontrollers can be put into various operational and sleep modes, how many of these sleep modes exist, how long it takes and how much energy it costs to switch between these modes.

- Currently used radio transceivers include the RFM TR1001 or Infineon or Chipcon devices; similar radio modems are available from various manufacturers. Typically, ASK or FSK is used, while the Berkeley PicoNodes employ OOK modulation. Radio concepts like ultra-wideband are in an advanced stage (e.g., the projects undertaken by the IEEE 802.15 working group). A crucial step forward would be the introduction of a reasonably working wake-up radio concept, which could either wake up all SNs in the vicinity of a sender or even only some directly addressed SNs. A wake-up radio allows a SN to sleep and to be wakened up by suitable transmissions from other SNs, using only a low-power detection circuit. Transmission media other than radio communication are also considered, e.g., optical communication or ultra-sound for underwater-applications.

- Batteries provide the required energy and an important concern is whether and how energy scavenging can be done to recharge batteries in the field. Also, self-discharge rates, self-recharge rates and lifetime of batteries can be an issue.

- The operating system and the run-time environment is a hotly debated issue in the literature. TinyOS [Tinyos] is the open source free event-driven platform is a popular operating system for

WSNs. On one hand, minimal memory footprint and execution overhead are required while on the other, flexible means of combining protocol building blocks are necessary, as Meta information has to be used in many places in a protocol stack.

• If each cluster is covered by more than one subset of SNs all the time, then some of the SNs can be put into sleep mode so as to conserve energy while keeping full coverage of each cluster and the area. The use of a second smaller radio has been suggested for waking up the sleeping sensor [Miller2005], thereby conserving the power of main wireless transmitter.

11.6.2 Network Lifetime

As discussed earlier, WSN architecture needs to cover a desired area both for sensing coverage and communication connectivity point of view. Therefore, density of the WSN network is critical for the effective use of the WSN. There is no well-defined measure of life-time of a WSN. Some assume either the failure of a single sensor running out of battery power, is taken as life-time of the network. Perhaps a better definition is if certain percentage of sensors stops working, may define the life-time as the network continues to operate. The percentage failure may depend on the nature of application and as long as the area is adequately covered by the working SNs, a WSN may be considered operational. Here you could also have some quantitative measure such as the monitored area is 95% covered.

The SNs are yet to become inexpensive to be deploying with some degree of redundancy. For example, it is good to say that a region 3 in Figure 11.3 can be monitored by three sensors simultaneously. But this is still a theoretical concept as coverage of a region by a single sensor is currently adequate. In addition, the degree of data reduction by collaborative aggregation plays a vital role in minimizing the energy consumption. A denser deployment of sensor and transmission of sensed data may cause more energy consumption and increased delay due to collisions. On the other hand, transmitting data between two far apart sensors may cause increased energy consumption due to increased energy consumption in wireless transmission (Figure 11.8). Therefore, there is an optimal distance between two sensors that would maximize the sensor lifetime [Bhardwaj2002]. So, if the density of sensors is high, then some of the sensors can be put into sleep mode to have close to optimal distance between the sensors. The network architecture as a whole has to take various aspects into account including:

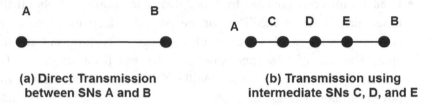

**(a) Direct Transmission
between SNs A and B**   **(b) Transmission using
intermediate SNs C, D, and E**

Figure 11.8 – Transmission strategies between two SNs A and B

- The protocol architecture has to take both application- and energy-driven point of view.
- QoS, dependability, redundancy and imprecision in sensor readings have to be considered.
- The addressing structures in WSNs are likely to be quite different: scalability and energy requirements can demand an "address-free structure" [Estrin2001]. Distributed assignments of addresses can be a key technique, even if these addresses are only unique in a two-hop neighborhood. Also, geographic and data-centric addressing structures are required.
- A crucial and defining property of WSNs will be the need for and their capability to perform in-network processing. This pertains to aggregation of data when multiple sensor readings are converge-casted to a single or multiple sinks, distributed signal processing, and the exploitation of correlation structures in the sensor readings in both time and space. In addition, aggregating data reduces the number of transmitted packets.
- As these services are, partially and eventually, invoked by agents outside the system, a BS concept is required: How to structure the integration of WSNs into larger networks, where to bridge the different communication protocols (starting from physical layer upwards) are open issues.
- More specifically, integration of such ill-defined services in middleware architectures like CORBA [CORBA] or into web

services is also not clear: how to describe a WSN service such that it can be accessed via a Web Service Description Language [WSDL] and Universal Description, Discovery and Integration [UDDI] description?

- Other options could be working with non-standard networking architectures, e.g., the user of agents that "wander" around a given network and explore the tomography or the "topology" of the sensed values.

- From time to time, it might be necessary to reassign tasks to the BS, i.e., to provide all its SNs with new tasks and new operating software.

11.6.3 Physical Layer

Very little work has been done on protocols that suits well to the needs of WSNs. With respect to the radio transmission, the main question is how to transmit as energy efficiently as possible, taking into account all related costs (possible retransmissions, overhead, and so on). Some energy efficient modulation techniques have been discussed in [Schurgers2001]. In [Gao2001], the hardware aspect for CDMA in SNs is considered and modulation issues are described. A discussion of communication protocol design based on the physical layer can be found in [Shih2001b]. Given the work being done at the IEEE level (e.g., the IEEE 802.15.4 standard) and also given the limited research in this area, we have chosen not to go into the details of the physical layer for WSNs. We note, however, that this is a very important issue that needs careful consideration.

11.7 MAC Layer

The MAC and the routing layers are the most active research areas in WSNs. Therefore, an exhaustive discussion of all schemes is impossible. However, most of the existing work addresses how to make SNs sleep as long as possible. Consequently, these proposals often tend to include at least some aspects of TDMA. The wireless channel is primarily a broadcast medium. All SNs within radio range of a SN can hear its transmission. This can be used as a unicast medium by specifically addressing a particular SN and all other SNs can drop the packet they receive. There are two types of schemes that could allocate a single broadcast channel among competing SNs: Static Channel Allocation and Dynamic Channel Allocation.

- Static Channel Allocation: In this category of protocols, if there are N SNs, the bandwidth is divided into N equal portions in frequency (FDMA), in time (TDMA), in code (CDMA), in space (SDMA) or in schemes such as OFDM or ultra-wideband. Since each SN is assigned a private portion, there is no or minimal interference amongst multiple SNs. These protocols work very well when there are only a small and fixed number of SNs, each of which has buffered (heavy) load of data;

- Dynamic Channel Allocation: In this category of protocols, there is no fixed assignment of bandwidth. When the number of active SNs changes dynamically and data becomes bursty, it is better to use dynamic channel allocation scheme. These are contention-based schemes, where SNs compete for the channel when they have data while minimizing collisions with other SNs' transmissions. When there is a collision, the SNs are forced to retransmit data, thus leading to increased wastage of energy and increased delay. Example protocols are: CSMA (persistent and non-persistent) [Tanenbaum1996], MACAW [Bharghavan1994], IEEE 802.11 [Crow1997], etc.

As we will see shortly, in a hierarchical clustering model, once clusters have been formed, it is desirable to keep the number of SNs in each cluster fixed and due to hierarchical clustering; the number of SNs per cluster is kept small, encouraging using a static channel allocation scheme. The use of TDMA for WSNs has been studied in [Heinzelman2000a, Intanagonwiwat2000], where each SN transmits data in its own slot to the CH and at all other times, its radio can be switched off, thereby saving energy. When it is not feasible to use TDMA, the SNs can use non-persistent CSMA since the data packets are of fixed size.

TDMA is suitable for either proactive or reactive type of networks. In proactive networks, as we have the SNs transmitting periodically, we can assign each SN a slot and thus avoid collisions. In reactive networks [Manjeshwar2001], when a sudden change takes place in some attribute being sensed, many SNs noticing abrupt difference will respond immediately. This may lead to collisions and it is possible that the data may never reach the user on time. For this reason, TDMA is employed so that each SN is given a slot and they transmit only in that slot. Even though many slots might be empty and could increases the delay, it is better than the energy consumption incurred due to dynamic channel allocation schemes.

CDMA can be used to avoid collisions due to use of the same channel by different clusters, although this means that more data needs to be transmitted per bit, it allows for multiple transmissions using the same frequency. A number of advantages have been pointed out for using TDMA/CDMA combination to avoid intra/inter cluster collision in and WSNs [Heinzelman2000b].

11.7.1 Design Issues

As with MAC protocols for traditional MANETs, WSNs have their own inherent characteristics that need to be addressed. Below we discuss some of the most important ones in the design of MAC protocols for WSNs.

Coping up with SN Failure

When many SNs have failed, the MAC and routing protocols must accommodate formation of new links and routes to other SNs and the BS. This may require dynamically adjusting transmit powers and signaling rates on the existing links, or rerouting packets through regions of the network with higher energy level.

Sources of Resource Consumption at the MAC Layer

There are several aspects of a traditional MAC protocol that have negative impact on wireless WSNs including:

• Collisions,
• Overhearing;
• Control packets overhead; and
• Idle.

Measures to Reduce Energy Consumption

One of the most cited methods to conserve energy in WSNs is to avoid listening to idle channels, that is, neighboring SNs periodically sleep (radio off) and auto synchronize as per sleep schedule. It is important to note that fairness, latency, throughput and bandwidth utilization are secondary in the WSNs.

Comparison of Scheduling & Reservation-based and Contention-based MAC Design

One approach of MAC design for WSNs is based on reservation and scheduling, for example TDMA-based protocols that conserve more energy as compared to contention-based protocols like the IEEE 802.11 DCF. This is because the duty cycle of the radio is increased and there is no contention-introduced overhead and collisions. However, formation of cluster, management of inter-cluster communication, and dynamic adaptation of the TDMA protocol to variation in the number of SNs in the cluster in terms of its frame length and time slot assignment are still the key challenges.

11.7.2 MAC Protocols

WSNs are designed to operate for long time as it is rather impractical to replenish the batteries. However, SNs are in idle state for most time and similar level of energy in idle mode as in receiving mode [Stemm1997]. Therefore, it is important that SNs are able to operate in low duty cycles. As far as the MAC layer, some of the relevant studies include [Bao2001, Kanodia2001, Woo2001], the PicoRadio MAC [Zhong2001], the S-MAC [Ye2002], the SMACS [Sohrabi2000], and the STEM [Schurgers2002]. As many of these protocols share common characteristics, in this section we discuss only those which are most prominent and necessary to understand the others. It is also important to note that more traditional MAC schemes such as FDMA, TDMA, CDMA, SDMA and a combination of these can also be employed. However, as these techniques are widely known, we do not discuss them.

11.7.3 The Sensor-MAC

The Sensor-MAC (S-MAC) protocol [Ye2002] explores design trade-offs for energy-conservation in the MAC layer. It reduces the radio energy consumption from the following sources: collision, control overhead, overhearing unnecessary traffic, and idle listening.

The basic scheme of S-MAC is to put all SNs into a low-duty-cycle mode –listen and sleep periodically. When SNs are listening, they follow a contention rule to access the medium, which is similar to the IEEE 802.11 DCF. In S-MAC, SNs exchange and coordinate on their sleep schedules rather than randomly sleep on their own. Before each SN starts the periodic sleep, it needs to choose a schedule and broadcast it to its neighbors. To prevent long-term clock drift, each SN periodically broadcasts its schedule as the SYNC packet. To reduce control overhead and simplify broadcasting, S-MAC encourages neighboring SNs to choose the same schedule, but it is not a requirement. A SN first listens for a fixed amount of time, which is at least the period for sending a SYNC packet. If it receives a SYNC packet from any neighbor, it will follow that schedule by setting its own schedule to be the same. Otherwise, the SN will choose an independent schedule after the initial listening period.

It is possible that two neighboring SNs have two different schedules. If they are aware of each other's schedules, they have two options:

• Follow the two schedules by listening at both scheduled listen time.
• Only follow its own schedule, but transmit twice as per both schedules when broadcasting a packet.

In some cases the two SNs may not be aware of the existence of each other, if their listen intervals do not overlap at all. To solve the problem, S-MAC let each SN periodically perform neighbor discovery, i.e., listening for the entire SYNC period, to find unknown neighbors on a different schedule. Figure 11.9 depicts the low-duty-cycle operation of each SN. The listen interval is divided into two parts for both SYNC and data packets. There is a contention window for randomized carrier sense time before sending each SYNC or data (RTS or broadcast) packet. For example, it SN A wants to send a unicast packet to SN B, it first perform carrier sense during B's listen time for data. If carrier sense indicates an idle channel, SN A will send RTS to SN B, and B will reply with CTS if it is ready to receive data. After that, they will use the normal sleep time to transmit and receive actual data packets. Broadcast does not use RTS/CTS due to the potential collisions on multiple CTS replies.

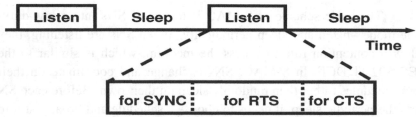

Figure 11.9 – Low-duty-cycle operation in S-MAC

Low-duty-cycle operation reduces energy consumption at the cost of increased latency, since a SN can only start sending when the intended receiver is listening. S-MAC developed an adaptive listen scheme to reduce the latency in a multi-hop transmission. The basic idea is to let the SN who overhears its neighbor's transmissions (ideally only RTS or CTS) wake up for a short period of time at the end of the transmission. In this way, if the SN is the next-hop SN, its neighbor is able to immediately pass the data to it, instead of waiting for its scheduled listen time. If the SN does not receive anything during the adaptive listening, it can go back to sleep mode.

Trade-offs on Energy, Latency and Throughput

With low-duty-cycle operation, S-MAC effectively reduces the energy waste due to idle listening. Experimental results show that an 802.11-like protocol without sleeping consumes 2–6 times more energy than S-MAC for traffic load with messages sent every 1–10 seconds in a 10-hop WSN. On the other hand, S-MAC with adaptive listen has about twice the latency as the MAC without sleeping. Periodic sleeping increases latency and reduces throughput. However, adaptive listening largely reduces such cost. It enables each SN to adaptively switch mode according to the traffic in the WSN. The overall gain on energy savings is much larger than the performance loss on latency and throughput [Ye2003].

11.8 The Self-Organizing MAC for WSNs and the Eaves-drop-And-Register Protocol

In this section a Self-Organizing MAC for WSNs (SMACS) and the Eaves-drop-And-Register (EAR) protocol [Sohrabi2000] is discussed. SMACS is an infrastructure-building protocol employed for network startup and link layer organization. EAR, in turn, enables seamless interconnection of SNs in the field of stationary wireless SNs, and represents the mobility management aspect of the protocol.

11.8.1 SMACS

The SMACS is an infrastructure-building protocol that forms a flat topology (as opposed to a cluster hierarchy) for WSNs. SMACS is a distributed protocol which enables a collection of SNs to discover their neighbors and establish transmission/reception schedules for communicating with them without the need for any local or global master SNs. In order to achieve this ease of formation, SMACS combines the neighbor discovery and channel assignment phases. Unlike methods such as the Linked Clustering Algorithm (LCA) [Baker1981] in which a first pass is performed on the entire WSN to discover neighbors and then another pass done to assign channels, or TDMA slots, to links between neighboring SNs, SMACS assigns a channel to a link immediately after the link's existence is discovered. This way, links begin to form concurrently throughout the network. By the time all SNs hear all their neighbors, they would have formed a connected WSN where there exists at least one multihop path between any two distinct SNs.

Since only partial information about radio connectivity in the vicinity of a SN is used to assign time intervals to links, there is a potential for time collisions with slots assigned to adjacent links whose existence is not known at the time of channel assignment. To reduce this, each link is required to operate on a different frequency. This frequency band is chosen at random from a large pool of possible choices when the links are formed. This idea is described in Figure 11.11(a) for the topology of Figure 11.10. Here, SNs A and D wake up at times T_a and T_d. After they find each other, they agree to transmit and receive during a pair of fixed time slots. This transmission/reception pattern is repeated periodically every T_{frame}. SNs B and C, in turn, wake up later at times T_b and T_c, respectively. Similarly, after they find each other they assign another pair of slots for transmission and reception. Note that if all the SNs operate on the same frequency band, there is the possibility that some transmissions collide in the given schedule. For example, a transmission from D to A will collide in time with a transmission from B to C. On the other hand, if different frequency band is assigned to different link (e.g., f_x to AD link and f_y to BC link); the time schedule of Figure 11.11(a) can work without any collision. When there are many frequencies from which to choose, and frequencies are randomly chosen, there is a small probability that the same frequency is chosen by two links within earshot.

T_{frame} as described above, is fixed for all SNs, and is a parameter of the MAC. In other words, T_{frame} can be seen as the length of the MAC superframe. As new neighbors are found and new links formed, the superframe of each SN would be filled. From Figure 11.8(a) we see that T_{frame} epochs for SNs A and B, for example, do not coincide. Now if we call each transmission or reception period a slot, we see from the same figure that the protocol will result in slot assignments which need not to be aligned throughout the entire network. The ability to assign non-synchronous slots in the network is the key issue that enables the SNs to form links on the fly, and is called non-synchronous scheduled communication (NSC). This spontaneity enables a quick method of scheduling links throughout the network. After a link is established, a SN knows when to turn on its transceiver ahead of time to communicate with another SN. It will turn off when no communication is scheduled. This scheduled mode of communication enables energy savings for the SN. Since link assignment is accomplished quickly, without requiring accumulation of global connectivity information or even connectivity information that reaches farther than one hop away, the overall effect is significant energy savings.

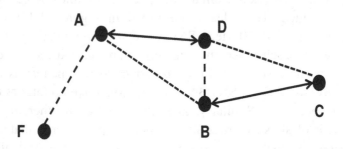

Figure 11.10 – Network topology

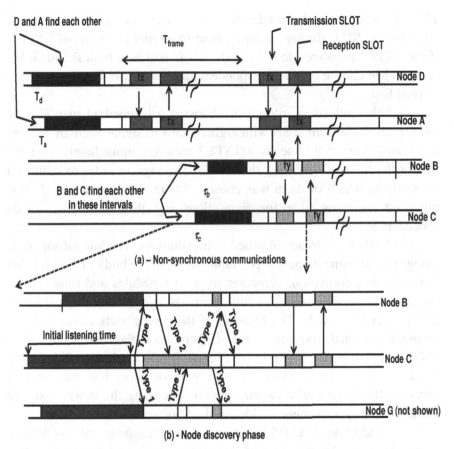

Figure 11.11 – Startup procedures in SMACS

The method by which SNs find each other and the scheme by which time slots and operating frequencies are determined constitute an important part of SMACS [Sohrabi1999]. To illustrate this scheme, consider SNs B, C, and D shown in Figure 11.11(b). These SNs are engaged in the process of finding neighbors and wake up at random times. Upon waking up, each SN listens to the channel on a fixed frequency band for some random time duration. A SN decides to transmit an invitation by the end of this initial listening time if it has not heard any invitations from other SNs. This is what happens to SN C, which broadcasts an invitation or TYPE1 message. SNs B and D hear this

TYPE1 message and broadcasts a response, or TYPE2, message addressed to SN C during a random time following reception of TYPE1. If the TYPE2 messages do not collide, SN C will hear both B and G. SN C must then choose only one respondent and selects SN B as its response arrived first.

Other selection criteria for choosing a respondent may also be used, such as choosing a SN with higher received signal strength or more neighbors. Next, SN C sends a TYPE3 message immediately after the end of the interval following the TYPE1 message in order to notify all respondents which of them was chosen. SN D, which was not chosen, turns off its transceiver for some time and then starts the search procedure again.

If SN C is already attached, it transmits its schedule information, along with the time its next superframe starts, in the body of TYPE3. SN B reads this information, compares the two schedules and time offsets, and arrives at a set of two free time intervals as the slots assigned to the link between C and B. The location of these time slots along with the randomly selected frequency band of operation is then sent by SN B to SN C in the body of a TYPE4 message. At this point, the two SNs have a pending link between them. Once a pair of short test messages is successfully exchanged between the two SNs using the newly assigned slots, the link is permanently added to the SNs' schedules.

In addition, SMACS has the concept of a *subnet* which is defined as a subset of SNs that form a connected graph and have coinciding superframe epochs. Therefore, there are at least two SNs in each subnet. For example, in Figure 11.11(a) SNs A and D form a subnet and B and C form another. As time goes on, these subnets grow in size by attaching new SNs. They will eventually become attached to other subnets, until finally almost all the SNs in the network are connected together. The case when two SNs find each other and attempt to form a link, while they are already members of different subnets, is a challenging scenario in the startup procedure. As long as the super frame of both SNs has enough overlap in unassigned regions to allocate a pair of slots for the new link, there is no need for the two SNs to re-organize their respective schedules in order to make room for the new link. If there is no room left, the two SNs simply give up and search for other SNs.

11.8.2 EAR

Mobility can be introduced into a WSN as extensions to the stationary WSN. Mobile connections are very useful to a WSN where small, low bit rate data packets can be exchanged to relay SNs to and from the SNs. At the same time, it cannot be assumed that each mobile SN is aware of the global network state and/or SN positions. The EAR protocol attempts to offer continuous service to these mobile SNs under both mobile and stationary constraints. As battery power is the primary concern of stationary SNs, the communication channels between stationary and few mobile SNs must be established with as few messages as possible. To avoid unnecessary consumption of energy associated with lost messages, the mobile SNs assume full control of the connection process. Furthermore, the overhead associated with acknowledgments can be eliminated due to proximity between SNs. To avoid connection handoff, the mobile SN keeps a registry of SNs, selecting a new connection only when absolutely necessary.

Since there will be few stationary SNs aware of the presence of the mobile SNs, the EAR protocol could be transparent to the existing stationary protocol. EAR assumes that stationary SNs use a TDMA-like frame structure, within which slots are designated for inviting neighboring SNs into the network. Since the stationary SN does not require a response to this message, the mobile SN is simply "eavesdropping" on the control signals of the stationary MAC protocol. It then decides the best course of action regarding the transmitting stationary sensor; hence, this invitation message acts as the trigger for the EAR algorithm.

In order to keep a constant record of neighboring activity, the mobile SN forms a registry of neighbors. The stationary SN simply registers mobile SNs that have formed connections and remove them when the link is broken. The EAR algorithm employs the following four primary messages:
- Broadcast Invite (BI).
- Mobile Invite (MI).
- Mobile Response (MR).
- Mobile Disconnect (MD.

11.8.3 The STEM

The Sparse Topology and Energy Management (STEM) protocol [Schurgers2002] is based on the assumption that most of the time the WSN is only sensing the environment, waiting for an event to happen. In other words, STEM may be seen as better suitable for reactive WSNs where the network is in the monitoring state for vast majority of time. One example of such application is a WSN designed to detect wildfire in a forest. These networks have to remain operational for months or years, but sensing only on the occurrence of a forest fire. Clearly, although it is desirable that the transfer state be energy-efficient, it may be more important that the monitoring state be ultra-low-power as the network resides in this state for most of the time.

The idea behind STEM is to turn on only a SN's sensors and some preprocessing circuitry during monitoring states. Whenever a possible event is detected, the main processor is woken up to analyze the sensed data in detail and forward it to the data sink. However, the radio of the next hop in the path to the data sink is still turned off, if it did not detect the same event. STEM solves this problem by having each SN to periodically turn on its radio for a short time to listen if someone else wants to communicate with it. The SN that wants to communicate, i.e., the *initiator SN*, sends out a beacon with the ID of the SN it is trying to wake up, i.e., the *target SN*. As soon as the target SN receives this beacon, it responds back to the initiator SN and both keep their radio on at this point. If the packet needs to be relayed further, the target SN will become the initiator SN for the next hop and the process is repeated.

Once both the SNs that make up a link have their radio on, the link is active and can be used for subsequent packets. However, the actual data transmissions may still interfere with the wakeup protocol. To overcome this problem, STEM proposes the wakeup protocol and the data transfer to employ different frequency bands as depicted in Figure 11.12. In addition, separate radios would be needed in each of these bands [Sensoria]. In Figure 11.12 we see that the wakeup messages are transmitted by the radio operating in frequency band f_1. STEM refers to these communications as occurring in the *wakeup plane*. Once the initiator SN has successfully notified the target SN, both SNs turn on their radio that operates in frequency band f_2. The actual data packets are transmitted in this band, called the *data plane*.

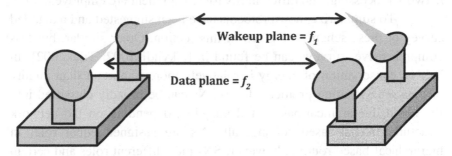

Figure 11.12 –SN configuration in STEM

11.8.4 Link Layer

Compared to the MAC and routing layers, very little work exists on the link layer for WSNs. The question of choosing suitable packet size for energy efficient operation is discussed in [Sankarasubramaniam2003a], while energy efficient issues at the link layer are also investigated in [Zorzi1997]. Finally, the use of FEC and transmission power variation on the energy spent per useful bit is studied in [Shih2001a].

11.9 Routing Layer

WSNs differ from traditional wireless networks as conventional flooding-based protocols widely employed in MANETs suffer from data explosion problem which wastes resources by sending and receiving duplicate data copies. In addition, as shown in Figure 8.1, flooding does not scale well in large networks due to multi-hop routing in WSNs. The main objective behind routing is to send data from a SN to a known destination SN, i.e., the BS with known location. A BS may be fixed or mobile, and is capable of connecting the WSN to an existing infrastructure (e.g., Internet) to let the user collect data. The task of finding and maintaining routes in WSNs is nontrivial since energy restrictions and sudden changes in SN status cause unpredictable topological changes. Thus, the main objective of routing techniques is to minimize the energy consumption in order to prolong WSN lifetime. To achieve this objective, routing protocols proposed in the literature employ some well-known routing techniques to WSNs. To preserve energy, strategies like clustering, data-centric methods, data aggregation and in-network processing, and different SN role assignment are employed.

To summarize, new protocols have been suggested and a detailed survey of these schemes is given in this section. Other similar, but less comprehensive, surveys can be found in [Akyildiz2002, Tilak2002]. In WSNs, conservation of energy is relatively more important than quality of data sent. Routing protocols for WSNs can be broadly classified into flat-based, hierarchical-based, and adaptive, depending on the network structure. In flat-based routing, all SNs are assigned equal role. In hierarchical-based routing, however, SNs play different roles and certain SNs, called cluster heads (CHs), are given more responsibility. In adaptive routing, certain system parameters are controlled in order to adapt to the current network conditions and available energy levels. Furthermore, these protocols can be classified into multipath-based, query-based, negotiation-based, or location-based routing techniques. In this section we use a classification according to the network structure and protocol operation (i.e., routing criteria), and is shown in Figure 11.13.

In majority of applications, SNs are expected to be stationary. Thus, it may be preferable to have table driven routing protocols rather than employing reactive schemes where a significant amount of energy is used in route discovery and setup. Another class of routing protocols is called the cooperative routing protocols wherein SNs send data to a CH for aggregation and reduce the energy use. Several other protocols, in turn, rely on timing and position information and are also covered in this section.

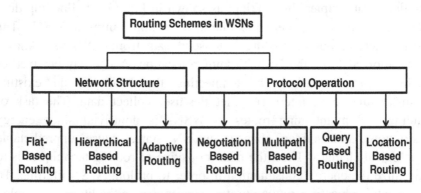

Figure 11.13 – Classification of Routing Protocols for WSNs

Network Structure Based

In this class of routing protocols, the network structure is one of the determinant factors. In addition, the network structure can be further subdivided into flat, hierarchical and adaptive depending upon its organization.

Flat Routing

In flat routing based protocols, all SNs play the same role. Here, we present the most prominent protocols falling in this category.

11.9.1 Directed Diffusion

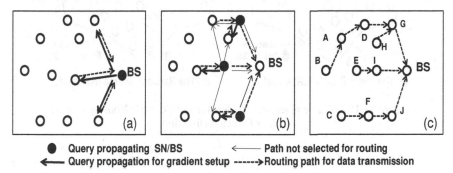

● Query propagating SN/BS   ◄——— Path not selected for routing
◄—— Query propagation for gradient setup ------►Routing path for data transmission

Figure 11.14 – Routing path set up by Directed Diffusion (a) Query from BS;
(b) Query propagation from three SNs; and (c) Routing path from each SN

Directed Diffusion [DD] is a data-centric data dissemination paradigm for WSNs [Intanagonwiwat2000]. DD is very useful for applications requiring processing of queries. As shown in Figure 11.14, in DD, BS and sensors create gradients of information in their respective neighborhoods. The BS requests data by broadcasting *query*, and interest is diffused through the network hop-by-hop, and is broadcast by each SN to its neighbors. As the interest is propagated throughout the network, gradients are setup to draw path satisfying shortest path from each SN towards the BS. This process continues until gradients are setup from each SN back to the BS. Figure 11.14(a) presents the propagation of interests; Figure 11.14(b) shows successive steps in the gradient construction; and Figure 11.14(c) depicts the data path from each SN to the BS. When multiple paths are formed from a SN, the best path is selected so as to prevent further flooding according to a local rule. In order to reduce communication costs, data is aggregated on the way to the BS when multiple SNs respond to a query such as in Figure 11.14(c), data from SNs H and D can be combined at SN G.

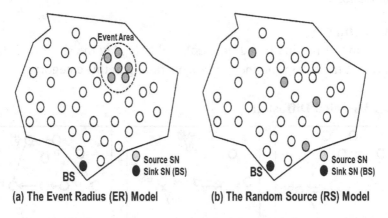

(a) The Event Radius (ER) Model   (b) The Random Source (RS) Model

Figure 11.15 – Source placement models for Data-Centric Routing schemes such as Directed Diffusion

DD-based WSNs are application-aware, which makes such type of information retrieval well suited for persistent queries where requesting SNs expect data that satisfy a query for a period of time. The performance of data aggregation is affected by the location of the source SN, the number of source SNs, and the WSN topology. In order to understand the impact of these factors, two models of event radius (ER) and random sources (RS) placement are shown in Figure 11.15. In the ER model, a single point in the network area is defined as the location of an event. All SNs within a sensing range of this event that are not sinks are considered to be data source SNs. In the RS model, K SNs that are not sinks are randomly placed act as sources and are not necessarily closed to each other.

11.9.2 Sequential Assignment Routing (SAR)

The routing scheme in SAR [Sohrabi2000] is dependent on three factors: energy resources, QoS on each path, and the priority level of each packet. To avoid single route failure, a multi-path with a tree rooted at the source SN to the destination SNs coupled with a localized path restoration scheme is employed. The paths of the tree are defined by avoiding SNs with low energy or QoS guarantees. For each SN, two metrics are associated with each path: delay (which is an additive QoS metric); and energy usage for routing on that path. SAR calculates a weighted QoS metric as the product of the additive QoS metric and a weight coefficient associated with the priority level of the packet. The goal of SAR is to minimize the average weighted QoS metric throughout the lifetime of the network. Also, a path re-computation is carried out if the topology changes due to SN failures or triggered by the BS. In addition, a handshake procedure based on a local path restoration scheme between neighboring SNs is used to recover from a failure.

11.9.3 Minimum Cost Forwarding Algorithm

The minimum cost forwarding algorithm (MCFA) [Ye2001] exploits the direction of routing, that is, towards fixed and predetermined external BS. Therefore, a SN need not have a unique ID nor maintain a routing table. Instead, each SN maintains the least cost estimate from itself to the BS. When a SN receives the message, it checks if it is on the least cost path between the source SN and the BS. This process repeats until the BS is reached. The BS broadcasts a message with the cost set to zero. Each SN, upon receiving the broadcast message originated at the BS, checks to see if the estimate in the message plus the cost of the link on which the message was received is less than the current estimate. If so, the current estimate and the estimate in the broadcast message are updated, and the message is re-broadcast. Otherwise, it is purged and nothing is done. To prevent multiple updates, the MCFA has been modified to run a back off algorithm at the setup phase. This algorithm mandates that a SN will not send the updated message until $a*l_c$ time units have elapsed from the last update, where a is a constant and l_c is the link cost from which the message is received.

Data processing is a major component in the operation of any WSN. In general, SNs cooperate with each other in processing different data flooded throughout the network. Two examples of data processing techniques are coherent and non-coherent data processing-based routing [Sohrabi2000] and are considered next.

11.9.4 Coherent and Non-Coherent Processing

In non-coherent data processing routing, SNs locally process the raw data before being sent to other SNs for further processing. The SNs that perform further processing are called the aggregators. In coherent routing, the data is forwarded to aggregators after minimum processing of time stamping and duplicate suppression. To perform energy-efficient routing, normally coherent processing is selected which generates long data streams and as such must achieve energy efficiency by path optimality. Non-coherent functions generate fairly low load and data processing is done three phases: (i) Target detection, data collection, and preprocessing; (ii) Membership declaration; and (iii) Central SN election.

A single and multiple winner algorithms have been proposed in [Sohrabi2000] for non-coherent and coherent processing, respectively. In the single winner algorithm (SWE), a single aggregator SN is elected for complex processing. The election of a SN is based on the energy reserves and computational capability of that SN. By the end of the SWE process, a minimum-hop spanning tree is constructed that completely covers the network. A simple extension to SWE is proposed to have a multiple winner algorithm (MWE). When all SNs are sources and send their data to the central aggregator SN, a large amount of energy is consumed and hence incurs a high cost. One way to lower the energy cost is to limit the number of sources that can send data to the central aggregator SN. Instead of keeping record of only the best SN, each SN keeps a record of up to n SNs of those candidates. At the end of the MWE process, each SN in the network has a set of minimum-energy paths to each source SN. After that, SWE is used to find the SN that yields the minimum energy consumption, which then serves as the central SN for the coherent processing. In general, the MWE process has longer delay, higher overheads, and lower scalability than that for non-coherent processing networks.

11.9.5 Energy Aware Routing

A destination initiated reactive protocol is proposed in [Shah2002] in order to prolong the WSN lifetime. This protocol is similar to directed diffusion with the difference that it maintains a set of paths through localized flooding instead of maintaining or enforcing one optimal path. These paths are chosen by means of a certain probability, which depends on how low the energy consumption of each path can be achieved and the probability is set to be inversely proportional to the path cost. By selecting different routes at different times, the energy of any single route will not deplete so quickly. With this scheme, the network degrades gracefully as energy is dissipated more equally amongst all SNs.

11.9.6 Hierarchical Routing

Hierarchical, or cluster-based, routing has its roots in wired networks, where the main goals are to achieve scalable and efficient communication. As such, the concept of hierarchical routing has also been employed in WSN to perform energy-efficient routing. In a hierarchical architecture, higher energy SNs (usually called cluster heads (CH)) can be used to process and send the accumulated information while low energy SNs can be used to sense in the neighborhood of the target and pass on to the CH. In these architectures, creation of clusters and assigning appropriate special tasks to CHs can contribute to overall system scalability, lifetime, and energy efficiency.

An example of a general hierarchical clustering scheme is depicted in Figure 11.7. As we can see from this figure, each cluster has a CH which collects data from its cluster members, aggregates it and sends it to the BS or there could be hierarchy among the CHs. For example in Figure 11.7, there are 5 SNs in cluster 2. There could be hierarchy among 4 CHs. For example, CH_2 could become a second level CH. This pattern can be repeated to form a hierarchy of CHs, with the uppermost level CH reporting directly to the BS. The BS forms the root of this hierarchy and supervises the entire network.

11.9.7 Cluster Based Routing Protocol (CBRP)

A simple cluster based routing protocol (CBRP) has been proposed in [Jiang1998]. It divides the network SNs into a number of overlapping or disjoint two-hop-diameter clusters in a distributed manner. The major drawback with CBRP is that it requires a lot of hello messages to form and maintain the clusters, and thus may not be suitable for WSN. Given that SNs are stationary this is a considerable and unnecessary overhead.

Scalable Coordination

In [Estrin1999], a hierarchical clustering method is discussed, with emphasis on localized behavior and the need for asymmetric communication and energy conservation in a WSN. In this method the cluster formation appears to require considerable amount of energy (no experimental results are available) as periodic advertisements are needed to form the hierarchy. Also, any changes in the network conditions or sensor energy level may result in re-clustering which may be not quite acceptable.

11.9.8 Low-Energy Adaptive Clustering Hierarchy (LEACH)

LEACH was introduced [Heinzelman2000b] as a hierarchical clustering algorithm for WSNs, called Low-Energy Adaptive Clustering Hierarchy (LEACH). LEACH is a good approximation of a proactive network protocol, with some minor differences which includes a distributed cluster formation algorithm. LEACH randomly selects a few SNs as CHs and rotates this role amongst the cluster members so as to evenly distribute the energy dissipation. In LEACH, the CH SNs compress data arriving from SNs that belong to the respective cluster, and send an aggregated packet to the BS in order to reduce the amount of information that must be transmitted. LEACH uses a TDMA and CDMA MAC to reduce intra-cluster and inter-cluster collisions, respectively. However, data collection is centralized and is performed periodically. Therefore, this protocol is better appropriate when there is a need for constant monitoring by the WSN. On the other hand, a user may not need all the data immediately, making periodic data transmissions unnecessary. After a given interval of time, a randomized rotation of the role of the CH is conducted so that uniform energy dissipation in the WSN is obtained. Based on simulation, it has been found that only 5% of the SNs actually need to act as CHs.

The operation of LEACH is separated into two phases, the setup phase and a longer steady state phase. In the setup phase, the clusters are organized and CHs are selected. In the steady state phase, actual data transfer to the BS takes place. During the setup phase, a predetermined fraction of SNs, say p, elect themselves as CHs as follows. A SN chooses a random number, say r, between 0 and 1. If this random number is less than a threshold value, say $T(n)$, the SN becomes a CH for the current round. The threshold value, in turn, is calculated based on an equation that incorporates the desired percentage to become a CH, the current round, and the set of SNs that have not been selected as a CH in the last $(1/p)$ rounds, denoted by G. As a result, $T(n)$ is given by:

$$T(n) = \frac{p}{1 - p\left(r \bmod \frac{1}{p}\right)} \quad if \ n \in G$$

where G is the set of SNs that are involved in the CH election. All the non-CH SNs, after receiving advertisement from newly elected CH, decide on the cluster to which they want to attach to. In LEACH, this decision is based on the signal strength of the advertisement. The non-CH SNs then inform the corresponding CH of their decision to be a member of its cluster. Based on the number of SNs in the cluster, the CH SN creates a TDMA schedule and assigns each SN a time slot. During the steady state phase, SNs begin sensing and transmitting data to their respective CHs. Once the CH receives the data from all of its members, it aggregates before relaying data to the BS. After a period time, which is determined a priori, the network goes back into setup phase and initiates another round of selecting new CHs.

Although LEACH is able to increase the network lifetime, there are still a number of issues regarding many assumptions. For example, LEACH assumes that all SNs can transmit with enough power to reach the BS if needed, and that every SN has enough computational power to support different MAC protocols. It also assumes that SNs always have data to send, and SNs located close to each other have correlated data. Also, it is not obvious how the number of the predetermined CHs (p) is going to be uniformly distributed through the WSN. Therefore, there is the possibility that the elected CHs be concentrated in one part of the network. Thus, some SNs will not at all find CHs in their proximity. Finally, the protocol assumes that all SNs begin with the same amount of energy capacity in each election round. LEACH could be extended to account for SNs with non-uniform energy. An extension has been introduced in [Heinzelman2000b] with the goal of preceding data transfers with negotiations similar to meta-data descriptors used in the SPIN protocol that ensures transmission of only data to the CHs that actually provides new information.

11.9.9 Threshold-sensitive Energy Efficient (TEEN)

In [Manjeshwar2001], a Threshold-sensitive Energy Efficient WSN (TEEN) protocol has been introduced its time line as depicted in Figure 11.16. In this scheme, at every cluster change time, the CH broadcasts the following to its members in addition to the attributes:

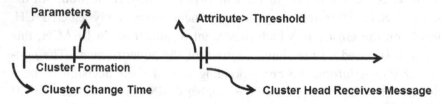

Figure 11.16 – Time line for TEEN [Taken from Manjeshwar2001]

- Hard Threshold (H_T): This is a threshold value for the sensed attribute. It is the absolute value of the attribute beyond which, the SN sensing this value must switch on its transmitter and report to its CH.
- Soft Threshold (S_T): This is a small change in the value of the sensed attribute which triggers the SN to switch on its transmitter and transmit once the H_T has been crossed.

In TEEN, SNs sense their environment continuously, thereby making it appropriate for real time applications. The first time a parameter from the attribute set reaches its hard threshold value, the SN switches on its transmitter and sends the sensed data. The sensed value is also stored in an internal variable in the SN, called the sensed value (SV). The SNs will next transmit data in the current cluster period, only when both the following conditions are true:

- The current value of the sensed attribute is greater than the hard threshold;

- The current value of the sensed attribute differs from SV by an amount equal to or greater than the soft threshold.

Whenever a SN transmits data, SV is set equal to the current value of the sensed attribute. Thus, hard threshold tries to reduce the number of transmissions by allowing the SNs to transmit only when the sensed attribute is in the range of interest. The soft threshold further reduces the number of transmissions by eliminating all data transfer which might have otherwise occurred when there is little or no change in the sensed attribute once the hard threshold is reached. The main features of this scheme are:

- Time critical data reaches the BS almost instantaneously. So, this scheme is eminently suited for time-critical data sensing applications.
- Message transmission consumes much more energy than data sensing. So, even though the SNs sense continuously, the energy consumption in this scheme can potentially be much less than in the proactive network, because data transmission is done less frequently.
- The soft threshold can be varied, depending on the criticality of the sensed attribute and the target application.
- A smaller value of the soft threshold gives a more accurate picture of the network, at the expense of increased energy consumption. Thus, the user can control the trade-off between the energy efficiency and the accuracy.
- At every cluster change time, the parameters are broadcast afresh and so, the user (BS) can change them as required.

On the other hand, the main drawback in TEEN is that if the thresholds are not reached, the SNs will never communicate and the user will not get any data from the network at all and will never be able to know even if SNs are working properly or have died. Thus, this scheme is not well suited for applications where the user needs to get data on a regular basis. Another possible problem with this scheme is that a practical implementation would have to ensure that there are no collisions within a cluster. TDMA scheduling of the SNs can be used to avoid this problem. This will, however, introduce a delay in reporting of time-critical data. CDMA/CA may be another possible solution to this problem. As we can see, TEEN is best suited for time critical applications such as intrusion and explosion detection.

11.9.10 Adaptive Periodic TEEN (APTEEN)

In some applications, the user wants not only time-critical data, but also wants to have an ability to query the network for analysis on conditions other than collecting time-critical information. In other words, the user might need a network that reacts immediately to time-critical situations and also gives an overall picture of the network at periodic intervals, so that it is able to answer the queries. The Adaptive Periodic TEEN (APTEEN) protocol [Manjeshwar2002a, Manjeshwar2002b] is an enhancement over the TEEN protocol which is able to combine the best features of proactive and reactive networks while minimizing their limitations to create a hybrid network. In this scheme, SNs not only send data periodically, they also respond to sudden changes in attribute values. This uses the same model as TEEN with the following changes. In APTEEN, once the CHs are decided the following events take place, in each cluster period. The CH broadcasts the following parameters:

- *Attributes* (A): This is a set of physical parameters which the user is interested in obtaining data about.
- *Thresholds*: This parameter consists of a hard threshold (H_T) and a soft threshold (S_T). H_T is a value of an attribute beyond which a SN can be triggered to transmit data. S_T, on the other hand, is a small change in the value of an attribute which can trigger a SN to transmit.
- *Schedule*: This is a TDMA schedule similar to the one used in [Heinzelman2000b], assigning a slot to each SN.
- *Count Time* (C_T): It is the maximum time period between two successive reports sent by a SN. It can be a multiple of the TDMA schedule length and introduces the proactive component in the protocol.

In APTEEN, SNs sense their environment continuously. However, only those SNs which sense a data value at or beyond the hard threshold transmit. Furthermore, once a SN senses a value beyond H_T, the next time it transmits data will be only when the value of that attribute changes by an amount equal to or greater than the soft threshold S_T. The exception to this rule is that if a SN does not send data for a time period equal to the count time, it is forced to sense and transmit the data irrespective of the sensed value of the attribute. Another issue in APTEEN is that SNs near to each other may fall into the same cluster hence sensing similar data at approximately the same time. If this happens, they may transmit their data simultaneously thus leading to collisions. To prevent this, a TDMA schedule is used and each SN in the cluster is assigned a transmission slot as shown in Figure 11.17. The main features of APTEEN are:

- It combines both proactive and reactive policies. By sending periodic data, it gives the user a complete picture of the network, like a proactive scheme. It also senses data continuously and responds immediately to drastic changes, thus making it responsive to time critical situations. It, thus, behaves as a reactive network also.

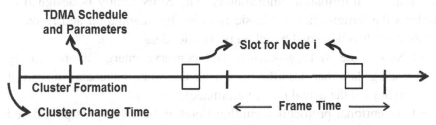

Figure 11.17 – Time line for APTEEN [Taken from Manjeshwar2002a]

- It offers a lot more flexibility by allowing the user to set the time interval (C_T) and the threshold values for the attributes.
- Changing the count time as well as the threshold values can control energy consumption.
- The hybrid network can emulate a proactive network or a reactive network, based on the application, by suitably setting the count time and the threshold values.

One drawback of APTEEN is the associated additional complexity to implement the threshold functions and the count time. However, this might be seen as a trade-off. By simulation, APTEEN [Manjeshwar2002a] is shown to surpass LEACH in prolonging the network lifetime.

11.9.11 SPIN-based Adaptive Routing

A family of adaptive protocols called Sensor Protocols for Information via Negotiation (SPIN) has been introduced in [Heinzelman1999, Kulik2002] that uses data negotiation and resource-adaptive algorithms. SPIN assigns a high-level name to appropriately describe their collected data, called meta-data, and perform meta-data negotiations before any data is transmitted. This ensures that no redundant data is transmitted throughout the network. The format of the meta-data is application-specific and is not specified in SPIN. For example, SNs might use their unique IDs to report meta-data if they cover a certain known region. In addition, SPIN has access to the current energy level of the SN and adapts the protocol it is running based on the remaining energy. In SPIN, all information at each SN is disseminated to every other SN in the WSN, with an implicit assumption that any SN is a potential BS. With this scheme, a user is able to query any SN and get the required information immediately. The SPIN family is designed to address the deficiencies of classic flooding by negotiation and resource adaptation. It is designed based on two basic ideas:

- SNs operate more efficiently and conserve energy by previously sending data that describe the sensor data (i.e., metadata) instead of sending all the actual data (e.g., image).
- Conventional protocols including flooding-based or gossiping-based routing protocols waste energy and bandwidth when sending extra and unnecessary copies of data through sensors covering overlapping areas.

SPIN is a three-stage protocol as SNs use three types of messages ADV, REQ and DATA. ADV is used to advertise new data, REQ to request data, and DATA is the actual message itself. The protocol starts when a SPIN SN obtains new data that it is willing to share. It does so by broadcasting an ADV message containing meta-data. If a neighbor is interested in the data, it sends a REQ message which is then sent to this neighbor SN. To disseminate the message, this neighbor SN repeats this process until the entire WSN receives a copy of the message.

The SPIN family of protocols includes two protocols, namely, SPIN-1 and SPIN-2, which incorporate negotiation before transmitting useful data. This mechanism helps in eliminating both implosion and overlapping. Also, each SN has its own resource manager which keeps track of resource consumption, and is polled by the SNs before data transmission. An extension to this, the SPIN-1 protocol, is SPIN-2, which incorporates threshold-based resource awareness mechanism in addition to negotiation. If the energy in the SNs is abundant, SPIN-2 communicates using the three-stage protocol of SPIN-1. However, when the energy in a SN starts approaching a low energy threshold, it reduces its participation in the protocol. This approach does not, however, prevent a SN from receiving and therefore spending energy on ADV or REQ messages below its low-energy threshold. However, it does not allow the SN from ever handling a DATA message when its energy level is below this threshold. SPIN-1 and SPIN-2 are simple protocols that efficiently disseminate data while maintaining no per-neighbor state. SPIN is very appropriate for an environment where the WSs are mobile. Other protocols of the SPIN family are:

- SPIN-BC: This protocol is designed for a SN to broadcast to all SNs. However, SNs have to wait for transmission if the channel is busy. SNs do not immediately send out REQ message when they hear the ADV message. Instead, each SN sets a random timer and upon expiration, the SN sends the REQ message. If other SNs, waiting for their timers to expire, overhear this message, they stop their timers, thereby preventing redundant copies of the same request to be sent.

- SPIN-PP: This protocol is used whenever two SNs can communicate in exclusive communication with each other without any interference from the other neighboring SNs. It is designed for a point to point communication, i.e., hop-by-hop routing. SPIN-PP assumes that energy is not a major constraint and that packets are never lost. Similar to SPIN-1, SPIN-PP is also a simple 3-way handshake protocol. One major advantage of SPIN-PP is its simplicity and the fact that each SN needs only know its single-hop neighbors while not requiring any other topology information;

- SPIN-EC: This protocol works similar to SPIN-PP, but with an energy heuristic added to ensure completion of all protocol stages without having its energy drop below a certain threshold.
- SPIN-RL: In SPIN-PP, it is assumed that packets are never lost, making it inappropriate for error prone channels. Instead, another protocol called SPIN-RL is used where two patches are added to SPIN-PP as to account for the lossy channel. First, each SN keeps track of all ADV messages it receives, with a provision of requesting data to be retransmitted if it did not get it within a specified window of time. Second, in order to adjust the rate of data retransmission, SNs wait for a certain pre-determined time before replying to the same REQ messages again. This procedure guarantees that data is retransmitted only after ensuring that the reply to the previous REQ message failed.

11.9.12 Power-Efficient Gathering in Sensor Information Systems (PEGASIS)

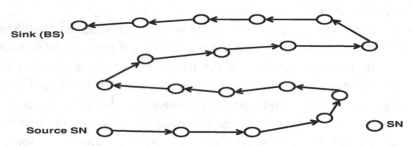

Figure 11.18 – Path established in PEGASIS algorithm

The Power-Efficient Gathering in Sensor Information Systems (PEGASIS) [Lindsey2002a] is a near optimal chain-based protocol which is an enhancement over LEACH. The algorithm works by finding links such that all SNs in a WSN are connected in the form of a single chain from the source SN to BS along the closest SNs as illustrated in Figure 11.18. To locate the closest neighbor, SNs use the signal strength to measure the distance to all of its adjacent SNs and then adjust the signal strength so that only one SN can be heard. The aggregated form of the data is sent to the BS by any SN in the chain, and SNs in the chain take turns in transmitting to the BS. This prolongs the network lifetime as SNs communicate with their closest neighbors only and so on till BS is reached. Then a new round starts which decreases power required to transmit data per round. As a result, PEGASIS has two main goals. First, it aims at increasing the lifetime of each SN by using collaborative techniques. Second, it only allows coordination between SNs that are close together, thus reducing the bandwidth consumed for communication. By simulation, it has been shown [Lindsey2002a] that PEGASIS increases the lifetime of the network twice as much as compared to when LEACH is used. On the other hand, PEGASIS makes some assumptions which may not always be true. First, PEGASIS assumes that each SN is able to communicate with the BS directly. In practical cases, however, SNs are expected to use multihop communication to reach the BS. In addition, it assumes that each SN maintains a complete database about the location of all other SNs in the network to form a single linked path. However, the method by which SN locations are obtained is not indicated. Finally, PEGASIS assumes all sensors to be immobile at all times, while this may not be true for certain applications.

11.9.13 Small Minimum Energy Communication Network (MECN)

The minimum energy communication network (MECN) [Rodoplu1999] protocol has been designed to compute an energy-efficient sub-network for a given WSN. On top of MECN, a new algorithm called Small MECN (SMECN) [Li2001a] has been proposed to construct such a sub-network. The sub-network (i.e., sub-graph G') constructed by SMECN is smaller than the one constructed by MECN if the broadcast region around the broadcasting SN is circular for a given power assignment. The sub-graph G' of graph G, which represents the WSN, minimizes the energy consumption satisfying the following conditions:

* The number of edges in G' is less than in G, while containing all the SNs in G.
* The energy required to transmit data from a SN to all its neighbors in the sub-graph G' is less than the energy required to transmit to all its neighbors in graph G.

The resulting sub-network computed by SMECN helps in the task of sending messages on minimum-energy paths. However, the proposed algorithm is local in the sense that it does not actually find the minimum-energy path; it just constructs a sub-network in which it is guaranteed to exist. Moreover, the sub-network built by SMECN leads to an increment in the probability that the path used is the one that consumes less energy.

11.9.14 Routing in Fixed-size Clusters

Routing in WSNs can also take advantage of geography-awareness. One such routing protocol called Geography Adaptive Fidelity (GAF) has been suggested in [Xu2001], where the network is firstly divided into fixed zones. Within each zone, SNs collaborate with each other to play different roles. For example, SNs elect one SN to stay awake for a certain period of time while the others sleep. This particular elected SN is responsible for monitoring and reporting data to the BS on behalf of all SNs within the zone. Each SN is assumed randomly positioned in a two dimensional plane. When a SN transmits a packet, the signal is strong enough for other SNs to hear it within the Euclidean distance r from the SN that originates the packet. Figure 11.19 depicts an example of fixed zoning applicable to WSN. In [Xu2001], both horizontal and vertical communications are guaranteed if the signal travels a distance $a = r / \sqrt{5}$. For a diagonal communication to take place, the signal has to span a distance $a = r / 2\sqrt{2}$. A CH can then ask the SNs in its own zone to switch on and start gathering data if it senses a target.

In GAF, one non-sleeping SN in each grid square is sufficient to maintain the connectivity of the original WSN. As connectivity is defined by the grid, selecting the active SN for each grid square does not require explicit exchange of connectivity information. Each SN transitions independently among three states: sleep, discovery, and active. SNs periodically wake up from the sleep state and transition to the discovery state. In the discovery state, a SN listens for other SNs' announcements and can announce its own grid position ID and residual energy status. If the SN hears no "higher ranking" announcement, it transitions to the active state; otherwise, it transitions back to the sleep state. After spending some time in the active state, a SN transitions back to the discovery state, allowing the active role to be rotated among the SNs in the grid square. It is assumed [Xu2001] that SNs obtain their location information through GPS, which may not be feasible for certain classes of WSN applications.

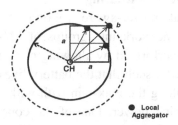

Figure 11.19 – Incorporating zones in wireless WSNs

Sensor Aggregates Routing

A number of algorithms for building and maintaining aggregates of SNs have been introduced in [Fang2003], with an objective to collectively monitor target activities in certain environments. A SN aggregate includes those SNs in a network that satisfy a grouping predicate for a collaborative task processing. SNs are divided into clusters according to their received communication signal strength. After that, local CHs are elected by exchanging information between neighboring SNs and a SN with highest connectivity among its one-hop neighbors is declared as a leader.

Three algorithms have been suggested. The first is a lightweight protocol called Distributed Aggregate Management (DAM) which forms aggregates of SNs for a target monitoring task with a predicate. The result of applying the predicate to the SN data as well as the information obtained from other SNs determines if a SN belongs to an aggregate or not and convergence of the process indicates aggregates are feasible. The second algorithm called Energy-Based Activity Monitoring (EBAM) is used to estimate the energy level at each SN by computing the area covered by the signal. The third algorithm called Expectation-Maximization Like Activity Monitoring (EMLAM) is used to remove the constant and equal sensors in target area assumption. EMLAM estimates the target positions and signal energy through received signals, and uses the resulting estimates to predict how signals from the targets may be mixed at each SN. This is an iterative process which is carried out until a good estimate is obtained. The system seems to work well in tracking multiple targets when these targets are not interfering.

Hierarchical Power-Aware Routing

A hierarchical power-aware routing scheme introduced in [Li2001b] divides the network into groups of SNs in their geographic proximity. Routing is performed by allowing each zone to decide how it routes across other zones, such that the battery lives are maximized. The messages are routed along the max-min path with the maximal-minimal fraction of the remaining power. One of the concerns is that traversal through the SNs with high residual power may be expensive as compared to minimal power consumption. To overcome this problem, an approximation algorithm called the *max-min zP_{min}* algorithm has been suggested [Li2001b] which is based on the tradeoff between minimizing the total power consumption using the Dijkstra algorithm and maximizing the minimal residual power of the network. It relaxes the minimal power consumption for the message to be equal to zP_{min} with parameter $z \geq 1$ to restrict the power consumption for sending one message to zP_{min}.

Another algorithm called zone-based hierarchical routing, relying on *max-min zP_{min}*, is also proposed in [Li2001b]. To send a message across the entire area, a global path from a zone to another zone is discovered. In this protocol, a global controller for message routing is assigned the role of managing the zones, for example, the SN with the highest power. If the network can be divided into a relatively small number of zones, the scale for the global routing algorithm is reduced. In order to represent connected neighboring zones, a zone graph has been employed. Here, each zone direction vertex is labeled by its estimated power level by a modified Bellman-Ford algorithm.

11.10 Flat versus Hierarchical

As we can see, hierarchical and flat approaches have their own advantages and disadvantages as an underlying routing organization for

WSN. To illustrate their differences and suitability for different applications, Table 11.1 compares the characteristics of these topologies.

Flat versus Hierarchical versus Adaptive

It is important to compare the flat, hierarchical and adaptive approaches for routing over WSN. As a general comparison of flat and hierarchical is given in Table 11.1, here we follow an approach where we compare them by contrasting the characteristics of one protocol under each category. To this end, Table 11.2 compares SPIN, LEACH, and the Directed Diffusion routing techniques according to different parameters. As we can see from the table, LEACH is shown to be a promising approach for energy-efficient routing in WSNs due to the use of in-network processing. We note, however, that the choice of the approach to be used is obviously application-dependent. There is no one solution that fits all problems.

11.11 Operation-Based Protocols

In this class of routing scheme, the protocol operation determines specific classification. Altogether, the protocols in this class can be classified as negotiation-based, multipath-based, query-based and location-based.

Negotiation-based Routing

Negotiation-based routing protocols use high level data descriptors in order to eliminate redundant data transmissions. The SPIN family protocols (discussed earlier) and the scheme presented in [Kulik2002] are examples of negotiation-based routing protocols. The motivation here is that the use of flooding to disseminate data produces implosion and data overlap, leading to scenarios where SNs receive duplicate copies of the same data, consuming considerable amount of energy. As we have seen before, the SPIN protocols are designed to disseminate the data of one sensor to all other SNs assuming these SNs are potential BSs. Thus, the main idea behind negotiation-based routing is to suppress duplicate information and is done by conducting a series of prior negotiations.

Table 11.3 – Hierarchical versus Flat topologies for WSN

Hierarchical	Flat
Reservation-based scheduling	Contention-based scheduling
Collisions avoided	Collision overhead present
Reduced duty cycle due to periodic sleeping	Variable duty cycle by controlling sleep time of SNs
Data aggregation by cluster head	SN on multi-hop path aggregates incoming data from neighbors
Simple but non-optimal routing	Routing is complex but optimal
Requires global and local synchronization	Links formed in the fly, without synchronization
Overhead of cluster formation throughout the network	Routes formed only in regions that have data for transmission
Lower latency as multi-hop network formed by cluster heads is always available	Latency in waking up intermediate SNs and setting up the multi-hop path
Energy dissipation is uniform	Energy dissipation depends on traffic patterns
Energy dissipation cannot be controlled	Energy dissipation adapts to traffic pattern
Fair channel allocation	Fairness not guaranteed

Table 11.4 – Comparison of SPIN, LEACH and Directed Diffusion

Parameter	SPIN	LEACH	Directed Diffusion
Optimal Route	No	No	Yes
Network Lifetime	Good	Very Good	Good
Resource Awareness	Yes	Yes	Yes
Use of Meta-Data	Yes	Yes	Yes

Multipath-based Routing

Network performance, and possibly lifetime, in WSNs can be significantly improved if the routing protocol is able to maintain multiple, instead of a single, paths to a destination, and protocols in this class are called multipath protocols. By employing multipath protocols, the fault tolerance (resilience) of the network is considerably increased. The fault tolerance of a protocol is increased if we maintain multiple paths between the source and the destination at the expense of an increased energy consumption and traffic generation (i.e., overhead), as alternate paths are kept alive by sending periodic messages. We would also like to note here that multipath routes between a source SN and the destination BS can be node-disjoint. Multiple paths between a source and destination are said to be SN-disjoint or not when there is no SN overlap amongst them. For the purpose of our discussion here, we refer to alternate routes as not being node disjoint, i.e., their routes are partially overlapped.

Packets are routed through a path with largest residual energy in the algorithm proposed in [Chang2000]. Here, the path is changed whenever a better path is discovered. The primary path is used until its energy falls below the energy of the backup path, at which time the backup path is used. By employing this mechanism, the SNs in the primary path do not deplete their energy resources through continual use of the same route, hence prolonging their lifetime. One issue with this scheme is the cost associated with switching paths, and how to deal with packets which are en-route. As we have seen before, there is a tradeoff between minimizing the total power consumed and the residual energy of a network. As a result, routing packets through paths with largest residual energy may turn out to be very energy-expensive. To minimize this effect, it has been proposed in [Li2001b] a scheme in which the residual energy of the route is relaxed a bit in order to select a more energy efficient path. To increases the network lifetime, suboptimal paths can be occasionally employed [Rahul2002] which are chosen by means of a probability which depends on how low the energy consumption of each path is.

Multipath routing was employed [Dulman2003] to enhance the reliability of unreliable environments. Here, reliability is enhanced by using several paths in sending the same packet from source to destination. Obviously, by using this technique, traffic increases significantly. Tradeoff between the amount of traffic and the network reliability is investigated in [Dulman2003] by splitting the original data packet into sub-packets and then send each sub-packet through one of the available multipath. It has been found that even if some of these sub-packets are lost, the original message can still be reconstructed. It has been found that for a given maximum SN failure probability, using multipath degree higher than a certain optimal value optimizes the failure probability.

An extension of the multi-path algorithm is described in [Jain2003a, Jain2003b, Jain2005b] that contains several important characteristics. The idea is to reduce the complexity of finding the paths by defining the rectangular region bounded by the responding SN S_1 and the BS as the routing region and defining the paths passing through cross-diagonal sensors as multiple paths. One such example for a rectangular mesh-based WSN is shown in Figure 11.20. This identifies many paths from S_1 to BS, with different path lengths in terms of number of intermediate SNs in the path and hence, reduces the delay between the responding SN S_1 and the BS by the process of data store-and-forward along the selected path. The path along the diagonal is shortest in length and if this path is used all the time in responding the persistent query, the energy of the sensors lying on this path, could get depleted at a much faster rate then rest of the network.

The energy consumption for such responses could be distributed throughout the rectangular region by using different paths and that decision could be made by responding SN based on residual energy of the SNs within the region. So, the strategy to be used is to transmit critical responses along the shortest path while non-critical periodic data updates need to be sent along longer paths and hence could support service differentiation. This enhances the lifetime of the WSN as it will take longer time to get any SN to run out of battery energy or to have the network disconnected or partitioned. The same phenomenon is observed for multiple queries in a mesh-based WSN or a randomly placed WSN [Jain2005b]. The multi-path algorithm has a very little control overhead [Jain2005a] and can support differential service. Therefore, it is more suited to applications such as explosion detection, intrusion detection, forest fire monitoring, and so on, that require service differentiation to associate a high priority level with the time-critical queries and a low priority level with non-critical queries proactively collecting data at the BS from on a regular basis. Essentially, this algorithm classifies the paths during multipath routing based on their route length, and route the critical queries through a set of paths with minimal route length while the rest of the traffic is spread with the objective of uniform SN utilization in the network.

Another good candidate for robust partially disjoin multipath routing and delivery is Directed Diffusion and is presented in [Ganesan2001]. The idea is that the use of multipath routing provides a viable alternative for energy efficient recovery from failures in WSN. The motivation of using these braided paths is to keep the cost of maintaining the multiple paths low. In this scheme, the costs of alternate paths are comparable to the primary path as they tend to be much closer to the primary path.

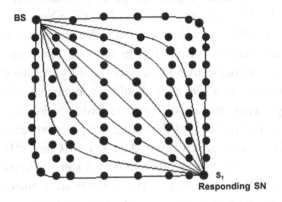

Figure 11.20 – Routing Region and Paths from S_1 to BS through other SNs

Query-based Routing

In query-based routing, the BS propagates a query for data (sensing task) from each SN. A SN having the data matching the query sends it back to the BS. Usually, these queries are described in natural language or in high-level query languages. For example, a BS may submit a query to SN S_1 inquiring: "Are there moving vehicles in battlefield region 1?". In query-based routing, all the SNs have tables consisting of the sensing tasks queries that they received, and send back data matching these tasks whenever they receive it. Whenever the SN S_1 has data matching the interest, it sends the data to the BS.

The rumor routing protocol [Braginsky2001] uses a set of long-lived agents to create paths towards the events they encounter. Here, each SN may generate an agent in a probabilistic fashion, and agents have a lifetime of a certain number of hops after which they die. Whenever an agent comes across a shorter path leading to an event that it has not encountered so far, it creates a path state. In this protocol, each SN maintains a list of its neighbors and is updated whenever new events are encountered. If there is no route available, the SN transmits a query in a random direction. The SN then waits for a certain period of time and if response has been received, it floods the network.

11.12 Location-Based Routing

In location-based routing, SNs are addressed by means of their locations. Here, the distance between neighboring SNs can be estimated on the basis of incoming signal strength [Bulusu2000, Capkun2001, Savvides2001]. Alternatively, the location of SNs may be available directly through GPS if we consider some SNs are equipped with a small low power GPS receiver [Xu2001]. In order to conserve energy, some location-based schemes demand that SNs should go to sleep if there is no activity. However, the active SNs should be connected, should cover the entire sensing region, and should provide basic routing and broadcasting functionalities. The problem of designing sleep period schedules for each SN in a localized manner has been addressed in [Chen2002, Xu2001]. In [Xu2001], the sensor field is divided into small squares in such a way as to ensure that two SNs in two neighboring squares are connected by making one SN in each square be active while all others will be in the sleep mode.

Transport Layer

Similar to the link layer, the issue of suitable transport protocols for WSNs has been given very little consideration so far. It ties obviously with the appropriate service definition of WSN, and also what level of dependability and QoS to provide, in saving the amount of energy. There are very few such proposals [Fu2003, Park2003, Sankarasubramaniam2003b, Stann2003, Wan2002] for WSN.

11.13 High-Level Application Layer Support

The protocols we have presented so far are also found, albeit in some different form in traditional wired, cellular, or ad hoc networks. For specific applications, a higher level of abstraction specifically tailored to WSN appears to be useful. In this section, we outline some of the activities in this direction.

11.13.1 Distributed Query Processing

The number of messages generated in distributed query processing is several magnitudes less than in centralized scheme. [Bonnet2000, Bonnet2001] discusses distributed query execution techniques that improve communication efficiency in WSNs. They discuss two approaches for processing sensor queries: warehousing and distributed. In the warehousing approach, data is extracted in a pre-defined manner and stored in a central database (e.g., the BS). Subsequently, query processing takes place on the BS. In the distributed approach, only relevant data is extracted from the WSN, when and where it is needed. A language similar to the Structured Query Language (SQL) has been proposed in [Madden2003] for query processing in homogeneous WSNs. They have developed a suite of techniques for power-based query optimization and built a prototype instantiation for the same, called TinyDB [TinyDB], which runs on Berkeley mica motes [MICA].

Sensor Databases

One can view the wireless WSN as a comprehensive distributed database and interact with it via database queries. This approach solves the entire problem of service definition and interfaces to WSNs by mandating, for example, SQL queries as the interface. The problem is in finding energy-efficiency ways of executing such queries and of defining proper query languages that can express the full richness of WSNs. A model for sensor database systems known as COUGAR [COUGAR] defines appropriate user and internal representation of queries. The sensor queries are also considered so that it is easier to aggregate the data and to combine two or more queries. In COUGAR, routing of queries is not handled. COUGAR follows three-tier architecture:

- The Query Proxy.
- A Front end Component.
- A Graphical User Interface.

Queries are formulated regardless of the physical structure or the organization of the WSN. Sensor data is different from the traditional relational data since it is not stored in a database server and it varies over time. Aggregate queries or correlation queries that give a bird eye's view of the environment also zoom on a particular region of interest. Each long running query defines a persistent view which it maintains during a given time interval. In addition, a sensor database should account for sensor and communication failures. Finally, it should be able to establish and run a distributed query execution plan without assuming global knowledge of the WSN.

Distributed Algorithms

WSNs are not only concerned with merely *sensing* the environment but also with interacting with the environment. Once actuators like valves are added to WSNs, the question of distributed algorithms becomes inevitable. One showcase is the question of distributed consensus, where several actuators have to reach a joint decision. This problem has been investigated to some degree for ad hoc networks [Malpani2000, Nakano2002, Srinivasan2003, Walter2001], but it has not been fully addressed in the context of WSNs where new scalability and reliability issues emerge.

11.13.2 In-Network Processing

In-network processing, requires data to be modified as it flows through the WSN. In-network processing is often very closely related to the distributed query processing, as the former takes place in the execution of the latter, the rationale being sensors close to the event sense similar data. Obviously, the number of SNs that sense attributes related to an event in a geographical region depends on the footprint of the event, also referred to as the *target region*. Therefore, it is possible to exploit correlation in the observed data both in time and in space (also called spatio-temporal correlation). Possibilities for in-network processing include compression [Petrovic2003] or aggregation [Boulis2003, Cristescu2003, Deb2003, Heinzelman2000b, Krishnamachari2002, Lindsey2002b, Petrovic2003, Zhao2003], the primary motivation being computation to be much cheaper in energy consumption than communication. Monitoring civil structures, machines, road traffic and environment are just a few applications that require spatio-temporal querying that could benefit from an in-network query processing architecture.

For aggregating data [Banerjee2005], some of the sensors need to have enhanced capabilities than the majority of the simple sensors and such resource rich wireless sensors (RRSN) make the WSN heterogeneous in nature, as illustrated in Figure 11.21. As the RRSNs SNs act as CHs, they also maintain partial network data. So, the next question is how many RRSN SNs need to be deployed and what the ratio with respect to simple WS SNs is. This would depend on the application and the type of desired query as response could also be provided by RRSNs, rather than getting information from individual SN. So, the queries can be broadly classified as [Biswas2005, Jain2005c]:

a) Simple Queries: This may require answer from a subset of SNs and could be provided by RRSN. An example could be, "What is the temperature in a given region?"

b) Aggregate Queries: This requires aggregation of currently sensed values by SNs in a given region.

c) Approximate Queries: This implies aggregation of data in the data form of a histogram, contour maps, or tables and the response could come from the RRSNs.

d) Complex Queries: This type of query would consist of several condition-based nested queries and one such example is, "Report the average temperature in a region has the highest wind velocity". This type of queries could be possibly responded by RRSNs.

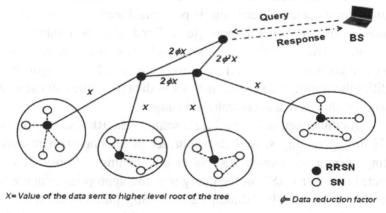

X= *Value of the data sent to higher level root of the tree*   φ= *Data reduction factor*

Figure 11.21 – Heterogeneous WSN with Resource Rich SNs for Data Aggregation

So, the query processing in a WSN need to be correlated to access data at RRW SNs as query tree need to be mapped to the flow of data along the routing tree between the RRSN and to the BS [Biswas2005, Buragohain2005, Hong2004]. The energy consumed in transmitting the query and receiving response from SNs and RRSNs could represent the cost of the query and hence minimization of power consumption is fairly involved. In-network query processing for multi-target regions is addressed by an energy aware routing scheme for spatio-temporal queries [Biswas2005, Jain2005c]. Queries are then evaluated based on a computation plan that is provided to the sink in the form of a *query tree*. Query trees are, in turn, defined as a logical representation of operator hierarchy in a given query with target regions as leaf SNs. A recent work addresses [Biswas2005, Jain2005c] the problem of translating a query tree at the sink to a corresponding *routing tree* such that the cost of transferring data from the target regions to the sink is minimized.

The problem of mapping a query tree to a routing tree is non-trivial. This query is usually specified in a declarative language like SQL containing operators such as selects, joins, projections and aggregations. Then, it has to be converted to a query tree (specifying the order of evaluation of operators) by using query optimization techniques based on power conservation [Madden2003]. Once the query tree is constructed, the task is of mapping the query tree to a routing tree by assigning query operators to individual SNs in an optimal way. Determination of the optimal query operator placement is performed such that the cost of data transfer from the target regions to a fixed sink is minimized. The algorithm in [Biswas2005, Jain2005c] achieves this goal by assigning query operators to RRSNs and by introducing an adaptive algorithm that modifies the routing tree to fluctuations in data properties and scarcity of network resources in a decentralized manner.

As illustrated in Figure 11.21, some of the RRSNs may not act as a CH in aggregating sensed data, but act as an important resource in routing information among RRSNs or maintaining a repository of data collected by other RRSN in the query tree and appropriately aggregating them by a factor Φ [Banerjee2005]. So, minimizing the energy consumed implies minimizing the amount of data transfer by each RRSN and the distance between the source RRSN and receiving RRSN or BS. The problem of operator placement has been first introduced in [Bonfils2003] for supporting in-network query processing in homogeneous WSNs. For optimal placement of operators in a query tree, the cost function employed includes a cumulative cost of the right and left sub-trees of an operator, but it is does not consider the outgoing data transfer rate to the parent operator which may be very important for minimizing local data transfer cost [Biswas2005, Jain2005c]. This is a nonlinear optimization problem and a two phase bottom-up from root SNs to BS and top-down from BS to root SNs based approach has been suggested [Biswas2005, Jain2005c] that provides close to optimal solution in selecting RRSN SNs as a part of the query tree. Finally, distributed algorithms based on non-linear optimization techniques have also been deployed in [Rabbat2004] for applications such as robust estimation, source localization, cluster analysis and density estimation in WSNs.

11.13.3 Data Aggregation

The need for data aggregation in a WSN has become extremely important in conserving energy as close by SNs usually satisfy spatial correlation whereby physical attributes exhibit a gradual and continuous variation over the two dimensional Euclidean space. The most common aggregation scheme advocated in the literature includes simple operators like sum, average, maxima, or minima. However, applying such a scheme to temperature maps will lose some useful information. Without compromising any relevant information, a scheme [Chen2006] determines optimal number of aggregator operators to be around 450 and 500 for energy-efficient data aggregation in a WSN with 10,000 sensors.

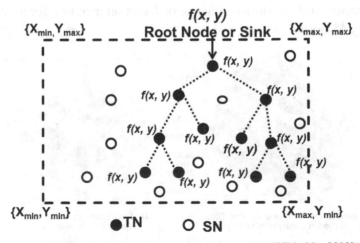

Figure 11.22 – TAG- Tiny Aggregation Tree in a WSN [Madden2002]

On the other hand, location information between multiple SNs in close proximity can be used to aggregate the attribute values. In the initial stage, a quad tree [Madden2002] is established as shown in Figure 11.22. Then, each sensor senses and reports its data to its nearest tree node (TN) acting as a CH. A data aggregation procedure [Banerjee2005] is used to create a polynomial, $f(x,y)$ as follows:

$$f(x,y) = \beta_0 + \beta_1 y + \beta_2 y^2 + \beta_3 x + \beta_4 xy + \beta_5 xy^2 + \beta_6 x^2 + \beta_7 x^2 y + \beta_8 x^2 y^2,$$

where, β_0, β_1,... β_8 are the nine coefficients of the regression polynomial for the attribute $f(x,y)$ sensed at coordinates (x, y). Four parameters x_{min}, y_{min}, x_{max}, and y_{max} are used to represent the rectangular area. Thus, each SN needs to send only two fields to its TN: one carrying the coefficient of polynomial, and the other for corresponding boundaries to its parent. At each level, parent TN calculates the new set of coefficients and covered area by combining with own reported readings and pass onto the next higher level TN. Once this procedure stops at the root (BS) with the final polynomial $f(x,y)$ in the given area and the root can get the attribute value at any point (x, y).

The data aggregation process has been simulated using *Mathematica* tool using the parameters of rooftop data [Wolf2001] at the University of Washington as shown in Figure 11.23(a). The temperature distribution of Figure 11.23(a) is used by 400 sensors in calculating synthetic data coefficients in the spread of 400x400 units. The temperature attribute shows a change of 1-unit temperature for a traversal of every 45-units distance.

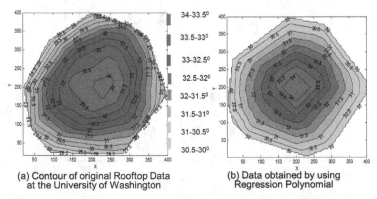

(a) Contour of original Rooftop Data at the University of Washington

(b) Data obtained by using Regression Polynomial

Figure 11.23 – Data Aggregation in a WSN using Regression Polynomial

Final relation at the BS [Banerjee2007] is obtained as:

$f(x,y)=26.1429+0.0427163y-0.000167934y^2+0.014x+0.000249xy$
$0.00000009231xy^2-0.0000181258x^2-0.000000860054x^2y+$
$0.00000000116143x^2y^2$.

and is drawn in Figure 11.23(b). The two figures look different. But, a closer analysis shows that the maximum error is limited to less than 5.64% and most of the time, it is less than 1.68% which may be acceptable in most WSN applications.

11.13.4 Mobile SNs and BSs

The enhancements in the field of electronics have made robots much smaller, lighter and more precise. Mobile WSNs have been suggested [Wang2004] to cover the area not reachable by static sensors. This has encouraged a BS to be placed on a robot so that they can move around and collect data from different SNs when it is in their neighborhood. This avoids unnecessary multi-hop forwarding of data to the BS. Such a scheme is extremely useful in defense and military applications where human time and life is very precious. Other applications, such as fire fighting, can also be envisioned where these robots would be carriers of fire extinguishers when triggered by SNs. Even in autonomous waste disposal, robots could play a major role as handling these wastes could be hazardous to human beings. But, this makes the WSN to be delay tolerant. Different SNs can be used to measure physical parameters of the surroundings, so that useful information can be procured. The collaboration between mobile robots and WSNs is a key factor towards achieving efficient transmission of data, network aggregation, quick detection of events and timely action by robots. Coordination between multiple robots for resource transportation has been explored for quite some time now. The time taken to detect an event depends entirely on the trail followed by the robots.

Typical resource-carrying are possible by robots by transferring its resources to another which has the capability to carefully transport their contents and transfer their resources. Once depleted of their resource, they may get themselves refilled from the sink which is a local reservoir of resources. The resource in demand could be water or sand (to extinguish fire), oxygen supply, medicines, bullets, clothes or chemicals to neutralize hazardous wastes, and so on. The target region that is in need of these resources is sometimes called an event location. If there are many more events than the number of resource rich robots, then priority has to be established to identify criticality of events.

11.14 Conclusions and Future Directions

WSN is one of the fastest growing areas in the broad wireless ad hoc networking field. The research in WSNs is flourishing at a rapid pace and is being considered as the revolutionary concept of this century. But, there are many challenges that need to be addressed such as, how to miniaturize the power source, how to have a self-power generating technology to provide indefinite power source and how to provide secured communication without exceeding the resource requirements. Another area that needs serious investigation is to come up with a killer non-defense civilian application so as to enhance its usefulness and general acceptance. The challenges are many. While we have partial answers or roadmaps to some of the above questions, there is still much to be done.

Homework Questions/Simulation Projects

Q. 1. Sensor networks sense physical parameters around its neighborhood. Assuming a set of 20bsensors are placed in the form of a linear array, can you compute energy consumed by a SN if:
   a. The sensing range of each sensor is limited to East/West neighboring SNs.
   b. The sensing range is extended to two neighbors in each direction.
   c. How much energy is consumed if data is sent to the BS located in one corner?
   d. How much energy is saved by all SNs if each SN aggregates transmitted data before forwarding to the BS?
   Assume the sensing range to be half that of communication range.

Q. 2. Can you show the S-MAC timing diagram for Question 1 (c)?

Q. 3. In Question 1, you are asked to use APTEEN protocol. Can you evaluate how much energy you can save as compared to MIT's LEACH approach if reactive data needs only 5% of bandwidth while proactive timings can be increased to 100 times? Make necessary assumptions and justify your answer.

Q. 4. 500 sensors are randomly deployed in a rectangular area of 40x40. Draw a Voronoi diagram and determine the following:
  a. What is the maximum distance between each SN pair?
  b. What is the average distance between any pair of SNs?
  c. What is the impact on the required sensing range and communication distance if you double or half the number of SNs?
  d. What is the impact on the performance if you double or half the area?
  e. If the SNs are divided in to two groups, how would you select the group so that there is minimum impact on the performance? Can you think of quick way of doing this? State clearly.
  f. In part e, if each set of sensors are allowed to sleep for 50% time, how much energy is consumed?
  g. If the sleep cycle is increased to 70%, what is the impact on the energy consumption?
  h. What will be minimum transmission radius requirement if you want to have full coverage by active sensors in part (f)?
  i. Derive an analytical model between the densities of nodes, coverage and sleep cycle.
  Assume appropriate parameters (if needed) as commonly used in the literature.

Q. 5. You are given 1,000 sensors and you need to spread them in the area of 100X100 units.
  a. What will be the average distance between two adjacent SNs if sensors are deployed randomly?
  b. Assume the average distance represents the sensing range of each sensor, what fraction of the area is covered by two adjacent SNs in situations a?
  c. Assume the radio transmission range to be double of the sensing range, how many sensors are present (average) within the transmission range of a single SN?
  d. If the sensing range of each sensor is reduced to half, what portion of the area remains uncovered?
  e. What will be the impact on wireless radio if the communication range is reduced to half?
  You need to come up with some quantitative measures.

Q. 6. In Question 3, show the use of directed diffusion scheme if the BS is located in one corner?

Q. 7. In Question 3, the source SN is at one corner while BS is placed at diagonally opposite corner. Can you identify multiple paths between the source SN and the BS? What is the minimum sensing and transmission range needed? Answer carefully.

Q. 8. In Question 3, can you illustrate how you can use PEGASIS algorithm to form a single chain of SNs, assuming the BS to be located at one corner?

Q. 9. Enumerate at least 5 different ways of conserving energy in WSNs. How do these energy conserving strategies affect the design of the network algorithms and the routing protocols?

Q. 10. In problem 3, you are allowed to group 5 sensors as a cluster. What is an easy way of forming a cluster? How effective it is as compared to an optimal solution? How do you select a CH in each cluster? Justify appropriateness of your answer.

Q. 11. Can you identify RRSNs that you can use? Justify your answer. If CHs are assumed to mobile units, how many events can be handled simultaneously? Justify your answer.

Q. 12. In Question 5, how much energy is saved if each CH does data aggregation before forwarding to the BS located in one corner of the field?

References

[Akyildiz2002] I. Akyildiz, W. Su, Y. Sankarasubramaniam, and E. Cayirci, "A survey on sensor networks," *IEEE Communications Magazine*, 40(8), August 2002.

[Aurenhammer1991] F. Aurenhammer, "Voronoi Diagrams- A survey of a fundamental geometric data structure," *ACM Computing Surveys* 23, pp. 345-405, 1991.

[Baker1981] D. Baker and A. Ephremides, "The Architectural Organization of a Mobile Radio Network via Distributed Algorithms," *IEEE Transactions on Communications*, no. 11, pp. 1694–1701, November 1981.

[Banerjee2005] T. Banerjee, K. Chowdhury, and D. P. Agrawal, "Tree Based Data Aggregation in Sensor Networks using Polynomial Regression," *The Eighth International Conference on Information Fusion*, Philadelphia, July 25-29, 2005.

[Banerjee2007] *Torsha Banerjee*, "Energy Efficient Data Representation and Aggregation with Event Region Detection in Wireless Sensor Networks," *PhD Dissertation*, University of Cincinnati, November 16, 2007.

[Bao2001] L. Bao and J. J. Garcia-Luna-Aceves, "A New Approach to Channel Access Scheduling for Ad Hoc Networks," In *Proc. ACM Intl. Conf. on Mobile Computing and Networking*, pp 210–220, Rome, Italy, July 2001.

[Bhardwaj2002] M. Bhardwaj and A. P. Chandrakasan , "Bounding the life time of sensor networks via optimal role assignment," *Proceedings IEEE INFOCOM 2002*.

[Bharghavan1994] V. Bharghavan, A. Demers, S. Shenker, and L. Zhang, "MACAW: A media access protocol for wireless LANs," *in ACM SIGCOMM*, August 1994.

[Biswas2005] R. Biswas, N. Jain, N. Nandiraju, and D. Agrawal, "Communication Architecture for processing Spatio-temporal continuous Queries in Sensor Networks," *Annals of Telecomm, 2005*.

[Bonfils2003] B. Bonfils and P. Bonnet, "Adaptive and Decentralized Operator Placement for In-Network Query Processing," *In Proceedings of Information Processing in Sensor Networks, Second International Workshop (IPSN)*, Palo Alto, CA, USA, April 2003.

[Bonnet2000] P. Bonnet, J. Gehrke, and P. Seshadri, "Querying the Physical World," *In IEEE Personal Communications*, October 2000.

[Bonnet2001] P. Bonnet, J. Gehrke, and P. Seshadri, "Towards Sensor Database Systems," *In 2nd Int. Conference on Mobile Data Management*, January 2001.

[Boulis2003] A. Boulis, S. Ganeriwal and M. Srivastava, "Aggregation in Sensor Networks: An Energy Accuracy Trade-off," *In Proceedings of the IEEE Intl. Workshop on Sensor Network Protocols and Applications (SNPA)*, Anchorage, AK, May 2003.

[Braginsky2001] D. Braginsky and D. Estrin, "Rumor Routing Algorithm For Sensor Networks," *International Conference on Distributed Computing Systems* (ICDCS), November 2001.

[Bulusu2000] N. Bulusu, J. Heidemann, and D. Estrin, "GPS-less low cost outdoor localization for very small devices," *Technical report 00-729, Computer Science department*, University of Southern California, April 2000.

[Buragohian2005] C. Buragohian, D. Agrawal, and S. Suri, "Power aware routing for sensor databases," *Proceedings INFOCOM 2005*.

[Capkun2001] S. Capkun, M. Hamdi, and J. Hubaux, "GPS-free positioning in mobile ad-hoc networks," *Proceedings of the 34th Annual Hawaii International Conference on System Sciences*, 2001.

[Chang2000] J.-H. Chang and L. Tassiulas, "Maximum Lifetime Routing in Wireless Sensor Networks," *Proceedings of the Advanced Telecommunications and Information Distribution Research Program (ATIRP)*, College Park, MD, March 2000.

[Chen2002] B. Chen, K. Jamieson, H. Balakrishnan, and R. Morris, "SPAN: an energy-efficient coordination algorithm for topology maintenance in ad hoc wireless networks," *Wireless Networks*, vol. 8, no. 5, September 2002.

[Chen2006] Y.P Chen, A.L. Liestman, Jiangchuan Liu "A hierarchical energy-efficient framework for data aggregation in wireless sensor networks," *IEEE Transactions on Vehicular Technology*, vol. 55, no. 3, May 2006, pp. 789–796.

[Chowdhury2005] K.R. Chowdhury, P. Chanda, D.P. Agrawal, and Q.A. Zeng, " DCA-A distributed channel allocation scheme for wireless sensor networks," *Proceedings PIMRC*, Sept. 2005.

[CORBA] The Common Object Request Broker Architecture (CORBA), http://www.corba.org/ .

[COUGAR] http://www.cs.cornell.edu/database/cougar/ , The Cornell Database Group.

[Cristescu2003] R. Cristescu and M. Vetterli, "Power Efficient Gathering of Correlated Data: Optimization, NP-Completeness and Heuristics," *In Proceedings of the ACM Intl. Symposium on Mobile Ad Hoc Networking and Computing (MobiHoc)*, Annapolis, MD, 2003.

[Crow1997] B. Crow, I. Wadjaja, J. Kim, and P. Sakai, "IEEE 802.11 Wireless Local Area Networks," *In IEEE Communications Maga*zine, September 1997.

[Dasgupta2003] Koustuv Dasgupta, Konstantinos Kalpakis, Parag Namjoshi "An Efficient Clustering--based Heuristic for Data Gathering and Aggregation in Sensor Networks". *In the Proceedings of the IEEE Wireless Communications and*

Networking Conference (WCNC), New Orleans, Louisiana, USA, March 16-20, 2003

[Deb2003] B. Deb, S. Bhatnagar, and B. Nath, "Multi-resolution State Retrieval in Sensor Networks," *In Proceedings of the IEEE Intl. Workshop on Sensor Network Protocols and Applications (SNPA)*, Anchorage, AK, May 2003.

[Dulman2003] S. Dulman, T. Nieberg, J. Wu, and P. Havinga, "Trade-Off between Traffic Overhead and Reliability in Multipath Routing for Wireless Sensor Networks," *WCNC Workshop*, New Orleans, Louisiana, March 2003.

[Estrin1999] D. Estrin, R. Govindan, J. Heidemann, and S. Kumar, "New Century Challenges: Scalable Coordination in Sensor Networks," *ACM Mobicom*, 1999.

[Estrin2001] D. Estrin, L. Girod, G. Pottie, and M. Srivastava, "Instrumenting the World with Wireless Sensor Networks," *In Proceedings Intl. Conf. on Acoustics, Speech and Signal Processing (ICASSP)*, Salt Lake City, Utah, May 2001.

[Fang2003] Q. Fang, F. Zhao, and L. Guibas, "Lightweight Sensing and Communication Protocols for Target Enumeration and Aggregation," *Proceedings of the 4th ACM international symposium on Mobile ad hoc networking and computing (MOBIHOC)*, 2003.

[Fu2003] Z. Fu, P. Zerfos, H. Luo, S. Lu, L. Zhang, and M. Gerla, "The Impact of Multihop Wireless Channel on TCP Throughput and Loss," *In IEEE INFOCOM*, San Francisco, CA, March 2003.

[Ganesan2001] D. Ganesan, R. Govindan, S. Shenker, and D. Estrin, "Highly-resilient, energy-efficient multipath routing in wireless sensor networks," *ACM SIGMOBILE Mobile Computing and Communications Review*, Vol. 5, No. 4, 2001.

[Gao2001] R. Gao and P. Hunerberg, "CDMA-based wireless data transmitter for embedded sensors," *In Proceedings 18th IEEE Instrumentation and Measurement Technology Conference*, 2001.

[Heinzelman1999] W. Heinzelman, J. Kulik, and H. Balakrishnan, "Adaptive Protocols for Information Dissemination in Wireless Sensor Networks," *ACM/IEEE Mobicom Conference*, Seattle, WA, August, 1999.

[Heinzelman2000a] W. Heinzelman, "Application-Specific Protocol Architectures for Wireless Networks," *PhD Thesis, Massachusetts Institute of Technology*, June 2000.

[Heinzelman2000b] W. Heinzelman, A. Chandrakasan and H. Balakrishnan, "Energy-Efficient Communication Protocol for Wireless Microsensor Networks," *Proceedings of the 33rd Hawaii International Conference on System Sciences*, January 2000.

[Hill2004] Jason Hill, M. Horton, R. King and L. Krishnamurthy, "The platform enabling wireless sensor networks," *Communications of the ACM*, vol. 47, no. 6, June 2004, pp 41-46.

[Ilyas2005] M. Ilyas and I. Mahgoub, Eds., *Handbook of Sensor Networks*, CRC Press, 2005.

[Intanagonwiwat2000] C. Intanagonwiwat, R. Govindan, and D. Estrin, "Directed Diffusion: A Scalable and Robust Communication Paradigm for Sensor Networks," *In ACM/IEEE MOBICOM*, August 2000.

[Jain2003a] N. Jain, D. Madathil and D. Agrawal, "Energy Aware Multi-Path Routing for Uniform Resource Utilization in Sensor Networks," *International Workshop on Information Processing in Sensor Networks (IPSN)*, Palo Alto, CA, April 2003.

[Jain2003b] N. Jain, D. Madathil and D. Agrawal, "Exploiting Multi-Path Routing to achieve Service Differentiation in Sensor Networks," *11th IEEE International Conference on Networks (ICON)*, Sydney, Australia, September 2003.

[Jain2005a] N. Jain, and D.P. Agrawal, "Current Trends in Wireless Sensor Network Design," *International Journal of Distributed Sensor Networks*, vol. 1, no. 1, 2005, pp 101-122.

[Jain2005b] N. Jain, D.K. Madathil and D.P. Agrawal, "MidHopRoute: A multiple routing framework for load balancing with service differentiation in wireless sensor network," *special issue on Wireless Sensor Networks of the International Journal of Ad-Hoc and Ubiquitous Computing*, 2005.

[Jain2005c] N. Jain, R. Biswas, N. Nandiraju, and D. Agrawal, "Energy Aware Routing for Spatio-temporal Queries in Sensor Networks," *Invited paper, IEEE Wireless Communications and Networking Conferences 2005*, March 13-17, 2005.

[Jiang1998] M. Jiang, J. Li, and Y. Tay, "Cluster Based Routing Protocol (CBRP) Functional Specification," *Internet Draft*, 1998.

[Kanodia2001] V. Kanodia, C. Li, A Sabharwal, B. Sadeghi, and E. Knightly, "Distributed Multi-Hop Scheduling and Medium Access with Delay and Throughput Constraints," *In Proc. ACM. Intl. Conf. on Mobile Computing and Networking*, pages 200–209, Rome, Italy, July 2001.

[Krishnamachari2002] B. Krishnamachari, D. Estrin, and S. Wicker, "The Impact of Data Aggregation in Wireless Sensor Networks," *In Proceedings of the IEEE Workshops of the Intl. Conference on Distributed Computing Systems*, pages 575–578, Vienna, Austria, July 2002.

[Kulik2002] J. Kulik, W. Heinzelman, and H. Balakrishnan, "Negotiation-based protocols for disseminating information in wireless sensor networks," *Wireless Networks*, Vol. 8, pp. 169-185, 2002.

[Li2001a] L. Li, and J. Y. Halpern, "Minimum-Energy Mobile Wireless Networks Revisited," *IEEE International Conference on Communications (ICC)*, 2001.

[Li2001b] Q. Li, J. Aslam, and D. Rus, "Hierarchical Power-aware Routing in Sensor Networks," *In Proceedings of the DIMACS Workshop on Pervasive Networking*, May 2001.

[Lindsey2002a] S. Lindsey and C. Raghavendra, "PEGASIS: Power-Efficient Gathering in Sensor Information Systems," *IEEE Aerospace Conference Proceedings*, 2002.

[Lindsey2002b] S. Lindsey and K. Sivalingam, "Data gathering algorithms in sensor networks using energy metrics," *IEEE Transactions on Parallel and Distributed Systems*, vol. 13, no. 9, pp. 924–934, 2002.

[Liu2004] B. Liu, and D. Towsley, "A study of the coverage of large-scale sensor networks," *Proceedings Mass 2004*, pp. 475-483.

[Liu2005] C. Liu, K. Wu, and J. Pei, "A dynamic clustering and scheduling approach to energy saving in data collection for wireless sensor networks," *Proceedings SECON*, Sept. 2005.

[Madden2002] Samuel Madden, Michael J. Franklin, Joseph M. Hellerstein, and Wei Hong, "TAG: a Tiny AGgregation Service for Ad-Hoc Sensor Networks," *5th Annual Symposium on Operating Systems Design and Implementation (OSDI)*, December, 2002.

[Madden2003] S. Madden, M. J. Franklin, J. M. Hellerstein, and W. Hong, "The Design of an Acquisitional Query Processor for Sensor Networks," *In ACM SIGMOD Conference*, San Diego, CA, June 2003.

[Malpani2000] N. Malpani, J. Welch, and N. Vaidya, "Leader Election Algorithms for Mobile Ad Hoc Networks," *In Proceedings of the Intl. Workshop on Discrete Algorithms and Methods for Mobile Computing and Communications*, Boston, MA, 2000.

[Manjeshwar2001] A. Manjeshwar and D. Agrawal, "TEEN: A protocol for Enhanced Efficiency in Wireless Sensor Networks," *Proceedings of the 1st Int. Workshop on Parallel and Distributed Computing Issues in Wireless Networks and Mobile Computing*, April 2001.

[Manjeshwar2002a] A. Manjeshwar and D. Agrawal, "APTEEN: A Hybrid Protocol for Efficient Routing and Comprehensive Information Retrieval in Wireless Sensor Networks," *Proceedings of the 2nd Int. Workshop on Parallel and Distributed Computing Issues in Wireless Networks and Mobile Computing*, April 2002.

[Manjeshwar2002b] A. Manjeshwar, Q.-A. Zeng, and D. Agrawal, "An analytical model for information retrieval in wireless sensor networks *using enhanced APTEEN protocol,*" *IEEE* Transactions on Parallel and Distributed Systems, vol. 13, no. 12, December 2002.

[Mayer2004] K. Mayer, K, Ellis and K. Taylor, "Cattle health monitoring using wireless sensor networks," *Proceedings 2nd IASTED International Conference on Communication and Computer Networks*, Cambridge, Massachusetts, Nov 8-10, 2004.

[Megerian2005] S. Megerian, F. Koushanfar, M. Potkonjak and M.B. Srivastava, "Worst and best case coverage in sensor networks," *IEEE Transactions on Mobile Computing*, Vol. 4, No. 1, Jan/Feb. 2005, pp 84-92.

[memsic] http://blog.memsic.com/mica2_mote/

[MICA]   MICASensorMote,   http://www.xbow.com/Products/WirelessSensor Networks.htm.

[microstrain] www.microstrain.com

[Miller2005] M.J. Miller and N.H. Vaidya, "A MAC protocol to reduce sensor network energy consumption using a wakeup radio," *IEEE Transactions on Mobile Computing*, May /June 2005, Vol. 4, No. 3, pp 228-242.

[moteiv] www.moteiv.com.

[Nakano2002] K. Nakano and S. Olariu, "A Survey on Leader Election Protocols for Radio Networks," *In Proceedings of the IEEE Intl. Symposium on Parallel Architectures, Algorithms and Networks*, pages 63–68, 2002.

[Nayak2010].A. Nayak and Ivan Stojmenović, *Wireless Sensor and Actuator Networks*, John Wiley, 2010.

[Olariu2006] Stephan Olariu and Ivan Stojmenović, "Design guidelines for maximizing lifetime and avoiding energy holes in sensor networks with uniform distribution and uniform reporting," *IEEE INFOCOM* 2006.

[Park2003] S.-J. Park and R. Sivakumar, "Sink-to-Sensors Reliability in Sensor Networks," In *Proc. ACM Intl. Symposium on Mobile Ad Hoc Networking and Computing (MOBIHOC)*, Annapolis, MD, June 2003.

[Petrovic2003] D. Petrovic, R. Shah, K. Ramchandran, and J. Rabaey, "Data Funneling: Routing with Aggregation and Compression for Sensor Networks," *In Proceedings of the IEEE Intl. Workshop on Sensor Network Protocols and Applications (SNPA)*, Anchorage, AK, May 2003.

[Pottie2000] G. Pottie, and W. Kaiser, "Wireless integrated Sensor Networks (WINS)," *Communications of the ACM*, Vol. 43, No. 5, May 2000.

[Rahul2002] C. Rahul and J. Rabaey, "Energy Aware Routing for Low Energy Ad Hoc Sensor Networks," *IEEE Wireless Communications and Networking Conference (WCNC)*, 2002.

[Rodoplu1999] V. Rodoplu and T. H. Meng, "Minimum Energy Mobile Wireless Networks," *IEEE Journal Selected Areas in Communica*tions, Vol. 17, No. 8, August 1999.

[Rabbat2004] M. Rabbat, and R. Nowak, "Distributed optimization in sensor networks," *In Proceedings of the Third International Symposium on Information processing in Sensor Networks*, pp. 20-27, 2004.

[Sankarasubramaniam2003a] Y. Sankarasubramaniam, I. Akyildiz, and S. McLaughlin, "Energy Efficiency Based Packet Size Optimization in Wireless Sensor Networks," *In Proceedings of IEEE Intl. Workshop on Sensor Network Protocols and Applications (SNPA)*, Anchorage, AK, May 2003.

[Sankarasubramaniam2003b] Y. Sankarasubramaniam, O. Akan, and I. Akyildiz, "ESRT: Event-to-Sink Reliable Transport in Wireless Sensor Networks," *In Proceedings of ACM MOBIHOC 2003*, Annapolis, Maryland, USA, June 2003.

[Savvides2001] A. Savvides, C-C Han, and M. Srivastava, "Dynamic fine-grained localization in Ad-Hoc networks of sensors," *Proceedings of the Seventh ACM Annual International Conference on Mobile Computing and Networking (MobiCom)*, July 2001.

[Schurgers2001] C. Schurgers, O. Aberthorne, and M. Srivastava, "Modulation Scaling for Energy Aware Communication Systems," *In Int. Symposium on Low Power Electronics and Design (ISLPED)*, pp. 96–99, August 2001.

[Schurgers2002] C. Schurgers, V. Tsiatsis, S. Ganeriwal, and M. Srivastava, "Optimizing Sensor Networks in the Energy-Latency-Density Design Space," *IEEE Transactions on Mobile Computing*, vol. 1, no. 1, 2002.

[Sensoria] Sensoria Corporation, http://www.sensoria.com.

[Shah2002] R. Shah and J. Rabaey, "Energy Aware Routing for Low Energy Ad Hoc Sensor Networks," *IEEE Wireless Communications and Networking Conference (WCNC)*, 2002.

[Shah2004] V. Shah, H. Deng, and D.P. Agrawal, "Parallel cluster formation for secured communication in wireless Ad hoc Networks," *IEEE international conference on networks 2004*, Nov. 16-19, 2004, pp 475-479.

[Shih2001a] E. Shih, B. Calhoun, S.-H. Cho, and A. Chandrakasan, "Energy-Efficient Link Layer for Wireless Microsensor Networks," *In Proceedings of the Workshop on VLSI (WVLSI)*, April 2001.

[Shih2001b] E. Shih, S.-H. Cho, N. Ickes, R. Min, A. Sinha, A. Wang, and A. Chandrakasan, "Physical-Layer Driven Protocol and Algorithm Design for Energy-Efficient Wireless Sensor Networks," *In Proceedings of 7th ACM Intl. Conf. on Mobile Computing and Networking (Mobicom)*, July 2001.

[Son2005] S. N. Son, M. Chiang, S. R. Kulkarni and S. C. Schwartz, "The value of clustering in distributed estimation for Sensor Networks," *Proceedings IEEE Wireless Com.*, June 2005.

[Sohrabi1999] K. Sohrabi and G. Pottie, "Performance of a Novel Self-Organization Protocol for Wireless Ad-Hoc Sensor Networks," *Proceedings of IEEE VTC*, Amsterdam, Netherlands, September 1999.

[Sohrabi2000] K. Sohrabi, J. Gao, V. Ailawadhi, and G. Pottie, "Protocols for self-organization of a wireless sensor network," *IEEE Personal Communications*, Vol. 7, No. 5, 2000.

[Srinivasan2003] V. Srinivasan, P. Nuggehalli, C. Chiasserini, and R. Rao, "Cooperation in Wireless Ad Hoc Networks," *In Proceedings of the IEEE Infocom*, San Francisco, CA, March 2003.

[Stann2003] F. Stann and J. Heideman, "RMST: Reliable Data Transport in Sensor Networks," In Proceedings of IEEE Intl. Workshop on Sensor Network Protocols and Applications (SNPA), Anchorage, AK, May 2003.

[Stemm1997] M. Stemm and R. H Katz, "Measuring and reducing energy consumption of network interfaces in hand-held devices," *IEICE Transactions on Communications*, E80-B(8):1125–1131, August 1997.

[Tanenbaum1996] A. Tanenbaum, *Computer Networks*, Prentice Hall PTR, 1996.

[Tilak2002] S. Tilak, N. Abu-Ghazaleh, and W. Heinzelman, "A taxonomy of wireless micro-sensor network models," *ACM SIGMOBILE Mobile Computing and Communications Review*, Volume 6, Issue 2, pp 28-36, April 2002.

[Tinyos] [http://www.tinyos.net/].

[TinyDB] TinyDB, http://telegraph.cs.berkeley.edu/tinydb/, University of California at Berkeley.

[UDDI] Universal Description, Discovery and Integration (UDDI), http://www.uddi.org/.

[Walter2001] J. Walter, N. Vaidya, and J. Welch, "A Mutual Exclusion Algorithm for Ad Hoc Mobile Networks," *Wireless Networks*, Vol. 9, No. 6, 2001.

[Wan2002] C.-Y. Wan, A. Campbell, and L. Krishnamurthy, "PSFQ: A Reliable Transport Protocol for Wireless Sensor Networks," *In Proceedings of ACM Intl. Workshop on Sensor Networks and Applications (WSNA)*, Atlanta, GA, September 2002.

[Wang2004] G. Wang, G. Cao, and T. L. Porta, "Proxy-based sensor deployment for mobile sensor network," *Proceedings of the MASS*, 2004.

[Wang2008] *Demin Wang*, "Wireless Sensor Networks: Deployment Alternatives and Analytical Modeling," *PhD Dissertation*, University of Cincinnati, November 17, 2008.

[Wolf2001] T. Wolf and S. Y. Choi, "Aggregated Hierarchical Multicast for Active Networks," MILCOM 2001., www.wolfran.com .

[Woo2001] A. Woo and D. Culler, "A Transmission Control Scheme for Media Access in Sensor Networks," In Proc. ACM Intl. Conf. on Mobile Computing and Networking, pages 221–235, Rome, Italy, July 2001.

[WSDL] Web Services Description Language, http://www.w3.org/TR/wsdl.

[xbow] www.xbow.com

[Xu2001] Y. Xu, J. Heidemann, and D. Estrin, "Geography-informed Energy Conservation for Ad-hoc Routing," *In Proceedings of the Seventh Annual ACM/IEEE International Conference on Mobile Computing and Networking*, 2001.

[Ye2001] F. Ye, A. Chen, S. Liu, and L. Zhang, "A scalable solution to minimum cost forwarding in large sensor networks," *Proceedings of the tenth International Conference on Computer Communications and Networks (ICCCN)*, pp. 304-309, 2001.

[Ye2002] W. Ye, J. Heidemann, and D. Estrin, "An energy-efficient MAC protocol for wireless sensor networks," In *Proceedings of the IEEE Infocom*, pages 1567–1576, New York, NY, June 2002.

[Ye2003] W. Ye, J. Heidemann, and D. Estrin, "Medium access control with coordinated, adaptive sleeping for wireless sensor networks," *Technical Report ISI-TR-567*, USC Information Sciences Institute, January 2003.

[Zhang2005] H. Zhang and J.C. Hou, "Maintaining Sensing Coverage and Connectivity in large Sensor Networks," *Ad hoc and Sensor Wireless Networks 2005*, pp 89-125

[Zhao2003] J. Zhao, R. Govindan, and D. Estrin, "Computing Aggregates for Monitoring Wireless Sensor Networks," In *Proceedings of the IEEE Intl. Workshop on Sensor Network Protocols and Applications (SNPA)*, Anchorage, AK, May 2003.

[Zhong2001] L. Zhong, R. Shah, C. Guo, and J. Rabaey, "An Ultra-Low Power and Distributed Access Protocol for Broadband Wireless Sensor Networks," In *IEEE Broadband Wireless Summit*, Las Vegas, NV, May 2001.

[Zorzi1997] M. Zorzi and R. R. Rao, "Error Control and Energy Consumption in Communications for Nomadic Computing," *IEEE Transactions on Computers*, Vol. 46, No. 3, 1997.

Chapter 12

Sensor Networks in Controlled Environment and Actuators

12.1 Introduction

In the previous Chapter, WSNs with randomly deployed SNs have been discussed. Now, we need to learn about how WSNs can be used for civilian applications where sensed area is easily accessible and SNs can be placed wherever desired. These include household and building automatic control, environmental observation for landfill gases, wastewater management, pollution effect, weather service monitoring, health and rehabilitation systems, structural health of tall buildings including cracks and large bridges management, habitat monitoring, factory assembly line synchronization, industrial monitoring, greenhouse monitoring [wikipedia], characterize presence of Chemical, Biological, Radiological, Nuclear, and Explosive (CBRNE) material, rural and urban environment in detecting changes in plains, forests, and oceans, security surveillance in shopping malls, parking garages, and other facilities, open spots indication in a large parking lot [wikipedia], safety of food and pharmaceutical industries [wapedia], and many other commercial and civilian applications [Akyildiz2004, Xia2007a, Xia2007b, Rezgui2007]. It may be noted that there are many civilian applications where access to the event area is either not possible or the occurrence region is unpredictable and SNs cannot be placed at predefined locations. Such applications include earthquake, cyclone and hurricane, excessive flooding and snowing, war region, forest wildfire, iceberg movement, landslide, etc.

First we define topologies that are appropriate for these applications and characterize associated performance issues. Most of these civilian applications can use utilize corrective action and is usually achieved by some mechanism known as actuator or a set of actuators [Ganek2003, Herrmann2005, Nayak2010]. For example, if temperature, humidity and other parameters are to be controlled in a house, then, based on current value detected and coordinated at the BS [Melodia2007, Shah2006], actuators such as air conditioner, humidifier/dehumidifier, etc., need to be activated [Arzen2006, Xia2006, Xia2007] and is illustrated by actuators by rectangular boxes, SNs, and BS in Figure 12.1 [*culler*]. The main objective of SNs is to sense different physical parameters and collect them at the BS which eventually issues control signals for corrective action to the actuators. This can be easily extended to a large office building or a shopping complex. Thus, the objective for having actuators is to compliment a WSN by simplifying the control, conserving the needed efforts, enhancing the convenience, increasing the efficiency, improving the reliability, adding it its flexibility, and making it safer [Bouyssounouse2005, El-Gendy2003, Li2007]. In most WSN applications, actuators are not shown explicitly, but thire existence and usefulness get unnoticed while they should not be ignored.

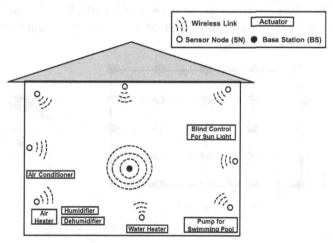

Figure 12.1 – SNs, BS and Actuators in a typical Home Control

For the application where the area is accessible, there is no need for deploying SNs randomly. Instead, SNs can be placed at predefined locations so as to cover the area adequately with minimum number of SNs. A simple strategy is to place the SNs in the form of 2-D grid as such cross-point and such configuration may be very useful for uniform coverage if the area to be deployed, is easily accessible and sensor can be placed anywhere. Such symmetric placement allows best possible regular coverage and easy clustering of the close-by SNs. Three such examples of SNs in rectangular, triangular and hexagonal tiles of clusters are shown in Figure 8.7 and 8.8. The first diagram shows clusters of square size 5x5, with a SN located at each intersection of lines. It may be noted that the square, triangle, or hexagonal placement of the SNs also dictates the minimum sensing area that need to be covered by each sensor.

12.2 Regularly Placed Sensors

Detailed views of SNs in three different configurations are shown in Figures 12.2-12.4. These three topologies are such that the basic scheme can be easily extended to a larger area by using the same type of tile and without leaving any gap, or having any overlap. A SN is placed at each cross point and the data packets are forwarded to the BS. As discussed in the previous chapter, if sensing range of a SN is assumed to be circular in nature, there will be overlapping areas in all three such WSNs. Therefore, it is easier to draw a Voronoi diagram and one can approximate that the Vononoi area to be covered by each SN is the sensing area for 1-coverage.

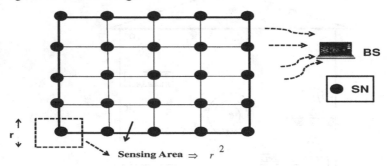

Figure 12.2 – SNs connected in the form of 2-D Mesh

Following the Voronoi diagram argument and for simplicity of calculation, the sensing area covered by rectangular placement is taken rectangular, while sensing areas by the two configurations are assumed hexagonal and triangular respectively. If the total area to be covered by N sensors is given by A, side of each rectangle/triangle/hex is given by r, then, the area covered by a rectangular placement can be given by r^2; while for triangular tile is $\frac{\sqrt{3}}{2} r^2$. Simple characteristics of three tiles are given in Table 12.1 [Wang2008a]. If a larger area is to be covered, then additional sensors are deployed as shown in Figures 12.5-12.7 [Wang2008a]. The three configurations also enable clustering of the SNs and the size of each cluster can be fixed as per application requirements, one such example is shown in these figures.

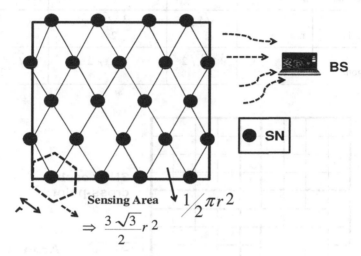

Figure 12.3 – SNs connected in the form of Triangles

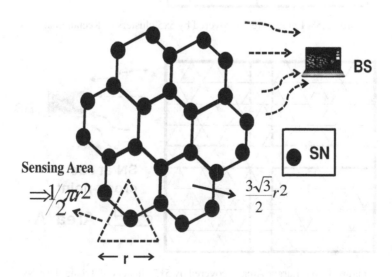

Figure 12.4 – SNs connected in the form of Hexagons

Table 12.1 Placement of Sensors and covered sensing area

Placement	Distance between Adjacent Sensors	Area of the topology	Sensing Area to be covered by each SN using Voronoi Diagram	Sensing area to be covered by N SNs
Rectangular	r	r^2	r^2	$N.r^2$
Triangular	r	$\dfrac{\sqrt{3}}{2}r^2$	$\dfrac{3\sqrt{3}}{2}r^2$	$N.\dfrac{3\sqrt{3}}{2}r^2$
Hexagon	r	$\dfrac{3\sqrt{3}}{2}r^2$	$\dfrac{\sqrt{3}}{2}r^2$	$N.\dfrac{\sqrt{3}}{2}r^2$

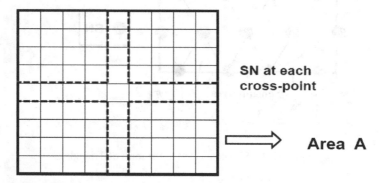

Figure 12.5 – Larger Area A covered by 5x5 clusters of Rectangular SNs

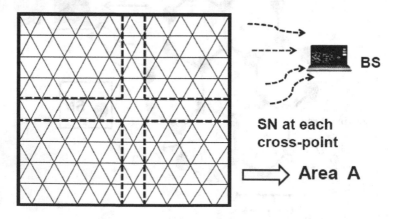

Figure 12.6 – Larger Area A covered by 5x5 clusters of Triangular SNs

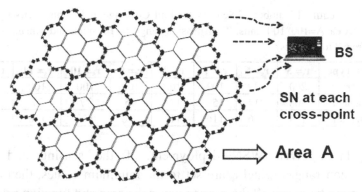

Figure 12.7 – Larger Area A covered by 5x5 clusters of Hexagonal SNs

Table 12.2 Required Number of Sensors for Full Coverage by three Lattices with different Sensing Ranges to cover an area A=100*100 units

Lattice Type	r_s=5	r_s=6	r_s=7	r_s=8	r_s=9	r_s=10	r_s=11	r_s=12
Square	200	139	102	79	62	50	42	35
Hexagonal	924	642	472	361	286	231	191	161
Triangular	154	107	79	61	48	39	32	27

Table 12.3 Required Number of Sensors for Full Connectivity by three Lattices with different transmission ranges for an area A=100*100 units

Lattice Type	r_c=5	r_c=6	r_c=7	r_c=8	r_c=9	r_c=10	r_c=11	r_c=12
Square	400	278	205	157	124	100	83	70
Hexagonal	924	642	472	361	286	231	191	161
Triangular	462	321	236	181	143	116	96	81

The area to be covered depends on the sensing range of each SN and the number of SNs required to cover an area of 100*100 is shown in Tables 12.2 for three placement patterns when the sensing range r_s of each SN is varied from 5 to 12 units. It may be note that the radio transmission distance between adjacent SNs need to be such that the sensors can receive data from adjacent sensors using wireless radio. The minimum communication distance thus becomes the distance between the SNs and is equal to r. A comparison between required numbers of SNs from the connectivity point of view is listed in Table 12.3 while Table 12.4 [Wang2008b] indicates the situation when both coverage and connectivity are considered together.

Table 12.4 Required Number of Sensors for Full Coverage and Connectivity by Three Lattices (Area A=100*100 units, Transmission range r_c=10 units, and sensing range r_s varied from 4 to 15 units)

Lattice Type	$r_s= 5$	$r_s= 6$	$r_s= 7$	$r_s= 8$	$r_s= 9$	$r_s= 10$	$r_s= 11$	$r_s= 12$
Square	200	139	102	100	100	100	100	100
Hexagonal	308	214	158	121	95	77	77	77
Triangular	154	116	116	116	116	116	116	116

In a given WSN deployment, if the sensing and radio transmission ranges r_s and r_c are set to the minimum values, then all the SNs need to be active all the time to cover the area and function properly. If the SNs density is increased, then the SNs can be divided into many sub-sets, with each sub-set covering the whole area such that other SNs belong to remaining groups can be allowed to go to sleep mode, thereby increasing the effective time of the WSN. Moreover, in many WSN applications, an event cannot be detected by a single SN and many SNs need to do so in identifying the event correctly. This may require each sub-region to be covered by more than one SN or k-covered, where $k>1$. An alternative is to increase the sensing range r_s so that the area covered can be larger than 1. Grouping of SN sub-sets in three different schemes are shown in Figure 12.8 as long as sensing and communication ranges r_s and r_c are adjusted according to the design requirements.

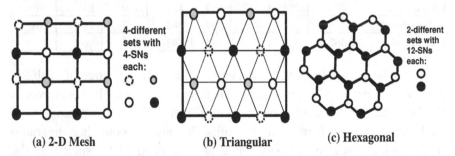

(a) 2-D Mesh   (b) Triangular   (c) Hexagonal

Figure 12.8 – Three Topologies divided into different sets

12.3 Design Issues

The controlled environment makes controlling SNs much easier. The distance between SNs is known and there is no need for location determination as relative distance between SNs is known ahead of time.

This also makes routing straightforward. In a similar way, it is relatively easier to form a cluster. Figure 12.9 shows clustering of SNs which are 1-hop away from the CH in three topologies. A similar clustering of SNs 2-hops away from CH is given in Figure 12.10 [Wang2008b]. These figures clearly indicate that clustering is relatively simpler in regular WSNs than random deployment.

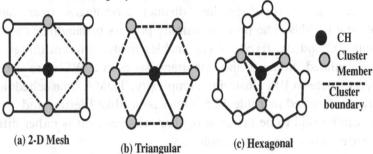

(a) 2-D Mesh   (b) Triangular   (c) Hexagonal

Figure 12.9 – 1-Clustering of SNs for three topologies

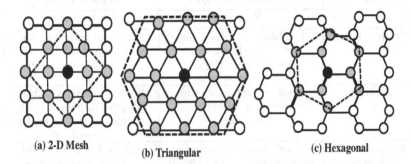

(a) 2-D Mesh   (b) Triangular   (c) Hexagonal

Figure 12.10 – 2-Clustering of SNs for three topologies

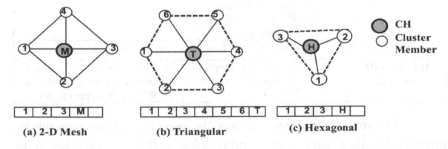

(a) 2-D Mesh   (b) Triangular   (c) Hexagonal

Figure 12.11 – TDMA Schedule for three topologies

Once the clusters have been formed, the CH can allocate TDMA schedule for each member to send data to the CH. One such time slot allocation is illustrated in Figure 12.11 where each SN transmits its data in an assigned slot only so as to avoid collisions.

12.4 Network Issues

A regular topology has distinct advantages over random deployment of SNs. The time consuming process of neighbor discovery is eliminated and the fixed pattern enables many different things to be easily calculated. Some of the important issues in a WSN are neighbor discovery, route to BS, clustering complexity, TDMA time schedule, etc and are summarized in Table 12.5. Other issues like lifetime and residual energy can be said to be better in regular schemes as it is rather difficult to quantize them for random topologies.

Table 12.5 Comparison of Random and Regular WSN Topologies

WSN Topology	Randomly Deployed	Regular Topology
SNs distribution	Random	Uniform
Neighborhood discovery	Use beacon signals to find nearest neighbors within communication range, an involved process	As SNs are uniformly placed, the location of neighboring SNs is built in the system
Routing	Have to find the path with known local neighborhood information	Routing is easy as each node with given address, is located at a location based on the basic topology
Formation of Clusters	Optimal clustering is complex, many heuristics defined based on SN connectivity	Easy to form clusters using topology information
TDMA schedule	Can be easily defined by CH if that has already done; otherwise too complex to implement	Can be easily defined by CH
Voronoi diagram	Fairly involved process	Easy as SNs are regularly placed
Impact of BS location	Routing to BS is a complex process	Does impact routing, but easy to do based on relative location of WSN with respect to BS

In terms of radio transmission, the energy efficiency depends on distances between adjacent SNs and path length from a source SN to the BS. This again favors regular schemes. Reactive, proactive or hybrid protocols are used for reporting sensed data to the BS. So far as grouping SNs for sleep-awake cycles, it is relatively easier to do so for regular topologies. The use of multi-path routing for reliability/fault-tolerance and equal consumption of energy across SNs seems straight forward in regular topologies as all alternate paths can be easily determined. This is much more involved in a random topology as single path itself requires a lot of efforts. Moreover, any discharged batteries or even failed SNs can be easily replaced as SNs in regular topologies are assumed to be easily accessible.

12.5 RF ID as a Passive Sensor

There has been a lot of discussion about RFID (Radio Frequency Identification Tag) for numerous tracking applications. This is being used in tracking valuable parcels; different products in superstores like Walmart, Sams Club; toll control in many express ways, parking lot control, and even in passports. The basic idea is to passively encode a number intelligently so that it can be read by RFID reader that gives the same number. This unique number is used to do other control operation or table look up to determine the balance in an associated account, charge money to that account or open the access bar in a parking lot. There is an increased use of such an RFID by placing them on trucks and when these trucks pass by RFID reader placed at pre-specified places, location of the truck is automatically determined. Many other applications of RFID are being explored. Further development has allowed having active RFIDs with very limited storage of few bytes.

12.6 Conclusions and Future Directions

WSNs deployed in a controlled environment possess some specific characteristics that make human life easy to do unmanned monitoring of an area and collected information at the BS can be used to control the associate actuator for necessary action. Applications of WSNs are being explored for new commercial applications. The challenges are many and one needs to use own imagination in finding use of WSNs in new areas. There are many open areas of applications and increased future utilization seems imminent.

Homework Questions/Simulation Projects

Q. 1. Assuming a set of sensors are placed in the form of a 2-D grid, can you compute energy consumed by a sensor if:

   a. The sensing range of each sensor is limited to Noth/South and East/West neighboring node.

   b. The sensing range is extended to all eight neighbors, including the diagonal neighbors.

Assume the sensing range to be half that of the communication range.

Q. 2. In the sensor network of Q. 1, a battery is connected to each sensor and it is important to optimize energy consumption. Therefore, it is critical to place them such that sensing can be done for all parts of the area and all sensors are able to communicate data to other sensors in the neighborhood. You are given 1,000 sensors and you need to spread them in the area of 50X50 units.

   a. What will be the average distance between two adjacent sensors in four directions (N-S-E-W) if the sensors are placed in the form of a regular grid?

   b. If the sensors are placed randomly?

   c. Assuming that average distance represents the sensing range of each sensor, what fraction of the area is covered by two adjacent sensors in situations 1 and 2?

   d. Assuming the radio transmission range to be double of the sensing range, how many sensors are present (average) within the transmission range of a single sensor both under 1 and 2?

   e. If the sensing range of each sensor is reduced to half, what portion of the area remains uncovered?

   f. What will be the impact on wireless radio if the communication range is reduced to half? You need to provide some quantitative measures.

Q. 3. 500 sensors equipped with wireless devices have been placed in a given rectangular area. The sensor nodes can be placed in a (i) rectangular, (ii) hexagonal, and (iii) triangular form. The 5x5 adjacent nodes are grouped together to form a cluster. Assume x units of energy is consumed for each data communication and y units of energy used for aggregation. The accuracy of aggregated data depends on the interval of aggregation and density of sensors.

   a. What is the average distance between each sensor node with respect to a cluster head?

   b. What is the average distance between the Cluster heads?

   c. What are the trade-offs if you double or halve the size of each cluster?

   d. If the sensors in each cluster are allowed to sleep for 50% time, how much energy is consumed?

   e. If the sleep cycle is increased to 70%, what is the impact on the energy consumption?

   f. What will be minimum transmission radius requirement if you want to have full coverage by active sensors in part (d)?

   g. Derive an analytical model between the densities of nodes, coverage and sleep cycle.

Assume appropriate parameters (if needed) as commonly used in the literature.

Q. 4. Repeat Q. 3 for bxb grid sensor structure with each cluster of axa sensors for integer values of a and b.

Q. 5. 500 sensors are deployed in a hexagonal fashion in an area of 40x40. Draw a Voronoi diagram and determine the following:

a. What is the maximum and average distance between each SN pair?

b. What is the impact on the required sensing range and communication distance if you double or halve the number of SNs?

c. What is the impact on the performance if you double or halve the area?

d. If the SNs are divided into two groups, how would you select the group so that there is minimum impact on the performance? Can you think of quick way of doing this? State clearly.

e. In part e, if each set of sensors is allowed to sleep for 50% time, how much energy is consumed?

f. If the sleep cycle is increased to 70%, what is the impact on the energy consumption?

g. What will be minimum transmission radius requirement if you want to have full coverage by active sensors in part (f)?

h. Assume appropriate parameters (if needed) as commonly used in the literature.

Q. 6. You are given 1,000 sensors and you need to spread them in the area of 100x100 units.

a. What will be the average distance between two adjacent SNs if sensors are deployed in a triangular topology?

b. Assuming the average distance represents the sensing range of each sensor, what fraction of the area is covered by two adjacent SNs in situations a?

c. Assuming the radio transmission range to be double of the sensing range, how many sensors are present (average) within the transmission range of a single SN?

d. If the sensing range of each sensor is reduced to half, what portion of the area remains uncovered?

e. What will be the impact on wireless radio if the communication range is reduced to half?

You need to come up with some quantitative measures.

Q. 7. In Question 3 with rectangular topology, the source SN is at one corner while BS is placed at the diagonally opposite corner. Can you identify multiple paths between the source SN and the BS? What is the minimum sensing and transmission range needed? Answer carefully.

Q. 8. In Question 6, can you illustrate how you can use PEGASIS algorithm to form a single chain of SNs, assuming the BS to be located at one corner?

Q. 9. In problem 3, you are allowed to group 5 sensors as a cluster. What is an easy way of forming a cluster? How effective it is as compared to an optimal solution? How do you select a CH in each cluster? Justify appropriateness of your answer.

Q. 10. In Question 5, how much energy is saved if each CH does data aggregation before forwarding to the BS located in one corner of the field?

References

[Akyildiz2004] I. F. Akyildiz and I. H. Kasimoglu, "Wireless sensor and actor networks: Research Challenges," *Ad Hoc Networks*, vol. 2, no. 4, pp. 351-367, 2004.

[Arzen2006] K.-E. Arzen, A. Robertsson, D. Henriksson, M. Johansson, H. Hjalmarsson, and K.H. Johansson, "Conclusions of the ARTIST2 Roadmap on Control of Computing Systems," *ACM SIGBED Review*, vol. 3, no. 3, pp. 11-20, 2006.

[Bouyssounouse2005] B. Bouyssounouse and J. Sifakis, (eds.), *Embedded Systems Design: The ARTIST Roadmap for Research and Development*, Lecture Notes in Computer Science 3436, Springer-Verlag, 2005.

[culler] *www.eecs.berkeley.edu/~culler/citris/sensorday/Arens.ppt*.

[El-Gendy2003] M. A. El-Gendy, A. Bose, and K. G. Shin, "Evolution of the Internet QoS and support for soft Real-time applications," *Proceedings of the IEEE*, vol. 91, no. 7, pp. 1086-1104, 2003.

[Ganek2003] A. G. Ganek and T. A. Corbi, "The dawning of the autonomic Computing Era," *IBM Systems Journal*, vol. 42, no. 1, pp. 5-18, 2003.

[Herrmann2005] K. Herrmann, G. Muhl, and K. Geihs, "Self management: the solution to complexity or just another problem," *IEEE Distributed Systems Online*, vol. 6, no. 1, pp. 1-17, 2005.

[Li2007] Y.J. Li, C.S. Chen, Y.-Q. Song, and Z. Wang, "Real-time QoS support in wireless sensor networks: a survey," *In Proc of 7th IFAC Int. Conf. on Fieldbuses & Networks in Industrial & Embedded Systems* (FeT'07), Toulouse, France, Nov. 2007.

[Melodia2007] T. Melodia, D. Pompili, V. C. Gungor, I. F. Akyildiz, "Communication and Coordination in Wireless Sensor and Actor Networks," *IEEE Transactions on Mobile Computing*, vol. 6, no. 10, pp. 1116-1129, 2007.

[Nayak2010] A. Nayak and I. Stojmenovic, *Wireless Sensor and Actuator Networks: Algorithms and Protocols for Scalable Coordination and Data Communication*, John Wiley, 2010.

[Rezgui2007] Rezgui, A.; Eltoweissy, M. Service-Oriented Sensor-Actuator Networks. *IEEE Communications Magazine*, 2007, vol. 45, no. 12, pp. 92-100.

[Shah2006] G.A. Shah, M. Bozyigit, O.B. Akan, and B. Baykal, "Real-Time Coordination and Routing in Wireless Sensor and Actor Networks," *In Proc. 6th Int. Conf. on Next Generation Teletraffic and Wired/Wireless Advanced Networking (NEW2AN)*, Lecture Notes in Computer Science, vol. 4003, pp. 365-383, 2006.

[Wang2008a] Yun Wang, "Application-Specific Quality of Service Constraint Design in Wireless Sensor Networks," *Ph.D. Dissertation*, University of Cincinnati, June 14, 2008.

[Wang2008b] Yun Wang and Dharma P. Agrawal, "Optimizing sensor networks for autonomous unmanned ground vehicles," *Optics/Photonics in Security & Defence*, 15-18 September 2008, Cardiff, Wales, United Kingdom.

[wapedia] http://wapedia.mobi/en/Wireless_sensor_network.

[wikipedia] http://en.wikipedia.org/wiki/Wireless_sensor_network.

[Xia2006] F. Xia, "Feedback scheduling of real-time control systems with resource constraints," *PhD thesis*, Zhejiang University, 2006.

[Xia2007] Xia, F.; Tian, G.S.; Sun, Y.X. Feedback Scheduling: An Event-Driven Paradigm. *ACM SIGPLAN Notices*, vol. 42, no. 12, pp. 7-14, 2007.

[Xia2007a] F. Xia, Y.C Tian, Y.J. Li, and Y.X. Sun, "Wireless Sensor/Actuator Network Design for Mobile Control Applications," *Sensors*, vol. 7, no. 10, pp. 2157-2173, 2007.

[Xia2007b] F. Xia, W.H. Zhao, Y.X. Sun, and Y.C. Tian, "Fuzzy Logic Control Based QoS Management in Wireless Sensor/Actuator Networks," *Sensors*, vol. 7, no. 12, pp. 3179-3191, 2007.

Chapter 13

Security in Ad Hoc and Sensor Networks

13.1 Introduction

As we have seen in the previous chapters, the advent of ad hoc and sensor networks brought with it a flurry of research primarily focused on communication and protocols in every layer of the protocol stack. Practical applications of this research range from simple chat programs to shared whiteboards and other collaborative schemes. Although intended for diverse audiences and contexts, many of these applications share a common characteristic: they are information–centric. The information transferred may be a trivial conversation between friends, confidential meeting notes shared among corporate executives, or mission–critical military information. Despite the deployment of information–driven applications such as these, the call for ad hoc and sensor network security remains largely unanswered.

Ad hoc and sensor networks security is not, however, a concern that has slipped through the cracks unnoticed: numerous research initiatives have been launched to surmount the challenge. Despite that, many questions remain open as many of the existing approaches have limited functionalities, unrealistic computational requirements, or inability to address core security issues. Security in ad hoc and sensor networks is an essential component for basic network functions like packet forwarding and routing and network operation can be easily jeopardized if countermeasures are not embedded into the basic network functions at the early stages of their design. In ad hoc and sensor networks, the basic functions are carried out by all available nodes.

If *a priori trust relationship* exists between the nodes of an ad hoc or a sensor network, entity authentication can be sufficient to assure the correct execution of critical network functions. A priori trust can only exist in a few special scenarios like military networks and corporate networks, where a common, trusted authority manages the network, and requires tamper-proof hardware for the implementation of critical functions. An environment where a common, trusted authority exists is called a *managed environment*. On the other hand, entity authentication in a large network raises key management requirements.

An ad hoc or a sensor network works like an *open environment* and any node can endanger the reliability of the network functions. The correct operation of the network also requires fair share of the functions by each participating node as power saving is a major concern. The considered threats are thus not just limited to maliciousness, a new type of misbehavior called selfishness should also be taken into account to prevent nodes that simply do not cooperate. With the *lack of a-priori trust*, a classical network security mechanism based on authentication cannot cope up with selfishness and collaborative security schemes. Therefore, security in ad hoc and sensor networks is a much harder task than in traditional wired networks and we delve into the specifics of security including key management schemes, secure routing algorithms, cooperation, and intrusion detection systems. In earlier part of this chapter, discussions are focused on ad hoc networks even though most approaches are equally applicable for sensor networks.

13.2 Distributed Systems Security

Threats are divided into three categories [Zwicky2000]: disclosure threats, integrity threats and denial of service (DoS) threats, even though this by no means covers all the possible threats. The disclosure threat involves the leakage of information from the system to a party that should not have seen the information and is a threat against the confidentiality of the information. Integrity threat involves an unauthorized modification of information. Finally, the DoS threat disables access to a system resource that is being blocked by a malicious attacker [Amoroso1994]. There are also other definitions that are important like authentication means ensuring the identity of another user while Non-repudiation ensures that a user that has sent a certain message cannot deny sending this message at a later time [Gollmann1999].

Another related process is called delegation. When a user, using a local access to login into a network, wants to execute a program on a remote machine, the program will need certain rights to use the resources on the remote machine. In such a case the user typically delegates access rights to the program, so that it can run on the remote machine. In distributed systems, there is always a possibility that the remote machine is weakly protected and a malicious user can exploit a legitimate user's rights. Another important parameter in distributed systems security is authentication that needs to be enforced centrally or locally. In centralized security enforcement, a Key Distribution Center (KDC), is used to store the keys of all the devices. The KDC acts as a Trusted Third Party (TTP) that users can use to authenticate themselves and other users. For example, Kerberos authentication and key exchange protocol can be found in [Schneier1996]. Here, the major limitation is the trustworthiness of the TTP. If it is compromised, all the secret keys become available for malicious use and the whole network collapses.

On the other hand, if the security enforcement scheme is to be local, other kinds of security measures are needed. Each user enforces security policy and trusts the machines one logs into. There could be a trusted Certification Authority (CA), which issues public key certificates and a Certification Distribution Center (CDC), which stores all the public/private certificates issued by the CA. The users have their own pair of keys and can certify their public keys with the CA. Then, if a user uses a private key to sign something, the signature can be verified to correspond with a public key. The public key, in turn, can be checked with the CDC to certify that in fact, it does belong to the user that originally did the signing. In this way, the security can be enforced locally and still have working authentication system with Public Key Infrastructure (PKI) [Gollmann1999].

13.3 Security in Ad Hoc and Sensor Networks

As we know, there is no fixed infrastructure in ad hoc and sensor networks and as the name implies, they are formed on the fly. The devices connect to each other in their own communication range via wireless links. Individual devices act as routers when relaying messages to other distant devices. The topology of an ad hoc network is not fixed either. It changes all the time when these mobile stations move in and out of each others transmission range. All this makes ad hoc and sensor networks very vulnerable to attacks. In this section we give an overview of the security issues over ad hoc and sensor networks.

13.3.1 Security Requirements

The security services of ad hoc and sensor networks are not altogether different than those of other network communication paradigms. Below we describe the requirements ad hoc networks must meet.

13.3.1.1 Availability

Availability ensures that the desired network services are available whenever they are needed. Systems that ensure availability seek to combat denial of service (DoS) and energy starvation attacks. As all the devices in the network depend on each other to relay messages, DoS attacks are easy to perpetrate. For example, a malicious user could try to jam or otherwise try to interfere with the flow of information. Or else, the routing protocol should be able to handle both the changing topology of the network and attacks from the malicious users. There are routing protocols that can adjust well to the changing topology, but there are none that can defy all the possible attacks [Deng2002, Zhou1999]. Another vulnerable point, which has no equivalence in traditional networks, is the limited battery power of wireless nodes. With battery exhaustion attacks, a malicious user can cause higher power consumption from other devices' battery, causing these devices to die prematurely [Stajano1999].

13.3.1.2 Authorization and Key Management

Authorization is another difficult matter in ad hoc and sensor networks. As there is little or no infrastructure, identifying users (e.g., participants in a meeting room) is not an easy task. There are problems with TTP schemes and identity-based mechanisms for key agreement. A generic protocol for password authenticated key exchange is described in [Asokan2000]. It has several drawbacks even though it is possible to construct very good authentication mechanisms for ad hoc and sensor networks. A password authenticated multi-party Diffie-Hellman key exchange seems to overcome many problems of the generic protocol.

13.3.1.3 Confidentiality and Integrity

Data confidentiality is a core security primitive for ad hoc and sensor networks. It ensures that the message cannot be understood by anyone other than the authorized personnel. With wireless communication, anyone can sniff the messages going through the air, and without proper encryption all the information is easily available. If a proper authenticity has been established, securing the connection with appropriate keys does not pose a big problem. Data integrity denotes the immaculateness of data sent from one node to another. That is, it ensures that a message sent from node A to node B was not modified during transmission by a malicious node C. If a robust confidentiality mechanism is employed, ensuring data integrity may be as simple as adding one–way hash to encrypted messages. In addition to malicious attacks, integrity may be compromised because of radio interference, etc., so some kind of integrity protection is definitely needed for ad hoc and sensor networks.

13.3.1.4 Non-Repudiation

Non-repudiation ensures that the origin of a message cannot deny having sent the message. It is useful for detection and isolation of compromised nodes. When a node A receives an erroneous message from a node B, non-repudiation allows A to accuse B using this message and to convince other nodes that B is compromised.

13.3.2 Security Solutions Constraints

Historically, network security personnel have adopted a centralized, largely protective paradigm to satisfy aforementioned requirements. This is effective because the privileges of every node in the network are managed by dedicated machines – authentication servers, firewalls, etc. – and the professionals who maintain them. Membership in such a network allows individual nodes to operate in an open fashion – sharing sensitive files, allowing incoming network connections – as any malicious user from outside world will not be allowed access. Although these solutions have been considered very early in the evolution of ad hoc and sensor networks, attempts to adapt similar client-server solutions to a decentralized environment have largely been ineffective ad the following characteristics have been identified for ad hoc and sensor networks:

- **Lightweight**: Solutions should minimize the amount of computation and communication to limit energy and computational resources.
- **Decentralized**: Attempts to secure must be done without reference to centralized, persistent entities and should levy cooperation of all trustworthy nodes.
- **Reactive**: MANETs are dynamic while sensor networks are static with many devices either run out of power or go to sleep mode, making the network dynamic. Therefore, security paradigms must react to changes in network state; they must seek to detect compromises and vulnerabilities.
- **Fault-Tolerant**: Wireless mediums are known to be unreliable; nodes are likely to leave or be compromised without warning.

Naturally, these are not stringent requirements: specific applications may relax some or all of the above based on their domain and the sensitivity of information involved. Moreover, many MANET applications do not require 2–party secure communication; instead, broadcast or group security may be needed. In sensor networks, devices are not mobile while all messages are directed towards the BS.

13.3.3 Challenges

The wireless links present in an ad hoc and a sensor network render them susceptible to many different attacks. Active attacks could range from message replay or deletion, injecting erroneous messages, impersonating a node, etc., thus violating availability, integrity, authentication and non-repudiation. Ad hoc nodes roaming freely in a hostile environment with relatively poor physical protection have good probability of being compromised. Hence, security solutions need to consider malicious attacks not only from outside but also from within the network. Therefore, security mechanism needs to be dynamic, and should be adequately scalable.

13.3.3.1 Key Management

Public key systems are generally recognized to have an upper hand in key distribution for a generic network. In a public key infrastructure, each node has a public/private key pair. A node distributes its public key freely to the other nodes in the network; however it keeps its private key to only itself. A CA is used for key management and has its own public/private key pair. The CA's public key is known to every network node. The trusted CA is responsible to sign certificates, binding public keys to nodes, and has to stay online to verify the current bindings. The public key of a node should be revoked if this node is no longer trusted or leaves the network. A single key management service for an ad hoc or a sensor network is probably not an acceptable solution, as it is likely to become Achilles' heel of the network. Hence, it may be more prudent to distribute the trust to a set of nodes by letting these nodes share the key management responsibility.

13.3.3.2 Secure Routing

The contemporary routing protocols designed for MANETs cope well with dynamically changing topology, but are not designed to provide defense against malicious attackers. In a sensor network, the devices are not dynamic, but the message forwarding is equally desirable. As for attackers, we can classify them into external and internal. External attackers may inject erroneous routing information, replay old routing data or distort routing information in order to partition or overload the network with retransmissions and inefficient routing. Compromised nodes inside the network are harder to detect and are far more detrimental. Routing information signed by each node may not work, as compromised nodes can generate valid signatures using their private keys. Isolating compromised nodes through routing information is also difficult due to the dynamic topology. Once the compromised nodes have been identified, the routing protocol should be able to bypass the compromised nodes by using alternate routes.

13.3.3.3 Intrusion Detection

Each MH in a MANET is an autonomous unit and is free to move independently and a node without adequate physical protection is susceptible to being captured or compromised. While intrusion prevention techniques such as encryption and authentication can reduce the risks of intrusion, they cannot be completely eliminated and can be used as a second line of defense to protect network systems.

13.3.4 Authentication

Authentication denotes an accurate, absolute identification of a user who wish to be a part of the network. Historically, authentication has been accomplished by a well-known central authentication server which maintains a database of entities, or users, and their corresponding unique IDs as a digital certificate, public key, or both. Unfortunately the ad hoc paradigm does not accommodate a centralized entity.

13.3.4.1 Trusted Third Parties

One of the most rudimentary approaches to authentication in ad hoc and sensor networks uses a TTP. Every node that wishes to participate obtains a certificate from a universally TTP. When two nodes wish to communicate, they first check to see if the other node has a valid certificate. Although popular, the TTP approach is laden with flaws. Foremost, it probably is not reasonable to require all devices to have a certificate. Secondly, each node needs to have a unique name. Although this is reasonable in a large internet, it is a bit too restrictive in an ad hoc and sensor setting.

13.3.4.2 Chain of Trust

The TTP model essentially relies on a fixed entity to ensure the validity of all nodes' identities. In contrast, the chain of trust paradigm relies on any node in the network to perform authentication and if you trust a friend, you tend to trust friend's friend. This paradigm fails if there are malicious modes within the network or the incoming nodes cannot be authenticated at all.

13.3.4.3 Location-Limited Authentication

Location-limited authentication levies on the fact that two nodes are close to one another and most ad hoc and sensor networks exist in a small area. Bluetooth and infrared are two of the most widely used protocols for this form of authentication. The authenticating node can be reasonably certain that the node it thinks is being authenticated is the node it is actually authenticating (i.e., there is no man-in-the-middle) by physical indications. Although location-limited authentication is well-suited for most applications with a single end-point, it is not feasible for large, group-based settings.

13.4 Key Management

This section provides a detailed description of the dominant key management paradigms that have been developed for ad hoc networks. The discussion is prefaced with an overview of key management terminology and the generalized Diffie–Hellman algorithm – the de facto standard for contributory key agreement algorithms. Although protocols had been developed for the wired network domain, their computational and communication requirements are shown to be prohibitive for ad hoc or sensor network scenarios. Notable improvements were made to these protocols by Burmester [Burmester1994], well–known CLIQUES, and the Tree-based Generalized Diffie-Hellman (TGDH) [Kim2000].

13.4.1 Conceptual Background

We first present definitions necessary to discuss and compare key management paradigms.

Definition 13.1: *A group key[1] is a secret that is used by two or more parties to communicate securely. Group keys are symmetric; that is, the same group key is used to encrypt and decrypt messages.*

Like most symmetric keys, group keys should be ephemeral in order to uphold key secrecy that guarantees the group key not to be discovered by a passive adversary within reasonable amount of time [Kim2000]. Group key secrecy assumes that the passive adversary has never been a member of the group. However, many group applications require that only current members of the group know the secret. This requirement necessitates key independence, and forward and backward secrecy are upheld by the key management protocol.

Definition 13.2: Key independence *ensures that a passive adversary*

---

[1]Group keys are sometimes referred to as "session keys" and use this in our discussions.

who knows a proper subset of group keys $\hat{K} \subset K$ *cannot discover any other group key* $\overline{K} \in (K - \hat{K})$.

Definition 13.3: Forward secrecy *ensures that a passive adversary (member or non-member) who knows a contiguous subset of old group keys cannot discover subsequent group keys.*

Definition 13.4: Backward secrecy *ensures that a passive adversary who knows a contiguous subset of group keys cannot discover preceding group keys.*

Definition 13.5: Key establishment *is the process, protocol, or algorithm by which a group key is created and distributed to the group.*

Key establishment is generally discussed as two discrete problems, namely, key agreement and key distribution.

Definition 13.6: Key agreement *is a protocol by which two or more parties contribute to the creation of a shared group key.*

Definition 13.7: Key distribution *is the process by which each group member is apprised of the group key.*

Some systems may necessitate key agreement and distribution protocols, while others may only have one or the other. Paradigms that only employ a key distribution protocol are often referred to as centralized. A single entity, typically the group controller or a TTP, is responsible for generating and distributing the keys. Similarly, paradigms that employ a key agreement protocol are often referred to as distributed among group members. Centralized key management technique is quite simple while such approach has a single point of failure. That means an active adversary needs only to compromise the key manager to change security of the entire group. Another notable drawback of centralized techniques is that they often require a secure channel to transmit the group key from the group controller to each member and is prohibitive.

Distributed key management techniques require that two or more group members contribute to the creation of the group key and one or more public values could be broadcast to the group. Upon receipt of the public value(s), each group member uses its own secret to calculate the actual group key. These are based on a generalization of the well-known Diffie-Hellman algorithm. Although distributed paradigms are appealing for their elegance and egalitarian treatment of group members, they are complex and often require synchronous group communication.

But, active adversaries may compromise security of the key by masquerading as a group member. The attack occurs during the key generation phase, when each member is contributing its share of the key. As the partial keys are being passed through the network, the adversary simply includes its own share. Distributed key management must account for such attacks by ensuring key integrity.

Definition 13.8: Key integrity *ensures that the group key is a function of all authenticated group members and no one else.*

One of the easiest ways for an adversary to sacrifice key integrity is by compromising prior keys of a group member. This closely related form of attack is known as a *known key attack.*

Definition 13.9: *A protocol is vulnerable to a* **known key attack** *if compromise of past session keys allows a passive adversary to compromise future group keys, or an active adversary to impersonate one of the protocol parties* [Steiner1996].

13.4.2 Diffie-Hellman Key Agreement

The Diffie-Hellman key agreement protocol [Diffie1976 is perhaps the largest publicly–known cryptographic breakthrough of the twentieth century. Unlike other cryptosystems, the Diffie-Hellman protocol provided a way for two parties to agree on a secret key and use it to communicate over an insecure medium in an ad hoc fashion.

Two constants p and α are chosen and are known to everyone.

A, B	Protocol participants
p	Large prime number
G	Unique subgroup of Z_*^p of order q with p, q prime
α	Exponentiation base – the generator in the group G
x	Random secret chosen by A such that $1 \leq x \leq p - 2$
y	Random secret chosen by B such that $1 \leq y \leq p - 2$
K	The shared key
K_i	The partial key created by member i

Key Agreement

1. A chooses its secret, x, creates $K_A = \alpha^x \bmod p$, and sends it to B.

$$A \xrightarrow{\quad K_A = \alpha^x \bmod p \quad} B$$

2. B chooses its secret, y, creates $K_B = \alpha^y \bmod p$, and sends it to A.

$$A \xleftarrow{\quad K_B = \alpha^y \bmod p \quad} B$$

3. A receives K_B and computes $K = K_{AB} = (\alpha^y)^x \bmod p$.
4. B receives K_A and computes $K = K_{BA} = (\alpha^x)^y \bmod p$.

13.4.2.3 Analysis

Despite the widespread acceptance of Diffie-Hellman, it is susceptible to the man-in-the-middle attack. For example, if an active adversary, Carol, is able to intercept the transmission of the Alice's public value, substitute her own, and send it to Bob. Similarly, Carol may intercept Bob's public value, substitute her own, and send it to Alice. The result of such an attack is that Bob's messages are actually understood by Carol, not Alice, and Alice's messages are understood by Carol as well, not Bob. The above attack is made feasible due to lack of authentication.

13.4.3 N-Party Diffie-Hellman Key Agreement

Several years after its inception, the Diffie-Hellman key agreement protocol was generalized to n participants. The revised protocol, henceforth referred to as Generalized Diffie-Hellman (GDH), is nearly identical to its predecessor: members agree on an *a priori* G and α; each member then generates its own secret $N_i \in$ G.

N	Number of protocols participants
i, j, k	Protocol participants; $i, j, k \in [1, n]$
M_i	i-th group member; $i \in [1, n]$
N_i	Random secret chosen by the member M_i
q	Order of the algebraic group
p	Large prime number
G	Unique subgroup of Z_*^p of order q with p, q prime
α	Exponentiation base – the generator in the group G
K_n	Group key shared by n members

Key Agreement

The Generalized Diffie-Hellman consists of two stages – up-flow and down-flow. Each member's contributions are collected during the up-flow stage, and the resultant intermediate values are broadcast to the group in the down-flow stage. The setup for GDH is identical to that of two-party Diffie-Hellman: all participants, $M_1, \ldots, M_n$ choose a cyclic group, G, of order q, and a generator, α in G; each member then chooses a secret share, $N_i \in G$.

Up-flow

During the up-flow, each member M_i performs a single exponentiation, appends it to the flow, and forwards the flow to M_{i+1}.as illustrated in Figure 13.1.

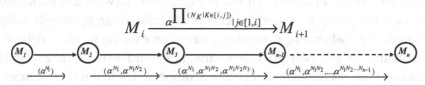

Figure 13.1 – Up flow in Diffie-Hellman Algorithm

The up-flow stage terminates and the down-flow commences when $M_n = M_i$ – when the last member has received the up-flow.

$$M_{n-1} \xrightarrow{(\alpha^{N_1}, \alpha^{N_1 N_2}, \ldots, \alpha^{N_1 N_2 \cdots N_{n-1}})} M_n$$

Upon receipt of the up-flow, M_n calculates the new group key, K_n, by exponentiation of the last intermediate value in the flow:

$$K = K_n = (\alpha^{N_1 N_2 \cdots N_{n-1}})^{N_n}$$

Once K_n has been calculated, M_n commences the down-flow.

Down-flow

The down-flow is initially comprised of n-1 intermediate values, $(\alpha^{N_1 N_n}, \alpha^{N_1 N_2 N_n}, \ldots, \alpha^{N_1 N_2 \cdots N_{n-2} N_n})$, exponentiated to the nth group member's secret, N_n. M_n sends the down-flow to M_{n-1}.

$$M_{n-1} \xleftarrow{(\alpha^{N_1 N_n}, \alpha^{N_1 N_2 N_n}, \ldots, \alpha^{N_1 N_2 \cdots N_{n-2} N_{n-1}})} M_n$$

Upon receipt of the down-flow, each member, M_i, removes its own intermediate value, $(\alpha^{N_1 N_2, \ldots, N_{i-1}, N_{i+1} \cdots N_n})$, calculates the group key, $K_n = (\alpha^{N_1 N_2, \ldots, N_{i-1} N_{i+1} \cdots N_n})^{N_i}$, exponentiates the remaining i-1 intermediate values in the flow, and forwards the flow to its predecessor, M_{i-1}.

$$\alpha^{N_1 N_n N_{n-1} \cdots N_{i+1} N_i},$$

$$M_{i-1} \leftarrow \frac{\alpha^{N_1N_2\cdots N_{i-2}N_{n-2}N_n \cdots N_{i+1}N_i}}{} M_i.$$

The down-flow terminates when $M_i = M_1$.

13.4.3.3 Group Mutation

The above protocol depicts the key agreement protocol when all group members are present at group genesis. Many contexts (e.g., ad hoc or sensor multicasting) necessitate group mutation after the initial group has been formed. In order to ensure key freshness, forward and backward secrecy, the group key must be changed whenever group membership changes (e.g., when a new member joins or an existing member departs, voluntarily or otherwise). A few straightforward extensions to GDH accommodate such mutation. For any addition and deletion of a member, the nth group member acts as a group controller similar to a group creation. For improved efficiency, the group controller caches the most recent up-flow message.

Member Addition

a) M_n generates a new secret $\hat{N}_n$. Note that this step may be removed from the protocol if backward secrecy is not required, as $\hat{N}_n$ simply prevents M_{n+1} from calculating the previous group key(s).

b) Using $\hat{N}_n$, M_n computes a new up-flow message of the form

c) $K = K_{n=1} = \alpha^{\prod (N_k | k \in [1,i] \wedge k \neq j)} \mid j \in [1,n], \alpha^{N_1N_2\cdots N_{n-1}\hat{N}_n}.$

d) M_n sends the new up-flow to the new member, M_{n+1}.

$$M_n \xrightarrow{(\alpha^{N_1}, \alpha^{N_1N_2}, \ldots, \alpha^{N_1N_2\cdots N_{n-1}})} M_{n+1}$$

e) M_{n+1} receives the up-flow and calculates the new group key, $K = K_{n+1}$ by exponentiating the previous cardinal value with its own share, N_{n+1}.

f) $K = K_{n+1} = (\alpha^{N_1N_2\cdots\hat{N}_n})^{N_{n+1}}$

g) M_{n+1} then calculate n intermediate values and commence the down-flow stage by sending the new intermediate values to M_n.

$$M_n \xleftarrow{\alpha^{N_1N_{n+1}}, \alpha^{N_1N_2N_{n+1}}, \ldots, \alpha^{N_1N_2\cdots N_{n-1}N_{n+1}}} M_{n+1}$$

h) Henceforth, M_{n+1} assume the role of the group controller.

Member Deletion

Deleting a member from the group requires that the excluded member will not be able decrypt subsequent group messages. Similar to [Steiner1996], let M_p be the member to be removed. The protocol works as follows:

a) The group controller, M_n, generates a new exponent, $\hat{N}_n$.

b) Using $\hat{N}_n$, Mn calculates a new set of $n-2$ sub-keys, such that N_p is absent, and broadcasts them to the group.

c) Each member receives the broadcast and calculates the new group key, $\hat{K}$ as:

$$\hat{K} = \alpha^{N_1 * \ldots N_{p-1} * N_{p+1} \ldots N_{n-1} * \hat{N}_n}$$

As N_p is absent from $\hat{K}$, M_p is unable to determine the new group key.

Perhaps the most notable strength of GDH is the ease with which it allows group members to generate and distribute a shared group key over an insecure medium. But, GDH is computationally complex, and does not scale in the context of a large group. Although GDH is often classified as a distributed paradigm, it relies heavily on a group control M_n. Compromising M_n eliminates integrity of the group. Although [Steiner1996] proved that GDH is as secure as 2-Party Diffie-Hellman, the proof assumes a passive adversary.

13.4.4 The Ingemarsson Protocol

Previously discussed generalized version of the Diffie-Hellman algorithm was published in 1996. A number of more restrictive variations on Diffie-Hellman have been presented; the Ingemarsson et al. protocol [Ingemarsson1982] was the first of these attempts. The performance and applicability of this protocol has since been surpassed by more effective protocols.

The design of the Ingermasson protocol requires that group members be organized as a logical ring as shown in Figure 13.2. Each member of the group receives the intermediate value from its predecessor, exponentiates using its own share, and forwards the result to its successor. After n-1 rounds, the protocol is complete, and each member calculates the group key Ingemarsson Protocol.

Setup: Prior to the first round of the protocol, all group members must synchronously form a ring. This step requires each member to be apprised of its successor and predecessor and the start time of first round.

Key Agreement

$$M_i \xrightarrow{\quad \alpha^{(\prod \{N_j | j \in [(i-k) \bmod n, i]\})} \quad} M_{i+1 \bmod n}$$

As discussed earlier, the Ingemarsson protocol requires n-1 rounds. At each round, n messages are sent – one by each member. This yields the complexity of this process (the input parameter, n, denotes the number of group members) as shown in Table 13.1.

Table 13.1 – Complexity of the Ingemarsson protocol

Characteristic	Complexity
Number of rounds	$n - 1$
Number of messages	$n(n - 1)$
Exponentiations per member	N
Total exponentiations	n^2
Total message size	$n(n - 1)$

M_n	$\{M_n\}$	$\{M_{n-1}\}$	....	$\{M_5\}$	$\{M_4\}$	$\{M_3\}$	$\{M_2\}$	$\{M_1\}$
j=1								{}
j=2							$\{N_1\}$	$\{N_2\}$
j=3						$\{N_1,N_2\}$	$\{N_1,N_3\}$	$\{N_2,N_3\}$
j=4					$\{N_1,N_2,N_3\}$	$\{N_1,N_2,N_4\}$	$\{N_1,N_2,N_4\}$	$\{N_2,N_3,N_4\}$
j=5				$\{N_1,N_2,N_3,N_4\}$	$\{N_1,N_2,N_3,N_5\}$	$\{N_1,N_2,N_4,N_5\}$	$\{N_1,N_3,N_4,N_5\}$	$\{N_2,N_3,N_4,N_5\}$

$\{z\}$ basic trigon with exponent base a and all numbers mod p

Figure 13.2 – Steps in Ingemarsson Protocol

13.4.4.1 Analysis

As the above complexity analysis indicates, the Ingemarsson protocol is quite slow – the slowest of all protocols discussed here. In addition, many of the design requirements make the resulting protocol quite restrictive as all members must join and form a ring synchronously and members need to maintain their predecessor and successor. The later requirement is especially prohibitive in fault-prone environments, where group members need to be constantly regrouping in ad hoc networks due to departure without notice or faults in the network.

13.4.5 The Burmester and Desmedt Protocol

The Burmester and Desmedt protocol [Burmester1994] presents a much faster variation of the generalized Diffie-Hellman. The setup phase of the Burmester and Desmedt protocol is identical to that of basic GDH; however, the group key construction is considerably different.

Each participant, M_i, $i \in [1, n]$ executes the following rounds:

a) M_i generates a secret, N_i, and broadcasts the new key, $z_i = \alpha^{N_i}$, to all group members.

b) Each member, M_i, computes and broadcasts $X_i = (\frac{z_{i+1}}{z_{i-1}})^{N_i}$.

c) $M_{i \bmod n}$ computes the new group key, K_n as:

$$K = K_n = z_{i-1}^{nN_i} * X_i^{n-1} * X_{i+1}^{n-2} ... X + i - 2 \bmod p$$

The Burmester and Desmedt protocol provides a considerable improvement over the Ingermasson protocol. The improved efficiency is achieved due to comparatively few rounds of messages. The complexity is shown in Table 13.2

Table 13.2 – Complexity of the Burmester and Desmedt protocol

Characteristic	Complexity
Number of rounds	2
Number of messages	$2n$
Exponentiations per member	$n + 1$
Total exponentiations	$n(n + 1)$
Total message size	$2n$

The Burmester protocol requires $n + 1$ exponentiations per group member; however, each exponentiation is significantly less complex. The implementation of the Burmester protocol is unlikely to achieve theoretical performance measures described above due to a heavy reliance on synchronous broadcast messaging. Although sequential broadcasts do not compromise the security of the protocol, they will most likely result in decreased performance.

13.4.6 The Hypercube Protocol

The high number of messages required by the Ingemarsson and Burmester protocols motivated Becker and Willie to define lower bounds on the communication complexity of the Diffie-Hellman based key agreement protocols. To minimize communication overhead, they developed the Hypercube and Octopus protocols [Becker1998]. The two protocols differ only in their logical arrangement of group members. The Hypercube Protocol requires 2^d participants, where d represents the dimensions of the cube. Each participant is logically positioned as a point in the cube; each edge of the cube depicts a key exchange. Because parallel edges represent exchanges that can be executed in parallel, a total of d rounds are required.

Notation

d	Dimensions of the cube
n	Number of group members ($n = 2^d$)
$\vec{v}$	Vector representing each participant; $\vec{v} \in GF(2)^d$
$\vec{b}_i$	Basis of $GF(2)^d$; i $\in [1, d]$
$r_{\vec{v}}$	Random secret generated by participant $\vec{v}$
q	Order of the algebraic group
p	Large prime number
G	Unique subgroup of Z_*^p of order q with p, q prime
α	Exponentiation base – the generator in the group G
φ	Bijection of the form $\varphi: G \to Z_*^p$
K	Group key shared by n members

Each participant in the d-dimensional vector space $GF(2)^d$ chooses a basis of $GF(2)^d$. Each participant, $\vec{v}$, then generates its secret, $r_{\vec{v}}$.

Key Agreement

    a)   Each participant, $\vec{v}$, performs a 2-Party Diffie-Hellman key exchange with $\vec{b}_1$.

b) During the ith round, each participant $\vec{v}$ performs a key exchange with $\vec{v} + \vec{b}_i$. The participant uses the result of i-1 round as the secret.

The protocol terminates when $i = d$, since $\vec{b}_1, \ldots, \vec{b}_d$ forms the basis of $GF(2)^d$. Thus, every participant can generate the shared key:

$$K = \alpha^{\varphi(\alpha^{r_{\vec{v}_1} \cdot r_{\vec{v}_2}}) \cdot \varphi(\alpha^{r_{\vec{v}_3} \cdot r_{\vec{v}_4}}) \cdot \ldots \cdot \varphi(\alpha^{r_{\vec{v}_{d-1}} \cdot r_{\vec{v}_d}})}$$

To clarify functioning of the protocol, examine the example shown in Figure 13.3. Note that there are $4 = 2^2$ members, and, therefore, $d = 2$ rounds of the protocol.

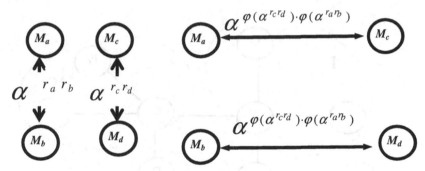

Figure 13.3 – Example of the Hypercube protocol ($d = 2$)

Complexity

Given the aforementioned characteristics, the complexity of the Hypercube protocol can be determined and is given in Table 13.3. The Hypercube protocol provides a reduction in the number of simple rounds if group members can be logically positioned as a cube.

Table 13.3 – Complexity of the Hypercube protocol

Characteristic	Complexity
Number of rounds	$\sqrt{n}$
Number of messages	$n\sqrt{n}$
Exchanges per round	n
Total exchanges	$n\sqrt{n}/2$

Unlike many of the other protocols, the Hypercube protocol assumes that group members can perform parallel key exchanges synchronously, which are difficult to achieve in both ad hoc and sensor networks. In addition, the Hypercube is extremely sensitive to network failures and member departure. Furthermore, if group semantics necessitate dynamic key management, as per the group membership changes, members must join and leave the group in powers of two so the cube structure can be maintained. Nonetheless, the Hypercube protocol provides an efficient, quite simple approach to key management for sensor networks where group membership is static.

13.4.7 The Octopus Protocol

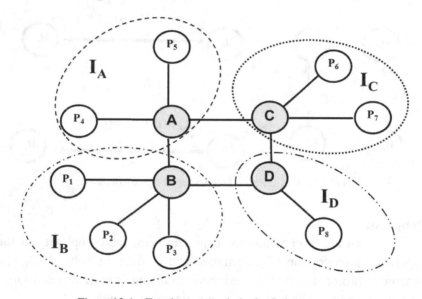

Figure 13.4 – Topology example in the Octopus protocol

The restrictive requirements of the Hypercube protocol to have 2^d group members motivated Becker and Willie to develop a more flexible protocol. The resulting protocol, named Octopus [Becker1998], allows an arbitrary number of group members to contribute to the group key. While minimizing the number of messages and rounds, the Octopus protocol has a desirable communication complexity. The topology of the Octopus protocol loosely models an octopus: 4 nodes comprise the core of the group and are positioned as a square; the remaining nodes are logically positioned as tentacles off the four nodes[2]. The four core nodes perform 2-Party Diffie-Hellman with each of the tentacles attached to it. Using the subkeys generated from their tentacles indicated by dotted lines, the four core nodes ABCD then perform the Hypercube protocol for $d = 2$. The group key is then sent to all tentacles in the octopus. Figure 13.4 shows an example of a topology in the Octopus protocol, with the corresponding elements described next.

Notation

A, B, C, D	Controlling group members
X	Central node; $X \in \{A, B, C, D\}$
I_X	Subgroup connected to x; I_A, I_B, I_C, I_D are pair-wise disjoint
P_i	Non-controlling participants; $P_i \in \{I_A, I_B, I_C, I_D\}$
r_i	Random secret generated by participant i
Q	Order of the algebraic group
P	Large prime number
G	Unique subgroup of Z_*^p of order q with p, q prime
α	Exponentiation base – the generator in the group G
φ	Bijection of the form $\varphi: G \rightarrow Z_*^p$
K	Group key shared by n members

Setup

All participants elect controlling nodes A, B, C, and D of Figure 13.4 and the remaining nodes establish membership in some I_X. A, B, C, D, P_i then choose a cyclic group, G, of order q, and a generator, α in G; each member then chooses a secret share, $r_i \in G$ as follows:

a) For all $X \in A, B, C, D$ and all $i \in I_X$, X and P_i perform 2-party Deffie-Hellman.

b) The controlling nodes, A, B, C, D perform the Hypercube protocol with $d = 2$. Each node uses the subkey generated during the previous steps as its secret for the Hypercube rounds of this step:

---

[2] A hybrid of the Hypercube and Octopus protocols whereby the square of the octopus is changed to a d dimensional cube has also been proposed in [Becker1998]. The resulting protocols is more scalable, but, unfortunately, more complex.

$r_a = K(I_A)$, $r_b = K(I_B)$, $r_c = K(I_C)$, $r_d = K(I_D)$). Thus, after performing the Hypercube rounds, A, B, C, D hold the group key:

$$K = \alpha^{\varphi(\alpha^{K(I_A \cup I_B)}) \cdot \varphi(\alpha^{K(I_C \cup I_D)})}$$

c) Each controlling node now sends the partial key to its subgroup:

    A. For all $j \in I_A$:

        i. A sends P_j: A sends P_j: $\alpha^{\frac{K(I_A) \cup K(I_B)}{j}}$

        ii. A sends P_j: $\alpha^{\varphi(\alpha^{K(I_C \cup I_D)})}$

        iii. P_j calculates $(\alpha^{\frac{K(I_A \cup I_B)}{j}})^{\varphi^{(K_j)}} = \alpha^{K(I_A \cup I_B)}$

        iv. P_j derives the group key:

$$K = (\alpha^{\varphi(\alpha^{K(I_C \cup I_D)})})^{\varphi(\alpha^{K(I_A \cup I_B)})}$$

    B. For all $j \in I_D$:

        i. D sends P_j: $\alpha^{\frac{K(I_C) \cup K(I_D)}{j}}$

        ii. D sends P_j: $\alpha^{\varphi(\alpha^{K(I_A \cup I_B)})}$

        iii. P_j calculates $(\alpha^{\frac{K(I_C \cup I_D)}{j}})^{\varphi^{(K_j)}} = \alpha^{K(I_C \cup I_D)}$

        iv. P_j derives the group key: $(\alpha^{\frac{K(I_C \cup I_D)}{j}})^{\varphi^{(K_j)}} = \alpha^{K(I_C \cup I_D)}$

$$K = (\alpha^{\varphi(\alpha^{K(I_A \cup I_B)})})^{\varphi(\alpha^{K(I_C \cup I_D)})}$$

Table 13.4 – Complexity of the Octopus protocol

Characteristic	Complexity
Messages	$3 \cdot (n - 2^d) + 2^d \cdot d$
Exchanges	$2 \cdot (n - 2^d) + 2^{d-1} \cdot d$
Simple rounds	$2 \cdot \left\lceil \dfrac{n - 2^d}{2^d} \right\rceil + d$
Synchronous rounds	$2 + d$

Complexity

    The complexity of the Octopus protocol is shown in Table 13.4. The Octopus protocol inherits many of the performance benefits of the Hypercube protocol, while relaxing the restrictions on the size and the topology of the group. Although not explicitly mentioned, the logical structure of the Octopus protocol can be easily extended to allow group mutation: tentacles can simply be removed, and missing controlling nodes can be replaced by a tentacle. Nonetheless, the Octopus is quite sensitive to network failures and node movement. Thus, it is more suitable to low mobility ad hoc or static sensor networks and is not suitable for highly mobile environments.

13.4.8 The CLIQUES Protocol Suite

The CLIQUES protocol [Steiner1998] remains one of the most effective and popular key management paradigms in use today. Unlike many of its predecessors, CLIQUES is a comprehensive key management paradigm, which includes well-defined group semantics, four key agreement protocols, an application programming interface (API), and an empirical analysis. Semantics for single and mass member join and leave, group merge and partition are included in the suite. The authors of CLIQUES were the first to formalize and prove the security of the GDH algorithm. Since its initial development, three additional variations of GDH have been developed: GDH.2, GDH.3, and STR. Each of these protocols is presented here.

13.4.8.1 Design

Unlike the protocols discussed thus far, CLIQUES is largely event-driven; that is, it uses membership events (e.g., join, leave, and merge) to trigger key regeneration. Because group events may occur concurrently, CLIQUES is designed above a synchronous networking layer that ensures total or causal ordering for broadcast messages, is employed at this layer. CLIQUES employs a group controller groups mutation events – add, remove, merge, partition, etc. However, the group controller is not solely responsible for key generation. Although it is strongly discouraged, the controller may be a trusted third-party. An alternative that appears very attractive is to make the newest or oldest member of the group as the controller.

Notation

These notation are common among all key agreement protocols in the CLIQUES suite – GDH.1 (i.e., base GDH discussed earlier), GDH.2, GDH.3, STR.

N, n	number of protocol participants
i, j	protocol participants; $i, j \in [1, j]$
M_i	i-th group memeber; $i \in [1, n]$
r_i	random secret chosen by member M_i
br_i	M_i's blinded session key; $br_i = \alpha^{r_i} \bmod p$
k_j	key shared by $M_1...M_j$
bk_j	blinded key shared by $M_1...M_j$; $bk_j = \alpha^{k_j} \bmod p$
Q	order of the algebraic group
P	large prime number
G	unique subgroup of Z_*^p of order q with p, q prime
α	exponentiation base – the generator in the group G
K_n	group key shared by n members

13.4.8.2 GDH.2

GDH.2 is a refinement of GDH.1, discussed earlier in this chapter. Like all CLIQUES protocols GDH.2 consists of a total of $(n-1)$ up-flow stages and corresponding down-flow broadcast of intermediate values.

Members select p, q, G, and with each member assigned a sequential identifier in $1...n$. M_n assumes the role of the group controller.

- Each M_i selects a random secret, $r_i \in Z_q^*$.

- M_i receives the up-flow, exponentiates each intermediate value, adds a new intermediate value that excludes its own contributions, updates the cardinal value, and forwards the up-flow to M_{i+1}:

$$M_i \xrightarrow{\quad \alpha^{\frac{r_1 \cdots r_i}{r_j}} \mid j \in [1,i], \alpha^{r_1 \cdots r_i} \quad} M_{i+1}$$

- M_n receives the up-flow from M_{n-1}, calculates the group key from the cardinal value, exponentiates all intermediate values, and broadcast the revised intermediate values to the group:

$$ALL \xleftarrow{\quad \alpha^{\frac{r_1 \cdots r_n}{r_i}} \mid i \in [1,n], \alpha^{r_1 \cdots r_n} \quad} M_n$$

Each member receives the broadcast message from M_n, extracts its intermediate value, and exponentiates it using its secret, r_i, to calculate the group key, K:

$$K = (\alpha^{r_1 \cdot r_2 \cdots r_{i-1} \cdot r_{i+1} \cdots r_{n-1} r_n})^{r_i}$$

Complexity

Table 13.5 illustrates the complexity of GDH.2.

Table 13.5 – Complexity of GDH.2

Characteristic	Complexity
Messages	n
Rounds	n
Message size	$(n-1)(\frac{n}{2}+2)-1$
Exponentiations per member	$(i+1) - O(n)$
Total exponentiations	$\frac{(n+3)n}{2}-1$

13.4.8.3 GDH.3

The GDH.3 protocol targets at reducing the number of exponentiations required by each member. Reducing the number and relative expense of exponentiations is of paramount importance in ad hoc networking context wherein the nodes have constrains on their computational capabilities. Members of the group select p, q, G, and α. In addition, each member is assigned a sequential identifier in $1...n$. M_n assumes the role of the group controller.

Key Agreement

a) The first state is identical to the up-flow of GDH.1: each member receives the up-flow, adds its contribution, and forwards it to its successor.

$$M_i \xrightarrow{\quad \alpha^{\prod \{r_k \,|k \in [1,i]\}} \quad} M_{i+1}$$

b) M_n receives the up-flow from M_{n-1}, calculates the group key from the cardinal value, exponentiates all intermediate values, and broadcast the revised intermediate values to the group:

$$M_i \longleftarrow \frac{\alpha^{\prod \{r_k | k \in [1,n-1]\}}}{} M_{n-1}$$

c) The third stage is the nth round. It entails a response from every group member.

$$M_i - \frac{\alpha^{\prod \{r_k | k \in [1,n-1]; k \neq i\}}}{- - - -} \rightarrow M_n$$

d) In the final stage, M_n collects all responses, exponentiates them to r_n, and broadcasts all intermediate values to the group.

$$M_i \longleftarrow \frac{\{\alpha^{\prod \{r_k | k \in [1,n]; k \neq i\}} | i \in [1,n-1]\}}{} M_n$$

Complexity

Table 13.6 – Complexity of GDH.3

Characteristic	Complexity
Messages	$2n - 1$
Rounds	$n + 1$
Message size	$3(n - 1)$
Exponentiations per member	4
Total exponentiations	$5n - 6$

Table 13.6 shows the complexity of GDH.3. The GDH.3 protocol provides considerable enhancements over GDH.1 and GDH.2 with its constant message sizes and comparatively fewer exponentiations per member while it does require the group controller. Although this requirement may be admissible in traditional, wired environments, it may be the source of a computational bottleneck.

13.4.8.4 STR

The GDH.1, GDH.2, and GDH.3 protocols have been designed for relatively small groups. Due to their computational requirements and high number of messages, they do not scale well in the context of larger groups. In order to accommodate these wide area group settings, the designers of CLIQUES developed a tree-based protocol called STR. Although the protocol can be used in small, proximal settings, it is best-suited for wide area networks.

The notation employed in GDH.1, GDH.2, and GDH.3 is used in STR with the following additions:

$N_{<j>}$	Tree node j
$IN_{<l>}$	Internal tree node at level l
$LN_{<i>}$	Leaf node associated with member M_i
$T_{<i>}$	Tree of member M_i
$BT_{<i>}$	Tree of member M_i including its blinded keys

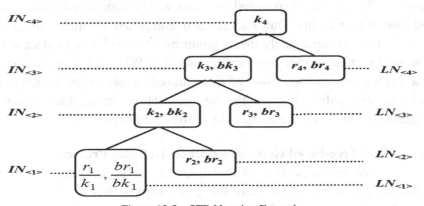

Figure 13.5 – STR Notation Example

Figure 13.5 presents an example of the notation employed in STR, depicting the various added notations.

Complexity

Table 13.7 – Complexity of STR

Characteristic	Complexity
Messages	2
Rounds	1
Exponentiations per member	2
Total exponentiations	$2n$

The complexity of STR is given in Table 13.7. CLIQUES provide a rather elegant solution to group key management. Key generation overhead is distributed, which makes it appealing for groups that are composed of a large number of nodes with limited resources. Unfortunately, CLIQUES is largely communication-centric: as many as

(M+1) messages may be sent for a single rekey event. In a wired, point-to-point context, each member will send and receive exactly one message for each rekey. In the context of an ad hoc and sensor networks, however, each member receives $M + 1$ broadcast message. Furthermore, protocol without address, often used, incur addition of member ids to each group; members will otherwise be unable to discern which intermediate value to exponentiation.

It is also important to recognize that the CLIQUES paradigm assumes that member leaves and evictions will be announced. Given that ad hoc networks are fault-tolerant and comprised of nodes in close proximity, it is quite likely that a group member will lose contact with the group without other members' knowledge. When these faults occur, the predecessor and successor of the removed member must be notified so they can unlink it from the key generation chain. Such updates significantly increase the complexity of the overall system.

13.4.9 The Tree-based Generalized Diffie-Hellman Protocol

One major problem with CLIQUES is the amount of time it takes to generate and distribute the group key M_n has to wait for n-1 members to exponentiate immediate and cardinal values. The Tree-based Generalized Diffie-Hellman (TGDH) key agreement protocol [Kim2000] improves this performance by hierarchically structuring the key generation. A tree-based key agreement algorithm (Diffie-Hellman) generates keys identical to CLIQUES while TGDH provides a secure tree structured group layer that supports wide-area members in a fault-prone network.

The Protocol

n	Number of group members
h	Height of the key tree
$<l, v>$	v-th node at level l in the tree
M_i	i^{th} group member, where $i \in [1, n]$
T_i	M_i's view of the key tree
$\hat{T}_i$	M_i's modified tree after the membership operation
p, q	Prime integers
α	Exponentiation base

Design

Each node in the key tree, $<l, v>$, has a key, $K_{l,v}$, a blinded key, $BK_{l,v}$, and member, M_i, associated with it. A blinded key is simply the modular exponentiation of the key in prime order groups (i.e., $BK_{l,v} = \alpha^k$ mod p). Each $K_{l,v}$ is computed recursively as follows:

$$K_{l,v} = \alpha^{K_{<l+1,2v>} * K_{<l+1,2v+1>}} \bmod p$$

Computing $K_{l,v}$ requires knowledge of one child's key and the other's blind key. It is important to note that the root node derives the group key. A cryptographically strong, one-way hash function is used to perform this derivation. In order to understand key computation by individual nodes, it is necessary to use the notion of *key-path* and *co-path*. The key path for a given node, i, is composed of every key that resides in the path from i to the root node, inclusive. The co-path is composed of all of the siblings of nodes that appear in the key-path. A node $<l, v>$ uses blind keys in its co-path and its own key to derive the group secret. Each node restructures its key tree after a join or leave operation.

13.5 Secure Routing

The MANET and sensor networks face a set of challenges specific to its environment. The insecurity of the wireless links, energy constraints, relatively poor physical protection of nodes in a hostile environment, and the vulnerability of statically configured security schemes have been identified [Zhou1999, Stajano1999]. In such an environment, there is no guarantee that a path between two nodes would be free of malicious nodes. For example, any node could claim that it is one hop away from the destination, causing all routes to pass through itself. Alternatively, a malicious node could corrupt any in-transit route request (or reply) and cause to be misrouted and intentionally falsified routing messages could cause DoS. The presence of even a small number of adversarial nodes could result in repeatedly compromised routes, and, as a result, the network nodes would have to rely on cycles of timeout and new route discoveries to communicate. Here we discuss how a secured routing can affect different parameters.

Infrastructure

Unlike traditional networks, there is no pre-deployed infrastructure or strict policy for supporting end-to-end routing in an ad hoc and a sensor network.

Frequent Changes in Network Topology

Nodes in an ad hoc network may frequently change their locations and frequently changing neighbors which a node has to use for routing. As we shall see, this has a significant impact on the implementation of secure routing over MANETs. A sensor network is fairly static and the topology may remain fixed.

Wireless Communication

As the communication is through a wireless medium, it is possible for any intruder to easily tap it. Wireless channels offer poor protection and routing related control messages can be tampered. The wireless medium is susceptible to signal interference, jamming, eavesdropping and distortion. An intruder may easily eavesdrop to find out sensitive routing information or jam the signals to prevent propagation of routing information. What is worse is that an intruder could even interrupt messages and distort them to manipulate routes. Secure routing protocols should be designed to handle such problems.

13.5.1 Problems with Existing Ad Hoc Routing Protocols

Existing ad hoc routing protocols possess many security vulnerabilities; routing security is very often peculiar to a specific protocol. It may happen that a given danger may be present in a given protocol, while it does not exist in another.

Implicit Trust Relationship amongst Neighbors

Most routing protocols are assumed cooperative and depend on trusting neighboring nodes to route packets. This naive trust model allows malicious nodes to paralyze an ad hoc or a sensor network by inserting erroneous routing updates, replaying old messages, changing routing updates or advertising incorrect routing information. While these attacks are possible in fixed network as well, ad hoc environment magnifies these issues and makes detection much harder.

Throughput

Ad hoc and sensor networks maximize total network throughput by using all available nodes for routing and forwarding. However, a node may misbehave by agreeing to forward packets and then fail to do so, because it is overloaded, selfish, malicious or out of service. Although the average loss in throughput due to a single misbehaving node may not be too high, it may be significantly high in the case of a group attack.

Attacks using Modification of Protocol Message Fields

Current routing protocols assume that nodes do not alter the protocol fields in the messages passing through them. Routing packets carry important control information that governs the behavior of data transmission. Since the level of trust in a traditional ad hoc and sensor network cannot be measured or enforced, enemy nodes or compromised nodes may participate directly in the route discovery and may intercept and filter routing protocol packets to disrupt communication. Malicious nodes can easily cause redirection of the network traffic and DoS attacks by simply altering these fields. These types of attacks can be classified as remote redirection attacks and DoS attacks, and are discussed next.

Remote Redirection with Modified Route Sequence Number (AODV)

Remote redirection attacks are also called black hole attacks [Deng2002]. In these attacks, a malicious node uses the routing protocol to advertise itself as the shortest path to those nodes whose packets it wants to intercept. As we have seen earlier, protocols such as AODV instantiate and maintain routes by assigning monotonically increasing sequence numbers in order to route packets towards a specific destination. In AODV, any node may divert traffic through itself simply by advertising a route to a node with a destination sequence number greater than the authentic (or actual) value. Suppose the malicious node M in Figure 13.6(a) receives a RREQ, originated from node S and destined to node X, after it is re-broadcast by node B during route discovery. Node M may redirect traffic towards itself simply by unicasting to node B a RREP containing a significantly higher destination sequence number for node X than the authentic value last advertised by X as illustrated in Figure 13.6(b).

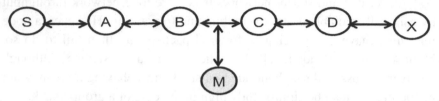

Figure 13.6 (a) – Malicious node M announces a shorter route to node X
(b) – Traffic destined to node X goes to node M

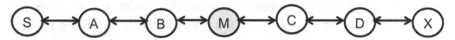

Figure 13.6(b) –Malicious node M keeps traffic from reaching node X

Redirection with Modified Hop Count (AODV)

A redirection attack is also possible in certain protocols, such as AODV, by modifying the hop count field of the route discovery messages. When routing decisions cannot be made by other metrics, AODV uses the hop count field to determine the shortest path. By setting the hop count field of the RREP to infinity, routes will tend to be created that do not include the malicious node. Once the malicious node has been able to insert itself between two communicating nodes, it is able to do nearly anything with the packets exchanged between them. It can choose to drop packets, to perform a DoS attack, or alternatively use its place on the route as a first step in man-in-the-middle attack.

DoS with Modified Source Routes

A DoS attack is also possible in the DSR routing protocol, which explicitly states routes in data packets. These routes lack any integrity checks and a simple DoS attack can be launched by altering the source routes in packet headers. Modification of source routes in DSR may also introduce loops in the specified path. Although DSR prevents looping during the route discovery process, there are insufficient safeguards to disable insertion of loops into a source route after it has been salvaged.

Attacks using Impersonation

Current ad hoc routing protocols do not authenticate the source IP address. Consequently, a malicious node can launch many attacks by altering its own MAC or IP address, with the goal of pretending to be some other node. This is called impersonation.

Attacks using Fabrication

Generation of false routing messages is termed as fabrication of messages. Such attacks are usually difficult to detect. Below we describe some attacks based on fabrication.

Falsifying Route Error Messages (AODV or DSR)

AODV and DSR have path maintenance measures to recover broken routes whenever nodes move. If the destination node or an intermediate node along an active path moves, the node upstream of the link breakage broadcasts a route error message to all active upstream neighbors. In addition, the node also invalidates the route for this particular destination in its routing table. The vulnerability here is that routing attacks can be launched by sending false route error messages. Suppose node S has a route to node X via nodes A, B, and C as in Figure 13.6(a). A malicious node M can launch a DoS attack against node X by continually sending route error messages to node B, hence spoofing node C by indicating a broken link between nodes C and X. Node B receives the spoofed route error message thinking that it came from C, deletes its routing table entry for X and forwards the route error message to node A, which also deletes its routing table entry. Whenever a route is established from node S to node X, node M listens and can broadcasts spoofed route error messages to prevent communications between S and X.

Route Cache Poisoning in DSR

This is a passive attack that can occur in DSR due to its optional promiscuous mode of updating routing tables. This occurs when information stored in the routing table at routers is deleted, altered or injected with the false information. In addition to learning routes from headers of packets being processed by a node along a path, routes in DSR may also be determined from promiscuously received packets.

A node overhearing any packet but not on the path, may add the routing information contained in that packet header to its own route cache. An attacker could easily exploit this vulnerability.

Routing Table Overflow Attack

In routing table overflow attack, the attacker attempts to create routes to non-existent nodes with an objective to have enough routes to prevent new routes from being created, or even overwhelm to flush out legitimate routes from the routing tables. Proactive routing algorithms continuously discover routing information, while reactive algorithms find routes only when needed. This makes proactive algorithms more vulnerable to routing table overflow attacks.

13.5.2 Detect and Isolate Misbehaving Nodes

As discussed earlier, misbehaving nodes can adversely affect the network throughput. Existing ad hoc routing protocols do not have any mechanism to identify these misbehaving nodes. On the other hand, a node appears to be misbehaving when it is actually encountering temporary overload problem or low battery. A routing protocol should be able to identify misbehaving nodes and isolate them.

13.5.3 Information Leaking on Network Topology

Ad hoc routing protocols (e.g., AODV and DSR) carry routes discovery packets which contain the routes to be followed in clear text. By analyzing these packets, any intruder can determine the network structure and use this information to determine the adjacency or physical location of a particular node. Such an attack can be carried out passively and intruders may use this to attack command and control nodes.

Lack of Self-Stabilization

An intruder should not be able to permanently disable the network by injecting a small number of malformed routing packets and routing protocols should be able to recover from an attack in finite time. For example, AODV is prone to self-stabilization problems as sequence numbers are used to verify route validity and incorrect state may remain in routing tables for a long period of time.

13.5.4 Secure Routing Protocols

Current efforts towards the design of secure routing protocols are mainly oriented to reactive (on-demand) routing protocols such as DSR or AODV, where a node attempts to discover a route to some destination only when it has a packet to send to that destination. These protocols are able to react quickly to topology changes, yet being able to reduce routing overhead or areas of the network with less frequent changes. In addition to active attacks, node selfishness also needs to be addressed. Furthermore, the prerequisite for all available solutions implies a *managed* environment and nodes wishing to communicate may be able to exchange initialization parameters beforehand. For example, session keys may be distributed directly or through a trusted third party. Major secure routing protocols for MANETs can be classified into four categories:

- Those using pre-deployed security infrastructures.
- Those aiming at concealing the network topology.
- Those targeting at mitigating node misbehavior.
- Other routing protocols.

13.5.4.1 Using Pre-Deployed Security Infrastructure

Here, ad hoc and sensor networks are called managed-open environment which assumes that these are pre-deployed. Nodes wishing to communicate can exchange initialization parameters beforehand, perhaps within the security of an infrastructure network where session keys may be exchanged or through a trusted third party like a certification authority.

13.5.4.2 The ARAN Protocol

The Authenticated Routing for Ad-hoc Networks (ARAN) [Dahill2002] detects and protects against malicious actions by third parties and peers. ARAN introduced for an ad hoc environment, provides authentication, message integrity and non-repudiation and could be easily applied to a sensor network. ARAN makes use of cryptographic certificates for the purposes of authentication and non-repudiation.

Route discovery in ARAN is composed of two distinct stages are follows.

Stage 1

The stage one contains a preliminary certification and a mandatory end-to-end authentication stage. This lightweight step does not demand too many resources.

a) **Preliminary Certification**: ARAN requires the use of a trusted certificate server, T that provides a certificate to each node before entering the ad hoc network. For example, the certificate is issued by T to a given node A would be:

$$T \rightarrow A: Cert_A = [IP_A, K_{A +} t, e]K_T-$$

Note that + sign indicates public key while − points out a private key. The certificate contains IP address of A (IP_A), public key of A (K_A-), a timestamp t indicating the creation time of certificate, and certificate expiration time e. These variables are concatenated and signed by T. All nodes must maintain fresh certificates with the trusted server T and must know its public key (K_{T+}).

b) **End-to-End Authentication:** The goal of stage one is to enable the source node to verify if the intended destination has been reached. The source trusts the destination to choose the return path. Two separate parts are: route request and route reply. Suppose a source node A wants to find a route to a destination node X. Node A initiates the route discovery procedure by broadcasting a route request (RREQ) packet to its neighbors in the following form:

$$A \rightarrow broadcast: [RREQ, IP_X, Cert_A, N_A, t]K_A-$$

The RREQ includes the packet type identifier ("RREQ"), the IP address of the destination (IP_x), node A's certificate ($Cert_A$), a nonce N_A, and the current time t, all signed with A's private key K_A-. Each time node A initiates a route discovery, it monotonically increases its nonce (i.e., N_A). Other nodes receiving the RREQ packet store the nonce they have last seen with its corresponding timestamp. Intermediate nodes receiving the RREQ record the neighbor from which the packet has been received. It rebroadcasts the RREQ packet to each of its neighbors, signing the contents of the packet. This signature prevents spoofing attacks that may alter the route or form loops. Let B be a neighbor of node A. Node B would rebroadcast the packet as:

$$B \rightarrow \text{broadcast: } [[\text{RREQ, IP}_X, \text{Cert}_A, N_A, t]K_A\text{-}]K_B\text{-}, \text{Cert}_B$$

Nodes do not rebroadcast duplicate packets for which they have already seen the (N_A, IP_A) tuple. Supposing node C is a neighbor of node B, it validates the packet signature with the given certificate upon receipt of node B's broadcast. Node C rebroadcasts the RREQ to its neighbors after removing node B's signature, resulting in:

$$C \rightarrow \text{broadcast: } [[\text{RREQ, IP}_X, \text{Cert}_A, N_A, t]K_A\text{-}]K_C\text{-}, \text{Cert}_C$$

Upon receipt of the first RREQ packet with the corresponding nonce at the destination node X, it replies back to the source with a route reply (RREP). Assume node D is the first hop in the reverse path from node X to node A. In this case, the RREP sent by node X takes the form:

$$X \rightarrow D: [\text{RREP, IP}_A, \text{Cert}_X, N_A, t]K_X\text{-}$$

Intermediate nodes receiving the RREP forward it to the predecessor node from which they received the corresponding RREQ packets signed by the sender. Let node D's next hop back to the source be node C. The RREP sent by node D is:

$$D \rightarrow C: [[\text{RREP, IP}_A, \text{Cert}_X, N_A, t]K_X\text{-}]K_D\text{-}, \text{Cert}_D$$

and this process continues till the source is reached. Nodes along the path check the signature of the previous hop as the RREP is returned to the source. This procedure avoids attacks where malicious nodes instantiate routes by impersonation or replay of node X's packet. When the source node A receives the RREP from the destination node X, it first verifies that the correct nonce has been returned by the destination as well as the destination's signature. Only the destination can answer the RREQ packet and other nodes already having paths to the destination cannot reply on its behalf. While some routing protocols allow this networking optimization (e.g., DSR and AODV), removing this feature cuts down the number of RREP packets received by the source. Since the destination is the only node that can originate a RREP, freedom from loops can be easily guaranteed.

c) **Disadvantages:** ARAN requires that nodes keep one routing table entry per source-destination pair that is currently active. This is certainly more costly than per-destination entries in non-secure ad hoc routing protocols.

Stage 2

Stage two is performed only after discovery of shortest path in Stage one as the destination certificate is required in this phase. Data transfer can be pipelined with the shortest path discovery operation employed in Stage two. Using the same example, the source node A initiates the shortest path discovery operation by broadcasting a Shortest Path Confirmation (SPC) message to its neighbors as:

$$A \rightarrow \text{broadcast: SPC, IP}_X, \text{Cert}_X, [[\text{IP}_X, \text{Cert}_A, N_A, t]K_A\text{-}]K_X\text{+}$$

The SPC message begins with the SPC packet identifier, followed by the destination node X's IP address and certificate. The source concatenates a signed message containing the IP address of X, its own certificate, a nonce and a timestamp. This signed message is then encrypted with node X's public key so that other nodes cannot modify its contents. Intermediate nodes receiving this message rebroadcast the same after including its own cryptographic credentials. For example, a node, say B, would sign the encrypted portion of the received SPC, include its own certificate, and re-encrypt with the public key of X obtained in the certificate forwarded by node A. Therefore, the message rebroadcast by node B would be of the form:

$$B \rightarrow \text{broadcast: SPC, IP}_X, \text{Cert}_X, [[[[\text{IP}_X, \text{Cert}_A, N_A, t]K_A\text{-}]K_X\text{+}]K_B\text{-}, \text{Cert}_B]K_X\text{+}$$

Similar to other non-secure routing algorithms, nodes receiving the SPC packet create entries in their routing table so as not to forward duplicate packets. In addition, this entry also serves to route the reply packet from the destination to the source along the reverse path. Upon receipt of the packet, the destination node X checks that all the signatures are valid. Node X replies to the first SPC it receives and also to any SPC with a shorter recorded path. Then, it sends a Recorded Shortest Path (RSP) packet to the source node A through its predecessor node, say D.

$$X \rightarrow D: [\text{RSP, IP}_A, \text{Cert}_X, N_A, \text{route}]K_X\text{-}$$

The source node A will eventually receive this packet and verify that the nonce corresponds to the SPC originally generated.

Advantages

The onion-ring like signing of messages prevents nodes in between source and destination from changing the path. First, to increase the path length of the SPC, malicious nodes require an additional valid certificate. Second, malicious nodes cannot decrease the recorded path length or alter it because doing so would break the integrity of the encrypted data.

Route Maintenance

ARAN is an on-demand protocol where nodes keep track of whether routes are active or not. When an existing route is not used after some pre-specified lifetime, it is simply deactivated (i.e., expired) in the route table. Nodes also use Error (ERR) packets to report links in active routes that are broken due to node movement. For a given route between source node A and destination node X, an intermediate node B generates a signed ERR-packet to its neighbor node C as follows:

$$B \rightarrow C: [ERR, IP_A, IP_X, Cert_C, N_B, t]K_B\text{-}$$

which prevents repudiation by malicious entities.

Key Revocation

ARAN attempts a best effort key revocation that is backed up with limited time certificates. In the event that a certificate needs to be revoked, the trusted certificate server node T sends a broadcast message to the ad hoc group announcing the revocation. The transmission of revoked certificate *Cert R* is sent as:

$$T \rightarrow \text{broadcast: } [revoke, Cert_R]K_T\text{-}$$

Any node receiving this message rebroadcasts it to its neighbors and need to be stored until the certificate has normally expired. Neighbors of the node with the revoked certificate may need to rebuild their routes so as to avoid paths passing through this un-trusted node. If an un-trusted node whose certificate is being revoked is in between two other nodes in the ad hoc network, it may simply not propagate the revocation message, thus leading to a partitioned network and is therefore not failsafe.

13.5.4 Concealing Network Topology

This is called a non-disclosure method (NDM) where a number of independent security agents (SA) are distributed over the network. Each of these agents SA_i owns a pair of asymmetric cryptographic keys K_{SAi+} and K_{SAi-}. Suppose node S wishes to transmit a message M to a receiver node R without disclosing its own location. Node S sends the message using a number of SAs, such as: $SA_1 \rightarrow SA_2 \rightarrow ... \rightarrow SA_N \rightarrow R$. The message is encapsulated N times using the public keys $K_{SA1+}...K_{San+}$ as follows:

$$M' = K_{SA1+}(SA_2, (K_{SA2+} (SA_3 (...(K_{SAN+}(R, M))...))))$$

In order to deliver the packet, node S sends it to the first security agent SA_1 which decrypts the outer most encapsulation and forwards the packet to the next agent. Each SA knows only the address of the previous and the next hop. The last agent finally decrypts the message and forwards it to node R. This scheme introduces a large amount of overhead and hence is not preferred for routing.

As we have discussed in an earlier chapter, the ZRP protocol is a hierarchical algorithm where the network is divided into zones which operate independently from each other. Such a hierarchical routing structure is favorable with respect to security since it should be able to confine certain problems to small portion of the hierarchy, leaving other portions unaffected. ZRP has some features that appear to make it somewhat less susceptible to routing attacks as its hierarchical organization hides some of the routing information within the zones. In addition, zoning provides some form of security against disclosing the network topology and limit their damage within the zone.

13.5.4.1 Identifying and Avoiding Misbehaving Nodes

In this scheme, wireless links are assumed to be bi-directional and is a realistic assumption in most MAC layer protocols including IEEE 802.11. Also promiscuous mode of operation is assumed that helps nodes to supervise each other's operation, which may not always be possible. The underlying routing protocol is considered as DSR, while it can be extended to other routing protocols.

The watchdog identifies misbehaving nodes, while the pathrater avoids routing packets through these nodes. When a node forwards a packet, the node's watchdog verifies that the next node in the path also forwards the packet. The watchdog does this by listening promiscuously to the next node's transmissions. If the next node does not forward the packet, then it is misbehaving. The pathrater uses this misbehavior knowledge to select a path that will most likely deliver packets.

Watchdog

As illustrated in Figure 13.7, the watchdog method is used to detect misbehaving nodes. Consider node S is transmitting packets to node D. Assuming node A's transmission cannot be heard by node C, but it can listen to node B's traffic. Thus, when node A transmits a packet to node B destined to node C, node A can often tell if node B retransmits the packet. If encryption is not performed separately for each link, then node A can also tell if node B has tampered with the packet payload or the header. Watchdog can be implemented by maintaining a buffer of recently sent packets and comparing each overheard packet with the packet in the buffer to determine if there is a match. If so, the packet in the buffer is removed and no longer monitored by the watchdog, as it has been forwarded on. If the packet has remained in the buffer for longer than a certain timeout, the watchdog increments a failure tally for the node responsible for not forwarding on the packet. If the tally exceeds a certain threshold bandwidth, it determines that the node is misbehaving and sends a message to the source notifying it of the misbehaving station.

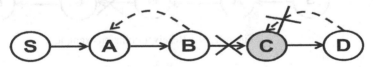

Figure 13.7 – Watchdog operation

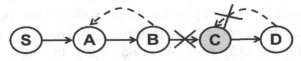

Figure 13.8 – Ambiguous collision example

a) Advantage: One advantage of the watchdog mechanism is that it can detect misbehaving nodes at forwarding level and not just the link level.

b)　Weaknesses: Watchdog might not detect misbehaving nodes in presence of:

- **Ambiguous collision**: The ambiguous collision problem prevents a given node, say A, from overhearing transmissions from another node, say B. As Figure 13.8 illustrates, a packet collision occurs at node A while it is waiting for node B to forward a packet. In this situation, node A is not able to figure out if the collision was caused by node B's transmission, or if node B never forwarded the packet and the collision was caused by other nodes in node A's neighborhood. Because of this uncertainty, node A should continue to watch node B over a longer period of time.

- **Receiver collision**: In the receiver collision problem, node A can only tell whether node B sends the packet to node C, but it cannot tell if node C receives it successfully. If a collision occurs at node C when node B first forwards the packet, node A can only determine that node B has forwarded the packet and assumes that node C has successfully received it. In this scenario, node B could skip the packet retransmission and evade detection. This situation is shown in Figure 13.9.

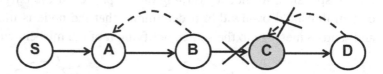

Figure 13.9 – Receiver collision example

- **False misbehavior**: False misbehavior can occur when nodes falsely report other nodes as misbehaving. A malicious node could attempt to partition the network by claiming that some following nodes in the path are misbehaving. For instance, in Figure 13.7, node A could report that node B is not forwarding packets when in fact it has done so. This will cause node S to mark node B as misbehaving when in fact node A is the culprit. Since node A passes messages onto node B (as verified by node S), then any acknowledgement from node D to node S will go from node A to node S. In this case, node S will wonder why it receives replies from node D when node B is supposedly dropping packets in the forward direction. In addition, if node A drops acknowledgements to hide them from node S, node B will detect this misbehavior and report it to node D.

- **Limited transmission power**: A misbehaving node that can control its transmission power can circumvent the watchdog. A node could limit its transmission power such that the signal is strong enough to be overheard by the previous node but too weak to be received by the true recipient.

- **Multiple colliding nodes**: Multiple nodes in collision can mount a more sophisticated attack. For example, nodes B and C in Figure 13.7 could collide so as to cause a mischief. In this case, node B forwards a packet to node C but does not report to node A when node C drops the packet. Because of its limitation, two consecutive untrusted nodes should not be allowed in a routing path.

- **Partial dropping**: A node can also circumvent the watchdog by dropping packets at a lower rate than watchdog's expected misbehavior threshold. Although the watchdog will not detect this node as misbehaving, this node is forced to forward at the threshold bandwidth. This way, watchdog serves to enforce this minimum bandwidth and to work properly, the watchdog ought to know where a packet should be in two hops.

13.5.4.2 Pathrater

Just like the watchdog, pathrater runs at each node, maintains a rating for every other node, and combines knowledge of misbehaving nodes with link reliability to pick the most reliable route. By averaging the node ratings over that path, it then calculates a path metric which provides a base for comparing overall reliability of different paths. It allows pathrater to emulate the shortest length path when no reliability information has been collected. In case multiple paths exist, the path with the highest metric is chosen. Since pathrater depends on the knowledge of the exact traversed path, it may be implemented on top of a routing protocol such as DSR. The pathrater assigns "neutral" rating of 0.5 when a node becomes known to the pathrater through route discovery. A node always rates itself as 1.0 so as to ensure that it picks the shortest length path if all other nodes are neutral nodes (rather than suspected misbehaving nodes). The pathrater increments the ratings by 0.01 at periodic intervals of 200 ms for all nodes along actively used paths. The maximum value a neutral node can attain is 0.8 while the lower bound is 0.0. The node's rating is decremented by 0.05 when a link breakage is detected during packet forwarding. The pathrater does not modify the ratings of inactive nodes. A special high negative value is assigned to nodes suspected of misbehaving by the watchdog mechanism. However, the values are slowly changed to avoid a permanent classification.

13.5.4.3 Security-Aware Ad Hoc Routing

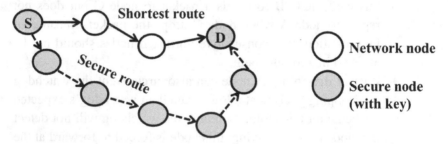

Figure 13.10 – Security-aware routing protocol

The Security-Aware ad hoc Routing (SAR) protocol [Yi2001] makes use of trust levels (security attributes assigned to nodes) to make informed, secure routing decisions. Existing routing protocols for ad hoc networks usually discover the shortest path between any two nodes. On the other hand, SAR does not take path length into account when establishing routes, while security attributes are the main criteria. An example of path setup in SAR is shown in Figure 13.10. A node initiating route discovery sets the required minimal trust level for nodes participating in the query/reply propagation. Nodes at each trust level share symmetric encryption keys. Only the nodes with the correct key can read the packet header, determine the destination node, and forward it. Intermediate nodes of different levels cannot decrypt in-transit routing packets and simply drop the packet. Thus, if a packet has reached the destination, it must have been propagated by nodes at the same level, as only these nodes can decrypt the packet, see its header and forward it.

SAR can be extended to any routing protocol. Here, we discuss SAR as an extension to AODV, and call the resulting scheme as Secure AODV (SAODV). Most of AODV's original behavior such as on-demand discovery using flooding, reverse path maintenance, and forward path setup via RREQ and RREP messages is retained. However, RREQ packet has an additional field called RQ_SEC_REQIREMENT that indicates the required security level for the route the sender wishes to discover. At each hop, a node performs 'AND' to its security guarantees RQ_SEC_GUARANTEE to existing RQ_SEC_REQUIREMENT value and write it back in the RREQ packet. At the destination, each RREP is initialized with the 'RQ_SEC_GUARANTEE' and the source can now choose the most secure route. On the other hand, SAODV protocol incurs a significant amount of encryption overhead.

13.5.5 Secure Routing Protocol

In this section Secure Routing Protocol (SRP) [Papadimitratos2002] is discussed which can be applied to existing *reactive* routing protocols. SRP combats attacks disrupting route discovery process and guarantees acquisition of correct topological information. A node initiating a route discovery is able to identify and discard replies with false routing.

13.5.5.1 Functional Algorithm

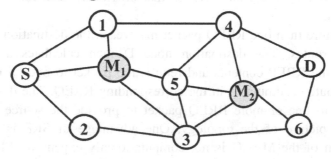

Figure 13.11 – Example of SRP functioning

First, it is assumed that a Security Association (SA) exists between the source node S and the destination node D by negotiating a shared secret key based on the public key of other end. The SA would be established by any of the group key exchange schemes. The existence of SAs with intermediate nodes is, however, unnecessary. In addition, similar to other routing protocols, SRP requires that end nodes be able to use non-volatile memory to maintain state information about relayed queries, so that previously seen route requests are discarded. It is also expected that a one-to-one mapping exists between MAC and IP addresses. Finally, broadcast nature of the radio channels requires transmissions to be received by all neighbors assumed to be in promiscuous mode.

To illustrate the functioning of SRP, consider the example shown in Figure 13.11 where node S wants to find a route to node D. It is assumed to have a SA (a shared key K_{SD}) between S and D. In this scenario, node S initiates the route discovery procedure by constructing a RREQ packet with a uniquely identified random query identifier (RND#) and a sequence number (SQ#). Node S then constructs a Message Authentication Code (MACO) which is a hash of the source, the destination, the random query identifier, the sequence number, and K_{SD}, i.e., MACO = hash(S, D, RND#, SQ#, K_{SD}). Intermediate nodes relay RREQ, and append their identifiers (IP addresses) to the RREQ packet. The entire path from source to destination is accumulated within the RREQ packet (such as in DSR). The intermediate nodes also maintain a limited amount of state information regarding relayed queries (by storing their random sequence number), so that previously seen route requests are discarded.

More than one RREQ packet may reach the destination through different routes. The destination node D then calculates a MACO covering the RREP contents and returns the packet to node S over the reverse route accumulated in the corresponding RREQ. The destination responds to one or more RREQ packet to provide the source with an adequate picture of the topology. One advantage of SRP is that the generation of the MACO is not computationally expensive. Moreover, SRP guarantees that the message integrity is preserved and data confidentiality can be implemented by encrypting the payload packet with the shared key K_{SD}.

Let M_1 and M_2 be two malicious intermediate nodes in the ad hoc network of Figure 13.11. We denote the query request as a list $\{Q_{SD}; n_1, n_2, ..., n_k\}$, where Q_{SD} denotes the SRP header for a search query for the destination node D, initiated by node S. In addition, n_i, $i \notin \{1, k\}$, are the IP addresses of the intermediate nodes, where $n_1 = S$, $n_k = D$. Similarly, a route reply is denoted by $\{R_{SD}; n_1, n_2, ..., n_k\}$, with similar parameters. The attacks are as follows:

- **Case 1**: When node M_1 receives $\{Q_{SD}; S\}$, it tries to mislead node S by creating $\{R_{SD}; S, M_1, D\}$ and faking that destination D is its neighbor. This is not possible in SRP as only node D can generate the MACO which is verified by node S.
- **Case 2**: If node M_1 discards request packets received from node S, it narrows the topology view of S. However, at the same time, it removes itself from node S's view and therefore cannot cause harm to data flows originating from node S that would exclude M_1.
- **Case 3**: When M_1 receives $\{R_{SD}; S, 1, M_1, S, 4, D\}$, it tampers its contents and relays $\{R_{SD}; S, 1, M_1, Y, D\}$, where Y can be any arbitrary sequence of nodes. SRP overcomes this attack given that node S will readily discard the reply due to the integrity protection provided by the MACO.
- **Case 4**: When node M_2 receives $\{Q_{SD}; S, 2, 3\}$, it corrupts the accumulated route and relays $\{Q_{SD}; S, X, 3, M_2\}$ to its neighbors, where X is a false IP address. This request arrives at node D and routes it over $\{D, M_2, 3, X, S\}$ towards the source node S. However, when node 3 receives the reply, it cannot forward it any further as X is not its neighbor. Therefore, the reply is dropped.
- **Case 5**: If node M_2 replays route requests in order to consume network resources, intermediate nodes will discard these as they maintain a list of query identifiers seen in the past. As we have seen, the query identifier is a random number so that it cannot be easily guessed by the malicious nodes.
- **Case 6**: If node M_1 attempts to forward $\{Q_{SD}; S, M_1^*\}$, i.e., M_1 attempts to spoof its IP address, node S would accept $\{R_{SD}; S, M_1^*, 1, 4, D\}$ as a possible route. However, the connectivity information

conveyed by such a reply is correct. In practice, a neighbor discovery scheme that maintains information on the binding of the MAC and IP address can strengthen the protocol. In this case, packets would be discarded when relayed by the same data link interface, i.e., same MAC address with more than one different IP address.

13.5.5.1 Attacks on SRP

We now describe possible attacks on the SRP for which there are no provisions in the protocol design.

Tunneling: During the route request and reply phases of a route discovery, if two nodes collide, then the protocol could be attacked. For example, if M_1 in Figure 13.11 received a route request, it can tunnel it to M_2, i.e., it can discover a route to M_2 and send the request encapsulated in a data packet. Then, M_2 would broadcast a request with the route segment between M_1 and M_2 falsified such as, for example, $\{Q_{SD}; S, M_1, Z, M_2\}$. The destination node D will, in turn, receive the request and construct a reply which is routed as $\{D, M_2, Z, M_1, S\}$. M_2 then receives the reply and tunnels it back to M_1, which, in turn, returns it to the source node S. As a result, the connectivity information is only partially correct.

Replay: If node M_1 rewrites the RND# with some other random number, its neighbors will think that it is a genuine packet and will keep on forwarding it, thus wasting their resources. Only when the packet reaches the destination, can this misuse be detected using the MACO. If the network is large, this can severely compromise the network lifetime.

13.5.5.2 ARIADNE

Based on DSR, an *on-demand* secure ad hoc routing protocol called ARIADNE is proposed in [Hu2002a] that can withstand node-compromise and relies only on highly efficient *symmetric* cryptography. ARIADNE guarantees that the target node of a route discovery process can authenticate the initiator, that the initiator can authenticate each intermediate node on the path to the destination indicated by the RREP message and that no intermediate node can remove a previous node in the node list of the RREQ or RREP messages themselves. As for the SRP protocol, ARIADNE needs some mechanism to bootstrap authentic keys required by the protocol. In particular, each node needs a shared secret key ($K_{S,D}$ is the shared key between S and D) with each node it communicates with at a higher layer, an authentic Timed Efficient Stream Loss-tolerant Authentication (TESLA) [Perrig2000] key for each node in the network, and an authentic "Route Discovery chain" for each node for this node which will forward RREQ messages.

ARIADNE provides point-to-point *authentication* of a routing message using a MACO and a shared key between the two parties. However, for authentication of a broadcast packet such as RREQ, ARIADNE employs the TESLA broadcast authentication protocol which can cope up with attacks by *malicious* nodes. Selfish nodes are not taken into account. In ARIADNE, eight fields are used:

<ROUTE REQUEST, initiator, target, id, time interval, hash chain, node list, MAC list>

The initiator and target are set to the address of the initiator and the target nodes, respectively. As in DSR, the initiator sets the id to an identifier that it has not recently been used in a Route Discovery. The TESLA time interval is set at a pessimistic expected arrival time of the request accounting for clock skew. The initiator of the request then initializes the hash chain to $\text{MACO}_{K_{SD}}$(initiator, target, id, time interval) and the node list and MAC list. When any node A receives a RREQ for which it is not the target, the node checks its local table of <initiator, id> values from recent requests. If it has already seen a request, the node discards the packet as in DSR. The node also checks whether the time interval in the request is valid: that time interval must not be too far in the future, and the key corresponding to it must not have been disclosed yet. The node modifies the request by appending its own address (A) to the node list in the request, replacing the hash chain field with H [A, *hash chain*], and appending a MACO of the entire REQUEST to the MACO list. The node uses the TESLA key K_{Ai} to compute the MACO, where i is the index for the time interval specified in the request. Finally, the node rebroadcasts the modified RREQ, as in DSR.

When the target node receives the RREQ, it checks the validity of the request by determining that the keys from the time interval specified have not been disclosed yet, and that the hash chain field is equal to:

$H [\eta_n, H [\eta_{n-1}, H [..., H [\eta_1, MACO_{K_{S,D}} \text{ (initiator, target, id, time interval)}]...]]]$

Where η_i is the node address at position i of the node list, and n is the number of nodes in the node list. If the target node determines that the request is valid, it returns a RREP to the initiator containing eight fields:

<ROUTE REPLY, target, initiator, time interval, node list, MACO list, target MACO, key list>

Here, the target, initiator, time interval, node list, and MACO list fields are set to the corresponding values from the RREQ, the target MACO is set to a MACO computed on the preceding fields in the reply with the key $K_{D,S}$, and the key list is initialized to the empty list. The RREP is then returned to the request initiator along the route obtained by reversing the sequence of hops in the request.

A node forwarding a RREP waits until it is able to disclose its key from the time interval specified. Then, it appends its key to the key list field and forwards the packet according to the source route indicated by the packet. Waiting time delays the return of RREP, but does not consume extra computational power. When the initiator receives a RREP, it verifies that each key in the key list is valid, that the target MACO is valid, and that each in the MACO list is valid. If all of these tests succeed, the node accepts the RREP; otherwise, it is discarded. In order to prevent the injection of invalid route errors by any node, each node that encounters a broken link adds TESLA authentication data to the route error message for authentication of the return path. However, TESLA authentication is delayed, so all the nodes on the return path buffer the error, but do not consider it until it is authenticated. Later, the node that encounters the broken link discloses the key and sends it over the return path, which enables nodes on that path to authenticate the buffered error messages. ARIADNE is also protected from a flood of RREQ packets that could lead to the cache poisoning attack. Benign nodes can filter out forged or excessive RREQ packets using *Route Discovery chains*, a mechanism for authenticating route discovery, allowing each node to rate-limit discoveries [Hu2002a]. ARIADNE is immune to an advanced version of wormhole attack, using the TIK (TESLA with Instant Key disclosure) protocol that allows very precise time synchronization between the nodes and enables detection of anomalies in routing traffic flows.

13.5.5.3 SEAD

A *proactive* secure routing protocol in [Hu2002b] is based on the DSDV proactive protocol. Its enhanced version (DSDV-SQ) [Hu2002a] has inspired a Secure Efficient Ad hoc Distance (SEAD) vector routing protocol that can deal with attackers *modifying* routing information broadcasted during the update phase of the DSDV-SQ protocol. Routing can be disrupted if an attacker modifies the sequence number and the routing table update message including *Replay attacks*. In order to secure DSDV-SQ, SEAD uses an efficient *one-way hash chain* rather than relying on expensive asymmetric cryptography operation. However, like the other secure protocols, SEAD assumes some mechanism for distribution of the hash that can be used to authenticate all the other elements of the chain. The key distribution is done by a trusted entity that relays public key certificates for each node to use in a hash chain.

The basic idea of SEAD is to authenticate the sequence number and routing table updates using hash chains. In addition, receiver of SEAD also authenticates the sender, ensuring that the routing information originates form the correct node. To create a one-way hash chain, a node chooses a random initial value $x \in \{0, 1\}^{\rho}$, where ρ is the length of the output hash function, and computes the list of values h_0, h_1, h_2, h_3, ..., h_n, where $h_0 = x$, and $h_i = H(h_{i-1})$ for $0 < i \le n$, for some n. As an example, given an authenticated h_i value, a node can authenticate h_{i-3} by computing $H(H(H(h_{i-3})))$ and verifying that the resulting value equals h_i.

In each routing update, each node uses a specific authentic (i.e., signed) element from its hash chain (metric 0). Based on this initial element, the one-way hash chain provides authentication for the lower bound on the metric in other routing updates for that node. The use of a hash value on the sequence number and metric in a routing update entry prevents any node from advertising a route to some destination with a greater sequence number than destination's own current sequence number. Likewise, a node cannot advertise a route better than those for which it has received an advertisement, since the metric in an existing route cannot be decreased by the hash chain. When a node receives a routing update, it checks the authenticity of the information for each entry using the destination address, the sequence number and the metric of the received entry, together with the prior *authentic* hash value received from that destination's hash chain. Received elements are hashed correct number of times to ensure authenticity of the received information by matching the calculated hash value with the authentic hash value. The source of each routing update message in SEAD must also be authenticated, since an attacker may be able to create routing loops through the *impersonation* attack. Two different approaches are possible to provide node authentication: the first is based on a broadcast authentication mechanism such as TESLA, and the second can use Message Authentication Codes, assuming a shared secret key between each a pair of nodes. SEAD does not cope with *wormhole* attacks though the authors propose to use the TIK protocol to detect the threat, similar to ARIADNE protocol.

13.5.6 The Wormhole Attack

The wormhole attack is a severe threat where a malicious node can record packets (or bits) at one location in the network and tunnel them to another location through a private network shared with a colluding malicious node. A dangerous threat can be perpetrated if a wormhole attacker tunnels all packets through the wormhole honestly and reliably since no harm seems to be done. The attacker actually seems to provide a useful service in efficiently connecting the network. However, when an attacker forwards only routing control messages and not data packets, communication may be severely damaged. As an example, when used against an on-demand routing protocol such as DSR, a powerful wormhole attack can be started by tunneling each RREQ message directly to the destination target node of the request. This attack prevents routes more than two hops long from being discovered, as RREP messages would arrive at the source faster than any other replies or, worse, RREQ messages arriving from other nodes next to the destination than the attacker would be discarded since already seen.

Temporal leashes (using precise time synchronization) or *Geographical leashes* (using location information) are used to calculate delay bounds on a packet. These bounds when compared to actual values can help in anomaly detection. In some special cases, wormholes can also be detected through techniques that do not require precise time synchronization nor location information. As an example, it would be sufficient to modify the routing protocol used to discover the path to a destination so that it could handle multiple routes. A verification mechanism would then detect anomalies when comparing the metric (e.g., number of hops) associated with each route. Any node advertising a path to a destination with a metric considerably lower than all the others could be branded as a suspect of a wormhole. Furthermore, if the wormhole attack is performed only on routing information while dropping data packets, other mechanisms can be used to detect this misbehavior.

13.6 Cooperation in MANETs

In ad hoc networks, basic networking functions such as packet forwarding and routing are carried out by all available network nodes. There is no reason to assume that the nodes will eventually cooperate with one another since network operation consumes energy in a battery powered environment. The new type of node misbehavior that is specific to ad hoc networks is caused by the lack of cooperation and goes under the name of *node selfishness* [Yoo2006]. A selfish node does not directly damage other nodes but it simply does not cooperate to the network operation, saving battery life for its own communications. Damages provoked by selfish behavior can not be underestimated: a simulation study presented in [Michiardi2002a] shows the impact of a selfish behavior in terms of global network throughput and global communication delay when the DSR routing protocol is used. The simulation results show that even a small percentage of selfish nodes present in the network leads to severe performance degradation.

Schemes that enforce node cooperation in a MANET can be divided in two categories: the first is currency based [Yoo2005] and the second uses a local monitoring technique. The currency based systems are simple to implement but rely on tamperproof hardware. The main limitation of this approach is how virtual currency is to be exchanged. On the other hand, cooperative security schemes based on local monitoring of neighbors by each node, evaluating a metric that reflects nodes' behavior and using that, a selfish node can be gradually isolated from the network. The main drawback of the second approach is the absence of a well-accepted securely identification mechanism. Any selfish node could elude the cooperation enforcement mechanism and get rid of its bad reputation just by changing its identity.

13.6.1 CONFIDANT

The CONFIDANT (Cooperation Of Nodes, Fairness In Dynamic Ad hoc NeTworks) cooperation mechanism [Buchegger2002a, Buchegger2002b] detects malicious nodes by observing or reporting several types of attacks, thus allowing nodes to route around misbehaved nodes and thereby isolating them. CONFIDANT works as an extension to a routing protocol such as DSR. The nodes are provided with:

• A monitor for observations.

• Reputation records for first-hand and trusted second-hand observations about routing and forwarding behavior of other nodes.

• Trust records to control trust given to a received warning.

• A path manager to adapt their behavior according to reputation and to take action against malicious nodes.

In CONFIDANT, the nodes monitor their neighbors and accordingly change the reputation. They can inform other nodes by sending an ALARM message either directly or by promiscuously listening. It evaluates how trustworthy the ALARM is based on the source of the ALARM and the accumulated ALARM messages about the node in question. It can then decide whether to take action against the misbehaved node in the form of excluding routes containing the misbehaved node, re-ranking paths in the path cache, reciprocating by non-cooperation, and forwarding an ALARM about the node.

Simulations with nodes that do not participate in the forwarding function have shown that CONFIDANT can cope well, even if half of the network nodes are malicious. Further simulations on the effect of second-hand information and slander have shown that slander can effectively be prevented while still retaining a significant detection speed-up over using merely first-hand information. CONFIDANT assumes that the reputation is based on observance of events to be feasible. Reputation can only be meaningful if identity of each node is persistent; otherwise it is vulnerable to spoofing attacks.

13.6.2 Token-Based

In an approach presented in [Yang2002], each node of the ad hoc network has a token and its local neighbors collaboratively monitor any misbehaving nodes in routing or packet forwarding services. Upon expiration of the token, each node renews its token via its multiple neighbors: the period of validity of a node's token is dependent on how long it has stayed and behaved well in the network to accumulate its credit. The security scheme is composed of four related components:

- *Neighbor verification*: Describes how to verify whether each node in the network is a legitimate or malicious node.
- *Neighbor monitoring*: Describes how to monitor the behavior of each node in the network and detect occasional attacks from malicious nodes.
- *Intrusion reaction*: Describes how to alert the network and isolate the attackers.
- *Security enhanced routing protocol*: Explicitly incorporates security information collected by other components into the routing protocol.

In the token issuing/renewal phase, it is assumed to have a global secret key (SK)/public key (PK) pair, where PK is well known by every node of the network. SK is shared by k neighbors who collaboratively sign the token requested or renewed by local nodes. On the other hand, token verification follows three steps: 1) Identity match between node's ID and the token ID; 2) Validity time verification; and 3) Issuer signature. If the token verification phase fails, the corresponding node is rejected and both routing and data packets are dropped for that node. Security relies on the redundancy of routing information rather than cryptographic techniques enforced by suitably modifying the AODV protocol and the Watchdog technique described earlier.

However, the proposed solution possesses drawback like the bootstrap phase needed to generate a valid collection of partial tokens that create a final token. As the number of neighbors necessary to complete the signature of every partial token has to be at least k, the use of such security mechanism is useful only for large and dense ad hoc and sensor networks. On the other hand, validity period of a token increases proportionally to the time the node behaves well, and has less impact if node mobility is high. Frequent changes in the subset of nodes sharing a key for issuing valid tokens can cause high computational overhead besides high traffic by issuing/renewing a token. This suggests that the token-based mechanism is more suitable for ad hoc networks with low node mobility. Spoofing attacks are not taken into account where a node can request more than one token based on different identities.

13.7 Wireless Sensor Networks

Wireless sensor network (WSN) discussed in Chapters 10, 11 and 12, is a self-configured network composed of a large number of tiny SNs. In most case, a sensor node (SN) is powered by energy limited battery and only has low computing power and limited memory storage. Each SN can exchange information with other nodes over wireless communication links within a short range. Unlike traditional wired networks or cellular networks and similar to MANETS, WSNs do not need infrastructural support; therefore they can easily be deployed in any terrains or environments.

13.7.1 WSN Security

Security is an extremely important issue in many WSN applications. Strong security requirement for military applications are often combined with an inhospitable and physically unprotected environment. In commercial applications, privacy protection is as important as security and reliability. There are some constraints in WSNs. On SN level, the main constraints are the limited battery power, limited memory size and short radio transmission range. On the network level, the main constraints are the limited pre-configuration, ad hoc networking and the intermittent connectivity. These constraints must be considered when we design the security protocols for WSNs. In the following sections we discuss the main security issues for WSNs.

Key distribution and Management: Traditional key distribution protocols, such as public key cryptographic can not be used directly in WSNs due to the inherent properties and the limited recourses. Key distribution and management protocols used in WSNs should scale to a large number of SNs. In WSNs, data propagate via multi-hop mode and to protect data privacy, each SN needs to set up keys with its neighbors.

Secrecy and Authentication: To prevent eavesdropping, injection, and modification of data packets, cryptography is required to provide secrecy and authentication. Hardware cryptographic support may achieve computational efficiency but increases the system cost at the same time. With recent sensor technology, software cryptography achieves the same secrecy and authentication only result in 5%~10% performance overhead. The cryptographic computations can be overlapped with transmission, their effect on system latency or throughput is trivial.

Privacy: With the development of WSN technology, malicious parties or individuals can deploy secret surveillance networks to spy on unaware victims, which increase the privacy abuses. Societal norms, laws as well as technological responses need to be adopted to protect privacy.

Secure Routing: In WSNs, routing and data forwarding are the essential services. But most of current routing protocols are vulnerable. Attacks, such as denial-of-service launched on the routing protocol can easily prevent the network's communication.

Secure Data Aggregation: Due to the large number of wireless SNs, the sensed data must be aggregated before sending to the base station. Data aggregation can be taking place in any places, depending on the architecture of the corresponding network, and all aggregation locations should be secured.

Denial of Service (DoS): WSNs can be jammed by the adversaries with various methods. A simplest way for an adversary to disrupt a network is broadcasting a high enough energy signal to jam the entire system's communication. The adversary also can inhibit communication by violating the 802.11 MAC protocols and create DoS. Jamming-resistant networks can identify the affected region and routing packets around this jammed area if only a small portion of the network is jammed.

Resilience to Node Capture: SNs might be physically captured by the attacker since they are usually placed in an accessible location. Once a SN is captured, the attacker can extract its cryptographic secrets, or replace it with a malicious node. Tamper-resistant hardware may provide some level of security against node capture attack but it is expensive; algorithmic solutions are preferable in WSNs. The security issues in WSNs involve different layers, with the most important issue being the key distribution and management.

13.7.2 Key Distribution and Management

To provide security for WSNs, communication should be encrypted and authenticated. An open research problem is how to secure the communications among SNs in wireless sensor networks, in other words, how to set up trust among the participating entities? Cryptographic key computations can be used to establish trust between communication nodes, either using secret key or public key based techniques. Due to the resource constraints of WSNs, secret key mechanisms are preferred. There are three types of general key agreement schemes: trusted-server scheme, self-enforcing scheme, and key pre-distribution scheme.

13.7.2.1 Trusted Server Protocols

The trusted–server scheme depends on a trusted server for key agreement between nodes, e.g., Kerberos [Neuman1994]. Kerberos uses a series of encrypted messages, timestamps and ticket-granting service. However, this scheme is not suitable for WSNs since there is usually no trusted infrastructure.

13.7.2.2 Self-enforcing Protocols

The self-enforcing scheme depends on asymmetric cryptography, such as key agreement using public key certificates. However, limited computation and energy resources of SNs in WSNs often make it undesirable to use public key algorithms, such as Diffie-Hellman key agreement [Diffie1976] or RSA [Rivest1978].

13.7.2.3 Key Pre-distribution Protocols

In key pre-distribution protocols, key information is distributed among all SNs prior to deployment so that adjacent SNs can communicate with each other if they share a common key [Cheng2005]. There exist many key pre-distribution schemes. A naïve approach is let all SNs share a master secret key. Any pair of nodes can use this global master secret key to achieve key agreement and obtain a new pair-wise key to secure the communication between them. The main advantage of is it requires much smaller memory size and has low computation complexity. However, once the master secret key is compromised by any SN, the security of the entire network could be easily compromised. Tamper-resistant hardware may increase the security level, but it also increases the cost and energy consumption at each SN. Furthermore, tamper-resistant hardware might not be always safe [Anderson1996].

Another key pre-distribution scheme is to let each pair of nodes share a secret pair-wise key. Suppose there are N nodes in the WSN, then each SN needs to store N-1 key in its memory. This scheme can provide perfect security of the network since compromising a SN or few SNs cannot affect the security of communications among other uncompromised SNs. However, this scheme is impractical for WSNs due to the large size of the network and limited memory of each SN. Another drawback of this scheme is adding new SNs to the existed network is very difficult. Further enhancement in key-distribution scheme has recently been suggested [Cheng2005].

13.7.2.4 Eschenauer-Gligor's Random Key Pre-distribution Scheme

In this random key pre-distribution scheme [Eschenauer2002], proposed by Eschenauer and Gligor, each SN receives a random subset of keys from a large key pool before deployment. One such distribution from a pool of keys shown in Figure 13.12(a) is shown in Figure 13.12(b) for an example 4-node WSN. Two nodes can communicate securely if they share a common key within their key subsets. The authors used the random graph theory to show that if the probability that any two nodes share at least one common key satisfies a certain critical value, the whole network can form a completely connected graph with secure links. That means any node can find a secure path to communicate with any other nodes in the network. A common key between two adjacent nodes is illustrated in Table 13.8 for the WSN of Figure 13.12(b). The non-adjacent nodes have to go through other intermediate SNs to have secured communication. The same is applicable if adjacent SNs do not have any common key between them for symmetric encoding/decoding. The Eschenauer-Gligor scheme is further improved by Du, Deng, Han, and Varshney [Du2003]], and by Liu and Ning [Liu2003a].

13.7.2.5 Blom's λ-secure key pre-distribution scheme

This key pre-distribution method [Blom1984] allows any pair of nodes in a network to be able to find a pair-wise secret key by constructing a $(\lambda+1)*N$ matrix **G** over a finite field GF(q), where N is the size of the network. As long as no more than λ nodes are compromised, the network is perfectly secure. Initially, Blom's scheme was not developed for WSNs; Du et al. [Du2003] made modifications to the original scheme and made it suitable for them. Separating the single key space in Blom's scheme into multiple key spaces and using random key selection scheme to improve performance and resilience, Du et al.' scheme improved the performance of λ-secure key pre-distribution scheme.

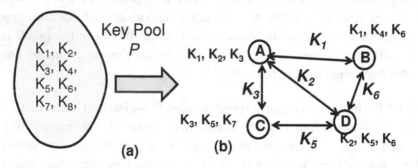

(a)   (b)

Figure 13.12(a) – Key Pool; (b) – Distribution of keys to SNs

Table 13.8 – Link between adjacent SNs and Common Key between them for the WSN of Figure 13.12 (b)

Link between two adjacent sensors	Keys at the first sensor	Keys at the second sensor	Common Key
A - B	A: k , k , k	B: k , k , k	k
A - C	A: k , k , k	C: k , k , k	k
A- D	A: k , k , k	D: k , k , k	k
B - D	A: k , k , k	D: k , k , k	k
B - D	B: k , k , k	D: k , k , k	k
C - D	C: k , k , k	D: k , k , k	k

13.7.2.6 Polynomial-based key pre-distribution scheme

A polynomial-based key pre-distribution technique [Blundo1993] was developed for group key pre-distribution scheme. Polynomial-based key pre-distribution schemes allow any group of t parties to compute a common key while being secure against collusion between some of them. These schemes focus on saving communication costs while memory constraints are not placed on group members. The polynomial-based key pre-distribution scheme discussed in [Blundo1993] can tolerate no more than t compromised nodes, where the value of t is limited by the storage capacity for pair-wise keys in a SN. Liu and Ning improved this [Liu2003b] and proposed a location-based pair-wise keys scheme using bivariate polynomials by taking advantage of sensors' expected locations information. This employs a threshold technique and provides a trade-off between the security against node capture and the performance of establishing pair-wise keys.

13.7.2.7 2-D Key Distribution Scheme for WSNs

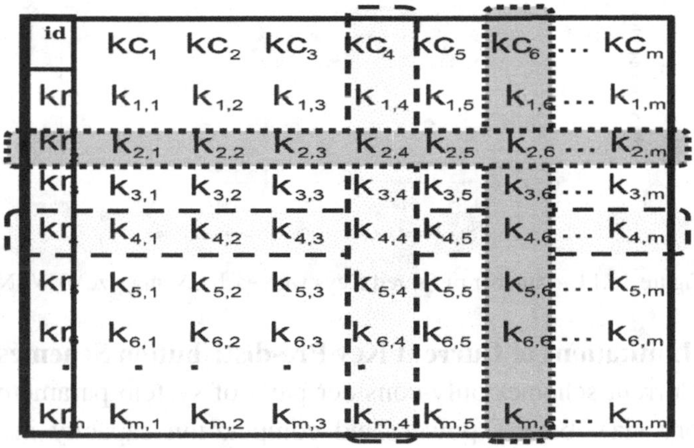

Figure 13.13 – 2-D organization of m^2 keys and assignment of $(2m - 1)$ keys to each SN for m^2 SNs (keys for SN 4,4 and 6,2 explicitly marked)

A novel 2-D key pre-distribution scheme has been introduced in [Cheng2005] where m^2 SNs are in a 2-D array. Similarily, m^2 *keys* are arranged in a two-dimensional array of size $m \times m$ and *each of* m^2 *sensors is allocated* a row and column of $(2m - 1)$ *keys*. After such keys pre-distribution, SNs can be deployed randomly, while remembering their row and column ids at the time of key distribution. This is illustrated in Figure 13.13 where it can be easily seen that there are at least two common keys between any two SNs with a given id. Presence of at least two common keys between any two SNs is guaranteed and compromise or destruction of a single SN does not impact keys for other SNs. The size of WSNs' required number of keys for random key distribution and 2-D distribution are compared in Figure 13.14. The later scheme is observed [Cheng2008] to have a better resiliency against node failure. Further enhancement has been done by using 3-D scheme of key distribution [Cheng2007]. A more detailed survey of secured communication in WSNs has been summarized in a recent article [Gaur2009].

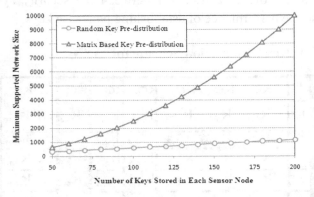

Figure 13.14 – Number of Stored Keys with each SN and size of WSN

13.7.2.8 Limitations of Current Key Pre-distribution Schemes

Current schemes only consider parts of system parameters, such as memory size, battery power, and computation capacity in general. Some important application based parameters, such as lifetime of network, required security level of different applications are not taken into account and not discussed sufficiently and carefully. There are still no systematic studies of all the different models for security of wireless sensor networks even though some important parameters have recently being considered in [Gaur2009].

13.8 Intrusion Detection Systems

The use of wireless links renders a MANET or WSN vulnerable to malicious attacks, ranging from passive eavesdropping to active interference. In wired networks, however, the attacker needs to gain access to the physical media (e.g., network wires, etc.) or pass through a plethora of firewalls and gateways. In MANETS and WSNs, the scenario is much different as there are no firewalls or gateways in place and attacks can take place from all directions. Every node in a MANET or WSN must be prepared for such encounters with the adversary. Each node in is an autonomous unit in itself, free to move independently. This means that a node without adequate physical protection is very much susceptible to being captured, hijacked or compromised. It is difficult to track down a single compromised node in a large network; attacks stemming from compromised nodes are far more detrimental and much harder to detect. Thus, every node in a MANET or WSN will be able to work in a mode wherein it trusts no peer.

As we know, MANET and WSN have a decentralized architecture, and many algorithms rely on cooperative participation of the member nodes. Adversaries can exploit this decentralized decision making architecture to launch new types of attacks aimed at breaking the cooperative algorithms. Furthermore, routing presents more vulnerabilities than one can imagine, since most routing protocols are designed based in their cooperative nature. The adversary who compromises a node could succeed in bringing down the whole network by disseminating false routing information and this could culminate into all nodes feeding data to the compromised node. Intrusion prevention techniques like encryption and authentication can reduce the risks of intrusion but cannot completely eliminate them (e.g., encryption and authentication cannot defend against compromised nodes).

13.8.1 Overview

The protocols and systems which are meant to provide useful services can be a prime target for attacks such as Distributed Denial of Service (DDoS). Intrusion detection can be used as a second line of defense to protect networks because once an intrusion is detected response can be put in place to minimize the damage or gather evidence for prosecution or take counter offensive measures. In general terms, "Intrusion" is defined as "any set of actions that attempt to compromise integrity, confidentiality or availability of a resource".

Intrusion detection assumes that "user and program activities are observable", which means that any activity which the user or an application program initiates gets logged somewhere into the system tables or some kind of a system log, and intrusion detection systems (IDS) have an easy access to these system logs. This logged system/user related data is called audit data. Thus, intrusion detection is all about capturing audit data, and on the basis of this audit data determining whether it is a significant aberration from normal system behavior. If yes, then IDS infers that the system is under attack. Based on the type of audit data, IDS can be classified into 2 types:

- **Network-based**: Network-based IDS sits on the network gateway and captures and examines network packets that go through the network hardware interface;
- **Host-based**: Host-based IDS relies on the operating system audit data to monitor and analyze the events generated by the users or programs on the host.

13.8.2 Unsuitability of Current IDS Techniques

Since almost all of current network-based IDS sit on the network gateways and routers and analyze the network packets passing through them, they are rendered ineffective for MANETS and WSNs due to lack of any fixed infrastructure. The only available audit data is restricted to the communication activities taking place within a node's radio range, and any IDS meant for these networks should be made to work with this partial and localized kind of audit data. Traditional anomaly detection models of IDS cannot be used for these networks, as the separating line between normalcy and anomaly is obscure. A node that transmits erroneous routing information can be either a compromised node or is currently out of sync due to volatile physical movement. Hence, it is relatively difficult to distinguish between false alarms and real intrusions.

13.8.3 An IDS Architecture for Ad Hoc and Sensor Networks

IDS should be both distributed and cooperative to suit the needs of MANETs and WSNs. What this means is that every node in the network should participate in possible intrusion detection. Each node is responsible for detecting intrusion locally and independently, but neighboring nodes have some mutual understanding and can collaboratively investigate in a broader range. Therefore, each node may have its own individual IDS agent and these agents monitor user and system activities as well as communication activities within their radio range. If an anomaly is detected in the local data or if the evidence is inconclusive, IDS agents on the neighboring nodes will cooperatively participate in a global intrusion detection scheme. These individual IDS agents constitute the IDS system employed to protect the wireless ad hoc network.

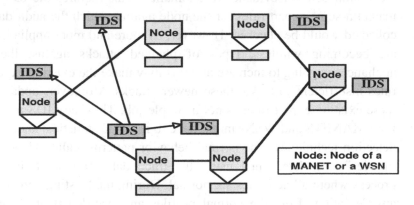

Figure 13.15 – An IDS architecture for a MANET and a WSN

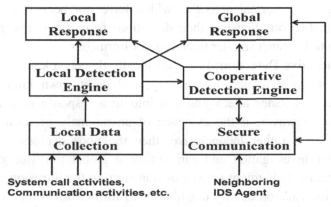

Figure 13.16 – Conceptual components of an IDS agent

Figure 13.15 depicts an IDS architecture, where the various IDS agents at different nodes communicate with each other to collectively detect the presence or absence of intrusions in the network. A Typical IDS agent, depicted in Figure 13.16, may include the following modules:

- **Local Data Collection**: The Local Data Collection module gathers streams of real time audit data from eclectic sources, which might include user and system activities within the node, communication activities by this node as well as any communication activities within the radio range of this node and observable to this node.

- **Local Detection Engine**: The Local Detection Engine analyzes the local audit data for evidence of anomalies. This requires the IDS to maintain some expert rules for the node against which the audit data collected would be checked. However, as more and more appliances are becoming wireless, types of planned attacks against these appliances is going to increase and this may make the existing expert rules insufficient to tackle these newer attacks. Moreover, updating these existing expert rules is not a simple job. Thus, any IDS meant for a MANETS and WSNs might have to resort to statistical anomaly detection techniques. The normal behavior patterns, called "Normal Profiles", are determined using the trace data from a "training" process where all activities are normal. During the "testing" process, any deviations from the normal profiles are recorded if at all any occur. A detection module is computed from the deviation data to distinguish anomalies from normalcy. Obviously, there will always going to be normal activities which have not been observed and recorded before, however their deviations from the normal profile is going to be much smaller than those of intrusions.

- **Cooperative Detection**: If a node locally detects a known intrusion with strong evidence, it can very well on its own infer that the network is under attack and can initiate a response or a remedial action. However, if the evidence of an anomaly or intrusion is a weak or is rather inconclusive, then the node decides it needs a broader investigation and can initiate a global intrusion detection procedure. This might consist of transmitting the intrusion detection state information among neighbors and further down the network if necessary.

The intrusion detection state information may be a mere level-of-confidence value expressed as percentage, for example:

- With p% confidence, node A after analyzing its local data concludes that there is an intrusion.
- With p% confidence, node A after analyzing the local data as well as that from its neighbors concludes that there is an intrusion.
- With p% confidence, nodes A, B, C, and so on, collectively conclude that there is an intrusion.

A more specific state that lists the suspects is also possible. For example, with p% confidence, node A concludes after analyzing its local data that node X has been compromised. A distributed consensus algorithm is then derived to compute the new intrusion detection state for the node under consideration, with the help of the state information recently received from other nodes in the network. The algorithm might involve a weighted computation, assuming that nearby nodes have greater effects than the far away ones. A majority-based Intrusion Detection Algorithm can include the following steps:

a) The node sends to its neighboring node an "intrusion state request".
b) Each node, including the one which initiates this algorithm, then propagates the state information indicating the likelihood of an intrusion to its immediate neighbors.
c) Each node then determines whether the majority of the received reports point towards an intrusion; if yes, then it concludes that the network is under attack.
d) Any node that detects an intrusion to the network can then initiate the remedial/response procedure.

As a rule of thumb, audit data from other nodes should not be trusted as compromised nodes might tend to send misleading data. However, sending audit data from a compromised node do not hold any incentives, as it might create a situation which would result in its expulsion from the network. Hence, unless the majority of nodes are compromised, remedial procedure will not be initiated.

13.8.3.1 Intrusion Response

The type of intrusion response for MANETs and WSNs depends on the type of intrusion, the type of network protocols and the confidence in the veracity of the audit trace data. The response might range from resetting the communication channels between nodes, or identifying the compromised nodes and precluding them from the network. The IDS agent can notify the end user to perform his/her own investigation and take the necessary actions. It also sends re-authentication requests to all the nodes in the network, to prompt their respective end users to authenticate themselves.

13.8.4 Anomaly Detection

In this section we discuss anomaly detection issues in IDS systems.

13.8.4.1 Detecting Abnormal Updates to Routing Tables

For ad hoc and sensor routing protocols, the primary concern is that false routing information generated and transmitted by a compromised node may be eventually used by other nodes in the network. Thus, a good candidate for audit data would be the updates of routing information. A legitimate change in the routing table is caused by physical motion of the nodes or changes in the membership of the network. For a node, its own movement and the change in its own routing table are the only data it can trust, and hence it is used as a basis of the trace data. The physical movement is measured by distance, direction and velocity. The routing table change is measured by Percentage of Changed Routes (PCR), and by the percentage changes in the sum of hops of all routes (PCH). Percentages are used as measurements as the number of nodes/routes is not fixed due to dynamic nature of the wireless ad hoc networks. During the "training" process, a wide variety of normal situations are simulated and the corresponding trace data is gathered for each node. The audit/trace data of all the nodes in the network are then merged together to obtain a set of all normal changes to the routing table for all nodes. The normal profile specifies the correlation of the physical movement of the node and the changes in the routing table. The classification algorithm divides available trace data into ranges. If the PCR and/or PCR values are beyond the valid range, then it is considered to be an anomalous situation and the necessary procedures are initiated.

13.8.4.2 Detecting Anomalous Activities in Other Layers

For medium access protocols, trace data could be in the form of total number of channel requests, the total number of nodes making those requests, etc., for last s seconds. The class can be the range of the current requests by a node. The classifier of the trace data describes the normal profile of a request. Anomaly detection model can then be computed on the basis of the deviation of the trace data from the normal profile. Similarly, the wireless application layer can use the service as the class and can contain the following features: for the past s seconds, the total number of requests to the same service, total number of services requested, the average duration of service, the number of nodes that requested service, the total number of service errors, etc. A classifier for each service then characterizes, for each service, a normal behavior for all requests.

13.9 Conclusions and Future Directions

Security has become a major concern in providing protected communication between network parties. Unlike wired networks, the unique characteristics of MANETs and WSNs pose a number of nontrivial challenges to security design, such as open peer-to-peer network architecture, shared wireless medium, stringent resource constraints, and the possibility of highly dynamic network topologies. These challenges clearly make a case for building multi-level security solutions that achieve both broad protection and desirable network performance. As we have seen throughout this chapter, the complete security solution should span all layers of the protocol stack, and encompass all three security components, namely, prevention, detection, and reaction. Although existing research has made considerable progress towards building end-to-end secure solutions, we still have a long way to go. The overhead of a foolproof solution would be prohibitive. Therefore, the research community is constantly trying to strike a balance between network performance and security robustness while such boundary is unclear. In any case, it is widely accepted that a lot more research is needed regarding security in ad hoc and sensor networks. While a perfect solution may be unrealistic to assume, the future looks bright as the basic foundation has already been established.

Homework Questions/Simulation Projects

Q. 1. Given a choice between an ad hoc network (MANET) and a wireless sensor network, which one would you prefer and why? Think about this and answer carefully.

Q. 2. In a given area, N sensor nodes are spread in a 2-D grid structure. To aggregate information and conserve energy, K x K (K<=sqrt(N)) sensors are grouped together to form a cluster. To reduce the complexity, an 8-bit Hash function is formed from the ID of the sensor node and is used to identify the symmetric key to be used for a source.

a.   What is the probability that two nodes within a cluster will have the same value of the symmetric key?
b.   Plot the probability when the cluster size in changed from K/2 x K/2 to 2K x 2K?
c.   What can you do to have a unique symmetric key within a cluster?
d.   Is it feasible to include location information in the symmetric key?
e.   If the distribution of sensor nodes is changed to random distribution (e.g., uniform distribution, Poisson distribution etc.), repeat the above questions.

Make necessary assumptions if needed.

Q. 3. Repeat Q. 2 if the sensors are placed as a triangular topology or a hexagonal topology.

Q. 4. Design a problem based on any of the material covered in this chapter (or in references contained therein) and solve it diligently.

References

[Amoroso1994] E. Amoroso, *Fundamentals of Computer Security Technology*, Prentice Hall, 403 pages, 1994.

[Anderson1996] R. Anderson and M. Kuhn, "Tamper resistance - a cautionary note," in *Proceedings of the Second Usenix Workshop on Electronic Commerce*, pp. 1-11, 1996.

[Asokan2000] N. Asokan and P. Ginzboorg, "Key Agreement in Ad-Hoc Networks," *Computer Communications*, 2000.

[Becker1998] K. Becker and U. Wille, "Communication complexity of group key distribution," in *Proceedings of the 5th ACM conference on Computer and communications security*, pages 1-6, 1998.

[Blom1984] R. Blom, "An optimal class of symmetric key generation system," Lecture Notes in computer science, Springer Verlag 1984, pp 335-338.

[Blundo1993] C. Blundo, A.D. Santis, A. Herzberg, S. Kutten, U. Vaccaro, and M. Yung, "Perfectly-secure key distribution for dynamic conferences," Lecture Notes in computer science, Vol. 740, pp 471-486, 1993.

[Buchegger2002] S. Buchegger and J-Y.L. Bodec, "Performance Analysis of the CONFIDANT Protocol. Cooperation of Nodes –Fairness in Dynamic Ad hoc Networks," Mobihoc 2002.

[Burmester1994] M. Burmester and Y. Desmedt, "A secure and efficient conference key distribution system," in *Advances in Cryptology – EUROCRYPT*, pages 275–286, 1994.

[Cheng2005] Y. Cheng and D.P. Agrawal, "Efficient pairwise key establishment and management in static Wireless Sensor Networks," 2nd IEEE International conference on Mobile Ad hoc and Sensor Systems (MASS 2005).

[Cheng2007] Y. Cheng, M. Malik, B. Xie, and Dharma P. Agrawal, "Enhanced Approach for Random Key Pre-Distribution in Wireless Sensor Networks," *International Conference on Communication, Networking and Information Technology*, Amman, Jordan, 6-8 December 2007.

[Cheng2008] Yi Cheng, "Security Mechanisms for Mobile Ad Hoc and Wireless Sensor Networks," PhD Dissertation, Advisor: Dharma P. Agrawal), University of Cincinnati, May 27, 2008.

[Dahill2002] B. Dahill, B. N. Levine, E. Royer, and C. Shields, "ARAN: A secure Routing Protocol for Ad Hoc Networks," *UMass Tech Report 02-32*, 2002.

[Deng2002] H. Deng, W. Li, and D.P. Agrawal, "Routing Security in Ad hoc Networks," IEEE Communications magazine, Vol. 40, No. 10, October 2002, pp 70-75.

[Diffie1976] W. Diffie and M. Hellman, "New directions in cryptography," *IEEE Transactions on Information Theory*, vol. 22, No. 6, pp. 644–654, 1976.

[Du2003] W. Du, J. Deng, Y. S. Han, and P. K. Varshney, "A pairwise key predistribution scheme for wireless sensor networks," in *Proceedings of the ACM CCS*, pp. 42-51, 2003.

[Eschenauer2002] L. Eschenauer and V. D. Gligor, "A key-management scheme for distributed sensor networks," in *Proceedings of the ACM CCS*, 2002.

[Gaur2009] Amit Gaur and Dharma P. Agrawal, "Secured Communication Strategies in Wireless Sensor Networks," Workshop on Security Risks in Internet, Indian Institute of Information Technology and Management, Gwalior, Dec. 18- 19, 2009, pp. 5-10.

[Gollmann1999] D. Gollmann, "Computer Security," *John Wiley & Sons Inc.*, 336 pages, 1999.

[Hu2002a] Y-C Hu, A. Perrig, and D. B. Johnson, "Ariadne: A secure On-Demand Routing Protocol for Ad Hoc Networks," in *Proceedings of Mobicom*, 2002.

[Hu2002b] Y-C Hu, D. B. Johnson, and A. Perrig, "SEAD: Secure Efficient Distance Vector Routing for Mobile Wireless Ad Hoc Networks," in *the Fourth IEEE Workshop on Mobile Computing Systems and Applications*, 2002.

[Ingemarsson1982] I. Ingemarsson, D. Tang, and C. Wong, "A conference key distribution system," *IEEE Transactions on Information Theory*, September 1982.

[Kim2000] Y. Kim, A. Perrig, and G. Tsudik, "Simple and fault-tolerant key agreement for dynamic collaborative groups," in *Proceedings of the 7th ACM Conference on Computer and Communications Security*, pages 235–244, 2000.

[Liu2003a] D. Liu and P. Ning, "Establishing pairwise keys in distributed sensor networks," in *Proceedings of the ACM CCS*, pp. 52-61, 2003.

[Liu2003b] D. Liu and P. Ning, "Location-Based Pairwise Key Establishments for Relatively Static Sensor Networks," in *Proceedings of ACM SASN*, 2003.

[Michiardi2002a] P. Michiardi and R. Molva, "Simulation-based Analysis of Security Exposures in Mobile Ad Hoc Networks," in *Proceedings of European Wireless Conference*, 2002.

[Neuman1994] B. Neuman, et al., "Kerberos: An Authentication Service for Computer Networks," *IEEE Communications Magazine*, vol. 32, No. 9, pp. 33-38, 1994.

[Papadimitratos2002] P. Papadimitratos and Z. Haas, "Secure Routing for Mobile Ad Hoc Networks," in *Proceedings of CNDS*, 2002.

[Perrig2000] A. Perrig, R. Canetti, J. D. Tygar, and D. Song, "Efficient authentication and signing of multicast streams over lossy channels," in *IEEE Symposium on Security and Privacy*, 2000.

[Rivest1978] R. L. Rivest, A. Shamir, and L. M. Adleman, "A method for obtaining digital signatures and public-key cryptosystems," *Communications of the ACM*, vol. 21, No. 2, pp. 120-126, 1978.

[Schneier1996] B. Schneier, *Applied Cryptography*, John Wiley & Sons Inc., 2nd Ed., 758 pages, 1996.

[Stajano1999] F. Stajano and R. Anderson, "The Resurrecting Duckling: Security Issues for Ad-hoc Wireless Networks," 7th *International Workshop on Security Protocols*, LNCS, Springer-Verlag, 1999.

[Steiner1996] M. Steiner, G. Tsudik, and M. Waidner, "Diffie-Hellman key distribution extended to group communication," in *Proceedings of the 3rd ACM conference on Computer and communications security*, pages 31–37, 1996.

[Yang2002] H. Yang, X. Meng, and S. Lu, "Self-Organized Network-Layer Security in Mobile Ad Hoc Networks," in *Proceedings of Wireless Security Workshop*, together with ACM Mobicom, September 2002.

[Yi2001] S. Yi, P. Naldurg, and R. Kravets, "Security-Aware Ad Hoc Routing for Wireless Networks," in *Proceedings of ACM MobiHoc*, 2001.

[Yoo2005] Y. Yoo, S. Ahn, and D.P. Agrawal, "A Credit-payment scheme for packet forwarding fairness in Mobile Ad hoc Networks," IEEE conference on communications, May 2005.

[Yoo2006] Y. Yoo and D.P. Agrawal, "Why it pays to be selfish in MANETs?," IEEE Wireless Communications Magazine, 2006, to appear.

[Zhou1999] L. Zhou and Z. Haas, "Securing Ad Hoc Networks," *IEEE Network*, November/December 1999.

[Zwicky2000] E. Zwicky, S. Cooper, and D. Chapman, *Building Internet Firewalls*, O'Reilly & Associates; 2nd Ed., 869 pages, January 2000.

Chapter 14

Integrating MANETs, WLANs, and Cellular Networks

14.1 Introduction

Popularity of the wireless communication systems can be seen almost everywhere in the form of cellular networks, WLANs and WPANs. Parallel with this evolution, increasing number of small portable devices are equipped with multiple communication interfaces, creating a heterogeneous environment in terms of access technologies. Resulting from diverse wireless communication systems and different access technologies future ubiquitous computing environment of the future has to exploit this multitude of connectivity alternatives to have guaranteed quality of service to the users. A recent trend that a ubiquitous computing environment should be capable of accessing information from different portable devices at any time and everywhere, has motivated researchers to integrate various wireless platforms such as MANETs, cellular networks, WPANs, WLANs, and WMN. Thus, this chapter envisions the architecture of state-of-the-art heterogeneous multi-hop networks, and identifies research issues that need to be addressed for a viable integration of the next generation wireless architecture.

As we can see in our daily lives, advances in wireless communications have expanded possible applications from simple voice connections in the early cellular networks (1G and 2G) to new integrated data applications. Wireless LANs based on the IEEE 802.11 family have recently become popular for allowing high data rates at relatively low cost. On the other hand, WLAN Access Points (AP) are expected to provide hot spot connectivity in the most common places, such as airports, hotels, shopping malls, schools, university campuses, homes, etc. It is expected that future advances in software defined radio (SDR) technology [SDRFORUM] will make multi-interface, multi-mode and

587

multi-band communication devices a commonplace. Such an integrated heterogeneous environment enables a user to access a cellular network, a WLAN, or a WMN, depending on the application needs and the types of radio access networks (RANs) available. For example, in a scenario, a student starts downloading a large video file using the cellular interface of a multi-mode phone and as higher data rate and lower cost connection through the home IEEE 802.11n AP becomes available, the connection could be automatically switched from the cellular network to the home AP. It is not unrealistic to expect automatic connection and seamless network migration for a single call.

The first step to provide effective and efficient support is to integrate WLANs (e.g., IEEE 802.11a/b/g/n and HiperLAN/2), Wireless WANs (e.g., 1G, 2G, 2.5G, 3G, GSM, the proposed IEEE 802.20, etc. [Agrawal2002]), Wireless MANs (e.g., IEEE 802.16 [Eklund2002]), and Wireless RANs (e.g., IEEE 802.22 [Cordeiro2005]) by observing a common characteristic of one-hop operation mode, where users access the system through a BS or an AP that is connected to a wired infrastructure. The second step is to extend this to multi-hop communication environment using the revolutionary paradigm of MANETs where router connectivity may change frequently, leading to the multi-hop communication that can provide connections inside hot-spot cells, and can also allow communication without the use of BS/AP.

Therefore, it is necessary to have global heterogeneous architectures and services that together provide seamless integration of one-hop networks (e.g., cellular, WLAN, WWAN) and multi-hop wireless systems. Furthermore, when all these technologies are integrated with the Internet, the possibilities are countless [Cavalcanti2005]. In this heterogeneous environment, users would have profiles such as price, data rate, battery life, service grade, and mobility pattern. A RAN has to be selected (e.g., SDR interface) for providing wireless connections and need to be done based on the user profile and the network state. If a node is not directly within the coverage area of a RAN, then we have to explore the possibility of reaching a RAN through multiple hops. Here, the node has to figure out what other nodes in its network can serve as the best gateway and provide access to RANs.

Internetworking of cellular systems with WLAN hot spots has also been investigated by standardization bodies, such as 3GPP (Third Generation Partnership Project) [Ahmavaara2003, Salkintzis2004] and 3GPP2 [Buddihikot2003]. The basic integration aspects considered in [Ahmavaara2003, Salkintzis2004] are the network selection, AAA (Authentication, Authorization and Accounting) and routing through the fixed infrastructure to the APs and the 3G network. The 3GPP2 group is also developing integrated solutions for WLANs and CDMA2000 based cellular systems [Buddihikot2003]. However, the architectures under investigation at 3GPP and 3GPP2 consider only infrastructure-based operation mode and does not address the issues related to multi-hop communications. The integration with the MANET paradigm is much more challenging and complex due to the dynamic nature of multi-hop communications and has many open research problems.

In this chapter, we discuss those issues that would enable us to integrate heterogeneous wireless networks with multi-interface devices. Firstly, we describe various components of a heterogeneous scenario and then we summarize design issues for each layer of the protocol stack and provide an overview of the existing integration models. Here, we give particular emphasis to the network layer issues, since it is intended to provide common base and hide the heterogeneity from the upper layers.

14.2 Ingredients of a Heterogeneous Architecture

As discussed before, a heterogeneous communication network provides transparent and self-configurable Wireless LANs, in both the infrastructure or the MANET mode, and Wireless WANs services. The basic components are Mobile Hosts (MHs), BSs/APs, and a Core IP Network (CN), with BSs and APs serve as the communication bridges for MHs (Figure 14.1). In a hot-spot area, multiple APs may overlay to some extent, as well as a BS and an AP may be co-located. MHs can arbitrarily move and at a given instant a particular MH can either be within or outside the coverage of BSs and/or APs. A connection from a MH to a BS/AP can be established by single hop or using multi-hop when the MH is out of the coverage of the corresponding BS/AP, as shown in Figure 14.1. Factors influencing the design of such a heterogeneous architecture include multi-interface MHs; transmit power and co-channel interference, topology and routing, mobility and handoff, load balance, interoperability and QoS service provisioning.

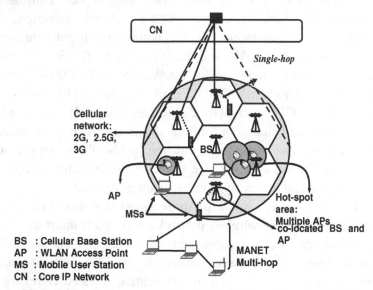

Figure 14.1 – Heterogeneous Network Architecture [Cavalcanti2005]

14.2.1 Mobile User Stations

Despite plethora of technology alternatives and future wireless interfaces, we can identify the following basic mobile stations types: single-mode cellular MH, single-mode WLAN MH or dual-mode MH. A single-mode cellular MH connects to a cellular network or a WMN through a BS. A single-mode WLAN can communicate through an AP or can connect to other WLAN equipped terminals in the ad hoc mode forming a MANET. A dual-mode MH can operate in both the cellular based mode or in the MANET mode using the WLAN interface. For example, in the UCAN architecture [Luo2003], each MH uses two air interfaces: a HDR (High Data Rate) [Esteves2000] interface for communicating with BS and an IEEE 802.11 based interfaces for peer-to-peer communication. Future MHs will be equipped with multiple interfaces based on different wireless technologies (e.g., HDR interface,

IEEE 802.11, IEEE 802.16, IEEE 802.15, IEEE 802.22, and so on) so that higher data speed rate could be supported for better connectivity.

14.2.2 Base Station and Access Point

The integration of cellular networks, WLANs and MANETs is not straightforward because of different communication requirements, different interface capabilities and mobility patterns of the MHs. Fixed network components, such as BSs and APs, can provide several services to MHs including:

a) Accessibility to the Internet.
b) Interoperability of existing networks and future networks.
c) Support of handoff between different wireless access networks.
d) Resources control.
e) Routing discovery.
f) Security management.

Both BS and AP should have the capability of interoperability with each other, and the possibility of integration with new emerging networks for supporting handoffs between them. The APs and BSs also have the responsibility to manage and control radio resources to the MHs. In fact, frequency allocation becomes more complicated, as different wireless technologies may possibly operate in the same frequency band. The high processing and power capacity of APs and BSs make them strategic components in selecting optimum routes between two MHs. Further, the APs and BSs can implement load balance functionalities by switching connections from the single-hop mode to the MANET mode or by diverting connections to free neighboring BS or AP by multi-hop communication.

14.2.3 Core IP Network (CN)

The Core IP Network (CN) serves as the backbone network with Internet connectivity and packet data services, but also supports seamless mobility, multi-hop cooperation, and security. The nodes in the CN may support Mobile IP [Tanenbaum1996] and Cellular IP [Valko1998] to provide continuous connectivity for MHs when they move between cellular networks and WLANs or change their points of attachment on cellular networks or WLANs. An integrated billing scheme and, possibly, a reward mechanism are required in the CN to encourage packet forwarding for multi-hop communication [Yoo2006]. Another important issue is security, and the CN plays an important role in preventing several types of attacks [Xie2004] by supporting authentication for all types of MHs.

14.2.4 Possible Communication Scenarios

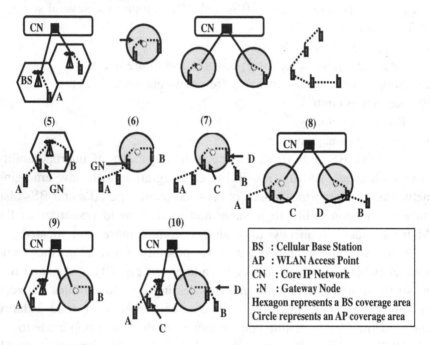

Figure 14.2 – Conection alternatives between two dual-mode MHs A and B
[Cavalcanti2005]

In the heterogeneous environment shown in Figure 14.1, different types of connections can be established between any two MHs [Cavalcanti2005]. For example, consider two dual-mode MHs A and B that try to establish a connection as shown in Figure 14.2. The MH A (B) can be under the coverage of an AP (cellular network). Another possibility is when both are using a WLAN interface, but B is operating in the ad hoc mode, while A is connected to an AP. When A and B are both single-mode cellular terminals, the only possibility is to use the cellular network. The most general case is when both end systems A and B have dual-mode capability. In this case, the following ten connection alternatives are possible:

a) A and B can communicate through the cellular interface, and the connection setup follows a typical procedure of the cellular network.

b) A and B can be connected through WLAN AP in a single-hop mode.

c) A and B can communicate with single-hop mode to their APs and these APs are interconnected through a fixed network.

d) A and B can use the WLAN interface in the ad hoc (i.e., MANET) mode to communicate directly or through multiple hops.

e) A can use the WLAN interface in the ad hoc mode to connect to a Gateway Node (GN), which can establish a connection to B through a cellular BS in the single-hop mode.

f) A can use its WLAN interface in the ad hoc mode to connect to a GN, which can establish a connection to B through a WLAN AP in the single-hop mode.

g) Both A and B are out of the coverage of an AP, but they can connect using the multi-hop ad hoc mode by identifying the corresponding GNs C and D that can communicate through the AP.

h) Both A and B are out of the coverage of WLAN APs, but they can connect using the multi-hop ad hoc mode by identifying corresponding GNs C and D that can communicate through the fixed infrastructure (i.e., CN) in the single-hop mode.

i) A and B are using the cellular and the WLAN interfaces, respectively, and the corresponding BS and AP are connected through the fixed network CN.

j) A is using its WLAN interface (via multi-hop) to access GN C that is connected to the BS, and this BS is connected through CN to the AP that provides connectivity to the destination B through the GN D.

14.2.5 Design Factors

Three unique features significantly affect the design of integrated solutions, namely: the availability of multiple interfaces for a MH, the integration of cellular networks and WLANs, and multi-hop communication (i.e., MANETs). Several questions need to be addressed in order to provide an integrated, transparent and self-configurable service. A fundamental question is what technology (or communication interface) to select to start a connection for a particular application. Another question is when to switch an ongoing connection from one interface to another (i.e., vertical handoff). Other important issues include transmission power selection for a given communication interface, co-channel interference, topology discovery and route creation, mobility and handoff management, and load balancing. Some of these questions have been addressed by researchers in several integration architectures. However, none of the existing models incorporate all these design factors. Clearly, the answer will depend on the MH's capabilities, on the connectivity options available at the current location, on the user's mobility profile, on the QoS expected, and on the service cost. For instance, the decision of what interface to use should be automatically taken by the system, as well as the selection of the end-to-end route for a particular connection may be based on the user's Service Level Agreement (SLA). Besides the goal of satisfying the user requirements, the selection of a given technology or the decision to perform a vertical handoff can also be used to enhance the overall system performance or to implement load balance functionalities.

14.3 Protocol Stack

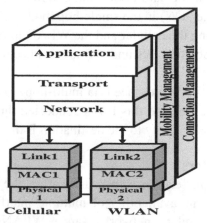

Figure 14.3 – Protocol stack of a dual-mode Mobile Terminal in a Heterogeneous
Network Environment [Cavalcanti2005]

In a homogeneous network, all network entities run the same protocol stack, where each protocol layer has a particular goal and provides services to the upper layers. In a heterogeneous environment shown in Figure 14.1, different mobile devices can execute different protocols for a given layer. The protocol stack of a dual mode MH is given in Figure 14.3. This protocol stack consists of multiple Physical, Data Link and MAC layers, and network, transport and application layers. Therefore, it is critical to select the most appropriate combination of lower layers (Link, MAC and Physical) that could provide the best service to the upper layers, which also implies the best communication interface. Also, some control planes such as mobility management and connection management can be added which could eventually use information from several layers to implement their functionalities.

As seen in Figure 14.3, the network layer has a fundamental role in this process, since it is the interface between available communications interfaces (or access technologies) that operate in a point-to-point fashion and the end-to-end layers (Transport and Application layers). In other words, the task of the network layer is to provide a uniform substrate over which transport (TCP and UDP) and application protocols can efficiently run, independent of the access technologies used in each of the point-to-point links in an end-to-end connection. However, this integration task is extremely complex and it requires the support of integration architecture in terms of mobility and connection management. Seamless handoffs for "out of coverage" terminals and resource management can usually be provided by the two vertical control planes shown in Figure 14.3.

14.3.1 The Physical Layer

MHs equipped with multiple network interfaces can access multiple networks simultaneously. Even though SDR based MHs are not fully capable of simultaneously accessing multiple wireless systems, discovering access networks available for a given connection need to be performed. If MH is connected to a cellular network that is also within the coverage area of an 802.11b AP; the network or the MH needs be able to switch from cellular to WLAN.

In a heterogeneous environment, different wireless technologies may be operating in the same frequency band, and it is critical that they must coexist friendly without degrading the performance of each other. Therefore, interference mitigation techniques are important. For MHs far away from their APs, for example, the multi-hop communication links may result in less interference than direct transmission to the AP. However, [Mengesha2001] shows that the capacity of AP decreases considerably, as a result of multi-hop, if MHs are close to the AP. In a direct transmission from a MH to a BS or an AP, only a single frequency channel is required, whereas in the multi-hop connection from the same MH to the same BS, several transmissions and receptions, and several frequency channels are assigned for each hop of the connection. Power control techniques have been applied to limit interference in CDMA/based cellular networks and MANETs, as well as coexistence analysis of IEEE 802.11 WLAN standard and Bluetooth have been published [Cordeiro2004].

14.3.1.1 Open Research Issues

The open research issues range from SDR based terminal design to power control techniques, and can be summarized as follows:

- Efficient design of SDR based MSs that switch between different technologies.
- Cognitive and agile radios [Cordeiro2005].
- Frequency planning schemes for BSs/APs that could satisfy resource constraints while increasing the spectrum utilization.
- Interference mitigation techniques between various wireless access technologies.

Note that modulation techniques and coding schemes that improve performance of a given technology will always be among the physical layer design challenges.

14.3.2 The Data Link Layer

The data link layer can be divided into logical link control (LLC) layer and MAC layer. The MHs will be able to use a centralized MAC, such as TDMA or CDMA, when connecting to a cellular network, or a distributed random access scheme, such as CSMA/CA, in an IEEE

802.11 WLAN. These access methods can provide different levels of service in terms of capacity and delay. The data rate in the cellular interface can reach up to 2.4 Mbps (the maximum in the current CDMA2000/HDR standard), while the 802.11b interface can provide up to 11 Mbps and the IEEE 802.11a/g standards support up to 54 Mbps. Also, achievable throughput and delay in CSMA/CA highly depends on the traffic load. Problems such as the hidden and exposed terminals are known to limit the capacity of IEEE 802.11. Furthermore, when two MHs are communicating through multiple intermediary hops in the ad hoc mode, the performance can be even worse due to random access problems. Moreover, as studied earlier, CSMA/CA scheme used in the IEEE 802.11 standard has performance limitations when used in a multi-hop environment. In addition, mechanisms such as power control MAC protocols [Jung2002] and MAC-based route selection [Cordeiro2002] can be used to improve the performance.

Although integration of heterogeneous technologies will not be performed at Link and MAC levels, these layers can provide useful information to upper layers. An end-to-end connection can involve a sequence of several different Link and MAC layer connections (scenario 8 in Figure 14.2) and the final end-to-end performance will be limited by the "weakest" link in this chain of connections. The cross-layer design approach can also be considered in heterogeneous networks. Security is also an important issue to be considered at the Link/MAC level. Although end-to-end security is considered in the application layer, some wireless access technologies provide a certain level of security at the lower layers, such as the WEP scheme that is part of the IEEE 802.11b standard and the IEEE 802.11i standard.

14.3.2.1 Open Research Issues

The Link and MAC layers in a dual mode MS may operate independently. Their operations have to be optimized to provide guaranteed service to the upper layers. The 802.11e, for example, is a supplementary to the MAC layer to provide QoS support for WLAN applications, and is applicable to 802.11a/b/g physical standards.

In summary, some of the open issues at these layers include:

- Design of efficient Link and MAC layers protocols for MANETs and WLANs that support different QoS levels.
- Channel management schemes in cellular networks that results in low call blocking and handoff failures in different types of traffic.
- Link/MAC layer security.

14.3.3 The Network Layer

The network layer seems to be the most challenging as it integrates many technologies. The multiple interfaces of MHs can have different physical and MAC layer protocols, and need to be taken into account in any integrated routing process. It also inherits multi-hop routing issues in MANETs, such as frequent route changes due to mobility, higher control overhead to discover and maintain valid routes, higher end-to-end delay, and limited end-to-end capacity due to problems at the lower layers (e.g., collisions at the MAC layer, and interference at the physical layer). Some existing solutions limit the number of multi-hops to a maximum of 2 or 3 [Aggelou2001, Luo2003, Wei2004]. However, it is not clear as to what extent multi-hop connections can enhance system's performance. Moreover, integrated solutions rely on high processing and power capabilities of BSs and APs.

The idea of integrating MANETs with single-hop networks is motivated not only by traffic load reduction in the BS/AP's, and improving the overall cell throughput [Luo2003], but also by providing connectivity to MHs that are out of fixed BS/APs coverage using GNs. Hence, the network layer must have mechanisms to allow these nodes in a MANET to find such gateways and to allow the MHs to correctly configure their IP addresses. Further, the nodes connected to the fixed infrastructure must be aware of the nodes in the MANET part (i.e., MH out of coverage) that can be reachable through the Gateways. In other words, the network layer has to discover the integrated topology and find the best route between any source and destination pair. Several metrics can be used to define the best route, including number of hops, delay, throughput, and so on. Furthermore, the network layer has to handle horizontal handoffs between BS/AP's of the same technology and vertical handoffs between different access technologies in a seamless way.

14.3.3.1 Integrated Architectures

Although there is no solution that considers all the possibilities described in Figure 14.2, several architectures and hybrid routing protocols have been proposed to integrate single-hop (cellular based model) and multi-hop (i.e., MANET) routing. The architectures and routing protocols discussed in this section include UCAN, Two-Hop-Relay, 1-hop and 2-hops Direct Transmission, HWN, MCN, iCAR, MADF, A-GSM, ODMA, SOPRANO, CAMA, and the Two-Tier Heterogeneous MANET Architecture.

UCAN: The Unified Cellular and Ad-Hoc Networks architecture (UCAN) [Luo2003] considers dual-mode MSs with a cellular CDMA/HDR interface and an IEEE 802.11b interface that can operate in the ad hoc mode. The UCAN architecture can be applied in a scenario similar to (5) in Figure 14.2, with all nodes assumed to be under the BS coverage (a unique cell). The basic goal is to use multi-hop routing to improve the throughput when the quality of the signal between the MS and the BS is poor. The system uses GNs (called proxy clients) with better downlink signal quality from the BS to relay packets towards the destination MS in the ad hoc mode. Thus, MSs have to discover the proxy clients that act as the interface between the ad hoc mode and the cellular network, as well as they have to decide when to execute vertical handoffs. Two proxy client discovery protocols have been proposed in [Luo2003], a proactive greedy scheme and an on-demand protocol. MHs monitor pilot bursts sent by the BS to estimate their current downlink channel conditions and it is used in the proxy discovery and routing process. The BS runs a scheduling algorithm to send data frames to the clients in the HDR downlink channel. Once a client, currently receiving data from the BS, experience degradation on the received data rate, it can send a route request (RTREQ) on the 802.11b interface trying to establish a new route (using a proxy client) to receive the data from the BS. The route request propagates through the ad hoc network (for a limited number of hops controlled by a TTL field) to find a proxy client.

Two-Hop-Relay: The Two-Hop-Relay architecture [Wei2004] also exploits availability of dual-mode terminals that can act as Relay Gateways (RG) between the single hop and the multi-hop domains (as shown in scenarios 5 to 8 in Figure 14.2) and it considers not only cellular BSs, but also WLAN APs. As suggested in [Wei2004], the relay gateways can be nodes placed by a wireless carrier or can be dual-mode MHs able to act as RGs. The two main goals in this architecture are to enhance the capacity of existing cellular network and extend the system coverage for WLAN terminals for up to two hops. As in the UCAN architecture, terminals with low downlink quality of signal from the BS can use multi-hop connections to achieve higher data rates. The RG can be used by WLAN terminals out of the AP coverage, only if they are properly registered with the cellular network. The RG periodically broadcasts the Relay Advertisements through its WLAN interface including its own identifier (GWid), the current BS/AP identifier (BSid), the bandwidth indicator type (BI-T) for QoS, the bandwidth indicator value (BI-V), and the registration method (RM).

Before establishing a connection, a dual-mode MH must decide whether to go directly through the single-hop (transmit direct to the BS) or to use the two-hop alternative. If the MH decides to go through the multi-hop routing, it sends a Relay Request to the RG. If the MH has only the WLAN interface it has to send the relay request message back to the RG after receiving the advertisement message. On its side, the RG stores the identification of the MH and forwards the Relay Request to the cellular network and waits for the authentication and authorization response from the cellular network by exchange several messages. Finally, when the cellular network sends the Relay Replay, the RG informs the MH the status of authorization of the connection.

1-hop and 2-hops Direct Transmission: In [Chang2003], it is introduced hybrid protocols for integrating single-hop and multi-hop operation in a WLAN environment, thereby combining the strengths of the two models to solve problems such as APs failures and handoff procedures in the single hop model, and weak connection under the ad hoc mode. These protocols can be used in a scenario similar to (8) in Figure 14.2. Several control messages are introduced for multi-hop operation as well as to allow MHs to discover GNs (called agents) to connect to an AP. If the receiver moves such that it can no longer directly hear the sender's signal and if the receiver can still receive data from one of the sender's neighbors, the connection can be switched to the 2-hop-direction transmission mode. However, if no sender's neighbor is accessible from the new receiver's position, they still can use the AP-oriented communication mode. Hence, two communicating MHs can be in one of the three modes.

Hybrid Wireless Network Architecture (HWN): The Hybrid Wireless Network (HWN) architecture [Hsieh2001] allows each cell (BS) to select the operation mode between a typical single-hop for sparse topologies or the ad hoc mode for dense topologies. It is assumed that all MHs have GPS capabilities and send periodically location information to the BS. The BS runs an algorithm to decide the operation mode that could maximize throughput. In the switching algorithm, the BS estimates and compares the throughput in the ad hoc mode with the current throughput in the single-hop mode. If the current operation mode is the ad hoc one, the BS compares the achieved throughput with $B/2N$, where B is the achievable bandwidth per cell and N is the number of MHs per cell. Furthermore, the BS periodically broadcasts minimum transmission power required to keep the network connected in the ad hoc mode. It has been suggested that the IEEE 802.11 PCF as MAC protocol for the cellular mode, and the DFC for the ad hoc mode [Hsieh2001]. Two drawbacks of the HWN architecture are: the minimum power used in the ad hoc mode can lead to disconnected topologies due to mobility, and ongoing connections are broken during the switching period. Also, the centralized selection of the mode for all the connections may not optimize the cell performance, and a better option may be to select the operation mode per connection.

Multi-Hop Cellular Network Architecture (MCN): In the Multi-hop Cellular Network MCN architecture [Lin2000], all cells use the same data and control channels. The MHs and BS data transmission power is reduced to half of the cell radius, to enable multiple simultaneous transmissions using the same channel. It is argued that this reduction factor of two represents a compromise between increasing the spatial reuse and keeping the number of wireless hops to a minimum. The transmission power in the control channel is corresponding to the cell radius and the MHs use this channel to send information about their neighbors to the BS. To ensure reliable connectivity information at the BS, the nodes recognize their neighbors using a contention-free beacon protocol and send their neighbors table to the BS. When a MH wants to connect to a given destination, it sends a route request to the BS in the control channel. Then, using the topology information, the BS finds the shortest path (using Djikstra's algorithm) between the source and the destination and sends back a route reply with the shortest path to the source node. Upon receiving the route reply, the source node inserts the route into the packet and begins its transmission. In addition, the nodes cache route information to eliminate the control overhead. When a node detects that the next hop is unreachable it sends a route error packet to the BS and buffers the current packet. The BS responds with a route reply to the node that generated the route error and also sends a correct route packet to the source node.

iCAR and MADF: In iCAR [Hu2001], the ad hoc relay stations (ARSs) are wireless devices deployed by the network operator and equipped with two interfaces, one to communicate with the cellular BSs and another to communicate in the ad hoc mode with other ARSs (WLAN based interface). Further, the ARSs nodes can have limited mobility controlled by the cellular MSC (Mobile Switching Center) in order to adapt to traffic variations. The iCAR architecture [Hu2001] uses ARS to balance the traffic load between overloaded cells to an un-congested one. Three basic relaying strategies can be used when a new call is originated in a congested cell. The primary relaying is when a MH A is originating a call in cell X and it can connect directly to a nearby ARS, which can divert the call to a neighboring cell Y. When no ARS is available to MH A, another ongoing call in cell X, let say MH C, can be transferred to a neighboring cell using another ARS (secondary relaying). If any of these two approaches can not be used, the third option, the cascade relaying, is to divert a call in cell Y to another cell Z, such that one of the ongoing call in the originating cell X, can be transferred to Y, and the new call is accepted in cell X. Besides load balancing, this scheme also increases the coverage of the cellular network, since MHs out of any BS coverage can access the system through the relay stations.

Like iCAR, the Mobile-assisted Data Forwarding (MADF) [Wu2002] achieves the load balance between cells by forwarding part of traffic in an overcrowded cell to some free cells. Unlike iCAR, which uses stationary ARS as relays, the traffic forwarding in MADF is achieved by using MHs as relaying nodes that are located between overloaded (hot) cells and free (cold) ones. Relaying MHs share a number of forwarding channels and they continuously monitor the delay of their packets. If the packet delay is high, the MH stops forwarding and requests BS for forwarding data packets to another neighboring cold cell. If the hot cell returns to low traffic, the MH may stop its MADF forwarding and then redirects packets back to its own cell. If MHs in a hot BS are far away from one another, the same forwarding-channels can be reused by two MHs to forward data to different neighboring BSs without interference. The implementation of MADF in the Aloha network and in the TDMA network shows that the throughput in a hot cell surrounding by some cold cells can be significantly improved [Wu2002]. The advantage of MADF over iCAR is that there is no need for additional devices.

A-GSM and ODMA: The A-GSM (Ad hoc GSM) architecture [Aggelou2001] allows GSM dual-mode terminals to relay packets in the MANET mode and provide connectivity in dead spot areas, thereby increasing the system capacity and robustness against link failures. The dual-mode terminals in [Aggelou2001] are equipped with a GSM air interface and a MANET interface and when one interface is being used, the other one can detect the availability of the alternative connectivity mode. The terminals have an internal unit called DIMIWU (Dual-mode Identity and Internetworking Unity), which is responsible for performing the physical and MAC layer protocols adaptation required for each air interface, i.e., GSM or MANET (A-GSM). At the link layer, the A-GSM mode uses an adaptation of the GSM Link Access Protocol for D channel (LAPDm) that advertises their capabilities of serving as relay nodes by beacon signals which could include BS as a relay node. The drawback of this proactive gateway discovery scheme is the high control overhead. A resource manager entity decides whether a relaying request should be accepted. A busy flag could reduce the number of beacon messages sent while the continuous transmission of beacons is used to keep information about the quality of the links between a node and its neighbors [Aggelou2001].

The basic idea in A-GSM is the same as in the ODMA (Opportunity Driven Multiple Access) scheme [Rouse2001]. Both solutions integrate multiple accesses and relaying function to support multi-hop connections. The ODMA breaks a single CDMA transmission from a MH to a BS, or vice versa, into a number of smaller radio-hops by using of other MHs in the same cell to relay the packets, thereby reducing the transmission power and co-channel interferences. However, the ODMA does not support the communications for the MHs that are outside the coverage of BSs as A-GSM does.

SOPRANO: The Self-organizing Packet Radio Ad hoc Network with Overlay (SOPRANO) [Zadeh2002] investigates some of the techniques by which the capacity of cellular network can be enhanced, including bandwidth allocation, access control, routing, traffic control, profile management, etc. The SOPRANO architecture advocates six steps of self-organization for the physical, data link, and network layers to optimize the network capacity: neighbor discovery, connection setup, channel assignment, planning transmit/receive mode, mobility management and topology updating, exchange of control and router information. Multi-user detection is also suggested for the physical layer since it is an effective technique to reduce the excessive interference due to multi-hop relaying. In the MAC layer, if transmissions are directed to a node with several intermediate nodes by multi-hop, clever frequency channel assignments for each node can significantly reduce the interference and could result in better performance. In the network layer, the system capacity can be enhanced by taking multi-hop routing, the interference and the energy consumption into account.

CAMA: The Cellular Aided Mobile Ad hoc Network (CAMA) architecture [Bhargava2004] has as underlying goal to enhance the performance of MANETs, with the aid of the existing cellular infrastructure. This concept is slightly different from most integrated solutions discussed so far, since the cellular network is used only to control the operation of the MANET by providing authentication, routing and security. Only control data is sent to the cellular network, i.e., the cellular channels can be viewed as an out-of-band signaling channels that are used by multi-radio MHs to connect to the CAMA agents located in the cellular infrastructure. The CAMA agents perform route discovery using a centralized position-based routing scheme, called multi-selection greedy positioning routing (MSGPR). All MHs are assumed to have GPS (Global Positioning System) capabilities, such that MHs can report their position to the CAMA agent through the cellular channels. All MHs are assumed to be under the cellular coverage, such that the CAMA agents are provided with information about the entire MANET. The authors in [Bhargava2004] claim that the low-cost, high-data-rate ad hoc channels are suitable for multimedia services and use this idea as motivation to forward all data traffic through the MANET. However, the dynamic nature of MANETs can not always assure QoS guarantees required by the multimedia services. For example, while the data rates are higher in MANETs, delay and jitter are affected by factors like interference and mobility. The connectivity between the MANET and the Internet is provided by special MHs (Gateways or APs) connected to the fixed network. Another point to be noted is that although no data traffic flows through the cellular network, the control traffic in the cellular network increases and this should not affect the QoS for cellular users.

As can be noted, CAMA is not a generic architecture, but uses the cellular infrastructure to improve the performance of MANETs. As suggested in [Bhargava2004], this concept could be extended to provide other control operations in the MANET, such as topology control and power management. Although CAMA is not designed to provide services to cellular users, it could be applied only to a scenario similar to (4) in Figure 14.2, if all MHs were inside the coverage of a cellular network.

Two-Tier Heterogeneous MANET Architecture: The Two-Tier Heterogeneous Mobile Ad hoc Network [Huang2004] considers WLAN MHs operating as a MANET and dual-mode MHs with WLAN and cellular capabilities, which are able to operate as gateways between the MANET and the Internet. The MHs operating in the MANETs form the lower tier, while the dual-mode gateways form the second tier of the architecture. Load balancing issues come into picture as the gateways relay on the limited bandwidth of the cellular channels to provide Internet connectivity to multiple MHs. Several gateway selection schemes are proposed in [Huang2004] that partition the network into clusters by associating MHs to gateways in a dynamic way according to different metrics. Similarly to CAMA, the Two-Tier Heterogeneous MANET architecture does not provide services to cellular users, as the main objective is to connect the MANETs with the Internet through the cellular network. The destination or the source node of a connection is always in the fixed network. Therefore, the architecture supports only a variation of the scenario (5) shown in Figure 14.2.

Network discovery: Network discovery is relatively simple in a single network. The wireless network interface only needs to scan the channel and frequency band operated by the network at the beginning of the connection or just before performing handover. In contrast, the network discovery problem becomes complicated while multiple networks are possibly available for connection. Periodic or reactive scanning multiple networks by multiple heterogeneous interfaces is required to provide efficient and economic connection. This is essentially required in a mobile environment if high HWN utilization is wanted. It needs careful design to coordinate the scanning on multiple interfaces. It is note that the scanning of multiple networks also consumes much more energy on MH than that in a single network. The available network may keep changing, which indicates frequently network discovery actions. The scanning of network usually consumes more energy than regular operation of radios. As the energy consumption is a critical issue for handheld, high energy consumption may constraint the development of the HWN on MH.

Inter-networking: When MH accesses multiple networks at the same time or performs handover from one network to another, the support for the fast information exchange among these networks is necessary. The information in general is through the gateways connecting different networks. Gateways between different networks need to covert data frame structure or perform "tunnel" function to make frames through one or several intermediate networks. Such functionality poses big challenges on the design of core networks. For example, to have fast handover between WiFi and cellular network is still not effectively solved. With the efforts on the next generation Internet, a robust and reliable inter-networking is expected to be formed in the near future.

14.3.3.2 Open Research Issues

Although several routing protocols have been proposed for the heterogeneous communication networks, the design of integrated and intelligent routing protocols is largely open for research:

- Routing capability in a heterogeneous environment that supports all communication alternatives described in Figure 14.2.
- Scalability in multi-hop routing without drastically increasing the overheads.
- The impact of additional routing constraints (co-channel interference, load-balance, bandwidth, and terminal interfaces), and requirement (services, speed, packet delay) needed by the MSs and the networks.

14.3.4 Transport Layer

As we studied earlier, performance degradation of the TCP protocol is the most important issue in any wireless transport layer as all losses are assumed due to congestion and factors such as channel errors, delay variations, and handoffs are ignored. We still see that several modified versions of TCP have been proposed to handle non-congestion related losses. In a heterogeneous scenario where MHs are equipped with multiple interfaces and several access networks are available, the transport protocol has also to handle the high delays involved in connection switching from one interface to another (vertical handoff procedures), server migration and bandwidth aggregation [Hsieh2003]. Since a TCP connection is identified by the tuple (IP address, port number) of both end points, the basic problem is how to maintain a TCP connection when a MH changes its IP address as it enters a new access network. Network layer solutions, such as Mobile IP, incur relatively high delay. Due to firewalls, server migration support may be required when the MH cannot access the original application server using the new access network address. Also, the overlap of coverage between different access technologies can be exploited to improve the aggregate connection's bandwidth. However, the MH must consider tradeoff between achieved throughput, power consumption and cost before using multiple active interfaces [Hsieh2003].

A RCP (Reception Control Protocol) is proposed in [Hsieh2003] which is TCP-compatible. Most approaches enhance TCP performance in wireless networks by providing feedback information about the causes of the errors at the wireless links to the protocol (basically the sender). A receiver centric approach is developed [Hsieh2003] such that the receiver closer to the wireless last-hop (where most errors occur) can obtain more accurate information about causes of the losses and avoids feedback overhead by taking proper actions. In case of server migration, overhead required to transfer connection state information from one sever to another is minimized, since the receiver has the control information.

An extension of the RCP protocol for host with multiple interfaces, called R2CP (Radial RCP), was also proposed in [Hsieh2003] to support seamless handoffs and bandwidth aggregation. R2CP aggregates multiple RCP connections into one abstract connection for the application layer. The protocol keeps multiple states at the host according to the number of active interfaces. Then, in a vertical handoff, an application can continue transmitting and receiving data in the old interface before the new connection is established. Note that the advantages of a receiver-centric transport protocol are highlighted when the sender is in the fixed network. When both ends are connected to wireless access networks (e.g., scenario 8 in Figure 14.2), the errors can occur not only close to the receiver, but also at the last wireless link at the sender side.

14.3.4.1 Open Research Issues

The main open problems at the transport level are:

- Design of new transport protocol or adapt the existing protocols (mainly the TCP) to take delays into account in vertical handoffs for end-to-end congestion control process.
- Implement server migration without interrupting ongoing connections, and support of bandwidth aggregation by exploiting the availability of multiple interfaces.

14.3.5 Application Layer

In a heterogeneous environment, the applications should have only an access point to the transport layer and network services as in the OSI network model, all the underlying complexity should be hidden. As discussed above, the network layer needs to exploit the availability of multiply access technologies and communication interfaces in the MHs to meet the QoS requirements. The multiple access networks available in a given location can also provide different kinds of application services to the users. For example, WLAN APs placed along the path to an airport can provide flight information service. In this case, WLAN capable MHs should be able to discover and inform the user availability of such service. In fixed networks, some particular nodes can be selected to store service availability information, while in MANETs decentralized service discovery schemes are required [Motegi2002]. A basic problem is how to provide information of services available in the fixed network part (through BS or AP) to MH participating in a MANET. Therefore, some kind of virtual service manager is needed, which can filter relevant information.

Due to multi-hop routing, design of charging and/or rewarding schemes in the application layer becomes a critical issue to encourage collaboration in packet forwarding [Luo2003, Salem2003, Yoo2006]. Another fundamental problem is end-to-end security. In an adversarial environment, heterogeneous network may suffer from variety security threats that may degrade the efficiency of packet relay, increase packet delivery latency, increase packet loss rate. In [Xie2004], a security framework has been investigated that traces each MH with unique ID to ensure the MH's Internet mobility and MANET routing security.

14.3.5.1 Open Research Issues

Some of the open issues at the application layer include:
- End-to-End security.
- Service Discovery mechanisms.
- Credit charging and rewarding mechanisms.

14.3.6 Mobility and Connection Management

As shown in Figure 14.3, mobility and connection management are two control planes that provide topology discovery, detect available

Internet access, as well as support vertical handoffs. The mobility and connection management functionalities cannot be clearly separated.

14.3.6.1 Internet Connectivity

Current cellular networks (2.5 and 3G) provide IP services based on the IPv4 protocol [Faccin2004] and support terminal mobility through Mobile IP. In WLANs, APs can act as Internet Gateways (IGW) to MHs. Future heterogeneous networks will be based on IPv6, and the main challenge is to provide Internet connectivity to nodes in the MANET. The APs can also provide Internet connectivity and perform mobility management functions for those "out of coverage" MHs. The gateway discovery process can be implemented in a proactive, reactive or hybrid fashion [Hsieh2001]. In the proactive scheme, the IGWs periodically broadcast advertisement messages, while in the reactive approach; nodes send gateway discovery requests as needed. The reactive discovery has a smaller control overhead, but results in higher delay. A hybrid discovery approach allows IGWs to broadcast only to nodes under a restricted area, while outside nodes have to perform reactive gateway discovery.

As described above, for some proposed integration architectures, such as Two-Hop-Relay and UCAN, the process of discovering an AP that acts as IGW, can also be considered as the process of discovering another MH (GN, Relay Gateways or proxy client) that can connect directly to the AP and forward control (registration messages) and data packets to and from nodes in the MANET.

14.3.6.2 Mobility and Connection Management

In a heterogeneous network, there are two types of handoffs, namely, horizontal handoffs (between AP/BS of the same technology) and vertical handoffs (between different interfaces or access networks). The horizontal handoffs can be handled by the cellular network components at the link layer and at the network layer, the Mobile IP (Mobile IPv4 and IPv6 [Tanenbaum1996]) protocol can be used to support macro and global mobility. The Mobile IP provides an effective solution for macro and global mobility management, but in a micro-mobility environment, inside the same cellular network, the latency in the handoff process increases significantly. Several mobility management schemes have being proposed to reduce the handoff latency in the micro-mobility environments (e.g., Cellular IP [Valko1998], Hawaii [Ramjee1999], TeleMIP [Das2000]), most of them introduce new Mobile IP agents and some of them, such as TeleMIP, uses DHCP [Tanenbaum1996] servers to discover the new network [Campbell2001].

In the case of vertical handoffs, two basic issues have to be considered: when to start the handoff process? How to redirect the traffic between interfaces? Note that the process of discovering the IGW for a given technology is also required to be supported for vertical handoffs. In order to use the functionalities of Mobile IP, the MH needs to register with a home agent running on a fixed IGW. There are some situations or trigger events where the traffic redirection can be performed [Montavont2004] (see Table 14.1).

The redirection process can be performed in a seamless or a reactive way [Montavont2004]. The first alternative is possible when the MH is under the coverage of the technologies corresponding to its interfaces and wants to start a vertical handoff to optimize QoS level or to perform load balance among its interfaces. For instance, MH currently downloading a file used the cellular interface can redirect the flow to the 802.11b interface if it is also under the coverage of the 802.11b AP. On the other hand, the reactive redirection is triggered by an interface down event (network failure), so that the packets can be received at a different interface. Clearly, packets will be lost if the node starts the redirection process only after detecting the interface is down. No integrated solution exists to handle all the possible types of vertical handoffs in a heterogeneous environment, as the shown in Figure 14.1.

In general, the vertical handoff latency can be characterized by three components [Aggelou2001]:

- Detection period is the time taken by the MH to discover an IGW.
- After detecting an IGW, address configuration interval is the time taken by a MS to update its routing table, and assign its interface with a new care-of-address (CoA) [Tanenbaum1996] based on the prefix of the new access network.

- Network registration time is the time taken to send a binding update to the home agent as well as the time it takes to receive the first packet on the new interface of the correspondent node.

When the discovery phase is based on IGW advertisements (reactive IGW discovery), some schemes have been proposed to reduce the handoff latency in a GPRS/WLAN scenario [Aggelou2001]: the fast router advertisements, route advertisements caching, binding update simulcasting, smart buffer management using a proxy in GPRS, a layer-3 soft handoff as briefed in Table 14.1.

Table 14.1 – Trigger events for a Vertical Handoff

Trigger Event	Description
Interface down	An interface currently used fails, so the MH cannot receive its flows anymore.
Interface up	A new interface is available, e.g., the node enters the coverage area of a new access network.
Horizontal handoff procedure	As a horizontal handoff procedure in the currently used interface is started, the node can redirect the traffic to another interface to reduce undesirable effects of the high delay involved in the handoff process.
Change in network capabilities	The QoS provided by one interface improves or degrades (coverage, data rate, power consumption, etc).

An extension of the Mobile IPv6 protocol is proposed in [Montavont2004] where the MMI (Multiple Interface Management) redirects traffic flows between two interfaces with two corresponding global IPv6 addresses (I_1, IP_1) and (I_2, IP_2). When MS wants to redirect its ongoing flows from the source interface I_1 to the target interface I_2, according to the MMI protocol, the node simple sends a Binding Update message to its home agent (HA_1) (for the source interface I_1) through the target interface I_2. The home address field in the Binding Update is set to IP_1 and the Care-of-Address field is set to IP_2. This allows the home agent for I_1 to register an association between IP_1 and IP_2 in its binding cache, upon the reception of the Binding Update. Hence, all the traffic previously forwarded to (I_1, IP_1) will be intercepted by the home agent

HA_1 and forwarded to the target interface (I_2, IP_2), without affecting the ongoing flows on the target interface. In case of horizontal handoff on the interface I_2 to a new subnet, where the MH gets the new CoA IP_3, both home agents HA_1 and HA_2 (for the target interface I_2) need to be informed of the new Care-of-Address IP_3. Finally, if the MN wants to use I_1 again, it only needs to invalidate the association between IP_1 and IP_2 in HA_1 binding cache.

14.3.6.3 Open Research Issues

In summary, some of the open research issues to be considered at the mobility and connection management layer in a heterogeneous scenario are:
- Efficient gateway discovery protocols to integrate MANETs with fixed network components.
- Development of new mobility and connection management approaches to reduce the delay during vertical handoffs.

14.4 Comparison of the Integrated Architectures

In this section we provide a comparison of existing architectures proposed for heterogeneous integrated networks. The first aspect to compare is the scenario considered by a particular architecture. The A-GSM, ODMA and iCAR proposals introduce the ad hoc mode (i.e., MANET) in a cellular system by exploiting the dual-mode terminals capabilities, but they do not consider the possibility of integrating WLAN APs. On the other hand, the 1-hop and 2-hop direct transmission protocols are especially designed to integrate single-hop (AP-based) and MANET mode in WLANs. The HWN and MCN architectures also focus on the integration of a generic single-hop mode (cellular or WLAN with AP) and MANETs, and they do not consider dual-mode terminals. The UCAN and Two-Hop-Relay architectures consider the most general scenario, including cellular network and WLAN in cellular-based mode and MANET mode. However, the results with these schemes presented in [Luo2003] and [Wei2004] assume that all nodes are dual-mode and are under a single cellular BS coverage and no WLAN AP is included in the evaluations.

As seen in Table 14.2, a common goal of the integration schemes is to improve the capacity in the cellular-based (BS or AP) systems by allowing some multi-hop transmissions. In A-GSM and ODMA, the reduction in the MH's transmission power is also identified as an advantage, since nodes far away from the BS do not need to increase their transmission power to reach the BS, but rather they can use a relatively low power level to connect to a close by relay MH in the ad hoc mode, which provides a path to the BS. The main goal in CAMA is slightly different from other solutions, as the cellular network is used solely to control MANET's operation by providing authentication, routing and security. Also, in the Two-Tier Heterogeneous architecture, the main focus is to provide Internet connectivity to MANET users. Therefore, the later two architectures cannot be considered as integrated solutions for the kind of heterogeneous scenario envisioned for future.

Table 14.2 – Comparison of Integrated Architectures

Architecture	Operation mode considered	Main Optimization goal	Mobile user Station	Supported Scneriosor in variation in Fig. 14.2	Support of out of coverage MHs	Connection Model Gateway Discovery
A-GSM	Cellular – MANET	Coverage, transmission power reduction and capacity	Dual-mode	(1), (5)	Yes	GNs send beacon (advertise) messages (proactive scheme)
ODMA	Cellular – MANET	Transmission power reduction and BS capacity	Dual-mode	(1), (5) but all MHs are under coverage of BSs	No	There is no concept of gateway, every node can relay packets, and the routing decision is based on the signal quality
ICAR	Cellular MANET	Load balance between BSs	Single-mode and Dual-mode	(1), (5)	No	The GN nodes are deployed in planned positions and the MH can use 1-hop away GNs.
UCAN	Cellular – WLAN - MANET	BS throughput and user downlink data rate	Single-mode and Dual-mode	(1), (2), (3), (5) but all MHs are under coverage of BSs	No	MHs search for gateway nodes when their transmission rate decreases below a given threshold (proactive or reactive discovery)
Two-hop Relay	Cellular - WLAN - MANET	BS throughput	Single-mode and Dual-mode	(3), (5)-(8)	Yes	The GNs sens advertising messages (proactive scheme)
1-hop and 2-hops Dir. Trans.	WLAN - MANET	Reliability to BS failures and handoffs	Single-mode	(3), (8)	Yes	Destination MH selects the connection mode, but MHs can act as gateways in case of APs failures
HWN	WLAN or Cellular – MANET	BS throughput	Single-mode	(1), (4)	No	BS selects the cell's operation mode
MCN	WLAN or Cellular- MANET	BS throughput	Single-mode	(1), (4)	No	BS/AP selects the operation mode and execute a centralized routing algorithm
MADF	Cellular- MANET	Load balance between BSs	Single-mode	(1), (5)	No	Ad hoc routing protocol for routing discovery
SOPRANO	Cellular- MANET	BS Capacity	Single-mode	(1), (5)	No	Routing decision is based on minimum interference and energy
CAMA	Cellular- MANET	Improve MANET performance	Single-mode and Dual-mode	(4) but all MHs under cellular coverage	No	Gateways are used only to provide access to CAMA agents in the cellular network
Two-Tier Heterogeneous MANET	Cellular- MANET	Connectivity MANET with the Internet	Single-mode and Dual-mode	(5)	No	Several gateway selection schemes are provided to have load balance among gateways

Other important aspect to consider is the connectivity support for MHs participating only in MANETs, i.e., out of the direct coverage of any BS or AP. Although the A-GSM scheme does not specifically consider the integration of isolated MANETs with the cellular network, one of its aims is to provide connectivity to terminals in dead spot areas, increasing the coverage of the cellular network. In ODMA [Rouse2002], iCAR, UCAN, HWN and MCN, all nodes are assumed to be under a BS or AP coverage.

The ODMA proposal achieves a capacity gain by reducing the interference level inside the cell, as some of the connections are established in the ad hoc mode. The iCAR architecture uses multi-hop only to transfer connections between BSs, performing load balance in the system. Different from the other schemes, the HWN and MCN approaches assume that the cell can operate only in the cellular-based or the ad hoc mode, i.e., they do not choose the transmission mode on a connection basis. By fixing a particular mode for all connections, the overall cell capacity will depend of the current topology. Hence, if the topology changes frequently, there can be a higher overhead and performance degradation in frequently switching between operation modes. Since different connections can have different QoS requirements, selecting the transmission mode for each connection request seems to be the most suitable approach. In HWN the routing protocol (BS-controlled or multi-hop routing) depends on the operation mode, while in MCN the route is selected by the BS even under ad hoc operation mode. In UCAN, the decision of the multi-hop route between the destination MS and the BS is based on the quality of the downlink transmission rate of the nodes in the path. Indeed, the GN (client proxy) for a given connection has to have a higher downlink rate then other terminals in the multi-hop path.

Although most architectures use some kind of GN as the interface between the cellular-based and the MANET mode, the GNs capabilities and responsibilities are not the same in all cases. For instance, the GNs can provide connectivity for MH out of the cellular-based infrastructure coverage, as in A-GSM and Two-Hop-Relay, or the GNs can be used only to improve the performance inside a cell (UCAN) or to perform load balancing (iCAR). Hence, in the former case, the "out of coverage" MHs have to find a gateway in order to join the network, which involves tasks as registration, authentication and addressing, while in the latter case, the MH uses the GN as the last hop in the multi-hop path to connect to a BS or AP and the performance metrics are used in the gateway selection. However, despite of the gateways functionalities, the network has to provide some means for the MHs to discover them.

Basically, the gateway discovery can be performed in a proactive or reactive fashion, and as in multi-hop routing for MANETs, each approach has its advantages and drawbacks. Proactive schemes generally provide a faster response time at the cost of more control traffic, while reactive discovery can reduce the amount of control traffic, but cannot achieve the same response time. The proactive scheme is used in A-GSM and Two-Hop-Relay the GNs periodically send advertise messages. In UCAN, the MH starts the search for the gateway as needed (reactive approach), but the authors have proposed two different discovery protocols, in a proactive approach, the MH keeps track of its neighbors capabilities of acting as gateway (downlink transmission rate from BS) and in a reactive scheme the request for gateway is forwarded until a candidate gateway is found, i.e., a MH with better downlink rate that the requesting MH. In iCAR, only 1-hop transmissions are allowed between a typical MH and a GN (called ASR), such that the MH can search for a gateway among its one-hop neighbors. In 1-hop and 2-hops Direct Transmission protocols, the MH can send a specific control message to find a gateway to connect to a nearby AP when its current AP fails. There is no concept of GN in ODMA, HWN and MCN, as all MHs are assumed to be under the cellular system coverage.

The general aim in most cases is to enhance users' throughput and improve the overall system performance, but no proposed architecture considers the applications' QoS requirements in the selection of the transmission mode, routing process or handoff procedures.

14.5 Conclusions and Future Directions

Future networking ought to integrate a myriad of heterogeneous terminals and access technologies, such as cellular, WLAN, WMAN and WPAN networks. Accessibility alternatives provide different QoS and coverage levels. The complexity of such a heterogeneous environment needs to be hidden not only from the end users, but also to be made transparent to the applications. The task of designing future adaptable heterogeneous network that provide QoS guarantees to users is extremely complex and challenging.

In this chapter, we have discussed open research issues that need to be addressed in order to integrate cellular, WLANs and MANETs. We have described the issues at each layer of the protocol stack as well as discussed various features and limitations of existing integration architectures proposed in the literature. The complexity in providing an integrated routing functionality increases with the necessity to consider isolated MHs or MHs operating in a MANET. Although the basic goal in several proposed integration architectures is to use multi-hop routing to enhance the performance in the cellular-based networks, the possibility of providing connectivity to users out of BS/AP coverage is also required to have a truly pervasive computing environment. The fixed infrastructure components, i.e., the BSs and APs, are expected to play an important role as their higher processing and power capabilities can be exploited to aid in the route decision process not only under their coverage, but also in the MANETs. Another fundamental task of the network layer is to support seamless horizontal and vertical handoffs so as to minimize the overall delay. Furthermore, service discovery and security mechanisms need to be provided.

Homework Questions/Simulation Projects

Q. 1. There are many types of wireless devices that are now commercially available. Some of the examples are Cellular systems, MANETs, WiFi, WPAN, WBAN, sensor networks, and recently WMAN in the form of WiMax. Very soon, we will also see WRAN chipsets. These are being used for different applications and so have differing characteristics such as coverage area, available bandwidth, power consumption, data and call rates, complexity, and associated security requirements.

    a. How many WLANs are needed to provide the data rate of a WPAN? Consider existing technologies.

    b. How many WiMax BSs are needed to provide the data rate of a WLAN?

    c. How do they compare in terms of channel bandwidth?

    d. What kind of communication hierarchy may be appropriate?

    e. What is the level of security in each device and how are they related as you move up in the hierarchy?

    f. Can you use multi-path routing to enhance higher bandwidth in ad hoc and personal area networks? What are the other alternatives?

Q. 2. Design a problem based on any of the material covered in this chapter (or in references contained therein) and solve it diligently.

References

[Aggelou2001] G. Aggelou and R. Tafazolli, "On the Relaying Capacity of Next-Generation GSM Cellular Networks," IEEE Personal Communications, Vol. 8, 2001.

[Agrawal2006] D. P. Agrawal and Q.-A. Zeng, "Introduction to Wireless and Mobile Systems," Thompson Publishing, Second Edition, 2006.

[Ahmavaara2003] K. Ahmavaara, H. Haverinen, and R. Pichna, "Interworking Architecture Between 3GPP and WLAN Systems," IEEE Communications Magazine, Vol. 41, No. 11, November 2003.

[Bhargava2004] B. Bhargava, X. Wu, Y. Lu and W. Wang, "Integrating Heterogeneous Wireless Technologies: A Cellular Aided Mobile Ad Hoc Network (CAMA)," Mobile Networks and Applications, Kluwer Academic Publishers, pp 393-408, September 2004.

[Buddihikot2003] M. Buddihikot, G. Chandranmenon, S. Han, Y. Lee, S. Miller, and L. Salgarelli, "Design and Implementation of a LAN/CDMA2000 Interworking Architecture," IEEE Communications Magazine, Vol. 41, No. 11, November 2003.

[Campbell2001] A. Campbell and J. Gomez-Catellanos, "IP Micro-mobility Protocols," ACM Mobile Computing and Communications Review, Vol. 2, No. 1, 2001.

[Cavalcanti2005] D. Cavalcanti, C. Cordeiro, D. Agrawal, B. Xie, and Anup Kumar, "Issues in Integrating cellular Networks, WLAN, and MANETs: A Futuristic Heterogeneous wireless Networks," IEEE Wireless Communications, June 2005, Vol. 12, No.3, pp 30-41.

[Chang2003] R. Chang, W. Yeh, and Y. Wen, "Hybrid Wireless Network Protocols," IEEE Transactions on Vehicular Technology, Vol. 52, No. 4, July 2003.

[Cordeiro2002] C. Cordeiro, S. Das, and D. Agrawal, "COPAS: Dynamic Contention-Balancing to Enhance the Performance of TCP over Multi-hop Wireless Networks," in IEEE IC3N, Miami, USA, October 2002.

[Cordeiro2004] C. Cordeiro, S. Abhyankar, R. Tonishwal, and D. P. Agrawal, "Bluestar: Enabling efficient integration between Bluetooth WPANs and IEEE 802.11 WLANs," ACM/Kluwer Mobile Networking and Applications (MONET) Special issue on Integration of Heterogeneous Wireless Technologies, Vol. 9, No. 4, August 2004.

[Cordeiro2005] C. Cordeiro, K. Challapali, D. Birru, and S. Shankar, "IEEE 802.22: The First Worldwide Wireless Standard based on Cognitive Radios," in IEEE International Conference on Dynamic Spectrum Access Networks (DySPAN), 2005.

[Das2000] S. Das, A. Misra, P. Agrawal, and S. Das, "TeleMIP: Telecommunications-Enhanced Mobile IP Architecture for Fast Intradomain Mobility," in IEEE Personal Communications, August 2000.

[Eklund2002] C. Eklund, R. Marks, K. Stanwood, and S. Wang, "IEEE Standard 802.16: A Technical Overview of the WirelessMANTM Air Interface for Broadband Wireless Access," in IEEE Communications Magazine, June 2002.

[Esteves2000] E. Esteves, "The High Data Rate Evolution of the cdma2000 Cellular System," in Multiaccess, Mobility, and Teletraffic for Wireless Communications, Vol. 55, Kluwer Academic Publishers, 2000.

[Faccin2004] S. Faccin, P. Lalwaney, and B. Patil, "IP Multimedia Services: Analysis of Mobile IP and SIP Interactions in 3G Networks," IEEE Communications Magazine, January 2004.

[Hsieh2001] H. Hsieh and R. Sivakumar, "Performance Comparison of Cellular and Multi-hop Wireless Networks: A Quantitative Study," in Proceedings of ACM SIGMETRICS, Cambridge, MA, USA, June 2001.

[Hsieh2003] H. Hsieh, K-H. Kim, Y. Zhu and R. Sivakumar, "A Receiver-Centric Transport Protocol for Mobile Hosts with Heterogeneous Wireless Interfaces," in Proceedings of the ACM MobiCom, September 2003.

[Hu2001] W. Hu, C. Qiao, S. De, and O. Tonguz, "Integrated Cellular and Ad Hoc Relaying Systems: iCAR," IEEE JSAC, Vol. 19, pp. 2105-15, 2001.
[Huang2004] C. Huang, H. Lee and Y. Tseng, "A Two-Tier Heterogeneous Mobile Ad Hoc Network Architecture and Its Load-Balance Routing Problem," Mobile Networks and Applications, Kluwer Academic Publishers, pp 379-391, September 2004.
[Jung2002] E.-S. Jung and N. Vaidya, "A Power Control MAC Protocol for Ad Hoc Networks," in Proceeding of the ACM Mobicom, 2002.
[Lin2000] Y. Lin and Y. Hsu, "Multi-Hop Cellular: A New Architecture for Wireless Communications," in Proceedings of the IEEE INFOCOM, Tel-Aviv, Israel, March 2000.
[Luo2003] H. Luo, R. Ramjee, R. Sicha, L. Li and S. Lu, "UCAN: A Unified Cellular and Ad-Hoc Network Architecture," in Proceeding of ACM Mobicom, September 2003.
[Mengesha2001] S. Mengesha, H. Karl, and A.Wolisz, "Improving Goodput by Relaying in Transmission-power-limited Wireless Systems," Informatik, Vienna, Austria, September 2001.
[Montavont2004] N. Montavont, T. Noel, and M. Kassi-Lahloy, "Description and Evaluation of Mobile IPv6 for Multiple Interfaces," in Proceedings of the IEEE Wireless Communications and Networking Conference (WCNC), Atlanta, USA, March 2004.
[Motegi2002] S. Motegi, K. Yoshihara, and H. Horiuchi, "Service Discovery for Wireless Ad Hoc Networks," in Proceedings of the 5th International Symposium on Wireless Personal Multimedia Communications, 2002.
[Ramjee1999] R. Ramjee, T. La Porta, S. Thuel, and K. Varadhan, "IP micro-mobility support using HAWAII," Internet draft, draft-ramjee-micro-mobility-hawaii-00.txt, February 1999 (work in progress).
[Rouse2001] T. Rouse, S. Maclaughlin, and Z. Hass, "Coverage-capacity Analysis of Opportunity Driven Multiple Access (ODMA) in UTRA-TDD," in Proceedings of the 2nd International Conference on 3G Mobile Commun. Technologies, 2001.
[Salem2003] N. Salem, L. Buttyán, J. Hubaux, and M. Jakobsson, "A Charging and Rewarding Scheme for Packet Forwarding in Multi-hop Cellular Networks," ACM Symposium on MobiHoc, 2003.
[Salkintzis2004] A. K. Salkintzis, "Interworking Techniques and Architectures for WLAN/3G Integration Toward 4G Mobile Data Networks," IEEE Wireless Communications, June 2004.
[SDRFORUM] Software Defined Radio (SDR) Forum, http://www.sdrforum.org/.
[Tanenbaum1996] A. Tanenbaum, Computer Networks, Prentice Hall, ISBN 0-13-349945-6, 1996.
[Valko1998] A. Valko, A. Campbell, and J. Gomez, "Cellular IP," Internet draft, draft-valko-cellularip-00.txt, November 1998 (work in progress).
[Wei2004] H-Y. Wei, and R. Gitlin, "Two-Hop-Relay Architecture for Next-Generation WWAN/WLAN Integration," IEEE Wireless Communications, April 2004.
[Wu2002] X.-X. Wu, S.-H. Chan, B. Mukherjee, and B. Bhargava, "MADF: Mobile-Assisted Data Forwarding for Wireless Data Network," Journal of Communications and Network, 2002.
[Xie2004] B. Xie and A. Kumar, "A Framework for Internet and Ad hoc Network Security," in IEEE Symposium on Computers and Communications, 2004.
[Yoo2006] Y.Yoo and D.P. Agrawal, "Why it pays to be selfish?," IEEE Wireless Communications magazine, 2006.
[Zadeh2002] A. Zadeh, B. Jabbari, R. Pickholtz, and B. Vojcic, "Self-organizing Packet Radio Ad Hoc Networks with Overlay (SOPRANO)," IEEE Communications Magazine, Vol. 40, No. 6, 2002.

Index

About the Authors

Carlos de Morais Cordeiro (Carlos.Cordeiro@intel.com) is a Chief Standards Architect in the Mobile Wireless Group of Intel Corporation, where he is responsible development and management of next generation wireless technologies. Before joining Intel, he worked for Philips Research North America and Nokia Research. Dr. Cordeiro has been involved in the wireless area including ad hoc and sensor networks, cognitive radios, millimeter wave technologies, MIMO, and wireless LANs and PANs for a number of years. Due to his contributions to this area, he received the *2007 IEEE New Face of Engineering Award* and was the recipient of the *IEEE Region 1's 2007 Technological Innovation Award*. Dr. Cordeiro was also part of the team that received the *2007 Frost & Sullivan Excellence in Research of the Year Award*. He is an Editor of IEEE Transactions on Wireless Communications and served as Guest Editor of various IEEE Journal on Selected Areas in Communications (JSAC) and ACM special issues. He is the co-founder of the IEEE ComSoc Technical Committee on Cognitive Networks, and is an active participant in standardization activities including IEEE 802.11, Wireless Gigabit Alliance (WiGig), Wi-Fi Alliance, IEEE 802.15, IEEE 802.22 and ECMA. Dr. Cordeiro is a Senior Member of the IEEE, has published over 80 papers in the wireless area alone, and holds dozen of patents.

Dharma P. Agrawal (<u>dpa@cs.uc.edu</u>) is the Ohio Board of Regents Distinguished Professor in the School of Computing Sciences and Informatics and the founding director for the Center for Distributed and Mobile Computing, University of Cincinnati, OH since Autumn 1998. He has been a faculty member at the Carnegie Mellon University (2006-2007), N.C. State University, Raleigh, NC (1982-1998) and the Wayne State University, Detroit (1977-1982). He has graduated 59 PhDs, and 45 MS students. His research interests include resource allocation and security in mesh networks, efficient deployment and security in sensor networks, use of Femto cells, and heterogeneous wireless networks. He has delivered 22 keynote talks and plenary speeches at major international conferences, 29 tutorials/intensive courses at different conferences and institutions, 580 papers in different journals and conference proceedings and holds 5 patents. His coauthored textbook on *Introduction to Wireless and Mobile Systems*, 3rd edition, published by Cengage has been adopted throughout the world. He has been the Program Chair and General Chair for numerous international conferences and meetings. He has received numerous certificates and meritorious service awards from the IEEE Computer Society. He was awarded a *"Third Millennium Medal,"* by the IEEE for his outstanding contributions. He has also been a Computer Science Accreditation Board visitor and an ABET team visitor. He is a Fellow of the IEEE, ACM, AAAS, and WIF and a recent recipient of IEEE CS Harry Goode award for his contribution to the wireless networking area.